FLORA ZAMBESIACA

Flora terrarum Zambesii aquis conjunctarum

VOLUME THREE: PART SEVEN

FLORA ZAMBESIACA

MOZAMBIQUE

MALAWI, ZAMBIA, ZIMBABWE

BOTSWANA

VOLUME THREE: PART SEVEN

Edited by
G.V. POPE, R.M. POLHILL & E.S. MARTINS

Published by the Royal Botanic Gardens, Kew,
for the Flora Zambesiaca Managing Committee
2003

ARRANGEMENT OF TRIBES IN VOLUME 3

LIST OF NEW NAMES PUBLISHED IN THIS PART

Tribe 12. **LOTEAE***

Loteae DC., Prodr. **2**: 115 (1825).
Coronilleae (Bronn) Boiss., Fl. Or. **2**: 176 (1872).

Herbs or shrublets. Leaves epulvinate, pinnately 3–5- to many-foliolate, sessile or with a very short petiole, the lowermost pair of leaflets often resembling stipules, rarely leaves simple; stipules minute or absent, often replaced by a gland; stipels very small or absent. Flowers in axillary usually pedunculate heads or umbels; bracts small, glandular, scarious, or absent, but peduncle often with a foliage leaf. Calyx campanulate or tubular with a hypanthium. Standard tapering or contracted to a claw, often with thickened and infolded margins. Stamen filaments all united into a sheath or more often the vexillary one free, all or at least some of them dilated upwards; anthers all similar. Fruits 2-valved or indehiscent, sometimes flattened or torulose, sometimes breaking up into 1-seeded articles. Seeds oblong-reniform to subglobular, elsewhere sometimes long and narrow.

A tribe of c. 22 genera if the *Coronilleae* are included, north temperate in distribution and centred mostly about the Mediterranean and in California, but extending into Africa, Australia and South America.

Loteae was formerly associated with the predominantly north temperate herbaceous tribes *Galegeae, Vicieae, Cicereae* and *Trifolieae,* but it lacks the inverted repeat of the chloroplast DNA characteristic of that group. It is now considered to be an independent derivative of the *Millettieae,* see Doyle in Adv. Leg. Syst. **7**: 21–22 (1995).

Fruit persistently straight, narrowly cylindrical, dehiscent as a whole; leaves 3–5-foliolate . . . . 83. **Lotus**
Mature fruit straight, but curving upwards during dehiscence, composed of spherical 1-seeded individually dehiscent articles; leaflets 5–11 . . . . 84. **Antopetitia**

83. LOTUS L.**

By E.S. Martins

Lotus L., Sp. Pl.: 773 (1753); Gen. Pl., ed. 5: 338 (1754). —Brand in Bot. Jahrb. Syst. **25**: 116–232 (1898). —J.B. Gillett in Kew Bull. **13**: 361–381 (1959).

Annual or perennial herbs, subshrubs or dwarf shrubs. Leaves sessile or with a very short petiole, usually with 5, rarely with 3–4, leaflets (or elsewhere occasionally more); leaflets petiolulate, the lower pair at the base of the rhachis and resembling stipules; stipules reduced to dark glands, sometimes inconspicuous, or absent; stipels absent. Flowers in axillary usually pedunculate umbels with a 1–3-foliolate foliage leaf, rarely umbels reduced to one flower and resembling solitary flowers. Calyx tubular or narrowly campanulate, subequally 5-toothed or the lowest tooth longest or the teeth in 2 lips. Corolla yellow, pink, cream or white, glabrous or rarely (in Macaronesia and Morocco) the standard and (very rarely) also the wings pubescent, free from the stamens; standard ± oblong-obovate, wedge-shaped at the base with infolded thickened margins; wings clawed, the blade oblong, auriculate, laterally pouched; keel curved, pointed or beaked with free claws. Vexillary stamen free, the other 9 with the filaments united into a sheath; free parts of at least the 5 longer filaments widened at the tip. Ovary sessile or shortly stipitate, glabrous (outside the Flora Zambesiaca area sometimes strigulose or hairy), subcylindrical, multi-ovulate; style bent upwards at the base, glabrous, but papillose, rigid. Pod usually cylindrical,

* *Lotus* by E.S. Martins. *Antopetitia* by B. Verdcourt.
** The author of this Flora account is grateful to Dr. D.D. Sokoloff for comments on the first draft, and for co-authorship of the full account of the new species *L. robsonii.*

straight, linear, usually septate within, dehiscent into 2 valves (except in *L. benoistii*, from Morocco, with an indehiscent fruit, and in the Mediterranean *L. edulis* with the fruit dehiscent ventrally), the valves often becoming twisted. Seeds ellipsoid, subreniform or subglobular, with a small, median hilum.

A genus of about 100 species mainly in temperate Europe and temperate Asia, extending to mountainous and drier places in Africa and Australia.

1. Leaves 3-foliolate · 2. *robsonii*
– Leaves 5-foliolate (the basal pair often resembling stipules) ·2
2. Annual herbs usually growing below 1000 m; leaf rhachis prolonged 0.5–1.5 mm beyond the insertion of the lateral leaflets; seeds usually mottled · · · · · · · · · · · · · · · · 7. *arabicus*
– Perennial herbs or subshrubs with a woody rootstock, growing above 1000 m; leaf rhachis not prolonged beyond insertion of the lateral leaflets; seeds not mottled · · · · · · · · · · · 3
3. Basal leaflets similar in shape to the other leaflets, symmetrical or only slightly asymmetrical, widest at or above the middle; leaves with or without red dots or markings · · · · · · · · · · 4
– Basal leaflets differing in shape from the other leaflets, ± asymmetrical and widest at or below the middle; leaves usually densely red-dotted or with reddish markings on upper and lower surfaces · 5
4. Leaf rhachis 1.5–2 mm long; leaflets with purple dots; axillary glands large, conspicuous · 3. *wildii*
– Leaf rhachis up to 6(10) mm long; leaflets without or with purple dots; axillary glands very small or absent · 4. *discolor*
5. Basal leaflets (2)2.5–4.5 times as long as wide; flowers 5–8; stems herbaceous, epidermis not purple; all leaflets usually very acute; hairs (at least in the old herbarium specimens) brownish-yellow, stiffly spreading · 6. *namulensis*
– Basal leaflets up to twice as long as wide; flowers 2–8; stems woody or herbaceous, epidermis often purple; leaflets obtuse to subacute · 6
6. Hairs laxly spreading, at least some of them up to 1.5 mm long; calyx tube with ± sparse hairs up to 1.5 mm long; flowers 2–5 · 1. *goetzei*
– Hairs ± appressed; calyx tube almost glabrous; flowers 3–8 · · · · · · · · · · · · · 5. *mlanjeanus*

1. **Lotus goetzei** Harms in Bot. Jahrb. Syst. **30**: 324 (1901). —E.G. Baker, Legum. Trop. Africa: 88 (1926). —Brenan, Check-list For. Trees Shrubs Tang. Terr.: 431 (1949). —J.B. Gillett in Kew Bull. **13**: 369 (1959); in F.T.E.A., Leguminosae, Pap.: 1045, fig. 146/1–12 (1971). —Brummitt in Wye Coll. Malawi Proj. Rep.: 65 (1973). —Lock, Leg. Afr. Check-list: 345 (1989). —Agnew, Upland Kenya Wild Fl., ed. 2: 149, t. 52 (1994). Type from Tanzania.

Lotus discolor var. *microcarpus* Brand in Bot. Jahrb. Syst. **25**: 214 (1898). Type from Tanzania.

Lotus nyikensis Baker f., Legum. Trop. Africa: 89 (1926). Type: Malawi, Nyika Plateau, *Crawshay* s.n. (BM, holotype).

Lotus oehleri Harms in Notizbl. Bot. Gart. Berlin-Dahlem **10**: 79 (1927). Type from Tanzania.

Lotus sp. nr. *oehleri* Harms sensu Brenan in Mem. New York Bot. Gard. **8**: 250 (1953).

Lotus discolor sensu auct. non E. Mey. —Thulin in Fl. Ethiopia **3**: 225 (1989).

Subshrub up to 50 cm tall from a woody rootstock. Stems erect, ascending or prostrate and up to 1(2) m long and up to 3 mm in diameter, woody below, pilose with ± dense white spreading hairs up to 1 mm long, the epidermis dark purple when dry. Leaves very variable in size, only the leaf base and rhachis persisting on older branches; basal leaflets asymmetrical, up to 15 × 8 mm, rhombic with the basal angle usually 70–90°; lateral and apical leaflets up to 18 × 7 mm, the apical one a little longer than the lateral ones, cuneate-obovate; all leaflets with margins revolute, pilose on upper and lower surfaces with spreading hairs up to 1.5(3) mm long, or subglabrous but with some hairs on the midrib and margins, subcoriaceous, densely marked with purple dots on upper surface and purple dots or wavy lines beneath; rhachis 1.5–2.5(?6) mm long, usually less than half the length of the basal leaflets; stipular glands purple, minute. Umbels 2–5-flowered; peduncle 0.5–4 cm long, pubescent; foliage leaf on the peduncle usually 3-foliolate; pedicels 1–2 mm long, pubescent. Calyx reddish, with ± sparse, spreading hairs up to 1.5 mm long; hypanthium c. 1.5 mm long; calyx tube c. 2 mm above the insertion of the stamens; calyx lobes 2.5–3.5 × 1.2–1.5 mm, triangular-acuminate. Corolla white or cream, the standard streaked with red and, at least when dry, the tip of the keel reddish or purplish; standard 8.5–10 × 4–5.5 mm, slightly violin-shaped, wedge-shaped at the base, retuse; wings 8.5–9 mm long including the 3 mm long claw, the blade oblong, laterally pouched, auriculate, the auricles of both wings often adhering to each other; keel c. 10 mm long including claws 3–3.5 mm long, curved, pointed. Filament sheath c. 6 mm long, often papillose near the base; free parts of the shorter filaments c. 2 mm long, those of the longer filaments c. 3 mm long. Ovary c. 7 mm long, glabrous, c. 26-ovulate; style 4 mm long, slightly curved upwards. Pod brownish-red, up to 25 × 2.2 mm, terete. Seeds 16 or less, c. 1.1 mm long, ovoid.

Malawi. N: Chipita Distr., 2 km from Mzengapakweru, 2225 m, fl. & fr. immat. 30.viii.1972, *Synge* WC337 (K; PRE; SRGH).

Also in Ethiopia, Uganda, Kenya and Tanzania. Montane grasslands; 2200–2350 m in the Flora Zambesiaca area, from 1500–3700 m in East Africa.

L. goetzei belongs to a taxonomically very difficult complex endemic to the African mountains. The *L. discolor* complex includes *L. goetzei, L. robsonii, L. wildii, L. discolor* sensu stricto, *L. mlanjeanus, L. namulensis, L. subdigitatus* Boutique and *L. becquetii* Boutique. There are no constant distinctions between the species in floral and fruit morphology except sometimes for the indumentum of the calyx. Principal diagnostic features are leaflet shape, rhachis length, indumentum density and hair type. However, these features are also variable, and transitional forms occur, especially between the three basic species, *L. goetzei, L. discolor* and *L. becquetii*. It is uncertain whether these transitional forms are hybrids; they are at least fertile and easily produce fruits with mature seeds. In Ethiopia, transitional forms are so common that it was impossible to segregate separate species (Thulin, 1989, loc. cit.). Sometimes it is difficult to regard unusual forms simply as hybrids between two or three basic species, and a number of specimens from Ethiopia cannot be effectively identified. However, in other regions of Africa intermediate forms are less frequent and specific boundaries are more evident. It is reasonable therefore to adopt here a narrow specific concept in the *L. discolor* complex. D.D. Sokoloff also considers that the complex needs a revision based on the study of variability in natural populations throughout their range.

2. **Lotus robsonii** E.S. Martins & D.D. Sokoloff, sp. nov.* Type: Malawi, Nyika Plateau, 40 km SE of Rest House on Nchenachena road, 2400 m, 28.x.1958, *Robson & Angus* 430 (K, holotype + isotype; MW, isotype).

Perennial herb with creeping rhizome, probably sometimes suffrutescent; rhizome sympodial, up to 4 mm in diameter. Year-old shoots with leaves well spaced, almost erect (up to 40 cm tall) or with prostrate presumably subterraneous lower part up to 10–15 cm long which later becomes a part of the rhizome; stems c. 1 mm in diameter, with epidermis cream-coloured to reddish when dry, with ± dense indumentum of spreading curled hairs in the upper and younger part to subglabrous in the lower and older part but then usually with a tuft of hairs in the leaf axils where buds occur. Lower leaves (1 to several on erect shoots; up to 11 on ascending ones) small, c. 1.5 mm long, scale-like, with margins in lower part glandular and with apex 3-lobed; the rest of the leaves sessile, 3-foliolate; rhachis up to 1 mm long or almost absent, sparsely hairy, persisting after the leaflets fall; basal leaflets 8–13 × 4.5–21 mm, usually 1.7–2 times longer than wide, asymmetrically ovate-oblong with a basal angle often more than 100°, subacute to obtuse at the apex; apical leaflet 12–22 × 6–12 mm, obovate, symmetrical, rounded or cuneate at the base, acute to obtuse at the apex; all leaflets glabrous on the upper surface, glabrous beneath or with a few hairs only on the midrib and sometimes basally near the margin, subcoriaceous, densely red-dotted above and beneath; stipular glands blackish, minute, usually persistent after leaflets fall. Umbels 2–5-flowered; peduncle (30)50–100 mm long, with indumentum of spreading curled hairs; foliage leaf on the peduncle usually 1-foliolate, small, with 2 blackish small glands (stipules) at the base; pedicels 1–2.5 mm long, densely spreading hairy; glands (bracts) at the base of the pedicels small, inconspicuous or

* **Lotus robsonii** E.S. Martins & D.D. Sokoloff, sp. nov., *L. goetzei* Harms et *L. namulensi* Brand affinis sed foliis trifoliolatis differt; a *Loto simonae* Maire, Weiller & Wilczek foliis inaequifoliolatis subglabris, a *L. ononopside* Balf.f. umbellis 2–5-floris longe pedunculatis, foliis inaequifoliolatis etc., a *L. molli* Balf.f. umbellis 2–5-floris longe pedunculatis, foliolis subglabris etc., a *L. garcinii* DC. dentibus calycini triangulari-acuminatis umbellisque 2–5-floris longe pedunculatis bene differt. Typus: Malawi, Nyika Plateau, 40 km SE of Rest House on Nchenachena road, 2400 m, 28.x.1958, *Robson & Angus* 430 (K, holotypus et isotypus; MW, isotypus).

lacking. Calyx reddish, with sparse weak hairs up to 1.3 mm long; hypanthium c. 1 mm long, calyx tube 2–2.5 mm long above the insertion of the stamens; calyx teeth 3–3.5 mm long, triangular-acuminate, the 2 upper ones wider than the others. Corolla pink, the standard crimson on the back, and the keel with deep purple tip; standard 12–13 × 7–8 mm, slightly violin-shaped, wedge-shaped at the base, retuse, the margins thickened and infolded over c. 3.5 mm from the base; wings 12–13 mm long including the 2.5–3 mm long claw, the blade 9.5–10 × 5.5–6 mm, oblong, strongly auriculate; keel 10.5–11.5 mm long, slightly curved, pointed, the claws c. 2.5 mm long. Filament sheath 6–7 mm long, papillose near the base; free parts of the shorter filaments 2–2.5 mm long, those of the longer filaments 3–4.5 mm long. Ovary 6–6.5 mm long, glabrous, 22–25-ovulate; style 4.5–5.5 mm long. Mature pod unknown.

Malawi. N: Rumphi Distr., Nyika Plateau, 40 km SE of Zambian Government Rest House, on Nchenachena road, 2400 m, fl. 28.x.1958, *Robson & Angus* 430 (K; MW).

Known only from the type collection. Montane scrub, beneath *Erica* (*Phillipia*) bushes; c. 2400 m.

Leaves with five leaflets are extremely characteristic for the genus *Lotus* sensu stricto, and there are only a few exceptions. Species with three leaflets are usually separated into distinct sections, *Ononidium* Boiss. and *Stipulati* Maire, Weiller & R. Wilczek, emend. (see Kramina & Sokoloff in Adansonia, sér. 3, **19**: 321–328 (1997)). *L. robsonii*, however, undoubtedly belongs to the *L. discolor* complex where it seems to be especially close to *L. goetzei* and *L. namulensis*. *L. robsonii* resembles *L. goetzei* and *L. namulensis* in the stem indumentum of spreading, curled hairs, in a somewhat hairy (not glabrous) calyx, in asymmetrical basal leaflets and in other characters. Besides the number of leaflets, *L. robsonii* may differ from *L. goetzei* and *L. namulensis* by having a creeping rhizome and longer style. *L. goetzei* is usually described as an undershrub and *L. namulensis* is a herb with numerous annual shoots from a woody and usually massive rootstock, whereas *L. robsonii* is a rhizomatous herb. However, the basal parts of *L. goetzei* are seldom preserved in herbarium specimens and its life form should be studied more critically. The life form of *L. namulensis* seems to be very characteristic, but it is uncertain whether this is the result of grass fires in the area of its distribution (cf. Gillett 1959, op. cit.: 372). *L. robsonii* has a style 4.5–5.5 mm long, whereas in other *Lotus* species of the Flora Zambesiaca region the style is 3–4 mm long. Such a distinction is significant for some groups of African *Lotus* (see Gillett, 1959, op. cit.). However, there are some specimens of the *L. discolor* sensu lato from Ethiopia with long styles (e.g., *Fries et al.* 1623 (K); *Scott* 160 (K)). On the other hand, the variability of the *L. discolor* complex in Ethiopia seems to be unusual as a whole (see above).

L. robsonii can easily be separated from other species of *Lotus* with constantly trifoliolate leaves. *L. simonae* Maire, Weiller & R. Wilczek (a poorly known species from Morocco) differs from *L. robsonii* in having all three leaflets similar to each other and in an indumentum of appressed, sericeous hairs, *L. mollis* Balf.f. (Socotra) has a dense indumentum on the leaflets and 1-flowered almost sessile umbels, *L. ononopsis* Balf.f. (Socotra) has yellow flowers, all three leaflets similar to each other, 1-flowered umbels and short peduncles without a foliage leaf. In a few species of *Lotus*, some leaves have three leaflets, others four or five, as in *L. garcinii* DC. (West Asia and East Africa) which differs from *L. robsonii* in 1-flowered umbels with shortened peduncle, the peculiar shape of the long calyx teeth and in many other characters. *L. robsonii* has definitely arisen from mountainous species of African *Lotus* with five leaflets and has no close relation to other trifoliolate species.

3. **Lotus wildii** J.B. Gillett in Kew Bull. **13**: 370 (1959). —Goodier & Phipps in Kirkia **1**: 56 (1961). —Drummond in Kirkia **8**: 223 (1972). —Corby in Kirkia **9**: 324 (1974). —Lock, Leg. Afr. Check-list: 348 (1989). TAB. 3,7: **1**. Type: Zimbabwe, Nyanga Distr., Matenderere (Mtenderere) R. source, 5.ix.1954, *Wild* 4590 (K, holotype).

Lotus discolor subsp. *discolor* sensu Goodier & Phipps in Kirkia **1**: 56 (1961). —sensu Drummond in Kirkia **8**: 223 (1972) non E. Mey.

Undershrub from a woody rootstock. Stems erect and up to 60 cm tall, or procumbent and up to 150 cm long, up to 2.5 mm in diameter, woody below, with ± densely appressed or ascending white or yellowish hairs up to 0.5 mm long, glabrescent, the epidermis at first cream-coloured, later dark purple or blackish when dry; internodes usually up to 35 mm long (longer in plants growing in shade). Leaves 5-foliolate, usually only the leaf base with stipules and rhachis persisting on older branches; rhachis 1.5–2 mm long (very rarely up to 3 mm), always less than half the length of the basal leaflets; basal leaflets shorter than the lateral and apical ones, symmetrical or slightly asymmetrical at the base, 5–10 × 2–3.5 mm, oblong-elliptic or obovate, obtuse at the base; lateral and apical leaflets up to 15 × 8 mm, the apical one a little narrower than the lateral, cuneate-obovate; all leaflets obtuse or rounded and apiculate at the apex, pubescent on upper and lower surfaces, sometimes sparsely so,

Tab. 3,7: **1**. LOTUS WILDII. 1, habit (× $^1/_9$); 2, rootstock (× 1), 1 & 2 from *Torre & Pereira* 12407; 3, fruiting and flowering branch (× 1); 4, leaves (× 2); 5, flower (× 4); 6, standard (× 4); 7, wing (× 4); 8, keel, opened out (× 4); 9, flower, petals removed (× 4); 10, androecium (× 4); 11, gynoecium (× 4), 3–11 from *Torre* 5944. Drawn by Judi Stone.

with appressed or ascending short hairs, densely marked with purple dots on upper surface and purple dots or wavy lines beneath; stipular glands large, conspicuous, dark purple or blackish. Umbels 3–5(7)-flowered; peduncle 10–50(65) mm long, strigose; foliage leaf on the peduncle 3-foliolate; pedicels 1.5–2 mm long, strigose; glands (bracts) at the base of the pedicels blackish, pronounced. Calyx appressed pubescent; hypanthium c. 0.7 mm long, calyx tube c. 2 mm above the insertion of the stamens; calyx lobes 1.8–2.5 mm long, triangular-acuminate. Corolla white or cream-coloured, occasionally pale mauve, the standard with reddish or purplish veining, the keel with purplish tip; standard c. 8 × 4.5 mm, slightly violin-shaped, wedge-shaped at the base, the margins thickened and infolded over 3–4 mm from the base; wings with a curved claw 2–2.5 mm long, the blade 6–7 × 3.5–4 mm, oblong, laterally pouched, auriculate; keel as long as the wings, curved, pointed, the claw c. 2 mm long. Filament sheath c. 5 mm long, often papillose near the base; free parts of the shorter filaments c. 2 mm long, those of the longer filaments c. 3 mm long. Ovary c. 5 mm long, glabrous, 18–24-ovulate; style 3.5 mm long, slightly curved upwards. Pod up to 28 × 2.5 mm, reddish-brown to greyish-brown. Seeds 11 or less, c. 1.5 mm long, ovoid, dark grey or greyish-brown.

Zimbabwe. E: Nyanga Distr., near Troutbeck Hotel, c. 2300 m, fl. 14.v.1959, *R.L. Williams* 108 (K; PRE; SRGH). **Mozambique**. MS: Gorongosa Distr., Serra de Gorongosa, mte. Nhandore, c. 1840 m, fl. & fr. 18.x.1965, *Torre & Pereira* 12407 (K; LISC; LMU; PRE; WAG).
Not known elsewhere. Montane grasslands; 1500–2590 m.

4. **Lotus discolor** E. Mey., Comment. Pl. Afr. Austr.: 92 (1836). —Harvey in F.C. **2**: 157 (1862). —Brand in Bot. Jahrb. Syst. **25**: 213 (1898). —E.G. Baker, Legum. Trop. Africa: 88 (1926). —Burtt Davy, Fl. Pl. Ferns Transvaal, pt. 2: 401 (1932). —Brenan in Mem. New York Bot. Gard. **8**: 249 (1953). —Boutique in F.C.B. **4**: 302, fig. 20 A, B (1953). —Hepper in F.W.T.A., ed. 2, **1**: 552 (1958). —J.B. Gillett in Kew Bull. **13**: 371 (1959). —White, F.F.N.R.: 159 (1962). —J.B. Gillett in F.T.E.A., Leguminosae, Pap.: 1047, fig.146/14 (1971). —Ross, Fl. Natal: 200 (1972). —Brummitt in Wye Coll. Malawi Proj. Rep.: 65 (1973). —Compton, Fl. Swaziland: 258 (1976). —Lock, Leg. Afr. Check-list: 344 (1989). —Agnew, Upland Kenya Wild Fl., ed. 2: 149 (1994). —Retief & Herman in Strelitzia **6**: 460 (1997). Type from South Africa (Eastern Cape).

Perennial herb with many herbaceous stems from a woody rootstock. Stems erect and 30–90 cm tall, or sometimes scrambling and up to 2 m long, usually up to 2.5 mm in diameter, the epidermis yellowish-green when dry, whitish or yellowish appressed pubescent, sometimes sparsely so, or pilose with spreading and arching or curled whitish or yellowish hairs, grooved, sometimes the prostrate stems woody and persisting, tinged purple. Leaves usually well spaced, often remote towards the base, 5-foliolate; leaf rhachis usually 1.5–4.5 mm long, less often up to 6(10) mm long, usually shorter than half the length of the basal leaflets or sometimes longer, indumentum as for the stems; basal leaflets 5–15 × 2.2–7 mm, 2–3 times as long as wide, oblanceolate to obovate, similar to the apical ones but smaller, or sometimes asymmetrically elliptic-oblong, subacute to rounded but apiculate at the apex, cuneate to obtuse at the base; apical leaflets 10–18 × 3–7 mm, 2.5–4 times as long as wide, cuneate-oblanceolate to cuneate-obovate, acute to rounded but apiculate at the apex; all leaflets appressed pubescent or spreading hairy on both surfaces, sometimes sparsely so; stipular glands very small or absent. Umbels 2–8-flowered, usually 5–6-flowered; peduncle 2–9 cm long, occasionally up to 15 cm long; foliage leaf on the peduncle sometimes inserted up to 5(8) mm above the tip, 1-foliolate or 3-foliolate with leaflets up to 14 mm long but often much shorter, oblanceolate; pedicels 1.5–3 mm long; glands (bracts) minute or absent. Calyx light green often purple-tinged, appressed pubescent or spreading hairy; hypanthium c. 1 mm long, calyx tube 2–2.5 mm above the insertion of the stamens; calyx teeth 1.5–3 mm long, triangular-acuminate. Corolla white or cream, the standard sometimes veined with purple, the keel with purple tip; standard 7–9 × 3.5–4.5 mm, usually 0.5–1.5 mm shorter than the wings, ± violin-shaped, wedge-shaped at the base, retuse, the margins infolded over 4–5 mm from the base, the upper third wider and abruptly bent upward; wings 9–11 mm long including the 3–3.5 mm long claw, the blade c. 4 mm wide, oblong, laterally pouched, strongly auriculate; keel c. 1 mm shorter than the wings, curved, pointed, the claws c. 3 mm long. Filament sheath 5–6 mm long, papillose near the base; free parts of the shorter filaments c. 2 mm long, those of the longer filaments c. 3 mm long. Ovary 5–6 mm long, glabrous, 25–29-ovulate; style 3–4 mm long, curved upwards. Pod up to 35 × 3 mm, often much shorter, reddish-brown to greyish-brown. Seeds 17 or less, c. 1.5 mm long, ovoid or globose, dark grey.

Hairs of the stems, leaves and inflorescences ± appressed, short, straight · · · · · subsp. *discolor*
Hairs of the stems, leaves and inflorescences spreading, longer, denser, ± curled · · subsp. *mollis*

Subsp. **discolor** —Gillett in Kew Bull. **13**: 371 (1959). —Gonçalves in Garcia de Orta, Sér. Bot. **5**: 96 (1982). —Lock, Leg. Afr. Check-list: 344 (1989).

Lotus tigrensis Baker in F.T.A. **2**: 61 (1871). —Taubert in Engler, Pflanzenw. Ost-Afrikas **C**: 213 (1895). Type from Ethiopia.

Lotus brandianus Harms in Bot. Jahrb. Syst. **28**: 401 (1900). —Brenan, Check-list For. Trees Shrubs Tang. Terr.: 431 (1949). Type from Tanzania.

Lotus discolor var. *cacondensis* Hiern ex Baker f., Legum. Trop. Africa: 878 (1930). —Torre in C.F.A. **3**: 81 (1962). Type from Angola.

Indumentum of the stems, leaves and inflorescences of whitish or yellowish, short, straight appressed or ascending hairs.

Zambia. N: Isoka Distr., Mafinga Mts., fl. & fr. 24.v.1973, *Fanshawe* 11995 (K; SRGH). E: Nyika Plateau, upper slopes of Kangampande, 2130 m, fl. 7.v.1952, *White* 2746 (K). **Malawi**. N: Rumphi Distr., Nyika Plateau, Sangule Kopje, 8 km southwest of Chelinda Camp, 2280 m, fl. 15.v.1970, *Brummitt* 10746 (K; LISC; MAL; SRGH). C: Dedza Distr., Dedza Mt., 1950 m, fl. 5.iv.1978, *Pawek* 14253 (K; MO; MAL). S: Zomba Distr., Zomba Plateau, by Chingwe's Hole, 1890 m, fl. 16.iii.1970, *Brummitt* 9148 (K; LISC; PRE; SRGH). **Mozambique**. N: Ngauma Distr., serra de Massangulo, 1600 m, fl. 5.iii.1964, *Torre & Paiva* 11034 (COI; LISC; LMU; PRE; SRGH; WAG). T: Macanga Distr., Furancungo, east slope of mte. Furancungo, 1265 m, fl. 17. iii.1966, *Pereira, Sarmento & Marques* 1844 (LMU).

Also in eastern Nigeria, Cameroon, eastern Dem. Rep. Congo, Ethiopia, Uganda, Kenya, Tanzania, Angola, Swaziland and South Africa. Montane grassland and evergreen forest margins, miombo woodland and plateau grasslands; 1230–2280 m.

Subsp. **mollis** J.B. Gillett in Kew Bull. **13**: 371 (1959). —Drummond in Kirkia **8**: 223 (1972). —Corby in Kirkia **9**: 324 (1974). —Lock, Leg. Afr. Check-list: 344 (1989). Type: Zimbabwe, Mutare Distr., Himalayas, Engwa Farm, *Wild* 4632 (K, holotype).

Indumentum of the stems, leaves and inflorescences of white, weak, ± curled spreading hairs.

Zimbabwe. E: Mutare Distr., Banti North, 1830 m, fl. 3.x.1962, *Chase* 7821 (K; LISC; SRGH).

Also in South Africa. Montane grassland; 1060–1980 m.

Salubeni 569 (K; LISC; PRE; SRGH) from Malawi, Dedza Distr., Chongoni Forestry School, "in *Brachystegia* woodland on western side near the meteorological screen", fl. & fr. 4.iii.1967, belongs to *L. discolor* subsp. *mollis*. It is probable that this collection is an escape from cultivation or an accidental introduction.

Gillett (1959, op. cit.) treated subsp. *mollis* and subsp. *discolor* as allopatric and reported subsp. *mollis* from Zimbabwe only. Later, however, both subspecies were found in KwaZulu-Natal and Mpumalanga, see Sokoloff in Kew Bull. **56**: 715 (2001). Subsp. *mollis* has been distinguished from subsp. *discolor* on the basis of patent (not appressed) indumentum only. Since no distinction in other features has been found and the geographic separation is not complete, Sokoloff suggests that a special taxonomic study of the whole *L. discolor* complex should be conducted to establish the taxonomic position of plants with spreading indumentum.

5. **Lotus mlanjeanus** J.B. Gillett in Kew Bull. **13**: 371 (1959). —Lock, Leg. Afr. Check-list: 346 (1989). Type: Malawi, Mt. Mulanje (Mlanji), *Mahon* s.n. (K, holotype).

Crotalaria minor C.H. Wright in Bull. Misc. Inform., Kew **1901**: 121 (1901). Type as above.

Lotus minor (C.H. Wright) Baker f., Legum. Trop. Africa: 89 (1926) non Bishop (1826) nec L.M. Clav. (1884).

Lotus discolor sensu Brenan in Mem. New York Bot. Gard. **8**: 249 (1953) pro parte quoad spec. *Brass* 16400, 16647 & 16791 non E. Mey.

Undershrub or perennial herb with numerous straggling herbaceous or rarely woody stems from a woody rootstock. Stems procumbent, up to 1 m long and

1.5(2.5) mm in diameter, red or light green becoming reddish with age, with an indumentum of sparse, ± appressed or short arching hairs. Leaves usually well spaced, 5-foliolate; rhachis 2–5(6) mm long, canaliculate, pubescent; basal leaflets smaller than the others, 4.5–10 × 3–6.5 mm, asymmetrically ovate, obtuse or subacute at the apex, obtuse at the base with a basal angle commonly 90–100°; apical leaflets up to 20 × 7.5 mm, the terminal one larger than the others, cuneate-obovate to oblanceolate, acute to obtuse at the apex; all leaflets sparsely pubescent to subglabrous, sometimes only with a few short hairs on the midrib and petiolules, densely red-dotted above; stipular glands blackish, usually easily visible. Umbels 3–8-flowered; peduncle 15–70 mm long, strigose to subglabrous; foliage leaf on the peduncle 1–3-foliolate, leaf-like; pedicels 1.5–3 mm long, subglabrous; glands (bracts) at the base of the pedicels blackish. Calyx subglabrous, light green tinged red, often the tube red overall; hypanthium c. 1 mm long, calyx tube 2–2.5 mm long above the insertion of the stamens; lobes 2.5–3.5 mm long, triangular-cuspidate. Corolla white or pale pink, the standard red-veined or sometimes reddish or brownish on back; standard 9–10 × 5–7 mm, slightly violin-shaped, wedge-shaped at the base, truncate to retuse at the apex, the margins thickened and infolded over c. 4 mm from the base; wings with a claw 3–3.5 mm long, curved, the blade 7.5–9 × 3.5–4.5 mm, oblong, laterally pouched, auriculate; keel as long as the wings, curved, pointed, the claws c. 3 mm long. Filament sheath c. 6 mm long; free parts of the shorter filaments c. 2.5 mm long, those of the longer filaments c. 4 mm long. Ovary c. 6 mm long, glabrous, c. 23-ovulate; style c. 4 mm long, slightly curved upwards. Pod c. 30 × 3.7 mm, terete, brown. Seeds c. 2 mm long, ovoid, brown.

Malawi. S: Mulanje Mt., beside Thuchila/Chambe path, 1820–2130 m, fl. 10.vi.1957, *J.D. Chapman* 463 (BM; K; LISC; PRE; SRGH).

Only known from Mt. Mulanje where it is the only *Lotus* species. Common in open montane grasslands, *Uapaca* woodlands, and *Widdringtonia* forests, usually in stony places; 1800–2580 m.

6. **Lotus namulensis** Brand in Bot. Jahrb. Syst. **25**: 213 (1898). —E.G. Baker, Legum. Trop. Africa: 87 (1926). —Gillett in Kew Bull. **13**: 372 (1959). —Drummond in Kirkia **8**: 223 (1972). —Corby in Kirkia **9**: 324 (1974). —Lock, Leg. Afr. Check-list: 346 (1989). Type: Mozambique, Zambézia, Namuli, 1887, *Last* s.n. (B, holotype, n.v.; K, isotype) "Namuli, Makua Country".

Lotus eylesii Baker f. in J. Bot. **73**: 294 (1935). Type: Zimbabwe, Stapleford, *Eyles* 6379 (SRGH, holotype).

Perennial herb with numerous annual stems from a woody rootstock. Stems erect or ascending, up to 40 cm tall but usually shorter, 0.5–1.5 mm in diameter, rarely up to 2.5 mm, ± quadrangular and grooved, spreading hairy or sometimes sparsely so. Leaves usually well spaced but sometimes very distant on lower stem, mainly on more vigorous stems; rhachis 1.5–3 mm long (rarely up to 7 mm), broad, canaliculate, hairy to almost glabrous; basal leaflets 4.5–9 × 1.3–3.5 mm, usually 2.5–4.5 times longer than wide, asymmetrically ovate with a basal angle often 70–80°, rarely up to c. 100°, acute to subacute at the apex; apical leaflets 7–15 × (1.5)2–3.5 mm, oblanceolate, narrower than the basal ones, acute to acuminate; all leaflets sparsely hairy to glabrous on upper surface, glabrous beneath or with long hairs only on the midrib and margins, often with reddish markings above and beneath; stipular glands small, inconspicuous. Umbels (4)5–8-flowered; peduncle 35–110 mm long, rarely longer, laterally compressed towards the apex, densely to sparsely covered with fulvous or yellowish spreading hairs, rarely almost glabrous; foliage leaf on the peduncle 1-foliolate or 3-foliolate, sometimes 2–3-lobed; pedicels 1.5–3.5 mm long, often densely spreading hairy, rarely subglabrous; glands (bracts) at the base of the pedicels small, oblong, purplish. Calyx light green, often with reddish or purplish spots, with sparse long weak hairs, sometimes subglabrous save on the teeth margins; hypanthium c. 0.8 mm long, calyx tube 1.8–2.5 mm above the insertion of the stamens; calyx teeth 2–4 mm long, triangular-acuminate. Corolla white with pink, mauve or cream spots, the keel with purplish tip; standard 9–11 × 4.5–5.5 mm, slightly violin-shaped, wedge-shaped at the base, retuse at the apex, the margins thickened and infolded over 3–4 mm from the base; wings 9–11 mm long with the claw 2.5–3 mm long, the blade 6.5–8 × 4–5 mm, laterally pouched, auriculate; keel 8–10 mm long, curved, rostrate, the claws c. 2.5 mm long. Filament sheath 5–6 mm long; free parts of the shorter filaments 2–2.5 mm long, those of the longer filaments 3.5–4 mm long. Ovary 6–7 mm long, glabrous, 29–30-ovulate; style c. 4 mm long. Pod up to 38 × 2.5 mm, usually shorter, terete, purplish or purplish-grey. Seeds 24 or usually fewer, c. 1.8 mm long, ovoid to subglobular, dark brown, dull.

Zimbabwe. E: Chimanimani Distr., Tarka Forest Reserve, 1520 m, fl. vi.1968, *Goldsmith* 97/68 (K; LISC; PRE; SRGH). **Malawi**. N: Chitipa Distr., Nyika Plateau, Domwe Peak, 2310 m, fr. immat. 16.iv.1975, *Pawek* 9270 (SRGH). **Mozambique**. Z: Gurué Distr., Namuli, Makua Country, 1887, *Last* s.n. (B, n.v.; K). MS: Sussundenga Distr., Rotanda, Serra de Messambuzi, 1600 m, fr. 26.xi.1965, *Torre & Correia* 13310 (LISC).

Also in South Africa (Northern Prov.). Mountain grasslands; 1520–2400 m.

7. **Lotus arabicus** L., Mant. Pl.: 104 (1767). —Seringe in A. De Candolle, Prodr. **2**: 212 (1825). —Klotzsch in Peters, Naturw. Reise Mossambique, **6**, part 1: 52 (1861). —J.G. Baker in F.T.A. **2**: 62 (1871). —E.G. Baker in Trans. Linn. Soc. London, Bot. **4**: 9 (1894). —Brand in Bot. Jahrb. Syst. **25**: 219 (1898). —E.G. Baker, Legum. Trop. Africa: 90 (1926). —Hepper in F.W.T.A., ed. 2, **1**: 553 (1958). —J.B. Gillett in Kew Bull. **13**: 375 (1959). —Torre in C.F.A. **3**: 81 (1962). —J.B. Gillett in F.T.E.A., Leguminosae, Pap.: 1048, fig. 146, 16 (1971). —Drummond in Kirkia **8**: 223 (1972). —Corby in Kirkia **9**: 324 (1974). —Gonçalves in Garcia de Orta, Sér. Bot. **5**: 96 (1982). —Lock, Leg. Afr. Check-list: 342 (1989). Lectotype designated by Lassen: Herb. Linn. No. 931.10.

Lotus mossamedensis Welw. ex Baker in F.T.A. **2**: 62 (1871). —Brand in Bot. Jahrb. Syst. **25**: 222 (1898). —Retief & Herman in Strelitzia **6**: 460 (1997). Type from Angola.

Annual herb with few to many stems from the base. Stems spreading or erect, usually up to 65 cm long, rarely more, and up to 4.5(6) mm in diameter, yellowish, pilose with spreading or arching hairs. Leaves usually well spaced, 5-foliolate; leaf rhachis 2–8.5 mm long, rather broad; leaflets all similar, up to 29 × 16 mm, cuneate-obovate, obtuse to truncate at the apex, the lateral ones often inserted on the rhachis at different levels, pilose on both surfaces, the indumentum similar to that on the stem and branches. Umbels 1–3(4)-flowered; peduncle up to 50 mm long; foliage leaf on the peduncle 1-foliolate, or sometimes 2–3-foliolate; glands (bracts) at the base of the pedicels usually small, sometimes inconspicuous; pedicels 2–3 mm long. Hypanthium 1–1.3 mm, calyx tube 1.5–2.2 mm above the insertion of the stamens; calyx lobes 5–6 mm long, becoming 7–7.5 mm long in fruit, narrowly triangular or subulate. Corolla pink, sometimes with purple blotches or stripes, or sometimes reddish or purplish; standard c. 8 mm long, oblong-obovate, wedge-shaped at the base, the margins thickened and infolded near the base; wings 7–7.5 mm long with a narrow claw 2–2.2 mm long, the limb oblong, auriculate; keel as long as the wings, curved, pouched at the sides, the claws 2 mm long, free. Filament sheath c. 4 mm long; free parts of the shorter filaments ± two-thirds as long as those of the longer ones. Ovary c. 5 mm long, c. 22–23-ovulate, glabrous; style 3.5–4 mm long, the tip incurved. Pod 23–36 × 3–3.5 mm, terete to subtorulose, purplish-red, glabrous. Seeds usually c. 18, sometimes less, rarely more, elliptic to ± kidney-shaped, brown, usually mottled.

Zambia. S: Gwembe Distr., Lusitu, fl. 26.ix.1959, *Fanshawe* 5224 (SRGH). **Zimbabwe**. N: Hurungwe Distr., Sanyati R., near junction of R. Vureche (Fulechi), 1050 m, fl. & fr. 11.x.1957, *Phipps* 753 (K; LISC; PRE; SRGH). C: Kadoma Distr., Munyati R., Gowe Pool, c. 850 m, fl. viii.1947, *Whellan* 79 (K; SRGH). E: Chipinge Distr., Save (Sabi) R., Dotts Drift, fl. & fr. 16.xi.1959, *Goodier* 658 (SRGH). S: Beitbridge Distr., Bubye R., near Chikwarakwara Store, c. 4 km upstream from Bubye–Limpopo R. confluence, fl. & fr. 12.v.1958, *Drummond* 5787 (K; SRGH). **Malawi**. S: Nsanje Distr., Lower Shire Valley, Thangadzi R., c. 17.5 km west of Chiromo, 45 m, fl. 16.vii.1958, *Seagrief* 3081 (K; SRGH). **Mozambique**. N: Balama Distr., R. Lúrio, próximo de Niuaia, fl. & fr. 2.ix.1948, *Barbosa* 1973 (LISC). Z: Mocuba arredores, fl. & fr. 1943, *Torre* s.n. (LISC). T: Changara Distr., Boruma (Boroma), R. Zambeze, fl. & fr. 22.x.1942, *Mendonça* 351 (LISC). MS: entre Marrínguè (Marínguè) e Gorongosa (Vila Paiva de Andrada), fl. & fr. 5.x.1944, *Mendonça* 2352 (K; LISC). GI: R. Limpopo, entre Xai-Xai (Vila João Belo) e Chibuto, fl. & fr. 20.vii.1944, *Torre* 6801 (LISC).

Also in Senegal, Gambia, northern Nigeria, Mali, Egypt, Sudan, Ethiopia, Tanzania, Angola, South Africa (Northern Prov.) and in Arabia. River banks, in sand and mud, usually in hot low altitudes; 45–1050 m.

84. ANTOPETITIA A. Rich.

By B. Verdcourt

Antopetitia A. Rich. in Ann. Sci. Nat., Bot., Sér. 2, **14**: 261, t. 15, fig. 2 (1840). — Verdcourt in Kirkia **9**: 555 (1974).

Annual herb. Leaves imparipinnate; petiole absent or very short; stipels absent or very short; stipules reduced to glands. Flowers in axillary pedunculate umbels; bracts and bracteoles minute, soon deciduous. Calyx tube and hypanthium obconic, the teeth subequal, narrowly triangular. Corolla glabrous, the petals with claws about equalling their blades. Stamens diadelphous, the vexillary one entirely free; free parts of filaments flattened; anthers uniform. Ovary stipitate, inserted laterally on the hypanthium. Fruit curved, made up of 2–5 spherical 1-seeded articles each of which splits into 2 valves; dorsal suture persistent. Seeds spherical, minutely warty; hilum circular, minute.

A monotypic African genus closely allied to *Ornithopus* L. but differing in the fruit segments being dehiscent and in the dorsal suture being persistent.

Antopetitia abyssinica A. Rich. in Ann. Sci. Nat., Bot., Sér. 2, **14**: 261, t. 15, fig. 2 (1840); Tent. Fl. Abyss. **1**: 209, t. 39 (1847). —E.G. Baker, Legum. Trop. Africa: 274 (1929). —Robyns, Fl. Sperm. Parc Nat. Alb. **1**: 318 (1948). —J. Léonard in F.C.B. **5**: 179, fig. 10 (1954). —Hepper in F.W.T.A., ed. 2, **1**: 577 (1958). —White, F.F.N.R.: 143 (1962). —Gillett in F.T.E.A., Leguminosae, Pap.: 1049, fig. 147 (1971). —Drummond in Kirkia **8**: 216 (1972). —Verdcourt in Kirkia **9**: 555 (1974). —Gonçalves in Garcia de Orta, Sér. Bot. **5**: 63 (1982). —Lock, Leg. Afr. Check-list: 123 (1989). TAB. 3,7: **2**. Type from Ethiopia.

Ornithopus coriandrinus Hochst. & Steud. ex Fielding & Gardner, Sert. Pl. **1**: t. 49 (1844). —Oliver in F.T.A. **2**: 140 (1871); in Trans. Linn. Soc. London, Bot. **2**: 331 (1887). —Taubert in Engler, Pflanzenw. Ost-Afrikas **C**: 213 (1895). —Harms in Engler, Pflanzenw. Afrikas [Veg. Erde 9] **3**: 605 (1915). Type from Ethiopia.

Hormolotus johnstonii Oliv. in Johnston, Kilimanjaro Exped.: 337, 339 (1886) *nom. nud.*

Spreading or, more rarely, erect pubescent annual herb 0.2–1.2 m tall, with a well-defined taproot. Leaflets 5–11, alternate or subopposite, 4–20 × 0.5–4 mm, narrowly elliptic to linear-oblanceolate, rounded to acute at the apex, cuneate, glabrous on upper surface, pubescent beneath; rhachis 5–25 mm long. Umbels 2–8-flowered; peduncle 1.5–8 cm long with a bract resembling the leaves situated near the middle or, less often, towards the apex; pedicels 2–3 mm long, bearing small bracteoles towards the apex. Calyx pubescent, the hypanthium and tube each c. 1 mm long; teeth 2–3 mm long. Corolla yellow or orange; standard limb 1.5–2 × 2.5 mm, oblate, brown-veined, the claw 1.5–2 mm long. Ovary glabrous, c. 4 mm long including the 1 mm long stipe; style scarcely 1 mm long. Fruit c. 7–12 mm long, the articles 2–2.5 mm in diameter, shining, glabrous, with reticulate venation. Seeds dark brown, 1.5 mm in diameter.

Zambia. N: Mbala (Abercorn), fl. & fr. 18.iii.1962, *E.A. Robinson* 5038 (K; SRGH). **Zimbabwe**. E: Nyanga Distr., Nyanga (Inyanga), road cutting, fr. 10.iii.1964, *Corby* (K; SRGH). **Malawi**. N: Khondowe (Kondowe) to Karonga, fl. & fr. vii.1896, *Whyte* (K). C: Ntchisi (Nchisi) Mt., fl. & fr. 6.v.1963, *Verboom* 850 (K; LISC; SRGH). S: Zomba Mt., fl. & fr. 29.vi.1955, *Jackson* 1697 (K). **Mozambique**. T: Tete Distr., a 2 km do cruzamento para Dedza em direccao a Vila Coutinho, fl. & fr. 17.vii.1949, *Barbosa & Carvalho* 3661 (K; LMA).

Also on Cameroon Mt. and in Eritrea, Ethiopia, eastern Dem. Rep. Congo, Rwanda, Burundi, Uganda, Kenya and Tanzania. Montane and submontane grassland, rocky outcrops, roadsides; 1460–2220 m (up to 3200 m in East Africa).

Tribe 13. GALEGEAE

By Maria Adélia Diniz

Galegeae (Bronn) Torr. & Gray, Fl. N. America **1**: 292 (1838).

Herbs or shrubs; hairs basifixed or biramous. Leaves usually without a distinct pulvinus, usually distichous, imparipinnate (sometimes paripinnate outside the Flora

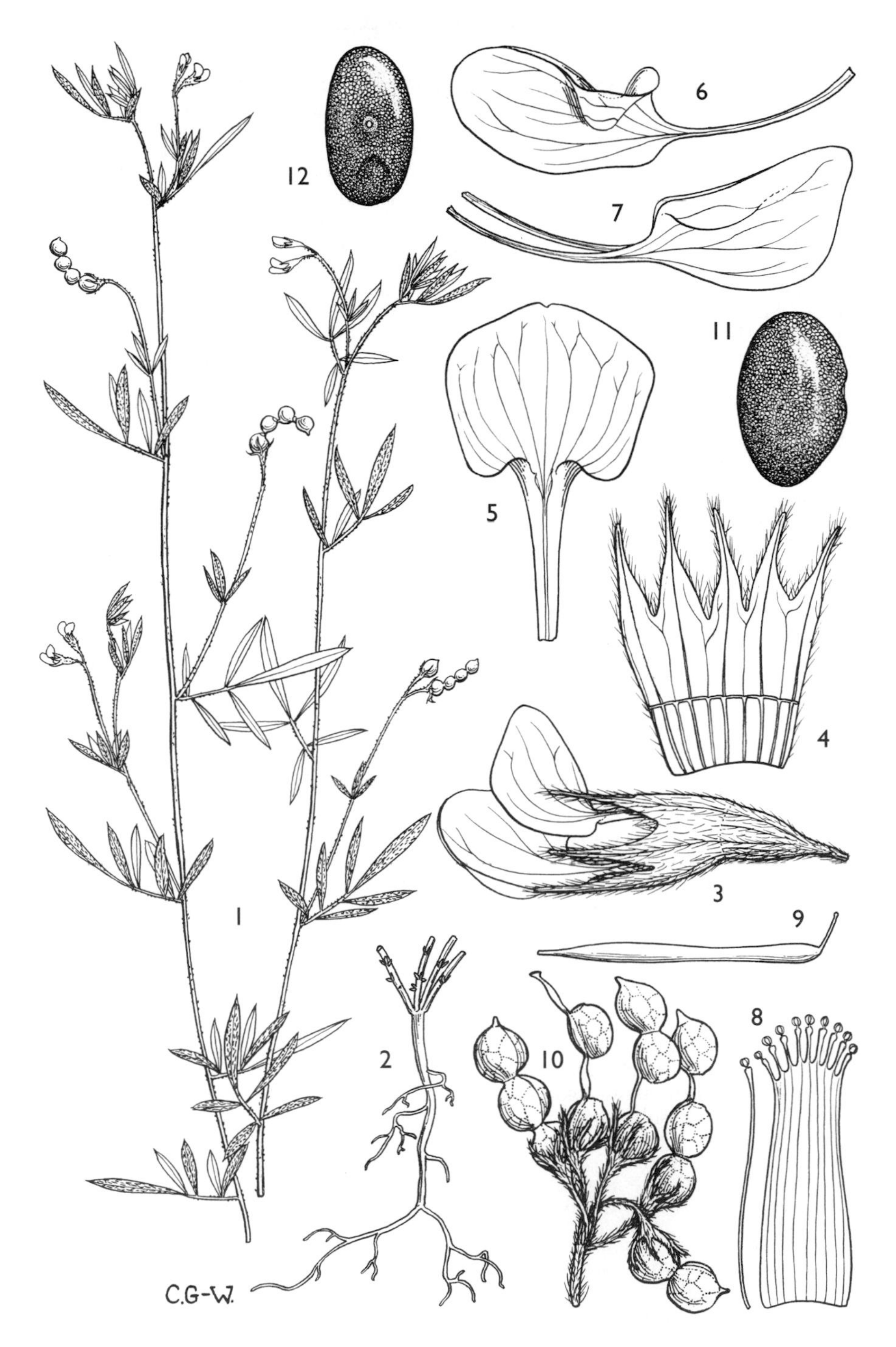

Tab. 3,7: **2**. ANTOPETITIA ABYSSINICA. 1, branches with flowers and fruits (× 1); 2, taproot with stem bases (× 1); 3, flower (× 10); 4, calyx, opened out (× 10); 5, standard (× 12); 6, wing (× 12); 7, keel (× 12); 8, androecium, spread out, with free vexillary stamen (× 10); 9, gynoecium (× 10), 1–9 from *Milne-Redhead & Taylor* 10752; 10, infructescence (× 4); 11, seed, side view (× 12); 12, seed, hilar view (× 12), 10–12 from *Chandler & Hancock* 2518. Drawn by Christopher Grey-Wilson. From F.T.E.A.

Zambesiaca area), usually with opposite leaflets (rarely 1–3-foliolate outside the Flora Zambesiaca area); stipels lacking; stipules usually adnate to the petiole and sometimes connate around the stem. Flowers in axillary extended or contracted racemes, rarely solitary; bracteoles present or absent. Calyx campanulate to tubular, with sinuses equal or the upper sinus deeper or shallower; hypanthium short and broad, sometimes oblique at the base. Standard clawed or narrowed to the base; wings usually well developed, rarely coarsely sculptured, usually with a fold interconnecting with the keel; keel blunt to apiculate. Stamen filaments all united into a sheath or more often the vexillary one free; anthers usually uniform. Ovary few to many-ovulate; style slender, tapered, bearded or not, with a terminal or lateral stigma. Pods compressed, angled or inflated, sometimes with the sutural margins intruded or longitudinally septate, occasionally torulose, dehiscent or indehiscent. Seeds oblong-reniform, with a small hilum.

A tribe of c. 20 genera mainly in Eurasia and North America, and extending into southern Africa, Australia and temperate South America.

The tribe *Hedysareae*, similar to the *Galegeae* but distinguished by having jointed fruits, has been recorded in the Flora Zambesiaca area by only one species in cultivation. *Hedysarum coronarium* L., "Spanish Sainfoin", native of southern Europe, has been grown in pasture trials at Marondera, Grasslands Research Station, fl. 29.v.1961, *Corby* 1015 (K; SRGH), and in Mutare, fl. 20.ix.1948, *Smithers* in *GHS* 21819 (K; SRGH). It is a perennial to 1 m tall; leaves imparipinnate with 7–11 leaflets; stipules free; racemes dense; flowers carmine, with the standard 12–15 mm long; fruits jointed and comprised of 2–4 spinulose articles.

1. Style glabrous ························ 87. **Astragalus**
– Style bearded or with a tuft of hairs below the stigma ························ 2
2. Standard exceeding the other petals; style with hairs along the upper side towards the apex, sometimes also with a tuft of bristles below, and in front of, the capitate stigma ························ 85. **Sutherlandia**
– Standard ± equalling the other petals; style without hairs along the upper side, but with a tuft of bristles below the capitate stigma mostly on the lower side but sometimes forming an almost complete ring ························ 86. **Lessertia**

85. SUTHERLANDIA R. Br. ex W.T. Aiton

Sutherlandia R. Br. ex W.T. Aiton, Hort. Kew., ed. 2, **4**: 327 (1812) *nom. conserv.* —E. Phillips & R.A. Dyer in Revista Sudamer. Bot. **1**: 69–80 (1934).

Shrubs up to 2 m high, rarely dwarf shrubs 10–20 cm high. Leaves imparipinnate; stipules small, ovate, joined. Flowers large, in 2–8-flowered racemes. Calyx campanulate, 5-lobed; lobes ovate. Standard oblong, shorter than the keel, curved when seen in profile, emarginate, slightly narrowed to the base, with the margins usually reflexed above; wings clawed, auriculate, the auricle shorter than to almost as long as the claw; keel clawed, oblong to oblong-lanceolate, auriculate, the claws connate except at the base. Stamens in a sheath open on the upper side; vexillary stamen free; stamen tube about 3 times as long as the free portion of the filaments; anthers equal, oblong. Ovary stipitate, many-ovuled, glabrous or rarely villous; style curved, laterally bearded along the upper side, sometimes with a tuft of hairs just below the small capitate stigma on one side. Pod membranous, bladdery. Seeds black, flattened, with well developed funicle.

A genus of 6 species mainly South African but also in Namibia. One or two species are cultivated as ornamentals in gardens.

Mature pods 1–2 times longer than broad; upper suture convex; corolla with the wings straight, obtuse at the apex ························ 1. *frutescens*
Mature pods 2.5–4 times longer than broad; upper suture straight; corolla with the wings oblique and acute at the apex ························ 2. *microphylla*

1. **Sutherlandia frutescens** (L.) W.T. Aiton, Hort. Kew., ed. 2, **4**: 327 (1812). —De Candolle, Prodr. **2**: 273 (1825). —E. Meyer, Comment. Pl. Afr. Austr.: 121 (1836). —Harvey in F.C. **2**:

Tab. 3,7: **3**. SUTHERLANDIA FRUTESCENS. 1, habit ($\times \frac{1}{6}$), from *Shantz* 58; 2, flowering and fruiting branch (× 1); 3, flower, side view ($\times 1\frac{1}{2}$); 4, calyx, opened out ($\times 1\frac{1}{2}$); 5, standard ($\times 1\frac{1}{2}$); 6, wings ($\times 1\frac{1}{2}$); 7, keel ($\times 1\frac{1}{2}$); 8, androecium and gynoecium ($\times 1\frac{1}{2}$); 9, tip of stigma, enlarged (× 16), 2–9 from *Harbor* 6612. Drawn by Judi Stone.

212 (1862). —E.G. Baker, Legum. Trop. Africa: 264 (1929). —E. Phillips & R.A. Dyer in Revista Sudamer. Bot. **1**: 73, fig. 1 (1934). —Rice & Compton, Wild Fls. Cape Good Hope: t. 38 (1951). —Schreiber in Merxmüller, Prodr. Fl. SW. Afrika, fam. 60: 113 (1970). —Lock, Leg. Afr. Check-list: 265 (1989). TAB. 3,7: **3**. Type from South Africa ("Cape").
Colutea frutescens L., Sp. Pl.: 723 (1753).

Shrub up to 1.7 m high. Stems usually erect, slightly ridged, densely appressed silvery-pubescent to glabrescent. Leaves 3–10 cm long; lateral leaflets in 3–8 pairs, alternate or opposite, 8–15 × 2.5–5 mm, linear-oblong to oblong, obtuse or rounded at the base, obtuse to slightly retuse, sometimes apiculate at the apex, glabrous or almost glabrous on the upper surface, usually thinly ± appressed silvery-pubescent on the lower surface; petiole 5–15 mm long, appressed-pubescent; rhachis sulcate on the upper side; stipules 1–2.5 × 0.3–0.5 mm, triangular and joined at the base, with a nerve near the middle. Racemes short, 3–6-flowered; peduncle 3–8 cm long, strigose; bracts 1.5–2.5 mm long, ovate to oblong; pedicels 5–10 mm long, strigose; bracteoles 2, at the apex of the pedicel, c. 1 mm long. Calyx tube 8–12 mm long, appressed pubescent or strigose outside, glabrous inside; lobes 1–3 mm long, ovate, sometimes acuminate, with short stiff sparse hairs outside. Corolla bright red or scarlet; standard 25–30 mm long including the claw, 8–14 mm wide, oblong, the claw c. 8 mm long, linear; wings 6–8 mm long, 1.5–2 mm wide, oblong, obtuse at the apex, obliquely clawed; keel longer than the standard, 30–35 mm long, c. 8 mm wide, oblong, curved, the petals auriculate with a claw 8–12 mm long, linear. Stamen tube 15–20 mm long; free parts of the filaments 5–8 mm long, filiform; anthers c. 0.8 mm long, oblong. Ovary stipitate, c. 15 mm long, oblong, semitranslucent, many-ovulate; style c. 15 mm long, bearded along the upper side towards the apex and with a tuft of hairs on the lower side just below the capitate stigma. Pod up to 5 × 2.5 cm, oblong to falcate, narrowed to the base into a stipe c. 5 mm long, and towards the apex to a beak terminated by the persistent curved style, inflated, membranaceous, indehiscent, 8–12-seeded. Seeds 2–3 mm long, subreniform-orbicular in outline, flattened, black.

Botswana. SW: Kgalagadi Distr., 1 km NE of Tshane, 24°01'S, 21°56'E, fl. & fr. 1.vii.1977, *Skarpe* S-188 (K; SRGH). SE: Kgatleng Distr., Mochudi, fl. i–iv.1914, *Harbor* in *Rogers* 6612 (K).
Also in South Africa, Lesotho and Namibia and introduced in Kenya. In low rainfall areas with low shrubs and sparse ground cover, on sandy soils, often on roadsides and disturbed areas; c. 1000 m.

2. **Sutherlandia microphylla** Burch. ex DC., Prodr. **2**: 273 (1825). —E. Phillips & R.A. Dyer in Revista Sudamer. Bot. **1**: 76, fig. 2 (1934). —Lock, Leg. Afr. Check-list: 265 (1989). Type from South Africa (Cape).
Sutherlandia frutescens var. *microphylla* (Burch. ex DC.) Harv. in F.C. **2**: 213 (1862).

Shrublet 0.6–1.3 m high. Stems few-branched, erect or ascending sometimes slightly ridged, strigose or more rarely almost glabrous, slender and hard. Leaves 2–9 cm long; lateral leaflets in 5–10 pairs, alternate to subopposite, 5–12 × 1–3 mm, elliptic to linear-oblong, obtuse at the apex, cuneate to rounded at the base, usually glabrous, more rarely puberulous on the upper surface, pubescent beneath; petiole 5–11 mm long; rhachis sulcate on the upper side, rounded on the back; stipules c. 2 mm long, ovate to oblong. Racemes 2–8-flowered; peduncle 7–17 mm long; rhachis 3–5 mm long; bracts 1–2 mm long, ovate; pedicels 3–6 mm long; bracteoles c. 1 mm long, at the base of the calyx, acute. Calyx tube campanulate, c. 10 mm long, strigose or setose without, glabrous within; lobes 3–4 mm long, ovate to linear-oblong, obtuse, pubescent or strigose without, pubescent to almost glabrous within. Corolla glabrous, scarlet; standard 16–29 mm long, elliptic to oblong, tapering at the base, with the margins usually reflexed above; wings 6–11 mm long, 1.5–1.8 mm wide, linear, oblique and acute at the apex, shortly auriculate, clawed; keel petals 25–38 mm long including the 6–13 mm long claws, the blades oblong to lanceolate. Filament sheath 17–24 mm long; free parts of the filaments 5–8 mm long, filiform; anthers 0.3 mm long, oblong. Ovary stipitate, 10–18 mm long, oblong, semitranslucent, glabrous; stipe 8–14 mm long; style 16 mm long, bearded along the upper side on the distal third. Pod 4–6 × 1–2 cm, membranaceous, inflated.

Botswana. SE: South-East Distr., Pharing, on old dam wall, at earthworks, 1190 m, fl. viii.1944, *O.B. Miller* 340 (PRE).
Also in South Africa. In low rainfall areas on dry rocky hillsides; c. 1070 m.

86. LESSERTIA DC.

Lessertia DC., Astrag.: 19 & 47 (1802) *nom. conserv.*; Prodr. **2**: 271 (1825).

Perennial or rarely annual herbs, subshrubs or shrubs, rarely spinescent. Leaves imparipinnate, with small stipules. Flowers in pedunculate axillary racemes, rarely solitary, terminal, bracteate. Calyx campanulate, 5-lobed; lobes small, subequal. Standard suborbicular, sessile or shortly clawed, sometimes bilobed, spreading or reflexed, exceeding the keel; wings usually auriculate, clawed; keel petals straight or incurved, obtuse at the apex, clawed. Stamens in a sheath open on the upper side; vexillary stamen free; anthers uniform. Ovary sessile or stipitate, 2–many-ovuled; style filiform, incurved, with a tuft of hairs below the terminal capitate stigma. Pod oblong or linear, membranaceous, somewhat inflated or compressed, the upper suture not impressed, at length dehiscing at the apex. Seeds reniform on a filiform funicle.

An African genus of more than 50 species, c. 35 of them from South Africa.

1. Pod compressed, oblong or linearly oblong, more than twice as long as broad, straight or curved; raceme shorter than the subtending leaf · 2
– Pod inflated, compressed or subcompressed, ± ovate, elliptic, oblong-elliptic or rounded; raceme longer or shorter than the subtending leaf · 3
2. Raceme 3–6-flowered; pod falcate · 3. *falciformis*
– Raceme 2-flowered; pod ± straight · 4. *pauciflora*
3. Raceme shorter than the leaf; pod with a stipe 2–3 mm long; undershrub, sparsely puberulous or almost glabrous · 1. *benguellensis*
– Raceme longer than the leaf; pod compressed with a stipe up to 2 mm long; shrublet, densely silky-villous · 4
4. Pedicel shorter than the calyx; pod elliptic-oblong, usually with small red or purple spots · 2. *macrostachya*
– Pedicel longer than the calyx; pod obliquely elliptic to ovate · · · · · · · · · · · · 5. *perennans*

1. **Lessertia benguellensis** Baker in F.T.A. **2**: 137 (1871). —Dinter in Repert. Spec. Nov. Regni Veg. **18**: 438 (1922). —E.G. Baker, Legum. Trop. Africa: 265 (1929). —E.P. Sousa in C.F.A. **3**: 186, t. 16 (1962). —Schreiber in Merxmüller, Prodr. Fl. SW. Afrika, fam. 60: 71 (1970). —Lock, Leg. Afr. Check-list: 261 (1989). TAB. 3,7: **4**. Type from Angola.
Lessertia emarginata Schinz in Verh. Bot. Vereins Prov. Brandenburg **30**: 166 (1888). —Dinter in Repert. Spec. Nov. Regni Veg. **18**: 438 (1922). —Drummond in Kirkia **8**: 223 (1972). —Lock, Leg. Afr. Check-list: 262 (1989). Type from Namibia.
Coluteastrum benguellense (Baker) Hiern, Cat. Afr. Pl. Welw. **1**: 232 (1896).

Erect robust many-stemmed perennial herb up to 1.5 m high, from a woody rootstock. Stems herbaceous and pithy above, branching, glabrous. Leaves 5–13.5 cm long; leaflets 2–9 on each side of the rhachis, opposite or distally subopposite, 6–25 × 2–9 mm, linear-oblong or obovate, truncate to emarginate and mucronulate at the apex, cuneate at the base, very sparsely white hairy beneath, glabrous above; petiole 1–1.5 cm long, with very sparse white hairs; rhachis 1.5–12 cm long, sulcate on the upper side; petiolules up to 1 mm long, with sparse white stiff hairs; stipules 1–3 mm long, lanceolate, with white stiff hairs at the margins. Racemes up to 4 cm long, 3–many-flowered; peduncle 3–8 mm long; bracts 1–1.5 mm long, ovate-lanceolate, with stiff white hairs at the margins; pedicels 7 mm long; bracteoles at the base of the calyx, minute, with stiff hairs. Calyx 2.5–3 mm long, campanulate, divided halfway into 5 equal teeth, with stiff hairs only at the margins. Corolla 8 mm long, yellow-greenish with the standard and wings streaked with purple-violet-bluish; standard c. 8 mm long and 6 mm wide, obovate, rounded and shortly emarginate at the apex, tapering at the base into a linear claw; wing petals c. 7 mm long including the 2 mm long filiform claws, the blades c. 2 mm wide, falcate-oblong, auriculate at the base; keel petals c. 8 mm long including the 3 mm long linear claws, the blades 2.5 mm

Tab. 3,7: **4.** LESSERTIA BENGUELLENSIS. 1, flowering and fruiting branch (× ²/₃); 2, flower (× 4); 3, standard (× 4); 4, wing (× 4); 5, keel (× 4); 6, androecium with free vexillary stamen (× 4); 7, gynoecium (× 4), 1–7 from *Skarpe* 285; 8, pod (× ²/₃); 9, pod, viewed from above (× ²/₃), 8 & 9 from *Mendes* 3843. Drawn by Pat Halliday.

wide, obovate, rounded at the apex, auriculate at the base. Filament sheath up to 5 mm long; free parts of the filaments 2.5 mm long; free stamen 6 mm long; anthers less than 1 mm long. Ovary c. 4 mm long, oblong, up to 7-ovulate, glabrous; style 2 mm long. Pod 1.5–4 × 1–2 cm, ovate or obliquely elliptic, sometimes with the dorsal suture arched, acute to obtuse or rounded at the apex, cuneate to rounded at the base and abruptly tapering into a 2–3 mm long stipe, membranaceous, translucent, brownish, glabrous. Seeds 1–7, obliquely oblong-reniform, 5 × 3.5 mm, dark brown.

Caprivi Strip. Okavango R., 19 km north of Shakawe, on Botswana border, fr. 16.iii.1965, *Wild & Drummond* 7090 (LISC; PRE; SRGH). **Botswana**. N: Ngamiland Distr., Nxaxumo, Gomoti R., 19°28.2'S, 23°25.1'E, fr. 13.v.1976, *P.A. Smith* 1732 (K; PRE; SRGH). SW: Ghanzi Distr., Kobe Pan, 23°14'S, 20°00'E, fl. & fr. 20.iii.1978, *Skarpe* S-285 (K; PRE; SRGH). **Zambia**. B: Sesheke, fr. 19.vi.1963, *Fanshawe* 7877 (BR; K). **Zimbabwe**. W: Umguza Distr., Nyamandhlovu, Redbank, fr. 31.i.1948, *Sharp* 2 (SRGH).

Also in Angola, Namibia and South Africa. In thickets and dry open woodland, on Kalahari Sand, often on floodplains and river banks; c. 900–1150 m.

This species was cultivated in Marondera, Grasslands Research Station, fl. & fr. 7.ii.1967, *Corby* 1738 (K; SRGH), from seeds from Okavango R. in Botswana N.

2. **Lessertia macrostachya** DC. in Ann. Sci. Nat. (Paris) **4**: 100 (1825); Prodr. **2**: 272 (1825). —Harvey in F.C. **2**: 216 (1862). —Schreiber in Merxmüller, Prodr. Fl. SW. Afrika, fam. 60: 73 (1970). —Lock, Leg. Afr. Check-list: 264 (1989). Type from South Africa (Cape).

Erect perennial herb up to 1 m high, from a woody taproot. Stems virgate, much branched from the base, herbaceous and pithy above, striate, green-greyish, covered by dense soft hairs. Leaves up to 9.5 cm long; leaflets 8–14 on each side of the rhachis, opposite or subopposite, 5–10 × 2–4 mm, elliptic to ovate, emarginate or rounded at the apex, cuneate to rounded at the base, thinly silky hairy on both surfaces; petiole 1–1.5 cm long, with appressed white soft hairs; rhachis 1.2–8 cm long, sulcate on the upper side; petiolules less than 1 mm long; stipules 2.5 mm long, subulate, ± silky. Racemes up to 25 cm long, laxly many-flowered, silky hairy; peduncle 3–4 cm long; bracts c. 1.5 mm long, ovate; pedicels 1.5 mm long. Calyx 3–4 mm long, campanulate, 5-toothed in the upper third, thinly silky hairy becoming puberulous. Corolla c. 6 mm long, yellow-purple or violet at the apex; standard c. 6 × 4 mm, elliptic, emarginate at the apex, tapering at the base into a short claw; wings 6 mm long including the 1.5 mm long claws, the blades 3 mm wide, falcate-oblong, auriculate at the base; keel petals c. 5 mm long including the 1.5 mm long claw, the blades c. 1.5 mm wide, rounded at the apex, auriculate at the base. Filament sheath up to 3.5 mm long; free parts of the filaments 1–2 mm long; free stamen 4 mm long; anthers c. 1 mm long. Ovary c. 3.5 mm long, oblong, few-ovulate, villous-pubescent or glabrous; style c. 2 mm long. Pod 15–22 × 8–10 mm, flattened, elliptic-oblong, rounded at the apex but with persistent curved style, tapering to the base into a 1–2 mm long stipe, thinly pubescent or glabrous, pale green, usually with small red or purple spots. Seeds 1–3 in each pod, 3 × 2 mm, reniform, blackish.

Botswana. SW: Kgalagadi Distr., Bohelabatho Pan, 23°50'S, 21°21'E, on pink sand dune south of the pan, fl. & fr. 2.xi.1978, *Skarpe* S-295 (K; PRE; SRGH).

Also in Namibia and South Africa. Low rainfall area in grassland with scattered trees and shrubs on Kalahari Sand, on sand dunes and sandy soils; also on roadsides; up to 1150 m.

3. **Lessertia falciformis** DC., Prodr. **2**: 272 (1825); Mém. Lég.: 292, t. 46 (1826). —Harvey in F.C. **2**: 221 (1862). —Schreiber in Merxmüller, Prodr. Fl. SW. Afrika, fam. 60: 72 (1970). —Drummond in Kirkia **8**: 223 (1972). —Lock, Leg. Afr. Check-list: 263 (1989). Type from South Africa (Cape).

Perennial herb with a long woody taproot. Stems 20–40 cm long, prostrate, covered by sparse small white appressed or ascending swollen simple hairs. Leaves up to 7.5 cm long; leaflets 4–8 on each side of the rhachis, opposite or subopposite, up to 15 × 2 mm, linear-oblong, with white hairs as on the stems and petiole on the lower surface, glabrous above; terminal leaflet much longer than the distal lateral pair; petiole 5–18 mm long, with ascending or ± appressed small white swollen hairs; rhachis 3.5–5 cm long, only slightly prolonged beyond the ultimate pair of leaflets; petiolules less than 1 mm long; stipules 2–4 mm long, narrowly triangular, curved, persistent. Racemes 3–6-flowered; peduncle (7)20–25 mm long; bracts 1–1.5 mm long, persistent; pedicels 2–3 mm long; bracteoles minute, linear, at the base of the calyx or often absent. Calyx c. 3 mm long, campanulate, divided to near the middle into 5 subequal teeth. Corolla 5–8 mm long, pink or purplish, glabrous; standard 5–8 × c. 6 mm, elliptic, emarginate at the apex, tapering at the base into a 1 mm long claw; wings 5–7 mm long, the blades c. 2 mm wide, oblong-falcate, auriculate at the base, the claws c. 2 mm long; keel petals 5–8 mm long, the blades c. 3 mm wide, obovate, rounded at the apex, auriculate at the base, the claws c. 2 mm long, linear. Filament sheath up to 5 mm long; free parts of the filaments c. 2 mm long; free stamen 5 mm long; anthers c. 0.2 mm long. Ovary shortly stipitate, 4–5 mm long, oblong, pluri-ovulate; style 2 mm long, persistent in fruit. Pod c. 40 × 8 mm, falcate, membranaceous, shortly stipitate, narrowed at the base, with a short beak at the apex terminated by the upcurved style, sparsely setulose. Seeds 9–15 in each pod, 3 × 1.5 mm, broadly reniform, yellowish-brown, with a long funicle.

Botswana. SE: Kweneng Distr., 6 km east of Motokwe, 24°06'S, 23°20'E, 1100 m, fr. 14.iii.1988, *Mithen* 530 (PRE). **Zimbabwe**. W: Matobo Distr., Matopos Research Station, fr. 9.i.1969, *Mangema* in *MRSH* 4418 (SRGH).

Also in Namibia and South Africa. In low rainfall areas in dry open scrubland, on Kalahari Sand and on calcrete pan margins; 1100–1350 m.

4. **Lessertia pauciflora** Harv. in F.C. **2**: 222 (1862). —E.G. Baker, Legum. Trop. Africa: 265 (1929). —Gillett in F.T.E.A., Leguminosae, Pap.: 1063 (1971). —Lock, Leg. Afr. Check-list: 264 (1989). —Agnew, Upland Kenya Wild Fl., ed. 2: 150 (1994). Syntypes from South Africa (? and Lesotho*).

Suberect or prostrate perennial herb or subshrub 5–35 cm high, from a woody taproot. Stems much branched from the base, ascending to erect, ± strigose canescent, becoming glabrescent. Leaves up to c. 7 cm long; leaflets (1)3–5 on each side of the rhachis, opposite or rarely subopposite, 3–25 × 1–4 mm, elliptic-oblong, obtuse to retuse and apiculate at the apex, obtuse or rounded at the base, with white hairs on the lower surface, glabrous above; terminal leaflet much longer than the others; petiole 5–15 mm long, with appressed small white hairs; rhachis 15–40 mm long, prolonged 1.5–7 mm beyond the ultimate pair of leaflets, sulcate on the upper side; petiolules short; stipules up to 5 mm long, obliquely subulate. Racemes 1–2-flowered; peduncle 7–22 mm long; bracts 1 mm long, ovate, persistent; pedicels 1.5 mm long, thinly puberulous, with 2 minute bracteoles at the apex. Calyx c. 3 mm long, campanulate, divided to just above the middle into 5 subequal teeth, puberulous. Corolla c. 7 mm long, purplish or reddish to pink, becoming very pale with age, caducous; standard c. 7 × 6 mm, elliptic, emarginate, tapering at the base into a short claw; wings 5 mm long including the 1.5 mm long claw, the blade 1 mm wide, falcate-oblong, auriculate at the base; keel petals c. 6.5 mm long including the 2 mm long linear claws, the blades c. 2 mm wide, obovate, rounded at the apex, auriculate at the base. Filament sheath up to 4 mm long; free parts of the filaments c. 1 mm long; free stamen 4 mm long; anthers minute. Ovary c. 4 mm long, oblong, many-ovulate, setulose along the upper suture; style 2 mm long, glabrous. Pod 30–43 × 5–8 mm, oblong-linear, straight or slightly falcate, cuneate at the base, rounded at the apex and terminated by the persistent style, membranaceous, glabrous or sparsely setulose. Seeds 7–12, 2.5 × 2 mm, oblong-reniform, dark brown.

Botswana. N: Central Distr., Boteti (Botletle) R. at Toromoja, fl. & fr. 25.iv.1975, *Ngoni* 439 (K; PRE; SRGH).

Also in Namibia and South Africa. In low rainfall areas on sandy soils with sparse ground cover; c. 900 m.

Gillett (loc. cit.) records this species from Kenya, Tanzania and Zimbabwe.

L. pauciflora and *L. falciformis* are very close and difficult to separate only by morphological features. Nevertheless *L. pauciflora* has racemes 2-flowered and the pods are straight or slightly falcate; *L. falciformis* has racemes usually 3–6-flowered and pods are always falcate. The habit, the shape of the leaflets, and the indumentum are more or less similar in both species.

---

*Harvey, loc. cit., refers to the collection *v. Schlicht,* from "Bassoutosland", for *L. pauciflora* var. *diffusa*. However, Jacot Guillarmod in Fl. Lesotho (1971) does not mention this species.

5. **Lessertia perennans** DC., Astrag.: 37 (1802); Prodr. **2**: 271 (1825). —Harvey in F.C. **2**: 216 (1862). —Jacot Guillarmod, Fl. Lesotho: 195 (1971). —Drummond in Kirkia **8**: 223 (1972). —Lock, Leg. Afr. Check-list: 264 (1989). Type from South Africa (Cape).

Erect perennial herb or subshrub up to 130 cm high, from a woody taproot. Stems simple or branched from the base, erect or divaricate, herbaceous and pithy above, striate, yellowish-brown, thinly silky-villous with scattered white hairs. Leaves 4–7.5 cm long; leaflets 3–10 on each side of the rhachis, opposite or subopposite, 7–15 × 3–6 mm, elliptic to ovate-elliptic, acute and mucronate at the apex, rounded at the base, silky-villous with scattered white hairs on both surfaces; terminal leaflet as long as the others; petiole 5–12 mm long, thinly silky-canescent; rhachis 1.4–5 cm long, sulcate on the upper side; petiolules 1–1.5 mm long; stipules 5–6 mm long, lanceolate-acuminate, silky-villous, membranaceous, persistent. Racemes up to 15 cm long in the upper axils, laxly many-flowered, silky-villous; peduncle 3.5–6 cm long; bracts 2 mm long, lanceolate-acuminate; pedicels 5–7 mm long; bracteoles minute, linear, very often absent. Calyx 3–4 mm long, campanulate, subequally 5-toothed; teeth 1–2 mm long, lanceolate-acuminate, with scattered white hairs. Corolla 7–8 mm long, pale purplish, mauve-red, blue or sometimes white, glabrous; standard 7–8 × 6–7 mm, suborbicular or broadly obovate, shallowly emarginate at the apex, abruptly tapering at the base into a 1–2 mm long claw; wings up to 7 × 2.2 mm, each obliquely ovate, rounded at the apex, auriculate above the 2 mm long claw; keel petals 6–7 mm long, the blades c. 3 mm wide, obovate, rounded at the apex, auriculate at the base, the claws c. 2 mm long. Free parts of the filaments up to 3 mm long; free filament 4–5 mm long, dilated at the base; anthers 0.2 mm long. Ovary shortly stipitate, c. 4 mm long, 2–5(6)-ovulate; style 2.5 mm long. Pod 15–20 × 9–13 mm, obliquely elliptic to ovate, mucronate at the apex with the upcurved persistent style, membranaceous; stipe c. 1 mm long. Seeds 2–5(6) in each pod, 2 × 1.5 mm, reniform, blackish.

Zimbabwe. E: Nyanga Distr., new Nyanga/Mutare (Inyanga/Umtali) road, at 1.6 km from Nyanga, fl. & fr. 5.iii.1969, *Corby* 2100 (BR; K; LISC; SRGH).
Also in Lesotho and South Africa. In submontane grassland and on river banks; 1620–1800 m.

87. ASTRAGALUS L.

Astragalus L., Sp. Pl.: 755 (1753); Gen. Pl., ed. 5: 335 (1754).

Herbs or subshrubs. Leaves imparipinnate, or paripinnate with a spine at the end of the rhachis (outside the Flora Zambesiaca area); leaflets entire; stipules entire, often foliaceous persistent; stipels absent. Inflorescence a many-flowered axillary raceme; bracts present; bracteoles present or absent. Calyx tubular, campanulate or inflated, divided into 5 subequal teeth or the upper two shorter. Standard much longer than wide, gradually narrowed to the base, glabrous, or rarely silky or tomentose (outside the Flora Zambesiaca area); wings oblong to oblong-ovate, clawed, the blades auriculate; keel petals ± as long as the wings, oblong, the lamina slightly pouched above the auricle and adhering to the wings. Vexillary stamen free (except in one Himalayan species), straight at the base; filament sheath long; free parts of the filaments curved upwards, filiform; anthers uniform. Ovary sessile or stipitate, 2–many-ovulate; style filiform, glabrous with a small terminal stigma. Pod sessile or stipitate, usually inflated, with the lower suture usually impressed, the upper one sometimes so, usually partly or wholly divided by a vertical false septum growing from the lower suture. Seeds reniform.

A genus of about 1500 species mainly in the Northern Hemisphere, particularly in temperate Asia.

Astragalus atropilosulus (Hochst.) Bunge, Gen. Astrag. Sp. Gerontogeae **1**: 6 (1868); **2**: 4 (1869). —Gillett in Kew Bull. **17**: 415 (1964); in F.T.E.A., Leguminosae, Pap.: 1054 (1971). —Troupin, Fl. Rwanda, Spermatophytes **2**: 159 (1983). —Wenninger in Mitt. Bot. Staatssamml. München **30**: 30 (1991). —Lock, Leg. Afr. Check-list: 253 (1989). Type from Ethiopia.
Diplotheca atropilosula Hochst. in Flora **29**: 596 (1846).

Subsp. **abyssinicus** (Hochst.) J.B. Gillett in Kew Bull. **17**: 417 (1964). —Wenninger in Mitt. Bot. Staatssamml. München **30**: 33, map 4 (1991). TAB. 3,7: **5**. Type from Ethiopia.

Diplotheca abyssinica Hochst. in Flora **29**: 595 (1846).

Astragalus abyssinicus (Hochst.) Steud. ex A. Rich., Tent. Fl. Abyss. **1**: 193 (1847). —Cronquist in F.C.B. **5**: 73, fig. 6 (1954).

Astragalus burkeanus Benth. ex Harv., Thes. Cap. **1**: 52 (1859). —E.G. Baker, Legum. Trop. Africa: 268 (1929). Type from South Africa (Gauteng).

Lessertia stipulata Baker f. in J. Bot. **37**: 430 (1899). Type: Zimbabwe, Harare (Salisbury), Sept. 1898, *Rand* 614 (BM, holotype).

Astragalus bequaertii De Wild. in Bull. Jard. Bot. État **8**: 126 (1923). Type from Dem. Rep. Congo.

Astragalus burkeanus var. *randii* Baker f., Legum. Trop. Africa: 268 (1929). Type as for *Lessertia stipulata.*

Astragalus atropilosulus subsp. *bequaertii* (De Wild.) J.B. Gillett in Kew Bull. **17**: 418 (1964); in F.T.E.A., Leguminosae, Pap.: 1055 (1971) with all its varieties.

Astragalus atropilosulus subsp. *burkeanus* (Harv.) J.B. Gillett in Kew Bull. **17**: 420 (1964); in F.T.E.A., Leguminosae, Pap.: 1056, fig. 149 (1971) with all its varieties. —Drummond in Kirkia **8**: 217 (1972).

Subshrub or perennial herb. Stems erect or ascending, usually glabrous. Leaves up to 20 cm long; leaflets 11–51, opposite or subopposite, up to 3 × 1.4 cm, lanceolate to narrowly elliptic, obtuse and mucronate at the apex, usually glabrous on the upper surface, with at least a few hairs on the midrib and margins on the lower surface and on the base of the petiolules; rhachis usually subglabrous, sometimes pubescent; petiole 1–2.5 cm long; stipules large, leaf-like, often larger than the leaflets, broadly triangular-ovate, acute, entire, free from the petiole. Racemes up to 10 cm long, many-flowered; peduncle and rhachis glabrous or subglabrous to densely covered with appressed hairs; bracts 1–2 mm long, linear; bracteoles absent; pedicels 2–3 mm long, pubescent, reflexed after anthesis. Calyx 3–5.5 mm long, oblique, glabrous or with appressed whitish, brownish or blackish hairs; tube 1.5–2 mm long, the upper 2 teeth c. 0.5 mm long, triangular, the lower ones up to 3 mm long, lanceolate. Corolla purplish, white or yellow, 6–12 mm long; standard and wings longer than, or sometimes equalling, the keel in length. Filament sheath 4–9 mm long, whitish, rather persistent. Ovary stipitate, divided by an almost complete longitudinal false septum, with 3–8 ovules on each side, glabrous or pilose; style 1.5–4 mm long. Pod stipitate, up to 40 × 7 mm, oblong-elliptic, narrowed to both ends, glabrous or brown strigulose, papery, conspicuously veined, with the lower suture impressed but not the upper one, entirely divided lengthwise by a pair of subhyaline septa and splitting into 2 halves at maturity. Seeds c. 2 × 1.5 mm, ± kidney-shaped in outline, dark brown.

Zambia. N: Mbala (Abercorn)–Mbeya road, about 9.5 km from Mbala (Abercorn), 1500 m, fl. & fr. 4.ix.1956, *Richards* 6106 (K). W: Copperbelt Distr., Katanino, dambo, fl. & fr. 3.xi.1955, *Fanshawe* 2560 (K; SRGH). C: Lusaka Distr., Mount Makulu Agricultural Research Station, fl. & fr. 14.iii.1969, *van Rensburg* 3126 (K; SRGH). E: Nyika Plateau, fl. & fr. 24.viii.1962, *Verboom* 665 (K). S: Kalomo Distr., Siantambo, Sichifula, fl. & fr. 19.xi.1962, *Mataundi* 15/61 (SRGH). **Zimbabwe**. N: Mazowe Distr., Glendale, fl. 13.vi.1951, *C.T. Carter* s.n. (K). C: Gweru Distr., 9.6 km SE of Gweru (Gwelo), fl. & fr. 16.vi.1966, *Biegel* 1271 (K; PRE; SRGH). E: Mutasa Distr., Nusa (Nuza) Plateau, fl. & fr. x.1934, *Gilliland* 889 (BM; K). S: Masvingo Distr., Great Zimbabwe National Park, fl. & fr. 27.iii.1973, *Chiparawasha* 609 (K; SRGH). **Malawi**. N: Nkhata Bay Distr., edge of road to Viphya, 41.5 km south west of Mzuzu, fl. & fr. 1.xi.1967, *Pawek* 1500 (MAL; SRGH). C: Kasungu Distr., Bua River Drift, road Kasungu–Nkhotakota, fl. & fr. 13.i.1959, *Robson & Jackson* 1144 (BM; K; PRE; SRGH). S: Zomba Distr., Mbulukuta Village, T.A. Mulumbe, fl. & fr. 31.viii.1989, *Balaka* 2161 (MAL; PRE). **Mozambique**. T: a 2 km do cruzamento para Dedza em direccao a Vila Coutinho, fr. 17.vii.1949, *Barbosa & Carvalho* 3660 (LMA).

Also in Yemen and Saudi Arabia, and southwards through the eastern Dem. Rep. Congo, southern Sudan, Ethiopia, Somalia and the East African highlands to South Africa (Northern Province, Mpumalanga, Gauteng and KwaZulu-Natal). In dambos and seasonally flooded grasslands, montane grasslands and in miombo woodlands, sometimes on termite mounds; also in cultivated lands and on roadsides; 900–2250 m (up to 3900 m outside the Flora Zambesiaca area).

Gillett op. cit. (1964 & 1971) considers numerous varieties within the species as well defined taxa. However, in the Flora Zambesiaca area, I have seen a continuous variation of the taxonomic features, difficult to delimit in varietal taxa. Therefore I have followed Wenninger (in loc. cit. 1991) and have treated all the material as a single subspecies.

The typical subspecies is distinguished from subsp. *abyssinicus* by having much of the inflorescence and ovary densely covered in appressed dark hairs, and is recorded only from Ethiopia.

Tab. 3,7: **5**. ASTRAGALUS ATROPILOSULUS subsp. ABYSSINICUS. 1, flowering branch (× 1); 2, part of leaf with subalternate leaflets (× 1); 3, bract (× 4); 4, flower (× 4); 5 standard (× 4); 6, wing (× 4); 7, keel (× 4); 8, androecium and gynoecium (× 4); 9, united stamens, spread out (× 4); 10, vexillary stamen (× 4); 11, anther (× 18); 12, gynoecium (× 4); 13, pod (× 2); 14, pod, with one valve displaced and part of septum cut away (× 2); 15, seeds (× 6). Drawn by Ann Webster from a plant grown at Kew. From F.T.E.A.

Tribe 14. **FABEAE**

By E.S. Martins

Fabeae Rchb., Fl. Germanica Excurs.: 525 (1832).
Vicieae (Bronn) DC., Prodr. **2**: 353 (1825).

Erect or mostly climbing herbs, with an indumentum of simple and short-stalked glandular hairs. Leaves epulvinate, distichous, paripinnate with the rhachis ending in a tendril or mucro, rarely imparipinnate or reduced to a tendril and stipules or a phyllode; leaflets usually entire, rarely toothed; stipels absent; stipules semi-sagittate or hastate or variously divided. Flowers in secund axillary racemes or single; bracteoles generally absent. Calyx with tube often oblique and asymmetrical, lobed with the upper lobes sometimes shorter. Standard clawed or not; wings superficially adnate to the keel by thumb-and-pocket configuration. Stamens joined in a tube, with the vexillary filament only lightly coherent; filaments slender or dilated at the apex; anthers uniform. Ovary 2–many-ovulate; style borne at right-angles to the ovary, variously flattened, thickened and bearded. Pod laterally compressed, generally dehiscent, sometimes winged or with "woolly" partitions between the seeds. Seeds compressed to spherical, with a long to short hilum.

A tribe of 5 genera, occurring in temperate regions and tropical mountains. A number of species are grown as crop plants or as garden ornamentals and food plants.

1. Pods short, ovate to broadly oblong, strongly compressed, with 1–2 lens-shaped seeds; calyx teeth subequal, long and slender ··· 89. **Lens**
– Pods narrowly oblong to linear-oblong, with several–many seeds; calyx with the upper teeth usually shorter than the lower ··· 2
2. Stipules large and foliaceous; style margin winged at the base ··· 91. **Pisum**
– Stipules not large and foliaceous; style margin not winged at the base ··· 3
3. Leaflets usually numerous, less often 2–4 (2 in *V. paucifolia* subsp. *malosana*), folded flat lengthwise in bud; staminal sheath oblique at the apex; style tapering apically, glabrous or distally hairy all round or on the leading edge only ··· 88. **Vicia**
– Leaflets usually 2 or 4, rarely numerous, inrolled in bud; staminal sheath truncate at the apex; style flattened or dorsiventrally compressed towards the apex, glabrous or bearded on the inner side ··· 90. **Lathyrus**

88. VICIA L.

Vicia L., Sp. Pl.: 734 (1753); Gen. Pl., ed. 5: 327 (1754).

Annual or perennial herbs, mostly climbing by means of tendrils, less often straggling or erect. Leaves usually paripinnate, the rhachis terminating in a tendril or bristle, rarely imparipinnate; leaflets often numerous, less often in 1–3 pairs, entire or toothed, folded flat lengthwise in bud; stipules usually small, semi-sagittate, often fimbriate or toothed, herbaceous; stipels absent. Flowers in axillary racemes or fascicles or solitary; bracts usually small, deciduous; bracteoles absent. Calyx 5-lobed; tube often asymmetrical; lobes subequal or the upper two shorter and partly joined. Corolla small to medium-sized, blue, mauve, purple, yellow or white; standard obovate or oblong, very often adhering to the keel; keel obtuse. Vexillary stamen free or ± united with the others; anthers uniform. Ovary stipitate or subsessile, 2–many-ovuled; style distally pubescent or pilose all round or on the lower side only, rarely glabrous; stigma terminal. Pod oblong to linear, compressed, dehiscent. Seeds globular or compressed.

A genus of about 160 species mainly in the northern temperate regions, and extending into South America and the highlands of tropical Africa.

Vicia faba L., the broad bean, is a widely cultivated food plant. A number of other species, subspecies and varieties have also been cultivated, usually as trial crops, at various agricultural experiment stations: *V. faba* (*Myre* 914), *V. ervilia* (L.) Willd. (*Myre* 962), *V. sativa* var. *macrocarpa* Moris (*Myre* 955, 960) and *V. villosa* subsp. *villosa* Roth (*Myre* 995) have been recorded from the Umbelúzi Experimental Station in Mozambique; *V. benghalensis* L. (*Corby* 952) cultivated at

Marondera, Grasslands Research Station in Zimbabwe; *V. villosa* subsp. *varia* (Host) Corb. cultivated at the Umbelúzi Experimental Station (*Myre* 982) and at the Agricultural Research Station in Zimbabwe (*Nicholls* in *GHS* 71431).

Flowers solitary or paired, subsessile in the leaf-axils; leaves 6–16-foliolate; stipules with a glandular median spot · 1. *sativa*
Flowers 1–5 in long pedunculate racemes; leaves 2-foliolate, rarely 3–4-foliolate; stipules without a glandular spot · 2. *paucifolia* subsp. *malosana*

1. **Vicia sativa** L., Sp. Pl.: 736 (1753). —Harvey in F.C. **2**: 233 (1862). —Burtt Davy, Fl. Pl. Ferns Transvaal, pt. 2: 404 (1932). —Torre in C.F.A. **3**: 238 (1966). —Verdcourt in F.T.E.A., Leguminosae, Pap.: 1068 (1971). —Corby in Kirkia **9**: 328 (1974). —C.C. Townsend in Fl. Iraq **3**: 534 (1974). Type from Europe.

Straggling or ascending annual herb up to 80 cm tall. Stems angular, sparsely pubescent to subglabrous. Leaves 6–16-foliolate, lower ones 2–4-foliolate; leaflets 9–35 × 1–11(15) mm, linear to broadly oblong, oblong-obovate or obcordate, acute, obtuse, rounded or emarginate at the apex and usually mucronate or shortly cuspidate, usually cuneate at the base, sparsely pilose on both surfaces or glabrescent; petiole up to 8 mm long; rhachis usually terminating in a branched tendril; petiolules c. 0.5 mm long; stipules up to 8 mm long, semi-sagittate, entire or dentate, with a dark glandular median spot. Flowers 1–2 in the axils, subsessile or very shortly pedicellate. Calyx sparsely pubescent to subglabrous; tube 3.5–7 mm long, slightly asymmetrical at the base, the mouth not oblique; lobes subequal, equalling or shorter than the tube, linear-lanceolate. Corolla 10–30 mm long, light reddish-purple, bluish or mauve; standard 6–11 mm wide, obovate, emarginate; wings 2–3 mm shorter than the standard, the lamina equalling or exceeding the claw and oblong-obovate with a long auricle; keel ± two-thirds as long as the wings, the lamina broadly oblong, shorter than the claw. Style dorsally compressed, pubescent all round towards the apex and with a tuft of longer hairs below the apex on the lower edge. Pod narrowly oblong, 25–50 × 4–9 mm, brownish-black, pubescent to glabrous. Seeds 2.5–6 mm in diameter, subglobose, smooth.

Var. **angustifolia** L., Fl. Suec., ed. 2: 255 (1755) in obs. —E.G. Baker, Legum. Trop. Africa: 346 (1929). —Verdcourt in F.T.E.A., Leguminosae, Pap.: 1069, fig. 153/13 (1971). —Troupin, Fl. Rwanda, Spermatophytes **2**: 164, fig. 47/4 (1983). Type from Switzerland.
Vicia angustifolia L., Amoen. Acad. **4**: 105 (1759). Type from England.
Vicia sativa var. *nigra* L., Sp. Pl., ed. 2: 1037 (1763). Type from Europe.
Vicia sativa subsp. *nigra* (L.) Ehrh., Hannover. Mag. **1780** (15): 229 (1780). —Drummond in Kirkia **8**: 228 (1972).
Vicia abyssinica Alef. in Bonplandia **9**: 72 (1861). Type from Ethiopia.
Vicia sativa var. *abyssinica* (Alef.) Baker in F.T.A. **2**: 172 (1871). —Robyns, Fl. Sperm. Parc Nat. Alb. **1**: 334 (1948). —Boutique in F.C.B. **6**: 80 (1954).

Leaves 6–12-foliolate; leaflets 9–35 × 1–6 mm, linear, narrowly oblong or narrowly obovate, obtuse to emarginate but usually apiculate or shortly cuspidate at the apex, cuneate at the base. Calyx tube 3.5–5.5 mm long; lobes 2.5–4.5 mm long, always shorter than the tube. Standard 10–18 × 7–10 mm, obovate; wings c. 4 mm wide; keel obtuse, sometimes greenish-white. Style 1.5–2 mm long. Pod 25–40 × 4–6 mm, brownish-black, glabrous or subglabrous. Seeds 2.5–4 mm in diameter.

Zimbabwe. N: Makonde Distr., fl. vi.1973, *Nicholls* in *GHS* 227143 (SRGH). C: Harare Distr., Spicers Farm, 4.8 km above Mukuvisi (Makabusi)–Manyame (Hunyani) confluence on Mukuvisi R., 1400 m, fr. 1.ix.1959, *Phipps* 2188 (BM; K; PRE; SRGH). E: Mutasa Distr., Imbeza Valley, 1097 m, fl. & fr. 14.ix.1956, *Chase* 6200 (K; LISC; PRE; SRGH).

Widely distributed in Europe, northern, eastern and southern Africa, Asia and the Middle East, cultivated or as an escape of cultivation, sometimes as a weed; c. 1000–1430 m.

Var. *macrocarpa* has been cultivated in Mozambique, see note above.

2. **Vicia paucifolia** Baker in F.T.A. **2**: 173 (1871). —E.G. Baker, Legum. Trop. Africa: 347 (1929). —Robyns, Fl. Sperm. Parc Nat. Alb. **1**: 336 (1948). —Boutique in F.C.B. **6**: 81 (1954). —Verdcourt in Kew Bull. **24**: 70 (1970); in F.T.E.A., Leguminosae, Pap.: 1069 (1971). Type from Ethiopia.

Delicate straggling climbing or suberect perennial herb up to 1 m tall. Stems angular, densely pilose when young, becoming sparsely pilose to subglabrous. Leaves 2–7-foliolate, the lower pairs opposite or alternate; leaflets 12–45 × 1–5.5(7) mm, linear to narrowly elliptic, narrowed at both ends, mucronulate, sparsely pilose on both surfaces; petiole obsolete or up to 8 mm long; rhachis terminating in a simple or branched tendril or sometimes a bristle; petiolules c. 1 mm long; stipules 5–13 mm long, semi-sagittate, sometimes with 1–2 additional teeth. Racemes 1–5-flowered; peduncle 20–35 mm long; pedicels 2–4 mm long, pilose, more densely so towards the apex. Calyx laxly pilose to subglabrous; tube 2–2.5 mm long; lobes subequal, 1–5 mm long, narrowly triangular-subulate to linear. Corolla blue, mauve or violet, 9–10 mm long; standard c. 10 × 6.5 mm, obovate, emarginate, sometimes greenish; wings as long as the standard or a little shorter; keel c. 6.5 mm long, obtuse. Style c. 1.5 mm long, densely pubescent all round towards the apex. Pod 24–30 × 4–7.5 mm, oblong, flat, glabrous. Seeds greenish-brown and densely mottled black to black, usually (3)6–10, c. 3 × 2.5 × 2 mm, irregularly subglobose, compressed; aril yellowish, produced into a tail.

Subsp. **malosana** (Baker) Verdc. in Kew Bull. **24**: 70 (1970); in F.T.E.A., Leguminosae, Pap.: 1070, fig. 153/12 (1971). —Lock, Leg. Afr. Check-list: 506 (1989). TAB. 3,7: **6**. Type: Malawi, Mt. Malosa, *Whyte* (K, holotype).

Lathyrus malosanus Baker in Bull. Misc. Inform., Kew **1897**: 261 (1897).

Vicia malosana (Baker) Baker f., Legum. Trop. Africa: 347 (1929).

Vicia paucifolia Baker var. *malosana* (Baker) Brenan in Mem. New York Bot. Gard. **8**: 256 (1953). —Binns, First Check List Herb. Fl. Malawi: 85 (1968).

Petiole obsolete or up to 2 mm long. Leaflets 2, rarely 3–4, the lower pair opposite. Tendrils mostly simple. Calyx lobes 1.5–2(3) mm long. Pods 4–5.5 mm wide.

Zambia. W: Chingola Distr., Mushishima, fl. & fr. 18.v.1968, *Mutimushi* 2609 (SRGH). E: Chama Distr., Nyika Plateau, 2100 m, fl. 3.i.1959, *Richards* 10418 (K). **Malawi**. N: Nkhata Bay Distr., South Viphya Plateau, by Luwawa Dam, 1560 m, fl. & fr. 8.v.1970, *Brummitt* 10494 (BR; K; LISC; MAL; P; PRE; SRGH). S: Zomba Distr., Zomba Plateau, fl. & fr. 25.v.1976, *Banda* 1237 (MAL; SRGH).

Also in southern Tanzania. Mushitu margins and wet or marshy places by rivers, and in montane grassland; 1220–2370 m.

The typical subspecies, from Ethiopia, Dem. Rep. Congo, Kenya and Tanzania, has petioles up to 6 mm long or petioles obsolete. Leaflets 3–4 with the lower pair often alternate; tendrils often branched. Calyx lobes 3–5 mm long. Pods mostly wider, 6–7.5 mm wide.

89. LENS Mill.

Lens Mill., Gard. Dict. abr. ed. 4 (1754) *nom. conserv.* —M.E. Ferguson et al. in Bot. J. Linn. Soc. **133**: 41–59 (2000).

Erect or spreading annual herbs. Leaves usually paripinnate, the rhachis terminating in a simple or rarely a branched tendril, or an awn, rarely imparipinnate; leaflets 2–several on each side of the rhachis, opposite or alternate, entire, folded flat lengthwise in bud; stipules linear to ovate or semi-sagittate, entire or dentate. Inflorescences axillary, few-flowered, racemose or flowers solitary; bracts small, deciduous; bracteoles absent. Calyx 5-lobed; lobes subequal, subulate, at least twice as long as the tube. Corolla small, white, bluish or violet; standard obovate to almost round, cuneate to the base; wings oblong or obovate, adnate to the keel; keel shorter than the wings. Vexillary stamen free; anthers uniform. Ovary subsessile, 2-ovuled; style somewhat flattened, distally pubescent on the inner side. Pod strongly compressed, 1–2-seeded. Seeds lenticular, compressed.

A genus of 5 species occurring in the Mediterranean region and western Asia, with 1 species probably indigenous in western Uganda and Ethiopia (*L. ervoides* (Brign.) Grande). According to some authors the genus comprises only 2 species: *L. culinaris* Medik. and *L. nigricans* Godr. Ferguson et al., in Bot. J. Linn. Soc. **133**: 41–59 (2000), recognized 4 species, one of them *L. culinaris* with 4 subspecies.

L. culinaris subsp. *culinaris*, the lentil, is a food crop cultivated since ancient times for its seeds (in Egypt, southern Europe and western Asia, and in India and much of China), see Purseglove, Tropical Crops, Dicotyledons part 1: 279-280 (1968). The split seeds are also known as dhal.

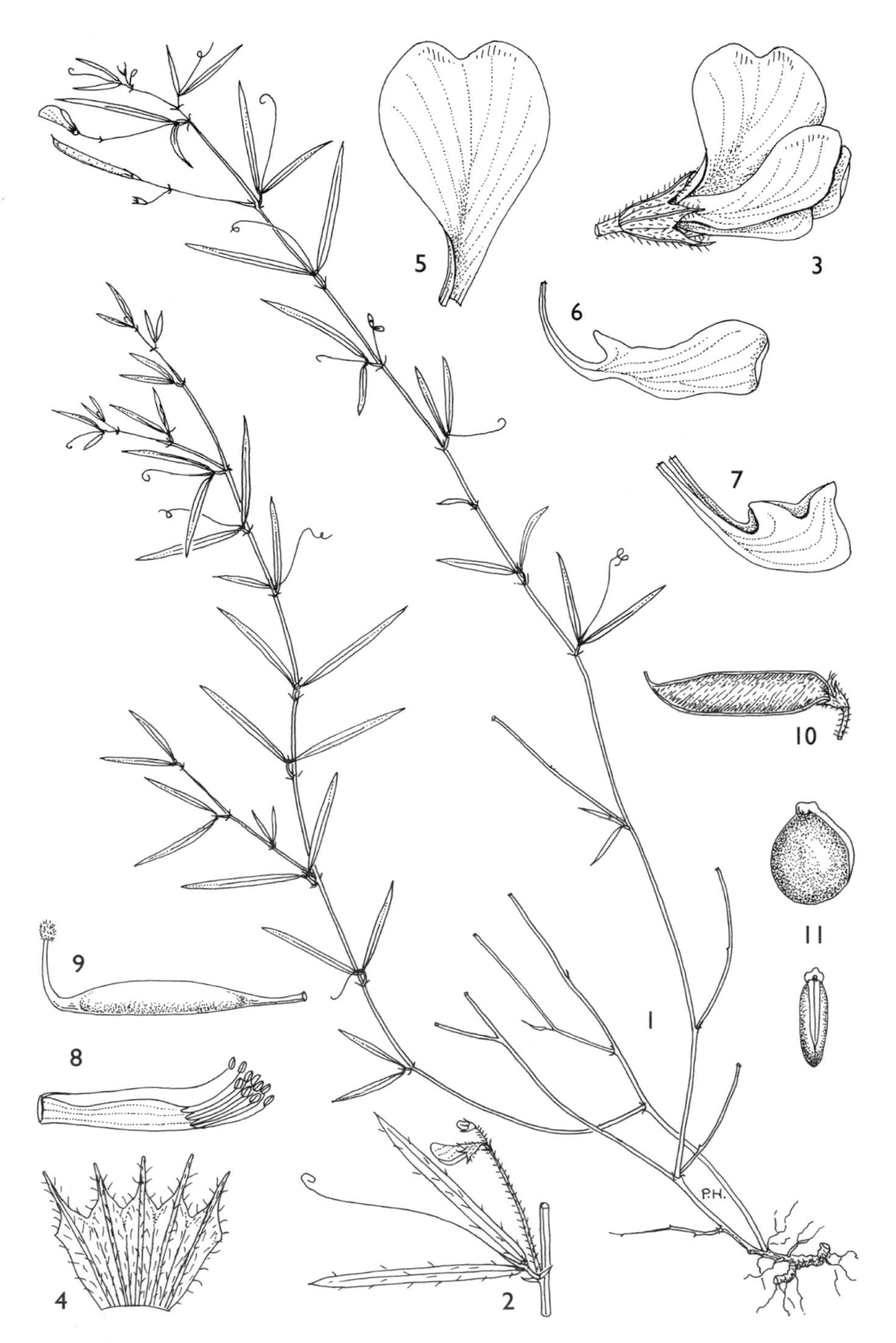

Tab. 3,7: **6**. VICIA PAUCIFOLIA subsp. MALOSANA. 1, habit (× $^{2}/_{3}$); 2, leaf with axillary raceme (× 1); 3, flower (× 4); 4, calyx, opened out, exterior (× 6); 5, standard (× 3); 6, wing (× 4); 7, keel (× 4); 8, androecium (× 4); 9, gynoecium (× 6), 1–9 from *Pawek* 2120; 10, pod (× 1); 11, seed, two views (× 4), 10 & 11 from *Brummitt & Patel* 15622. Drawn by Pat Halliday.

Lens culinaris Medik. in Vorles. Churpfälz. Phys.-Öcon. Ges. **2**: 361 (1787). —Verdcourt in F.T.E.A., Leguminosae, Pap.: 1074 (1971). —A.E. Gonçalves in Garcia de Orta, Sér. Bot. **5**: 94 (1982). —Lock, Leg. Afr. Check-list: 503 (1989). —Castroviejo & Pascual in Fl. Iberica **7**(1): 418, t. 113/f–i (1999). Type from France.
Ervum lens L., Sp. Pl.: 738 (1753). —Schinz, Pl. Menyharth.: 56 (1905). Type from Europe.
Lens esculenta Moench, Meth. Pl.: 131 (1794) *nom. illegit.*

Subsp. **culinaris**. TAB. 3,7: **7**.

Erect or suberect annual herb up to 50 cm tall. Stem slender, 4-angular, few-branched, sparsely softly hairy to subglabrous. Leaves shortly petiolate, paripinnate with 3–8 leaflets on each side; leaflets up to 18 × 5 mm, shortly petiolulate, oblong-linear to narrowly elliptic, ± rounded and apiculate at the apex, cuneate to rounded at the base, sparsely pilose on both surfaces; rhachis sparsely pilose, usually terminating in an unbranched tendril or an awn; stipules 3.5–7 mm long, asymmetrically oblong-lanceolate, entire. Racemes usually 2–4-flowered, ± equalling the leaves; rhachis terminating in a short bract-like awn. Calyx 6–9 mm long, sparsely pilose; tube c. 1.5 mm long, campanulate; teeth 4–6 times as long as the tube, subulate. Corolla usually white, rarely pink or violet, as long as the calyx or little shorter; standard c. 5 × 5 mm, almost round with a short and wide claw. Pod 12–16 × 6–12 mm, ovate-rhombic, glabrous. Seeds 1–2, lenticular, 2–3 mm in diameter.

Zimbabwe. C: Harare, fl. 31.i.1970, *Corby* 2153 (K; LISC; SRGH). **Mozambique**. M: Namaacha Distr., Umbelúzi, Estação Experimental, fl. & fr. immat. 29.ix.1950, *Myre* 954 (LMA).
It has been cultivated in the Flora Zambesiaca area, mainly in trials on agricultural experiment stations.

90. LATHYRUS L.

Lathyrus L., Sp. Pl.: 729 (1753); Gen. Pl., ed. 5: 326 (1754). —Kupicha in Notes Roy. Bot. Gard. Edinburgh **41**: 209–244 (1983).

Annual or perennial herbs, erect or straggling, climbing by means of tendrils. Stems angular or ± winged. Leaves paripinnate, rarely imparipinnate; rhachis terminating in a tendril or a bristle, rarely reduced to a tendril; petiole and rhachis sometimes dilated and leaf-like; leaflets usually few, inrolled in bud, rarely absent, entire; stipules foliaceous, often semi-sagittate, persistent, rarely entire. Flowers solitary or in axillary racemes; bracts usually minute, early caducous; bracteoles absent. Calyx 5-lobed, sometimes asymmetrical with the upper 2 lobes shorter. Corolla small to medium-sized, red, blue or yellow; standard oblong-obovate to transversely elliptic with a short broad claw; wings oblong to falcate-obovate, adherent to the keel or free; keel shorter than the wings, incurved, obtuse. Vexillary stamen free or connate with the staminal sheath; filament sheath truncate at the apex; anthers uniform. Ovary subsessile or stipitate, few–many-flowered; style incurved, dorsally compressed, often indurated towards the apex, mostly bearded on the upper side, or rarely glabrous; stigma capitate. Pods linear-oblong, terete or ± laterally compressed, dehiscent. Seeds globose or ellipsoid, sometimes compressed, smooth or rugulose, with a slender aril.

A genus of about 130 species in the temperate regions of the northern hemisphere and South America, a few species in Africa. A number of species have been cultivated as seed or fodder crops in the Flora Zambesiaca area, with the following recorded from the Umbelúzi Experimental Station in southern Mozambique: *L. clymenum* L. (*Myre* 1019), *L. sativus* L. (*Myre* 916) and *L. cicera* L. (*Myre* 920). *L. odoratus* L., the sweet pea, is widely cultivated as a garden ornamental, grown for its showy scented flowers.

Leaf tendrils unbranched; flowers solitary or rarely 2; peduncle up to 25(33) mm long; pods glabrous or only ciliate along the margins ························· 1. *hygrophilus*
Leaf tendrils 3–5-branched; flowers 1–3(4) in axillary racemes; peduncle 50–80 mm long; pods covered with tuberculate-based long hairs and small sessile glands ··········· 2. *hirsutus*

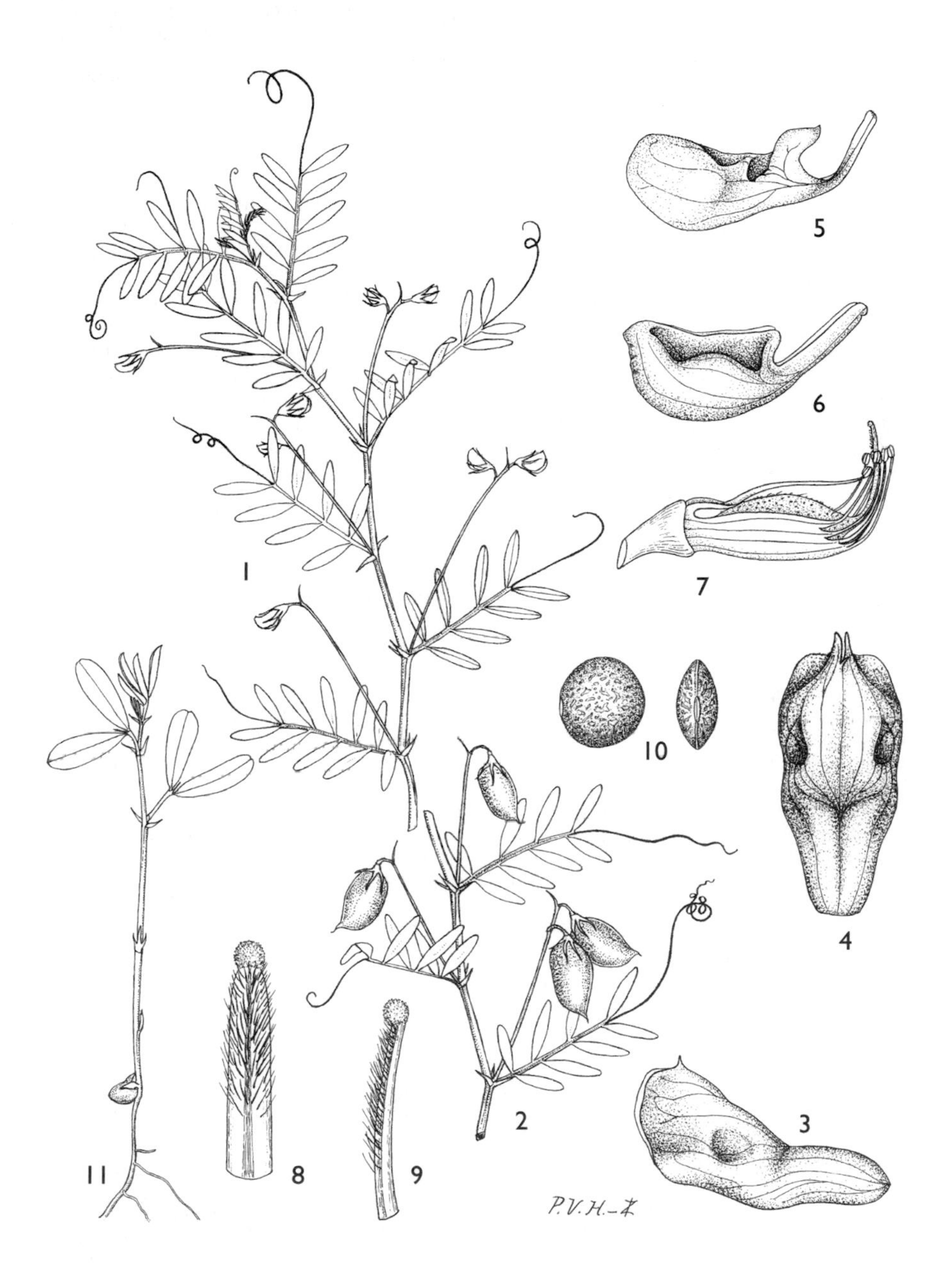

Tab. 3,7: **7**. LENS CULINARIS subsp. CULINARIS. 1, flowering branch ($\times \frac{5}{6}$), from *Westphal* WP7616; 2, fruiting branch ($\times \frac{5}{6}$), from *Westphal* WP7617; 3, standard, side view ($\times 7\frac{1}{2}$); 4, standard, back view ($\times 7\frac{1}{2}$); 5, wing ($\times 7\frac{1}{2}$); 6, keel ($\times 7\frac{1}{2}$); 7, androecium and gynoecium ($\times 7\frac{1}{2}$); 8 & 9, detail of style and stigma ($\times 30$), 3–9 from *Westphal* WP7616; 10, seed, two views ($\times 2\frac{1}{2}$), from *Westphal* WP2816A; 11, seedling ($\times \frac{5}{6}$), from *Westphal* WP7615. Drawn by Pamela Verheij-Hayes and Ike Zewald. From Pulses in Ethiopia. Reproduced with permission of the National Herbarium of the Netherlands.

1. **Lathyrus hygrophilus** Taub. in Engler, Pflanzenw. Ost-Afrikas **C**: 219 (1895). —R.E. Fries, Wiss. Ergebn. Schwed. Rhod.-Kongo-Exped. **1**: 92 (1914). —E.G. Baker, Legum. Trop. Africa: 349 (1929). —Robyns, Fl. Sperm. Parc Nat. Alb. **1**: 336 (1948) pro parte. — Boutique in F.C.B. **6**: 78, t. 6 (1954). —Hedberg in Symb. Bot. Upsal. **XV.1**: 122 (1957). — Binns, First Check List Herb. Fl. Malawi: 83 (1968). —Verdcourt in F.T.E.A., Leguminosae, Pap.: 1077, fig. 155 (1971). —Brummitt in Wye Coll. Malawi Proj. Rep.: 65 (1973). — Troupin, Fl. Rwanda, Spermatophytes **2**: 166, fig. 47/2 (1983). —Lock, Leg. Afr. Check-list: 501 (1989). —Agnew, Upland Kenya Wild Fl., ed. 2: 151, t. 53 (1994). TAB. 3,7: **8**. Syntypes from Tanzania.

Lathyrus kilimandscharicus Taub. in Engler, Pflanzenw. Ost-Afrikas **C**: 219 (1895). —E.G. Baker, Legum. Trop. Africa: 350 (1929). Syntypes from Tanzania.

Lathyrus intricatus Baker in Bull. Misc. Inform., Kew **1897**: 261 (1897). —E.G. Baker, Legum. Trop. Africa: 350 (1929). Type: Malawi, Nyika Plateau, 6000–7000 ft, July 1896, *Whyte* (K, lectotype, chosen here).

Lathyrus hygrophilus var. *angustifolius* Baker f. in J. Bot. **62**: 333 (1924). Type from Uganda.

Straggling or climbing perennial herb; rootstock not seen but probably small. Stems numerous, much branched, 0.5–1.5 mm in diameter, at first pubescent, later glabrous or subglabrous. Leaves 2-foliolate; leaflets 13–65 × 2–12 mm, linear to elliptic, acute to rounded but mucronulate at the apex, cuneate at the base, glabrous to pilose on both surfaces or only beneath, sometimes with numerous small red ellipsoid glands; petiole 3–15 mm long, very narrowly winged, produced beyond the leaflets into an unbranched tendril; petiolules 0.5–1 mm long; stipules 5–12 mm long, narrowly semi-sagittate with the lobes narrow, sometimes with a small tooth between the lobes. Flowers solitary or less often paired; peduncle 2–25(33) mm long, pubescent; pedicels 4–6 mm long, pubescent; bract 1–1.3 mm long, hyaline, subulate, soon caducous or absent. Calyx pubescent or pilose to glabrous; tube 2–2.5 mm long; lobes 2–5.5 mm long, narrowly triangular-subulate. Corolla whitish, purplish or pinkish, 11–13 mm long; standard greenish-white, pinkish or yellowish, usually with purple lines inside, 12–13 × 9 mm, obovate, emarginate, narrowed to the base into a 4 mm long, broad claw; wings c. 11 mm long, the lamina falcate-obovate, with a short conjunctival flap above the auricle and adhering to the keel; keel a little shorter than the wings, incurved, obtuse, the lamina slightly pouched above the small auricles, the claw c. 4 mm long, straight. Vexillary stamen adhering to the others; filament sheath 7–7.5 mm long; free parts of the filaments 5–5.5 mm long. Style 4–5 mm long, pubescent towards the apex on the upper side. Pod 25–48 × 5–6 mm, linear-oblong, laterally compressed, glabrous to pubescent or ciliate on the margins; valves with a scarcely evident reticulate venation. Seeds usually 5–8, dark brown to purplish-brown, with maximum dimensions of 4.1 × 2.9 × 1.8 mm, subglobose to ellipsoid.

Malawi. N: Mzimba Distr., Mtangatanga Forest Reserve, fl. 14.viii.1971, *Salubeni* 1692 (K; LISC; SRGH). C: Ntcheu Distr., Kirk Range, Tsangano (Zangano) Hill, 1850 m, fl. & fr. 31.i.1959, *Robson* 1389 (BM; K; LISC; PRE).

Also in Dem. Rep. Congo, Sudan, Uganda, Kenya and Tanzania. Montane grasslands, evergreen forest edges, swampy places and stream banks; 1220–2300 m (up to 4100 m in East Africa).

2. **Lathyrus hirsutus** L., Sp. Pl.: 732 (1753). —P.W. Ball in Fl. Europaea **2**: 142 (1968). — Verdcourt in F.T.E.A., Leguminosae, Pap.: 1076 (1971). —C.C. Townsend in Fl. Iraq **3**: 563 (1974). —M.J. Gallego in Fl. Iberica **7**(1): 434, t. 115/e–g (1999). Type from Europe (Habitat inter Angliae, Galiae segetes).

Climbing annual herb sparsely pubescent on the young vegetative parts. Stem up to c. 80 cm long, winged, glabrous; wings 0.5–2.5 mm wide. Leaves 2-foliolate; leaflets

---

Tab. 3,7: **8**. LATHYRUS HYGROPHILUS. 1, flowering and fruiting branch (× 1), from *D. Davis* 40; 2, leaflet (× 1½), from *Haarer* 413; 3, leaflet, showing variation (× 1½), from *Liebenberg* 1646; 4, flower (× 3); 5, calyx, opened out (× 3); 6, standard (× 3); 7, wing (× 3); 8, keel (× 3); 9, androecium and gynoecium (× 3); 10, androecium, spread out (× 3); 11, gynoecium (× 3), 4–11 from *D. Davis* 40; 12 & 13, pods (× 1½); 14, seed (8), 12–14 from *Liebenberg* 1646. Drawn by Margaret Stones. From F.T.E.A.

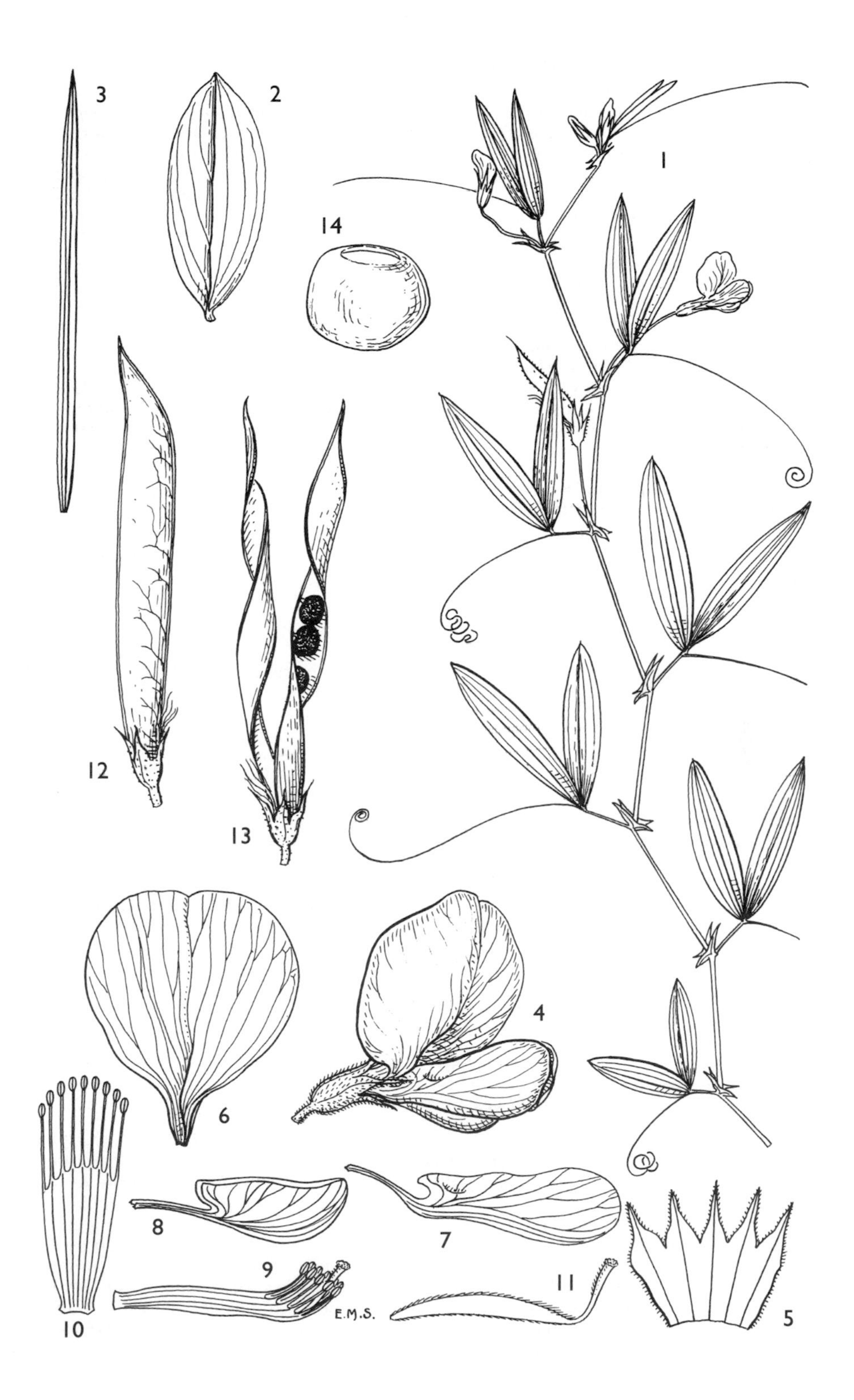
3
2
1
14
12
13
4
6
10
8
7
9
11
E.M.S.
5

15–80 × 3–20 mm, linear to oblanceolate, subacute to rounded and shortly acuminate or mucronate at the apex, cuneate at the base, glabrous, usually with 2–3 longitudinal veins from the base on each side of the midrib, prominent on both surfaces; petiole 13–30 mm long, narrowly winged, produced beyond the leaflets into a branched prehensile tendril; petiolules up to 0.5 mm long; stipules 10–20 mm long and up to 2 mm wide, semi-sagittate, the lobes diverging at a very wide angle, entire; rarely a small tooth present between the lobes. Flowers 1–3(4) in axillary racemes; peduncle 50–80(200) mm long, usually longer than the subtending leaf, slender; bracts up to 1.7 mm long, subulate; pedicels 4–6 mm long, glabrous or sparsely glandular-pubescent. Calyx 6–8 mm long, glabrous; tube c. 2 mm long, lobes 3.5–4 mm long, subequal, ovate-lanceolate, acuminate. Corolla 10–15 mm long; standard red, 10–15 × 8–12 mm, the lamina suborbicular, truncate-retuse, abruptly narrowed to a broad claw c. 4 mm long; wings blue, 8–12 mm long, the lamina twice as long as the claw, oblong-obovate, obtuse, with a very long conjunctival flap above the acute auricle; keel usually c. 2 mm shorter than the wings, the lamina arcuately ascending, deltoid-subquadrate, as long as the claw, with a short obtuse auricle. Vexillary stamen free; filament sheath c. 3.5 mm long. Pod 25–50 × 6–8 mm, linear-oblong, abruptly narrowed at both ends, laterally compressed, brown or yellowish with tubercle-based long silky hairs and small sessile glands. Seeds 3–10, globular or subglobose, tuberculate, brownish or blackish.

Zimbabwe. C: Marondera Distr., Grasslands Research Station, fl. 25.x.1961, *Corby* 1023 (K).
Native to central and southern Europe, probably introduced into the Flora Zambesiaca area as a forage plant. In pastures.

91. PISUM L.

Pisum L., Sp. Pl.: 727 (1753); Gen. Pl., ed. 5: 324 (1754).

Annual or perennial herbs, spreading or climbing by means of tendrils. Leaves paripinnate, the rhachis terminating in a prehensile tendril or a bristle; leaflets in 1–3 pairs; stipules small to very large, foliaceous, semicordate, usually equalling or exceeding the leaflets, toothed at least towards the base. Flowers solitary or in few-flowered axillary racemes; bracts small, deciduous; bracteoles absent. Calyx 5-partite; tube asymmetrical, slightly gibbous at the base; teeth subequal or the upper 2 shorter and broader. Corolla purple, pink or white, medium-sized to large; standard broadly ovate to suborbicular with a short broad claw; wings shorter than the standard, adhering to the keel, with the lamina asymmetrical, ascending, and the claw curved; keel shorter than the wings, oblong-falcate, often with a wing on the outer upper surface. Vexillary stamen free, at least in part; filament sheath truncate; filaments somewhat dilated towards the apex; anthers uniform. Style dorsally compressed, folded longitudinally with margins meeting abaxially, pubescent on the upper side towards the apex; stigma capitate. Pod oblong, little compressed, dehiscent. Seeds globular, numerous, with a slender aril.

A genus of 3 species, occurring in the Mediterranean region and western Asia. *Pisum sativum*, the garden pea, is widely cultivated as a garden vegetable and as a fodder crop.

Pisum sativum L., Sp. Pl.: 727 (1753). —Schinz, Pl. Menyharth.: 56 (1905). —A.E. Gonçalves in Garcia de Orta, Sér. Bot. **5**: 101 (1982). —Troupin, Fl. Rwanda, Spermatophytes **2**: 161, fig. 47/1 (1983). —Lock, Leg. Afr. Check-list: 503 (1989). —Romero Zarco in Fl. Iberica **7**(1): 483 (1999). Type from Europe.

Climbing annual herb up to 2 m tall (in cultivation), glabrous. Stems ± terete. Leaves 2–6(8)-foliolate, the leaflets usually opposite; leaflets 15–70 × 7–40 mm, ovate to elliptical, obtuse to emarginate and sometimes apiculate at the apex, cuneate at the base, entire to dentate; petiole up to 60 mm long; rhachis terminating in a branched prehensile tendril; petiolules 0.5–1 mm long; stipules foliaceous, up to 80 × 40 mm, usually larger than the leaflets, semicordate, semiamplexicaul, dentate towards the base or rarely subentire, glaucous or sometimes with a violet spot at the base. Flowers solitary or up to 3 in axillary racemes; peduncle 5–190 mm long. Calyx tube 4–8 mm long, campanulate; lobes as long as or longer than the tube, unequal, lanceolate, acute. Corolla white, pinkish or purplish; standard 15–30 × 23–45 mm,

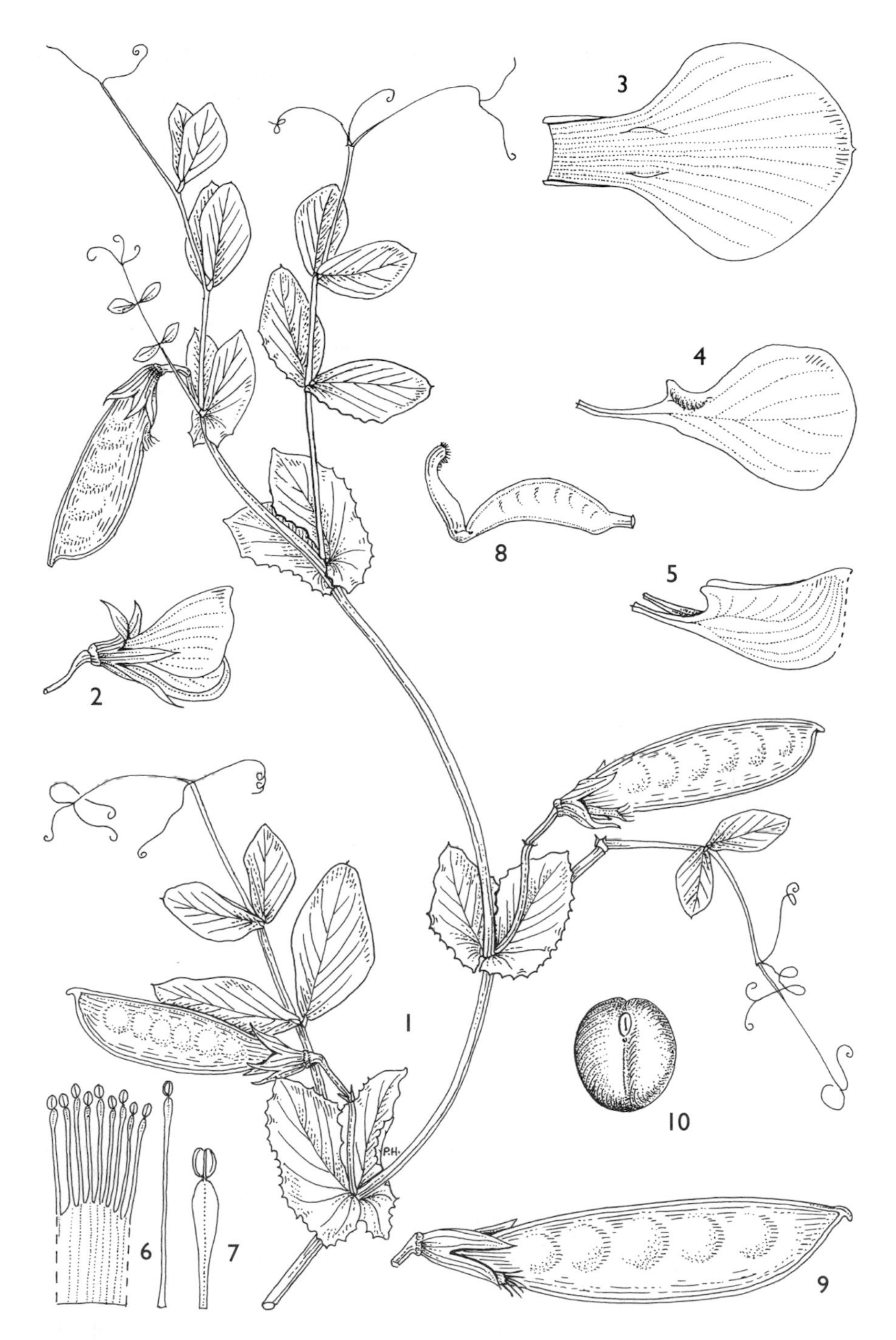

Tab. 3,7: **9**. PISUM SATIVUM var. SATIVUM. 1, fruiting branch (× 2/3), from *Torre & Paiva* 10907; 2, flower (× 1), from *Roscoe* s.n.; 3, standard (× 2); 4, wing (× 2); 5, keel (× 2); 6, androecium with free vexillary stamen (× 2); 7, anther and top of expanded filament, enlarged; 8, gynoecium (× 2); 9, pod (× 1); 10, seed (× 2), 3–10 from *Torre & Paiva* 10907. Drawn by Pat Halliday.

with the lamina broadly ovate, emarginate-apiculate, plicate and subappendiculate at the base; wings purplish or whitish, a little shorter than the standard, the lamina orbicular-obovate, abruptly narrowed above the auricles; keel coloured as the standard, much shorter than the wings, subacute at the apex. Style c. 7 mm long. Pod up to 100 × 25 mm, oblong-obovate, abruptly narrowed to both ends, whitish or yellowish when ripe. Seeds 6–10 in each pod, globular.

A very polymorphic species, widely cultivated for thousands of years for its edible seeds and for fodder; some varieties and cultivars are also cultivated for their edible fresh pods.

Var. **sativum**. TAB. 3,7: **9**.

Stipules glaucous, without a violet spot at the base. Racemes shorter than the leaves. Flowers uniformly white. Seeds c. 7 mm in diameter, smooth.

Mozambique. N: Lago Distr., 50 km de Lichinga (Vila Cabral) para Maniamba, 1350 m, fl. & fr. 29.ii.1964, *Torre & Paiva* 10907 (COI; K; LISC; LMU; SRGH; WAG).

Native or spontaneous in the Mediterranean Region and widely cultivated in most temperate and highland areas of the world.

Tribe 15. **CICEREAE**

By Maria Adélia Diniz

Cicereae Alef. in Oesterr. Bot. Z. **9**: 345 (1859) as "*Ciceridae*". —Kupicha in Bot. J. Linn. Soc. **74**: 158 (1977).

Annual or perennial herbs, sometimes spiny, with an indumentum of glandular hairs. Leaves epulvinate, distichous, imparipinnate or paripinnate, with the rhachis ending in a tendril or spine; leaflets dentate; stipels absent; stipules free from the petiole, toothed or spiny. Flowers solitary or in axillary racemes, similar to *Vicieae*, but wings free from the keel, vexillary stamen free and the style glabrous and curved away from the ovary. Pods elliptic, obovate or elongate-ellipsoid, inflated, with 1–10 seeds. Seeds beaked and bilobular to subglobose.

A tribe of only one genus, ranging from Morocco to central and western Asia.

92. **CICER** L.

Cicer L., Sp. Pl.: 738 (1753); Gen. Pl., ed. 5: 327 (1754). —M.G. Popov in Bull. Appl. Bot. Gen. Pl. Breed. **21**: 3–240 (1929). —van der Maesen in Meded. Landbouwhoogeschool **72**, part 10: 1–342 (1972).

Erect or prostrate annuals, or perennial herbs with annual stems from woody rootstocks, pubescent with stalked glandular hairs or hairs eglandular, sometimes the leaf rhachis, peduncles and stipules spinose. Leaves 3-foliolate or more usually pinnate, imparipinnate or paripinnate with the rhachis ending in a tendril or spine; leaflets conspicuously dentate; stipules small or large, foliaceous, dentate or incised. Flowers solitary or in 2–5-flowered axillary racemes; bracts small; bracteoles absent. Calyx nearly regular, 5-lobed; tube short, oblique or dorsally gibbous; lobes subequal. Corolla white, pink, purplish or blue, veined; standard obovate, narrowed into a broad claw, without appendages; wings oblong-obovate, free from the keel. Vexillary stamen free; anthers all similar. Ovary sessile, 1–10-ovuled; style filiform, incurved, glabrous; stigma dilated or not, terminal. Pods sessile, elliptic, obovate or elongate-rhomboid, acuminate, inflated. Seeds subglobose to oblong-obovoid, beaked; hilum without an aril.

A genus of about 40 species mostly in central and western Asia, 1 endemic in Ethiopia. *Cicer arietinum*, the chick pea, grown for its seeds, is widely cultivated in Central Asia, India, Iran, Mediterranean countries, Ethiopia, Mexico, Peru and Chile.

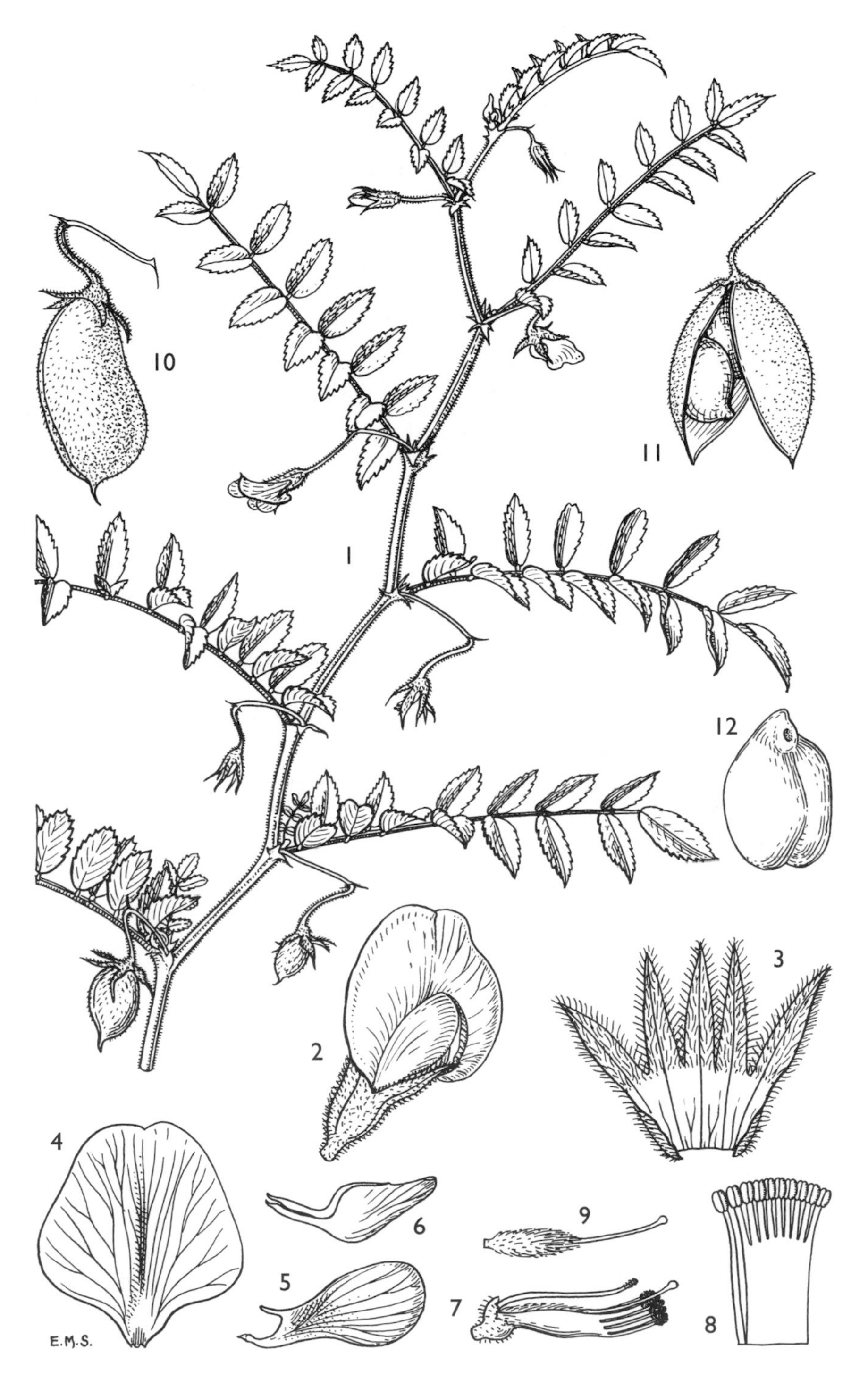

Tab. 3,7: **10**. CICER ARIETINUM. 1, flowering and fruiting branch (× 1); 2, flower (× 4); 3, calyx, opened out (× 4); 4, standard (× 4); 5, wing (× 4); 6, keel (× 4); 7, androecium and gynoecium (× 4); 8, androecium, spread out (× 4); 9, gynoecium (× 4); 10 & 11, pods (× $1^1/_2$); 12, seed (× 3). Drawn by Margaret Stones. From F.T.E.A.

Cicer arietinum L., Sp. Pl.: 738 (1753). —J.G. Baker in F.T.A. **2**: 172 (1871) pro parte. —E.G. Baker, Legum. Trop. Africa: 345 (1926). —Popov in Bull. Appl. Bot. Gen. Pl. Breed. **21**: 177, figs. 15, 16 (1929). —Purseglove, Tropical Crops **1**: 246, fig.37 (1968). —Verdcourt in F.T.E.A, Leguminosae, Pap.: 1065, fig. 152 (1971). —van der Maesen in Meded. Landbouwhoogeschool **72**, part 10: 30 (1972). —Lock, Leg. Afr. Check-list: 123 (1989). TAB. 3,7: **10**. Type from southern Europe.

Erect or prostrate annual herb, pubescent on all parts except the corolla. Stems simple or branched from the base, up to 1 m high. Leaves imparipinnate, with (3)5–7(8) pairs of leaflets; leaflets subsessile, 7–19 × 3–10 mm, elliptic, mucronate or aristate, cuneate at the base, with the upper two-thirds of the margins conspicuously dentate, glandular pubescent above and beneath; rhachis 25–60 mm long, grooved above; stipules 3–5 × 2–4 mm, ovate to triangular, 2–4-fid. Flowers solitary, axillary; peduncle 0.7–2(3) cm long, ending in a small arista 0.2–4 mm long; pedicels 5–12 mm long, recurved in fruit; bracts up to 3 mm long, linear to triangular. Calyx tube 3–4 mm long, dorsally gibbous at the base; teeth 4–5 mm long, lanceolate, with prominent midribs. Corolla white, or pinkish to purplish, the standard darker; standard 8–10 × 7–10 mm, obovate, glabrous or loosely eglandular pubescent; wings 6–9 × c. 4 mm, obovate, obtuse, asymmetrical at the base, auriculate, the auricle c. 1 mm long; keel 6–8 mm long, the petals adnate for two-thirds of the frontal side of the ventral margin and with a 2–3 mm long claw. Stamens 9, united into a sheath 4–5 mm long, free above for 2–3 mm, and 1 stamen free from the base; anthers basi-dorsifixed. Ovary 2–3 × 1–1.5 mm, ovate, 1–2(4)-ovulate; style 3–4 mm long; stigma widened after pollination. Pod 20–35 × 10–20 mm, elongate-ellipsoid, inflated, densely glandular pubescent. Seeds 1–2(4), 7–10 × 5–8 mm, subglobose or oblong-obovoid, with a median groove and a curved beak overhanging the hilum, whitish, yellowish or pale brownish, surface smooth, wrinkled or tuberculate.

Zimbabwe. N: Mazowe Distr., Archie Henderson Research Station, fl. & fr. 1.ii.1953, *Wild* 4005 (K; LISC). C: Marondera Distr., Grasslands Research Station, nursery, fl. & fr. immat. 31.iii.1965, *Corby* 1296 (K).

Also in Tanzania where it is considered to have become semi-naturalized; and widely cultivated in dry regions of the Mediterranean countries and Asia, and introduced into America, Africa and the Orient. *Cicer arietinum* (the 'chick pea') is recorded from agricultural research stations, but also introduced as a food crop, and would be expected to be found semi-naturalized as an escape.

Tribe 16. **TRIFOLIEAE**

By Maria Paula Vidigal

Trifolieae (Bronn) Benth. in Gen. Pl. **1**: 442 (1865).

Herbs or small shrubs. Leaves epulvinate, distichous, pinnately or digitately 3-foliolate (elsewhere rarely 1- or 5–7-foliolate); leaflets toothed, the nerves usually extending into the teeth or to the margin; stipels absent; stipules ± adnate to the petiole (free in *Parochetus*). Flowers few to numerous, in axillary or rarely terminal heads, spikes or short racemes, rarely solitary; bracteoles absent. Calyx campanulate, usually with subequal lobes, sometimes inflated. Standard clawed or narrowed to the base; wings sometimes interlocking with the keel, but without sculpturing. Vexillary stamen free or all filaments united below to form a tube; free portion of the filaments slender or dilated at the apex; anthers uniform or dimorphic. Ovary sessile or stipitate, 1–many-ovuled; style glabrous, straight or upcurved; stigma terminal. Fruits various, straight, falcate, spirally coiled or ovate, included in or exserted from the calyx, dehiscent or not. Seeds small, mostly rounded, with a small hilum.

A tribe of 6 genera in Eurasia, extending thinly to Africa, the Americas and Australia. Several species are widely cultivated as pasture plants.

1. Keel ± acute; stipules not adnate to petiole; creeping herbs, rooting at the nodes · 93. **Parochetus**

– Keel obtuse; stipules at least basally adnate to petiole (but also sheathing the stem in *Trifolium*); not creeping herbs (except for rare species in *Trifolium*) · · · · · · · · · · · · · · · 2
2. Petals ± adnate to the stamens, usually persisting in fruit; filaments dilated at the apex; fruit mostly included within the calyx; leaves digitately 3-foliolate or rarely pinnately so · 97. **Trifolium**
– Petals free from the stamens, not persisting in fruit; filaments not dilated; fruit much exceeding the calyx; leaves pinnately 3-foliolate · 3
3. Flowers with explosively tripping pollination mechanism; keel and wings very tightly interlocked; staminal tube strongly arched at the apex; stigma mushroom-shaped with a ring of long papillae at the base, apparently free of papillae at the apex; fruits usually coiled, very rarely falcate, often spiny · 96. **Medicago**
– Flowers lacking explosively tripping pollination mechanism; keel and wings not or loosely interlocked; staminal tube truncate or slightly arched at the apex; stigma not mushroom-shaped, lacking a ring of long papillae at the base, with short papillae at apex; fruits not coiled, never spiny · 4
4. Pods less than 3 times as long as the calyx, globose, subglobose to ovoid, indehiscent, 1–2-seeded, rarely more; flowers in extended racemes · · · · · · · · · · · · · · · · · · · 94. **Melilotus**
– Pods more than 3 times as long as the calyx, straight or rarely falcate, often beaked, dehiscent, usually many-seeded; flowers not in extended racemes · · · · · · · 95. **Trigonella**

93. PAROCHETUS Buch.-Ham. ex D. Don

Parochetus Buch.-Ham. ex D. Don, Prodr. Fl. Nepal.: 240 (1825).

Creeping herbs, usually rooting (and also forming small tubers in *P. communis*) at the nodes. Leaves digitately 3-foliolate, long-petiolate; stipules free from the petiole. Inflorescence axillary, 1–3-flowered, pedunculate; bracts present; bracteoles absent. Flowers pedicellate. Calyx with the 2 upper lobes connate almost to their tips. Petals not adnate to the stamens, caducous; standard shortly clawed; wings with the upper proximal corner of the blade extended into a short auricle; keel shorter than the wings, upcurved, rather acute. Stamens with free portions of filaments abruptly upcurved, not dilated at the apex; vexillary filament free; anthers uniform. Ovary sessile; ovules many; style slender, tapering, abruptly bent upwards; stigma minute, terminal. Pod 3–4 times as long as the calyx, oblong, acute, beaked, not septate inside, many-seeded. Seeds with tuberculate testa; funicle filiform.

A genus of 2 species, one from tropical Africa, the other from southern Asia.

Parochetus was formerly only known from the type species, *Parochetus communis* D. Don. The African material is now treated as a separate species *Parochetus africanus* Polhill. *Parochetus communis*, the name that has been widely used for these plants, is now seen to be correctly applied to plants mainly from the Himalayas, but also on higher mountains south to Sri Lanka, east to China and southeast to Java. Several characters allow the recognition of the African representatives as a second species. The two species differ in habit, phenology, flower structure and seed-pods, see Beckett & Polhill in Kew Mag. **8**: 54–58 (1991). Dr. John Grimshaw (notes in the Kew Herbarium) suspects that the colour plate in that article was inadvertently prepared from introduced Asiatic material — the African plant, in his experience, has paler blue flowers and lacks the red and white "chevron" on the leaflets, tending instead to have an irregular pattern of silvery markings. The line drawing, diagnosis and typification of the African species remain unaffected. A colour photograph is published in Blundell, Wild Fl. East Africa: pl. 830 (1987) under *Parochetus sp.*

Parochetus africanus Polhill in Kew Mag. **8**: 56, fig. on page 57 (1991). TAB. 3,7: **11**. Type: Malawi, Mt. Chiradzulu (Chiradzura), *Meller* s.n. (K, holotype).

Parochetus communis sensu J.G. Baker in F.T.A. **2**: 48 (1871). —sensu Gürke in Engler, Pflanzenw. Ost-Afrikas **C**: 208 (1895). —sensu Gillett in F.C.B. **4**: 289 (1953). —sensu Brenan in Mem. New York Bot. Gard. **8**: 249 (1953). —sensu Gillett in F.T.E.A., Leguminosae, Pap.: 1014, fig. 141 (1971) non Buch.-Ham. ex D. Don (1825).

Parochetus major sensu E.G. Baker, Legum. Trop. Africa: 70 (1926) non D. Don (1825).

Parochetus communis var. *grossecrenatus* Cufod. in Senckenberg. Biol. **39**: 295, t. 35/1 (1958). Type from Ethiopia.

Parochetus sp. Blundell, Wild Fl. East Africa: 114, pl. 830 (1987).

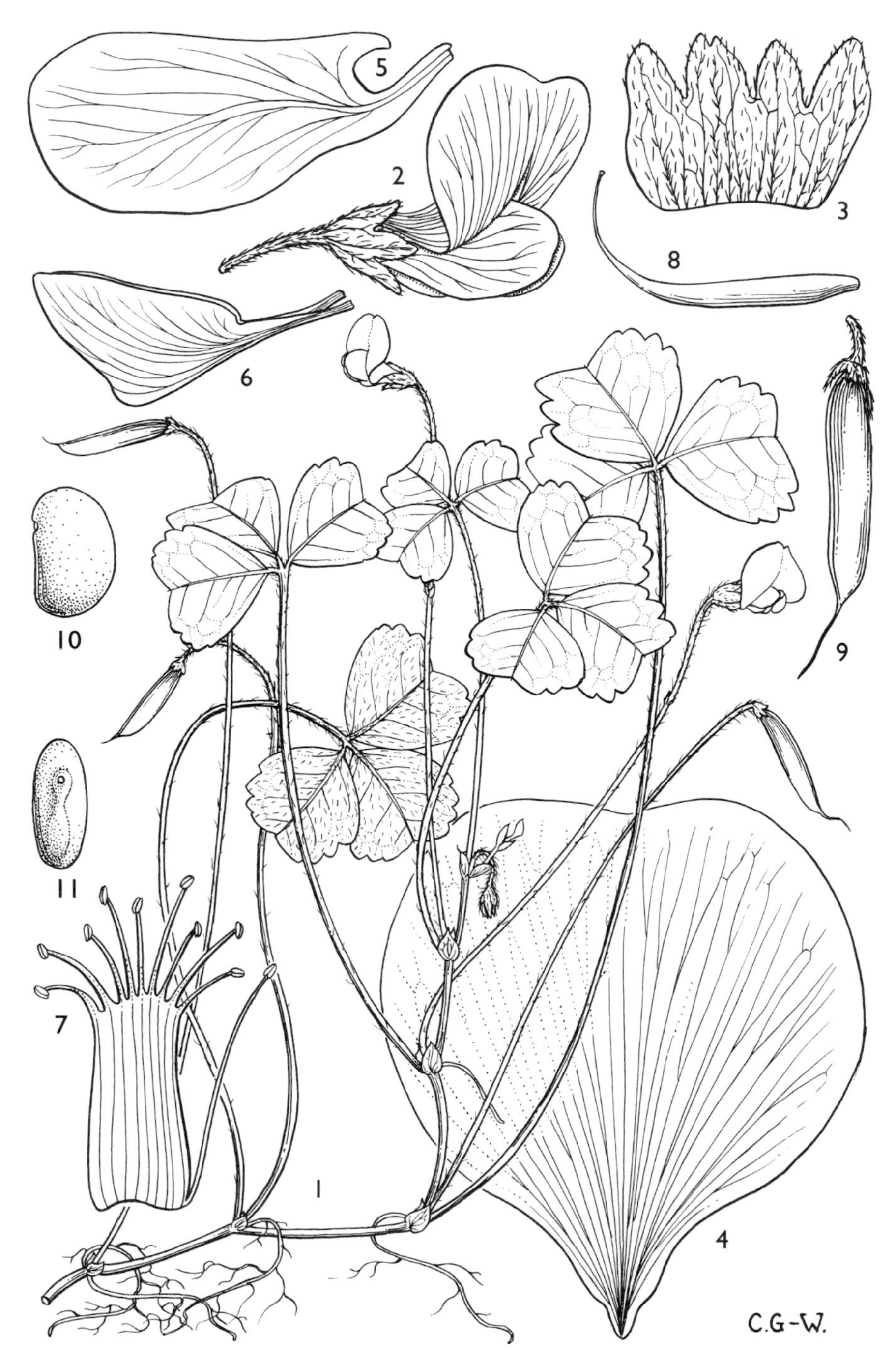

Tab. 3,7: **11**. PAROCHETUS AFRICANUS. 1, habit (× $^{2}/_{3}$); 2, flower (× 2); 3, calyx, opened out (× 4); 4, standard (× 4); 5, wing (× 4); 6, keel (× 4); 7, androecium, spread out (× 4); 8, gynoecium (× 4); 9, pod (× 1), 1–9 from *Verdcourt* 3265; 10, seed, side view (× 8); 11, seed, hilar view (× 8), 10 & 11 from *Tanner* 3859. Drawn by Christopher Grey-Wilson. From F.T.E.A.

Creeping prostrate herb, forming ground cover. Stems slender, rooting at the nodes, lacking tubers, glabrous to sparsely pilose. Leaves digitately 3-foliolate; leaflets up to 4 cm long and wide, obovate, emarginate at the apex, cuneate at the base, margins usually coarsely toothed towards the apex, with a distinct midrib and 4 primary lateral nerves on each side, green blotched with white in the centre, glabrous above, subglabrous to pilose beneath; petiole 3–30 cm long, pilose; stipules 6–10 mm long, ovate, acute, scarious. Inflorescence 1–2(3)-flowered; peduncle 1.5–25 cm long, ascending; bracts similar to stipules but smaller. Flowers sweetly scented, on pedicels 1.5–6 cm long, reflexed in fruit. Calyx c. 7 mm long, the 2 upper lobes connate almost to their tips, the lateral lobes triangular-lanceolate, the lowest lobe longer than the 2 lateral, glabrous to sparsely pilose. Corolla generally pale blue (rarely white), glabrous; standard 12–18 mm long, broadly ovate-elliptic, emarginate, contracted at the base into a short claw; wings c. 13 mm long, oblong-ovate, the auricle only 0.5–1.5 mm (2.5–4 mm in *P. communis*) long; keel c. 10 mm long, subangular about the middle of the blade (upcurved near the tip in *P. communis*), acute, almost straight along the upper margin, darker blue at the tip. Stamens glabrous. Ovary glabrous; style slender, glabrous; stigma minute, capitate. Pod 15–25 × 4–5 mm, narrowly oblong, shortly beaked, glabrous or rarely pilose, tardily dehiscent or sometimes forced open by seeds germinating inside. Seeds 2–2.5 × 1.8–2 mm, slightly reniform, brown, sometimes mottled darker.

Zimbabwe. C: Marondera (Marandellas), Grasslands Research Station, cultivated, fl. 15.ii.1967, *Corby* 1761 (K; PRE). **Malawi**. N: Rumphi Distr., Nyika Forest, above Nchenachena, c. 1950 m, fl. vii.1953, *Chapman* 182 (BM). S: Zomba Distr., Zomba Plateau, by Mlunguzi Stream near Trout Ponds, 1520 m, fl. 26.vii.1970, *Brummitt* 12244 (K; LISC; MAL; SRGH).

Also in central Ethiopia, eastern Dem. Rep. Congo, Rwanda, Burundi, western Uganda, Kenya and Tanzania. Submontane or montane, forest floor and on stream and river banks in damp shady places; 1500–2000 m.

Although Gillett, loc. cit. (1971), has recorded this species from Mozambique (Namuli) I have not seen any material from this area.

94. MELILOTUS Mill.

Melilotus Mill., Gard. Dict. abr. ed. 4 (1754). —O.E. Schulz in Bot. Jahrb. Syst. **29**: 682–735 (1901).

Annual or biennial herbs. Leaves pinnately 3-foliolate; leaflets toothed, the nerves ending in teeth; stipules basally adnate to the petiole. Flowers small, yellow or white, in extended pedunculate axillary racemes, lacking an explosively tripping pollination mechanism; bracts minute or absent; bracteoles absent. Calyx lobes subequal, ± as long as the tube. Petals not adnate to the stamens, glabrous, deciduous; standard subsessile, with 3 major basal veins; wings extended at the upper proximal corner into a small auricle; keel and wings not or loosely interlocked by a wing spur in a keel pocket (both spur and pocket not or weakly developed). Anther sheath straight at the apex; free portions of filaments mostly thin; vexillary filament free; anthers uniform, versatile. Ovary sessile or stipitate, few-ovuled; style long, incurved above, glabrous; stigma minute, terminal, with short papillae on the apex. Pod small, subglobose or ovoid, falling off with the calyx and pedicel, indehiscent or tardily dehiscent, variously veined, mostly indehiscent. Seeds ovoid, smooth or tuberculate.

A genus of about 20 species in temperate and subtropical Eurasia, some also in North Africa and extending into Ethiopia.

In addition to *Melilotus indicus* and *M. albus*, two other species have been recorded in cultivation in the Flora Zambesiaca area. However, the material is too poor for positive determination and further collections are needed. These specimens are represented by herbarium material: Mozambique. M: Umbelúzi, Estação Experimental, cultivated, 23.iii.1951, *Myre* 1075 (LMA). GI: Guijá, Posto de Culturas, Regadas do Vale do R. Limpopo, cultivated, 11.ix.1952, *Pedro* 4008 (LMA).

Flowers up to 2.5 mm long, yellow; pod prominently reticulate-veined · · · · · · · · · · 1. *indicus*
Flowers 3–6 mm long, white; pod weakly reticulate-veined · 2. *albus*

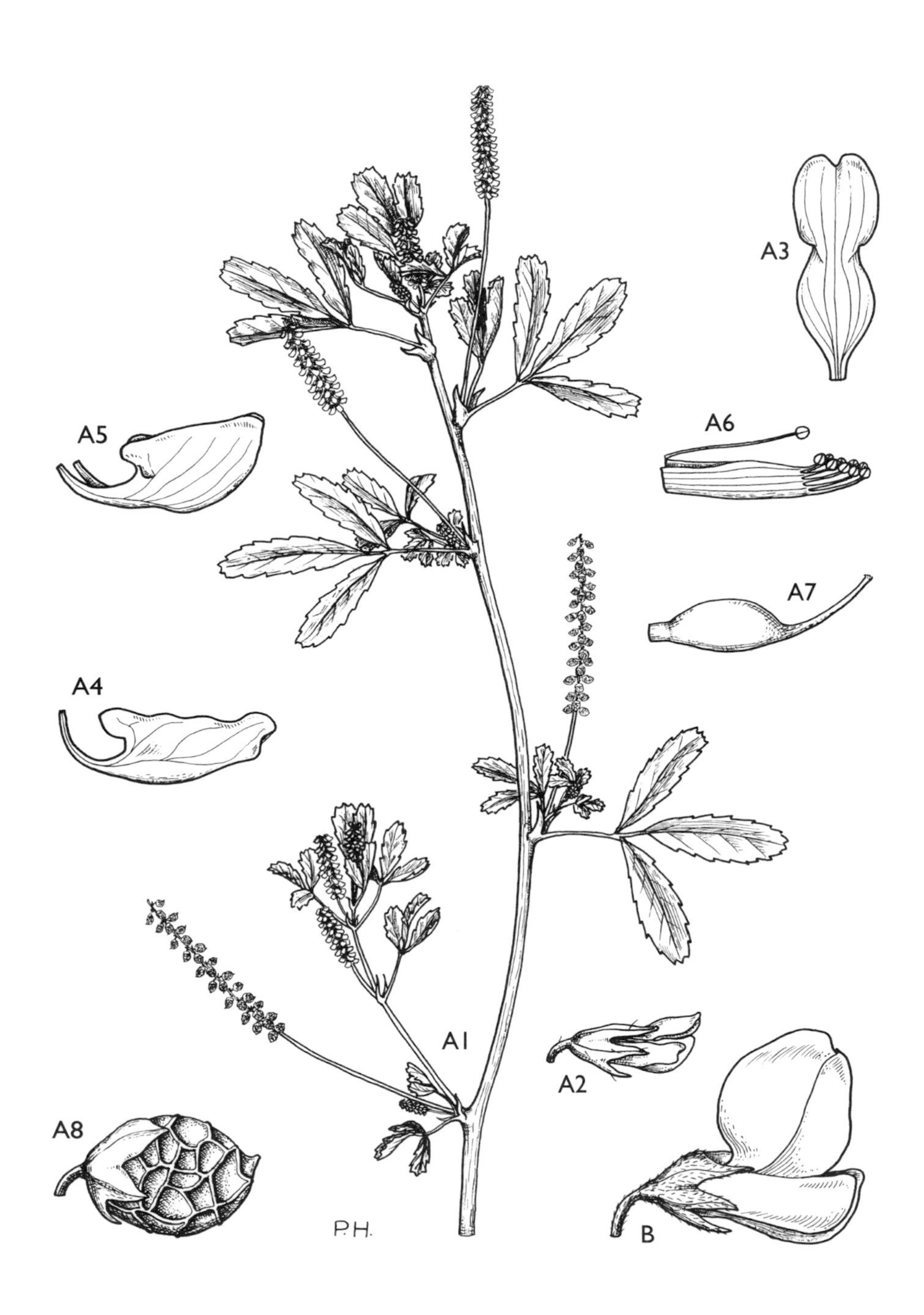

Tab. 3,7: **12**. A. —MELILOTUS INDICUS. A1, flowering and fruiting branch (× $^{2}/_{3}$); A2, flower (× 6); A3, standard (× 12); A4, wing (× 12); A5, keel (× 12); A6, androecium (× 12); A7, gynoecium (× 12), A1–A7 from *Newton* in *E.A.H.* 19156; A8, pod (× 6), from *Purseglove* 2315. B. —MELILOTUS ALBUS, flower (× 6), from *Bally* 6205. Drawn by Pat Halliday. From F.T.E.A.

1. **Melilotus indicus** (L.) All., Fl. Pedem. **1**: 308 (1785). —E.G. Baker, Legum. Trop. Africa: 74 (1926). —Gillett in F.T.E.A., Leguminosae, Pap.: 1039, fig 145/1–8 (1971) as "*indica*". —Drummond in Kirkia **8**: 224 (1972) as "*indica*". —Lock, Leg. Afr. Check-list: 481 (1989) as "*indica*". —Agnew, Upland Kenya Wild Fl., ed. 2: 149 (1994) as "*indica*". TAB. 3,7: **12**, fig. A. Type from India.
   Trifolium indicus L., Sp. Pl.: 765 (1753) as "*indica*".
   Melilotus parviflorus Desf., Fl. Atl. **2**: 192 (1799) as "*parviflora*". —J.G. Baker in F.T.A. **2**: 52 (1871) as "*parviflora*". Type from Algeria.

Erect, much branched, slender annual herb. Leaflets up to 25 × 9 mm, shortly cuneate-oblong or lanceolate, rounded at the apex, denticulate around the upper half or more; stipules lanceolate-subulate, entire, sometimes denticulate near the base. Racemes slender, many-flowered, at first compact and cylindrical, later elongating and laxer; peduncle up to 3 cm long. Flowers with pedicels c. 1 mm long, reflexed in fruit. Calyx c. 1 mm long; teeth narrowly deltoid, subequal, thinly hairy. Corolla yellow, ± twice as long as the calyx; standard c. 2.5 mm long, emarginate at apex; wings subequalling the keel, with small auricles; keel c. 2 mm long. Pod 2–3 mm long, globose-ovoid, strongly reticulately veined, 1-seeded. Seeds ovoid, finely verrucose.

Zimbabwe. N: Mazowe Distr., 1310 m, fl. & fr. 30.x.1926, *Eyles* 1202 (BM). W: Matobo Distr., Matopos, fl. & fr. 5.ix.1931, *J.M. Rattray* 396 (PRE).

Also in northern, central and eastern Africa, Socotra, Asia, Europe and the Middle East. Introduced as a fodder crop and often in cultivation on agricultural research stations it has become naturalized and is a weed of disturbed ground.

Drummond, in Kirkia **8**: 224 (1972), has recorded this species from Zimbabwe C and E, but I have not seen material from these areas.

2. **Melilotus albus** Medik. in Vorles. Churpfälz. Phys.-Okon. Ges. **2**: 382 (1787) as "*alba*". —Desrousseaux, Poiret & Savigny, Encycl. Méth. Bot. **4**: 63 (1796). —Burtt Davy, Fl. Pl. Ferns Transvaal, pt. 2: 402 (1932). —Gillett in F.T.E.A., Leguminosae, Pap.: 1041, fig. 145/9 (1971) as "*alba*". —Drummond in Kirkia **8**: 224 (1972) as "*alba*". —Lock, Leg. Afr. Check-list: 481 (1989) as "*alba*". —Agnew, Upland Kenya Wild Fl., ed. 2: 149 (1994) as "*alba*". TAB. 3,7: **12**, fig. B. Type: a plant cultivated in Paris from Siberian seed.

Erect, branched annual herb. Leaflets narrowly oblong-obovate to linear or elliptic, cuneate and entire at the base, denticulate around the upper half or more; stipules linear-subulate, entire. Racemes many-flowered, up to 17 cm long, lax; peduncle up to 4 cm long. Flowers small, with pedicels 1.5–2.5 mm long, becoming deflexed in fruit. Calyx c. 2 mm long; teeth narrowly deltoid, somewhat unequal (the lowest shortest), glabrous or somewhat appressed-hairy. Corolla white, ± twice as long as the calyx; standard c. 4 mm long, broadly oblong-ovate, distinctly emarginate at the apex; wings slightly exceeding the keel, with small, hooked auricles; keel c. 4 mm long, oblong. Pod 3–5 mm long, globose-ovoid or subglobose, with a short, curved beak at the apex, weakly reticulately veined, ± brown when mature, glabrous. Seeds ovoid, smooth.

Zimbabwe. N: Mazowe (Mazoe), Mazowe Citrus Estate, fl. & fr. 4.xi.1970, *Searle* 34 (LISC; PRE; SRGH). C: Marondera (Marandellas), fl. & fr. 15.ii.1962, *Corby* 1027 (K). E: Mutare Distr., Old Mutare (Umtali) Mission, near water furrow, 1100 m, fl. 26.ix.1955, *Chase* 5808 (BM). **Mozambique**. MS: Sussundenga Distr., Mavita, Rotanda, Campos Experimentais da Junta dos Cereais, 17.ix.1946, *Pedro & Pedrógão* 285 (PRE). M: Maputo (Lourenço Marques), Umbelúzi Experimental Farm, fl. 7.x.1965, *Mogg* 32586 (LISC).

Also in Algeria, Libya, Egypt, Ethiopia, Kenya, Tanzania, South Africa and Lesotho; in Asia; almost throughout Europe; Middle East. In grasslands, growing in profusion alongside irrigation channels, near water furrows, apparently in light sandy or loamy soil, abundant in sunny places.

95. TRIGONELLA L.

Trigonella L., Sp. Pl.: 776 (1753); Gen. Pl., ed. 5: 338 (1754).

Herbs, mostly annual, often strongly aromatic. Leaves pinnately 3-foliolate; leaflets often denticulate; stipules basally adnate to the petiole. Flowers in axillary racemes, heads or umbels, or flowers solitary or in pairs, lacking an explosive tripping pollination mechanism; bracts minute; bracteoles absent. Calyx campanulate, more rarely tubular; teeth subequal. Petals free from the stamens, mostly yellow, deciduous; standard obovate, with basal vein 3-branched near the base; wings oblong with the upper proximal corner extended into a small auricle (c. $^1/_{10}$–$^1/_5$ the length of the wing blade); keel obtuse, not or loosely adherent to the wing by a wing spur in a keel pocket (both spur and pocket not or weakly developed). Filament sheath truncate at the apex; free portions of filaments mostly thin; anthers uniform, versatile; vexillary filament free. Ovary sessile, few–many-ovuled; style long, incurved above, glabrous; stigma small, somewhat expanded from the style but not markedly capitate, with short papillae on the apex. Pod usually linear or oblong, straight or ± curved, subterete or compressed, often beaked, never coiled or spiny, dehiscent along the ventral suture or indehiscent, usually many-seeded. Seeds mostly verruculose, sometimes smooth.

A genus of about 80 species mostly in the Mediterranean region, extending into central Europe and central Asia, to the Canary Islands, and into the drier parts of southern Africa; one species indigenous to Australia.

One species native and two species known only in cultivation in the Flora Zambesiaca area.

1. Flowers 1–2, axillary; pods 6–15 cm long, long-beaked · · · · · · *foenum-graecum* (cultivated)
– Flowers several together; pods up to 1 cm long, hardly beaked · · · · · · · · · · · · · · · · · · 2
2. Pods undulate (wavy) · *anguina* (cultivated)
– Pods not undulate · *laciniata*

Trigonella foenum-graecum L.

Erect, annual herb, strongly aromatic. Leaflets obovate-oblong, obtuse at the apex, denticulate in the upper half; stipules triangular-lanceolate, acuminate, entire, glabrous. Flowers solitary or paired, subsessile in the leaf axils. Calyx tubular, pilose; teeth as long as the tube. Pod 60–150 × 4–6 mm, somewhat flattened, linear, straight or curved tapering into a 20–40 mm long beak, nerved lengthwise, usually 10–20-seeded.

Zimbabwe. N: Mazowe Distr., fr. ix.1975, *Fenwick* s.n. (K). **Mozambique**. N: Malema, fr. 21.vi.1948, *Pedro & Pedrógão* 4349 (LMU). GI: Guijá Distr., Posto de Culturas Regadas do Vale do R. Limpopo, 24.viii.1952, *Myre* 1227 (LMA).

Cultivated for its seeds, which when ground up are used as an ingredient of curry-powder, a spice used in cooking.

Trigonella anguina Del.

Decumbent annual herb, branched from the base, glabrous or thinly hairy. Leaflets small, cuneate-obovate, truncate to retuse at the apex, denticulate almost to the base, glabrous above, sparsely pubescent beneath; stipules lanceolate-acuminate, the lower ones dentate. Flowers small, yellow, 2–7-fasciculate, subumbellate, sessile or subsessile. Calyx campanulate, sparsely pubescent; teeth equalling or exceeding the tube. Pod 4–10 mm long, linear, wavy, shortly acuminate, with a very slender persistent style, 2–5-seeded. Easily recognized by its small eel-like fruits.

Zimbabwe. C: Marondera (Marandellas), Grasslands Research Station, 25.ii.1964, *Corby* 1081 (PRE).

Native of North Africa, the Middle East and western Asia, and recorded in cultivation at Grasslands, a pasture research station in Zimbabwe, where it was under trial as a drylands pasture crop.

Trigonella laciniata L., Sp. Pl., ed. 2: 1095 (1763). —J.G. Baker in F.T.A. **2**: 50 (1871). —E.G. Baker, Legum. Trop. Africa: 71 (1926). —F.W. Andrews, Fl. Pl. Anglo-Egypt. Sudan **2**: 243 (1952). —Gillett in Kew Bull. **19**: 387 (1965). —Lock, Leg. Afr. Check-list: 499 (1989). TAB. 3,7: **13**. Type from Egypt.

Prostrate annual herb, diffusely branched from the base. Leaves petiolate; leaflets 3–6 mm long, obtriangular-oblong, acutely and sharply dentate; stipules deeply and

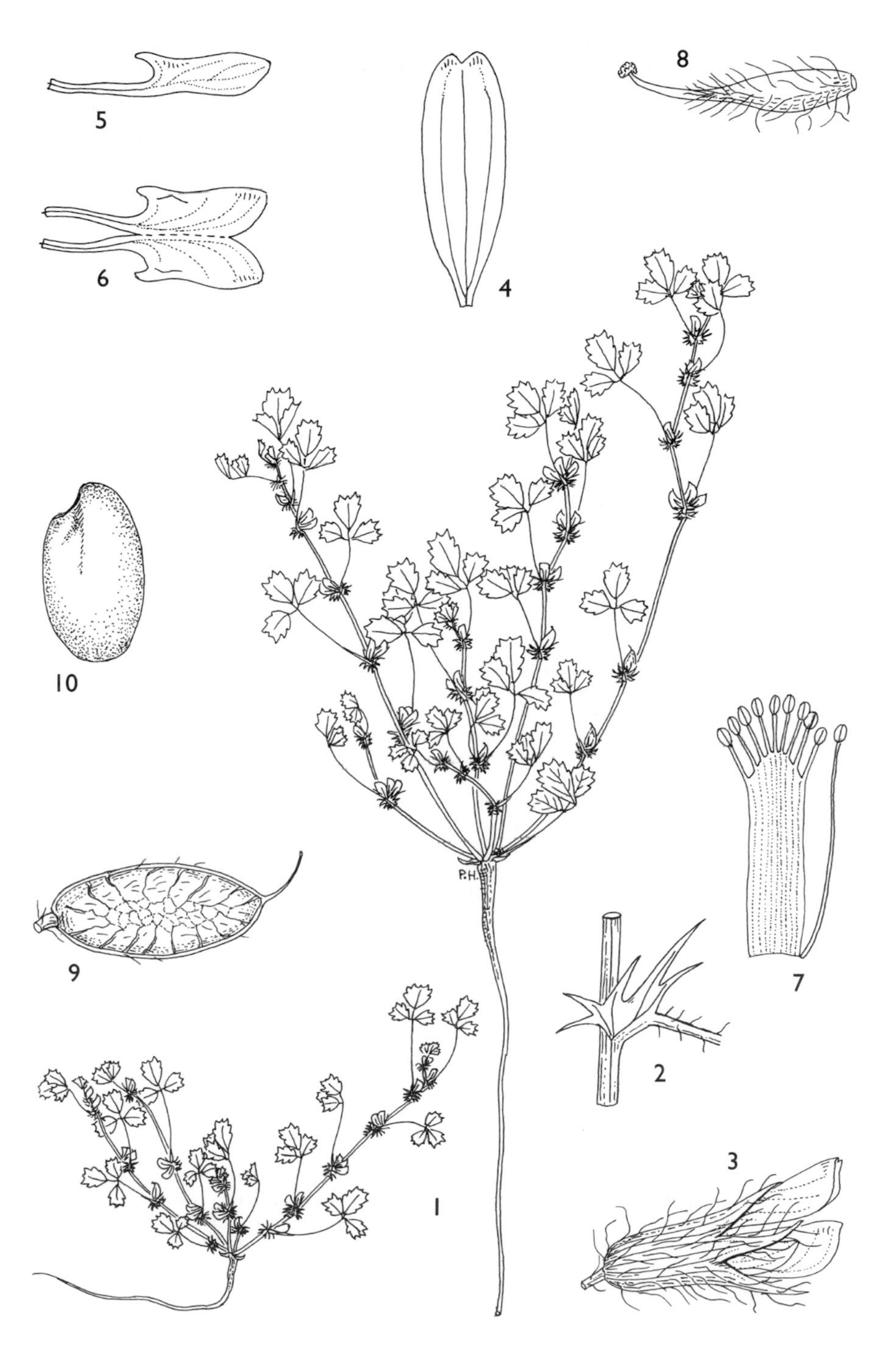

Tab. 3,7: **13**. TRIGONELLA LACINIATA. 1, habit (× 1); 2, stipules, much enlarged; 3, flower (× 10); 4, standard (× 10); 5, wing (× 10); 6, keel (× 10); 7, androecium (× 12); 8, gynoecium (× 12); 9, pod, calyx removed (× 6); 10, seed (× 10), 1–10 from *van Rensburg* 2536. Drawn by Pat Halliday.

sharply laciniate. Flowers in sessile axillary heads (heads pedunculate outside the Flora Zambesiaca area). Calyx 2–2.5 mm long; teeth triangular, ± half as long as the tube. Corolla twice or more as long as the calyx, yellow. Pods c. 7 × 3 mm, oblong, straight, with prominent longitudinal nerves crossed by transverse ones, subglabrous to sparsely hairy, many-seeded. Seeds c. 1 mm long, ovoid, brown, smooth.

Zambia. S: Namwala Distr., Kafue National Park, Ngoma, fl. & fr. 12.viii.1962, *Fanshawe* 7001 (K); Namwala Distr., Kafue River, Ibamba Ranch, between Katoshi and Kasenga, fl. & fr. 15.x.1963, *van Rensburg* 2536 (K; PRE; SRGH); Namwala Distr., along a stream emanating from hot springs east of Ngoma, fr. 1.ix.1964, *van Rensburg* 2953 (K; SRGH).

Also in Algeria, Egypt, Sudan, Saudi Arabia and the Middle East. Alluvial clay on river banks and in dry lagoons, in *Vossia* grassland.

Gillett, loc. cit. (1965) considers the gatherings *Fanshawe* 7001 and *van Rensburg* 2536 cited above, as belonging to the forma *arguta* (Vis.) Sick.

96. MEDICAGO L.

Medicago L., Sp. Pl.: 778 (1753); Gen. Pl., ed. 5: 339 (1754). —Heyn in Scripta Acad. Hierosolymitana, Sci. Rep. **12**: 1–154 (1963). —E. Small & Jomphe in Canad. J. Bot. **67**: 3260–3294 (1989).

Annual or perennial herbs. Leaves pinnately 3-foliolate; leaflets at least distally marginally toothed; stipules basally adnate to the petiole, deeply incised or laciniate (entire or basally toothed in *M. sativa*). Racemes short, axillary, pedunculate; bracts short. Flowers pedicellate, with an elaborate explosive tripping pollination mechanism (in connection with a syndrome of floral characteristics). Calyx short, 5-dentate; teeth subequal. Petals free from the stamens, mostly yellow or golden (blue, violet or mauve in *M. sativa*), glabrous, caducous; standard with basal vein usually more than 3-branched near the base; wings with the upper proximal corner extended into a large auricle (about one-third the length of the wing blade); keel and wings very tightly interlocked by an adaxial (ventral) prominent wing spur in an abaxial (dorsal) developed keel pocket. Staminal tube strongly arched at the apex; free portions of filaments relatively thick; vexillary filament free; anthers uniform. Ovary sessile, 1–few–many-ovuled; style always short, massive, glabrous; stigma terminal, large, fungiliform (mushroom-shaped), with a basal ring of papillae. Pod exserted from the calyx, usually spirally coiled, discoid to cylindrical, mostly indehiscent, the dorsal suture usually spiny or smooth or tuberculate. Seeds 1–several, small, ± reniform, ellipsoid-oblong or slightly curved, mostly smooth; hilum ± central.

A genus of about 50 species distributed mainly around the Mediterranean, and extending into Europe (absent in northernmost Europe), North Africa to Ethiopia, and central and western Asia, with a few species in tropical and southern Africa; much cultivated in pastures and for fodder.

Medicago sativa L. (lucerne, alfalfa), a perennial herb with blue flowers, and fruits without spines, is cultivated as a fodder crop in the Flora Zambesiaca area, e.g., Zambia (*Angus* 3782 (K), and for seeds in Mozambique (*Myre* 987 (LMA)). Native of Eurasia it is widely introduced and escaped from cultivation.

1. Pods flat, without spines; surface of coils without a submarginal vein · · · · · · 1. *orbicularis*
– Pods not flat, spiny, or if spineless, with a distinct submarginal vein · · · · · · · · · · · · · · · 2
2. Radial veins on surface of coils S-shaped, not anastomosing; grooves between the submarginal vein and dorsal suture of pod narrow (not visible in side view of pod) · · · · · · 2. *laciniata*
– Radial veins on surface of coils not S-shaped, anastomosing freely; grooves between the submarginal vein and dorsal suture of pod wide (visible in side view of pod) · · 3. *polymorpha*

1. **Medicago orbicularis** (L.) Bartal., Cat. Piante Siena: 61 (1776). —J.G. Baker in F.T.A. **2**: 51 (1871). —E.G. Baker, Legum. Trop. Africa: 73 (1926). —Heyn in Scripta Acad. Hierosolymitana, Sci. Rep. **12**: 135, fig. 36 (1963). —Drummond in Kirkia **8**: 224 (1972). —Lock, Leg. Afr. Check-list: 480 (1989). Type from England.

Medicago polymorpha var. *orbicularis* L., Sp. Pl.: 779 (1753).

Annual herb, usually procumbent. Leaflets obovate, cuneate at the base, usually retuse at the apex, finely denticulate in the upper half, glabrous; stipules deeply laciniate, glabrous. Inflorescence 1–3(5)-flowered; peduncle produced into a fine arista beyond the flowers. Flowers with pedicels longer than the calyx. Calyx glabrous or sparsely hairy; teeth almost as long as the tube. Corolla yellow, ± twice as long as the calyx. Pod lenticular or discoid, glabrous, without spines; coils 3–5, turning anticlockwise, thin, the middle coil 8–15 mm in diameter; surface of coils with a net of anastomosing radial veins entering into the dorsal suture; no submarginal vein present. Seeds not seen.

Zimbabwe. C: Marondera (Marandellas), Grasslands Research Station, fl. & fr. 23.ii.1962, *Corby* 1037 (SRGH).

Also in Morocco, Algeria, Tunisia, Libya, Egypt, Ethiopia, South Africa; and in Asia, Europe and the Middle East. Pastures, escaped from cultivation.

2. **Medicago laciniata** (L.) Mill., Gard. Dict. ed. 8, Medicago no. 5 (1768). —Harvey in F.C. **2**: 163 (1862). —J.G. Baker in F.T.A. **2**: 51 (1871). —E.G. Baker, Legum. Trop. Africa: 73 (1926). —Heyn in Scripta Acad. Hierosolymitana, Sci. Rep. **12**: 55, fig. 15 A–C (1963). —Schreiber in Merxmüller, Prodr. Fl. SW. Afrika, fam. 60: 86 (1970). —Gillett in F.T.E.A., Leguminosae, Pap.: 1037 (1971). —Jacot Guillarmod, Fl. Lesotho: 192 (1971). —Stirton in Bothalia **14**: 28, fig. 1 (1982). —Lock, Leg. Afr. Check-list: 479 (1989). —Agnew, Upland Kenya Wild Fl., ed. 2: 149 (1994). —Retief & Herman in Strelitzia **6**: 461 (1997). TAB. 3,7: **14**, fig. A. Type: plate 34 in Breyne, Exoticarum aliarumque minus cognitarum plantarum (1678), *fide* Heyn, loc. cit.

Medicago polymorpha var. *laciniata* L., Sp. Pl.: 781 (1753).

Medicago aschersoniana Urb. in Verh. Bot. Vereins Prov. Brandenburg **15**: 77, fig. 58 (1873). —Ross, Fl. Natal: 199 (1972). Type: the figure cited.

Spreading annual herb, branching from the base. Leaves petiolate; leaflets 4–15 × 2–7 mm, narrowly cuneate-obovate or cuneate-oblong, truncate to retuse and apiculate at the apex, denticulate in the upper half or deeply toothed or irregularly laciniate, glabrous above, sparsely pilose beneath; petioles up to 2.5 cm long; rhachis up to 5 mm long; stipules laciniate, with simple hairs beneath. Inflorescence 1–2-flowered; peduncle produced into an arista beyond the terminal flower. Flowers very small, with pedicels shorter than the calyx tube. Calyx 3–4 mm long, sparsely appressed-hairy; teeth equal, shorter than the tube, hairy. Corolla yellow, twice as long as the calyx; standard ovate, longer than the keel; wings shorter than the keel; keel blunt, oblong. Pod cylindrical to ellipsoid, usually glabrous, spiny; coils 3–7, turning anticlockwise, the broadest coil 3–5 mm in diameter, not strongly appressed to each other; surface of coils with numerous S-shaped radial veins, not anastomosing before joining the submarginal vein near to the dorsal suture; grooves between dorsal suture and submarginal vein of pod narrow (invisible in side view of pod); spines 2–4 mm long, with 2-rooted base, one root arising in the dorsal suture, the other in the submarginal vein, the longer ones hooked at the tip, divergent and intermeshing with those of adjoining coils. Seeds 5–10 per pod, c. 3 mm long, oblong to subreniform, yellowish, smooth.

Botswana. SW: Ghanzi Distr., Khomodimo (Kmodimo) Pan, 332 km northwest of Molepolole (Molepole), fr. 18.vi.1955, *Story* 4923 (PRE); Ghanzi Distr., small pan with Ncojane Ranches, 42 km from Ncojane, 1050 m, fl. & fr. 24.iii.1988, *Mithen* 573 (PRE).

Also in Morocco, Algeria, Tunisia, Egypt, western Sahara, Ethiopia, Somalia, Kenya, Tanzania, Namibia, South Africa (KwaZulu-Natal) and Lesotho; and in S Yemen, Asia and Middle East. Low rainfall areas, calcareous pans; 1000–1050 m.

Two varieties are generally recognized (Heyn 1963): var. *laciniata* with the peduncle much longer than petiole and pods with 5–7 coils; var. *brachyacantha* Boiss. with the peduncle shorter than the petiole and pods with 2?–4? coils. The specimens seen, *Mithen* 525 and 573 and *Story* 4923, from Botswana, show a combination of characters of the two varieties.

3. **Medicago polymorpha** L., Sp. Pl.: 779 (1753), emend. Shin in Rhodora **58**: 5 (1956). —Heyn in Scripta Acad. Hierosolymitana, Sci. Rep. **12**: 71, fig. 19 (1963). —Gillett in F.T.E.A., Leguminosae, Pap.: 1039 (1971). —Drummond in Kirkia **8**: 224 (1972). —Stirton in Bothalia **14**: 30 (1982). —Lock, Leg. Afr. Check-list: 480 (1989). —Retief & Herman in Strelitzia **6**: 461 (1997). Type from Europe.

Medicago hispida Gaertn., Fruct. Sem. Pl. **2**: 349, t. 55 (1791), emend. Urban in Verh. Bot.

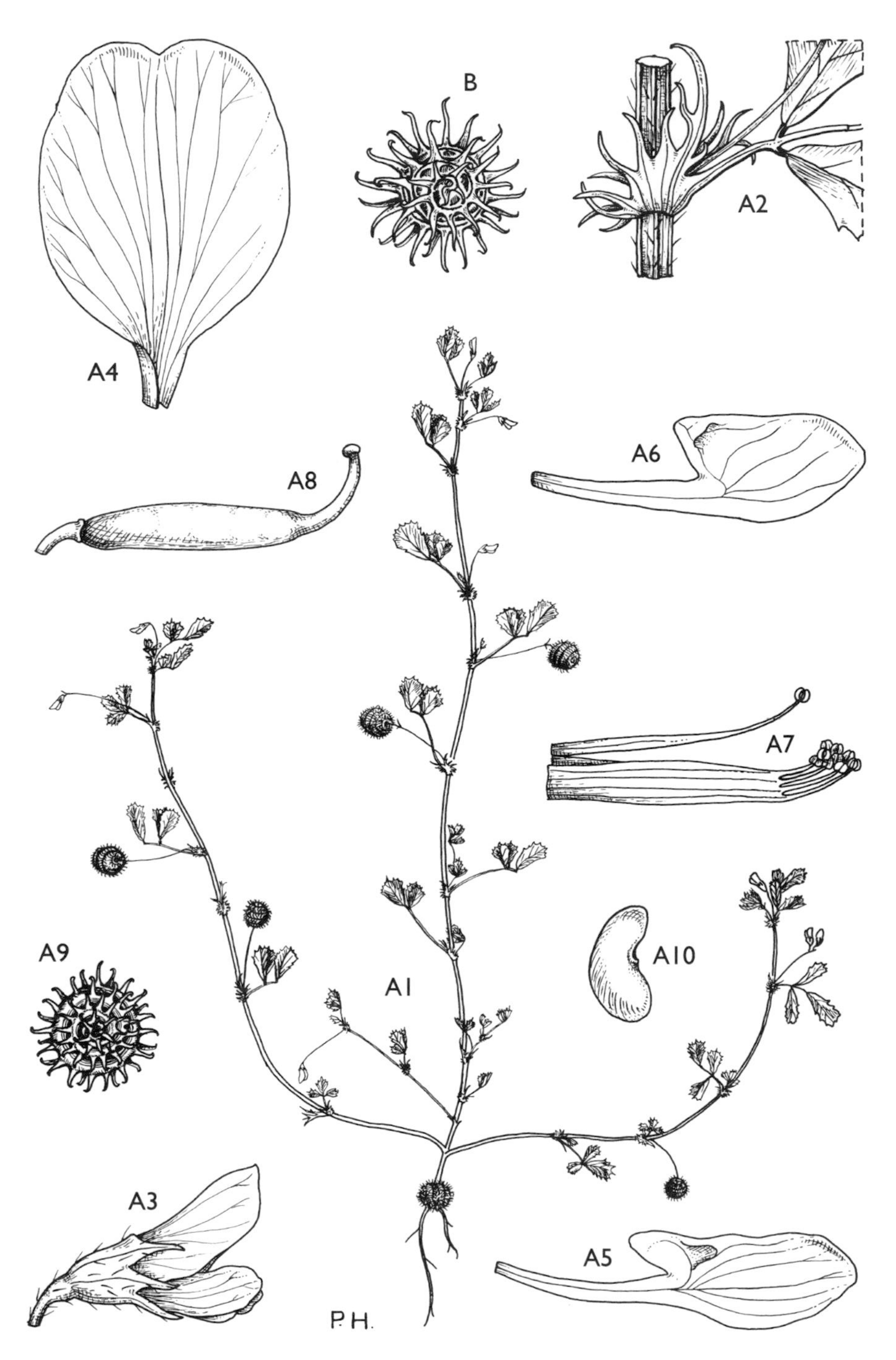

Tab. 3,7: **14**. A. —MEDICAGO LACINIATA. A1, habit (× $^{2}/_{3}$); A2, detail showing stipules (× 6); A3, flower (× 6); A4, standard (× 12); A5, wing (× 12); A6, keel (× 12); A7, androecium (× 12); A8, gynoecium (× 12), A1–A8 from *Bogdan* 3487; A9, fruit, viewed from above (× 2); A10, seed (× 6), A9 & A10 from *Ayre* in *A.D.* 2953. B. —MEDICAGO POLYMORPHA var. POLYMORPHA, fruit, viewed from above (× 2), from *Boonman* K.542333. Drawn by Pat Halliday. From F.T.E.A.

Vereins Prov. Brandenburg **15**: 74 (1873). —Ross, Fl. Natal: 199 (1972). Type is a fruit specimen of unknown origin.

Medicago denticulata Willd., Sp. Pl. **3**: 1414 (1802). —Harvey in F.C. **2**: 162 (1862). —J.G. Baker in F.T.A. **2**: 51 (1871). —E.G. Baker, Legum. Trop. Africa: 73 (1926). Type from southern Europe.

Annual herb, glabrous except for a few simple hairs on the petiolules, lower surface of leaflets, midribs, peduncles and calyces. Leaflets obovate to obovate-cuneate, obtuse, usually retuse and apiculate at the apex, serrate or dentate in the upper third; stipules ovate-oblong, laciniate, the lower side glabrous or with a few hairs along the margins. Inflorescence 2–5-flowered; peduncle usually not protruding beyond the terminal flower. Flowers with pedicels shorter than the calyx tube. Calyx sparingly hairy; teeth ± as long as the tube. Corolla yellow, usually less than twice as long as the calyx; wings longer than the keel. Pod discoid to cylindrical, glabrous, spiny, tuberculate or spineless; coils 1.5–6, turning anticlockwise, loosely appressed to each other, the broadest coil 3.5–8 mm in diameter, hardened at maturity; surface of coils with many radial veins curved, never S-shaped, anastomosing freely before entering the submarginal vein; grooves between dorsal suture and submarginal vein of pod prominent (visible in side view of pod); spines, when present, up to 4 mm long, c. 15 in each row, with 2-rooted base, one root arising in the dorsal suture, the other in the submarginal vein, hooked. Seeds 3 mm long, subreniform, yellowish to yellow-brown, smooth.

Also in Morocco, Algeria, Tunisia, Libya, Egypt, Ethiopia, Djibouti, Kenya, Tanzania, South Africa and S Yemen (Socotra); and in Asia, Europe and the Middle East.

Var. **polymorpha**. TAB. 3,7: **14**, fig. B.

Pods spiny.

Zimbabwe. W: Matobo Distr., Matopos Research Station, fl. & fr. 26.viii.1966, *R.P. Denny* in *MRSH* 4044 (SRGH). **Mozambique**. GI: Guijá, Posto de Culturas Regadas do Vale do R. Limpopo, cultivated, fl. & fr. 16.ix.1950, *Myre* 934 (LMA).

Only recorded from agricultural research stations.

Var. **brevispina** (Benth.) Heyn in Scripta Acad. Hierosolymitana, Sci. Rep. **12**: 77 (1963). Type as for *M. apiculata* Willd.

Medicago apiculata Willd., Sp. Pl. **3**: 1414 (1802). Type from southern Europe.

Medicago denticulata Willd. var. *brevispina* Benth., Cat. Pl. Pyrénées: 103 (1826) pro parte.

Pods spineless.

Zimbabwe. C: Harare Distr., Harare (Salisbury), Agricultural Experiment Station, fl. & fr. 2.vi.1956, *Whellan* 1110 (K).

A weed of cultivation; introduced. Apparently a rare weed in the Flora Zambesiaca area.

The specimen *Hopkins* in *GHS* 16640, from Matopos, Pasture Research Station in Zimbabwe is not in fruit and so cannot be determined as to variety.

97. TRIFOLIUM L.

Trifolium L., Sp. Pl.: 764 (1753); Gen. Pl., ed. 5: 337 (1754). —Gillett in Kew Bull. **7**: 367–404 (1952). —M. Zohary & D. Heller, The Genus Trifolium (1984).

Annual or perennial herbs. Leaves digitately or pinnately (*T. campestre*) 3-foliolate (rarely 5-foliolate outside the Flora Zambesiaca area); leaflets mostly denticulate; stipules well developed, basally adnate to the petiole often also sheathing the stem, herbaceous or membranaceous. Racemes axillary, less often ± terminal, usually contracted and capitate, or spicate or umbellate, pedunculate or sessile; bracts present or absent; bracteoles absent. Flowers pedicellate or sessile. Calyx tubular or campanulate, (5)11- or more-nerved, sometimes bilabiate, accrescent, inflated, the mouth open or closed by a callosity or by a ring of hairs. Corolla purple, pink, white or yellow, glabrous, usually persistent; standard free or connate at the base with the wings and keel; wings often longer than keel; keel obtuse. Stamens with free portions of filaments usually dilated at the apex; vexillary filament free; anthers uniform. Ovary small, sessile or shortly stipitate, 1- or few-ovulate; style straight, slightly incurved towards the tip; stigma small, punctate or capitate. Pod small, ± included in the persistent calyx and corolla, usually membranous, indehiscent or rarely dehiscent. Seeds 1–2, globular to ovoid, reniform or lenticular.

A genus of about 250 species mostly in temperate Eurasia, Africa and America, with main centre of distribution in the Mediterranean region.

Introduced species recorded from trial plantings on agricultural research stations include:

Trifolium resupinatum L. from Mozambique, Maputo, Umbelúzi, Estação Experimental, fl. 30.x.1950, *Myre* 1008 (LMA). "Trevo".

Trifolium nigrescens Viv. from Mozambique, Maputo, Umbelúzi, Estação Experimental, fl. 29.ix.1950, *Myre* 964 (LMA). "Trevo".

Trifolium pratense L. from Zimbabwe, Gweru (Gwelo), Mlezu School Farm, 16 km SSE of Kwekwe (QueQue), cultivated in grass nursery, fl. 2.xii.1965, *Biegel* 654 (LISC; PRE; SRGH); Mozambique, Guijá, Posto de Culturas Regadas do Vale do R. Limpopo, fl. 26.v.1952, *Myre & Carvalho* 1207 (LMA). "Trevo-dos-prados".

Trifolium alexandrinum L. from Zimbabwe, Mutare, fl. 3.xi.1954, *Seagrief* in *GHS* 48529 (K) and Mozambique, Maputo, Umbelúzi, Estação Experimental, fl. 5.ix.1950, *Myre* 927 (LMA). "Bersin"; "Trevo-de-Alexandria".

1. Flowers pedicellate, bracteate; calyx mouth without a ring of hairs or an annular callosity; ovules 2 or more · 2
– Flowers sessile, ebracteate; calyx mouth with a ring of hairs (or outside the Flora Zambesiaca area, with an annular callosity); ovules 1 (sect. *Trifolium*) · · · · · · · · · · · · · 12
2. Leaves pinnately 3-foliolate; calyx 5-nerved; corolla yellow; pod 1-seeded (sect. *Cronosemium*) · 1. *campestre*
– Leaves digitately 3-foliolate; calyx more than 5-nerved; corolla purple, white or pink; pod usually 2- or more-seeded · 3
3. Calyx bilabiate, inflated and conspicuously reticulate at maturity; corolla resupinate (sect. *Vesicaria*) · *resupinatum* (cultivated)
– Calyx ± regularly 5-toothed, not inflated at maturity, the transverse veins never conspicuous; corolla not resupinate (sect. *Lotoidea*) · 4
4. Petiole in all leaves adnate to the stipules throughout its length, or almost all its length · · 5
– Petiole free for the greater part of its length, at least in the lower leaves · · · · · · · · · · · · 7
5. Leaflets more than 5 times as long as wide; calyx more than 15-nerved; stems not rooting at nodes · 2. *simense*
– Leaflets up to 5 times as long as wide; calyx up to 15-nerved, usually 11-nerved; stems rooting at nodes · 6
6. Leaflets usually emarginate, truncate or rounded at the apex; standard less than 5 mm long · 3. *usambarense*
– Leaflets usually acute; standard 5 mm long or more · · · · · · 4. *polystachyum* var. *psoraleoides*
7. Inflorescences sessile in the axil of the two uppermost leaves · · · · · · · · · 8. *pseudostriatum*
– Inflorescences pedunculate · 8
8. Pedicels less than 2 mm long (c. 1 mm long), not reflexed after flowering; corolla purple, very rarely white · 9. *rueppellianum*
– Pedicels usually more than 2 mm long, conspicuously reflexed after flowering; corolla white or pinkish · 9
9. Annual herb; peduncle less than 5 times as long as leaflets · · · · · · · · · · · · · · · · · · 10
– Perennial herb; peduncle usually 5–10 times as long as leaflets · · · · · · · · · · · · · · · · · 11
10. Calyx teeth 2–3 times as long as tube · 7. *masaiense*
– Calyx teeth half as long as tube · *nigrescens* (cultivated)
11. Calyx, pedicels and lower surface of leaflets at least on the midrib pilose; calyx teeth at least 4 times as long as wide · 5. *semipilosum*
– Calyx, pedicels and leaflets glabrous; calyx teeth less than 4 times as long as wide · · · 6. *repens*
12. Heads sessile, involucrate with the stipules of the subtending leaves; calyx teeth filiform-subulate with long, spreading, swollen-based hairs and separated by broad sinuses · *pratense* (cultivated)
– Heads pedunculate, not involucrate; calyx teeth triangular-lanceolate to subulate, plumose with antrorse, silky, not swollen-based hairs · · · · · · · · · · · · · · · *alexandrinum* (cultivated)

1. **Trifolium campestre** Schreb. in Sturm, Deutschl. Fl. **1**: 16, t. 253 (1804). —Gillett in Kew Bull. **7**: 368 (1952). —Drummond in Kirkia **8**: 228 (1972). —Täckholm, Stud. Fl. Egypt, ed. 2: 241 (1974). —Jafri in Fl. Libya **86**: 228, fig. 86 (1980). —M. Zohary & Heller, Genus *Trifolium*: 336, t. 127 (1984). —Lock, Leg. Afr. Check-list: 491 (1989). —Retief & Herman in Strelitzia **6**: 472 (1997). —Boulos, Fl. Egypt **1**: 184, t. 49, fig. 8 (1999). Type from Europe.
Trifolium agrarium L., Sp. Pl.: 772 (1753) pro parte.
Trifolium procumbens L., Fl. Suec., ed. 2: 261 (1755) pro parte, nomen ambiguum, non L., Sp. Pl.: 772 (1753). —J.G. Baker in F.T.A. **2**: 57 (1871). —E.G. Baker, Legum. Trop. Africa: 76 (1926).

Annual herb. Stems erect or ascending, hairy. Leaves petiolate; leaflets 6–12 × 4–8 mm, obovate, emarginate or retuse at the apex, the midrib pilose beneath but otherwise glabrous, nervation straight; petioles 1–1.5 cm long in the lower leaves, reducing above; stipules ovate to oblong, long-acuminate. Inflorescence rounded or ovoid, many-flowered; peduncle usually exceeding the subtending leaves. Flowers with pedicels c. 1 mm long, becoming deflexed early. Calyx c. 2 mm long, white, 5-nerved, glabrous; tube membranous; teeth very unequal, the two upper very short, triangular or lanceolate, the others long, linear, twice as long as the tube or longer, long-subulate. Corolla 5–7 mm long, pale yellow fading to pale brown; standard orbicular, flattish or spoon-shaped when in fruit, entire or slightly retuse at the apex, denticulate at margin, markedly longitudinally sulcate, persistent, ± twice as long as the claw; wings sinuate-oblong, rounded at the apex, twice as long as the claw, the auricles separated from the claw by a broad, square sinus; keel oblong, obtuse or slightly acuminate, twice as long as the claw. Pod oblong, stipitate, enclosed within the keel, 1-seeded. Seeds c. 1 mm long, ovoid, chestnut-brown, smooth.

Zimbabwe. C: Harare Distr., Spicer's Farm, c. 5 km above Mukuvisi-Manyame (Makabusi-Hunyani) confluence, on Mukuvisi, c. 1400 m, fl. 1.ix.1959, *Phipps* 2187 (K; PRE; SRGH).
Also in Morocco, Algeria, Tunisia, Libya, Egypt, Sudan, Ethiopia, Djibouti and South Africa; Europe; Middle East. A rare weed on edges of irrigated oat fields.

2. **Trifolium simense** Fresen. in Flora **22**: 49 (1839). —J.G. Baker in F.T.A. **2**: 57 (1871). —E.G. Baker, Legum. Trop. Africa: 79 (1926). —Gillett in Kew Bull. **7**: 375, figs. 2/1–7 & 3 (1952). —F.W. Andrews, Fl. Pl. Anglo-Egypt. Sudan **2**: 242 (1952). —Gillett in F.C.B. **4**: 293 (1953); in F.W.T.A., ed. 2, **1**: 553 (1958). —Binns, First Check List Herb. Fl. Malawi: 85 (1968). —Moriarty, Wild Fl. Malawi: 132, t. 66, fig. 3 (1975). —Gillett in F.T.E.A., Leguminosae, Pap.: 1022, fig. 143 (1971). —M. Zohary & Heller, Genus *Trifolium*: 98, t. 17 (1984). —Lock, Leg. Afr. Check-list: 496 (1989). —Agnew, Upland Kenya Wild Fl., ed. 2: 147, t. 51 (1994). TAB. 3,7: **15**. Type from Ethiopia.

Perennial herb. Roots with elongated nodules. Stems up to 50 cm, erect or sometimes semi-decumbent but not rooting at the nodes. Leaves petiolate; leaflets glabrous or sparsely pilose, those of the lower leaves c. 1 cm long, narrowly cuneate-oblong, those of the upper leaves up to 6 cm long, linear-lanceolate, with well-marked, subulate teeth on the margins, the lateral nerves many on each side, at 5–15° to the midrib; petioles 1–3 cm long, adnate to the stem-sheathing stipules for all their length; stipules ending in bristly points, glabrous but pilose at the margins. Inflorescence hemispherical or ovoid, c. 13 mm in diameter; peduncle up to 5 cm long, pilose, especially at the top; bracts up to 1 mm long, white. Flowers with pedicels c. 1 mm long, erect in fruit. Calyx whitish, glabrous or sparsely pilose; tube c. 2 mm long, 17–20-nerved; teeth 2–3 mm long, subulate. Corolla nearly twice as long as the calyx, pink-purple, rarely white; standard c. 6 mm long. Stamens 3–5 mm long. Ovary 2-ovulate, glabrous; style 2.5 mm long, persistent. Pod 1–2-seeded. Seeds c. 1.3 × 1 mm, ovoid, pale brown.

Zambia. E: Nyika, fl. 8.vii.1962, *Lawton* 916 (K). **Malawi**. N: Rumphi Distr., Nyika Plateau, Kasaramba Viewpoint, 2345 m, fl. & fr. 14.v.1970, *Brummitt* 10689 (K; LISC; MAL; PRE; SRGH).
Also from Cameroon, Equatorial Guinea, Dem. Rep. Congo, Rwanda, Burundi, Sudan, Ethiopia, Uganda, Kenya and Tanzania. Montane grassland; 2000–2350 m.
The species shows great variation in the width and length of the leaflets.

3. **Trifolium usambarense** Taub. in Engler, Pflanzenw. Ost-Afrikas **C**: 208 (1895). —E.G. Baker, Legum. Trop. Africa: 79 (1926). —Robyns, Fl. Sperm. Parc Nat. Alb. **1**: 286 (1948). —Gillett in Kew Bull. **7**: 370, fig. 1/13–15 (1952); in F.C.B. **4**: 292 (1953); in F.W.T.A., ed. 2, **1**: 553

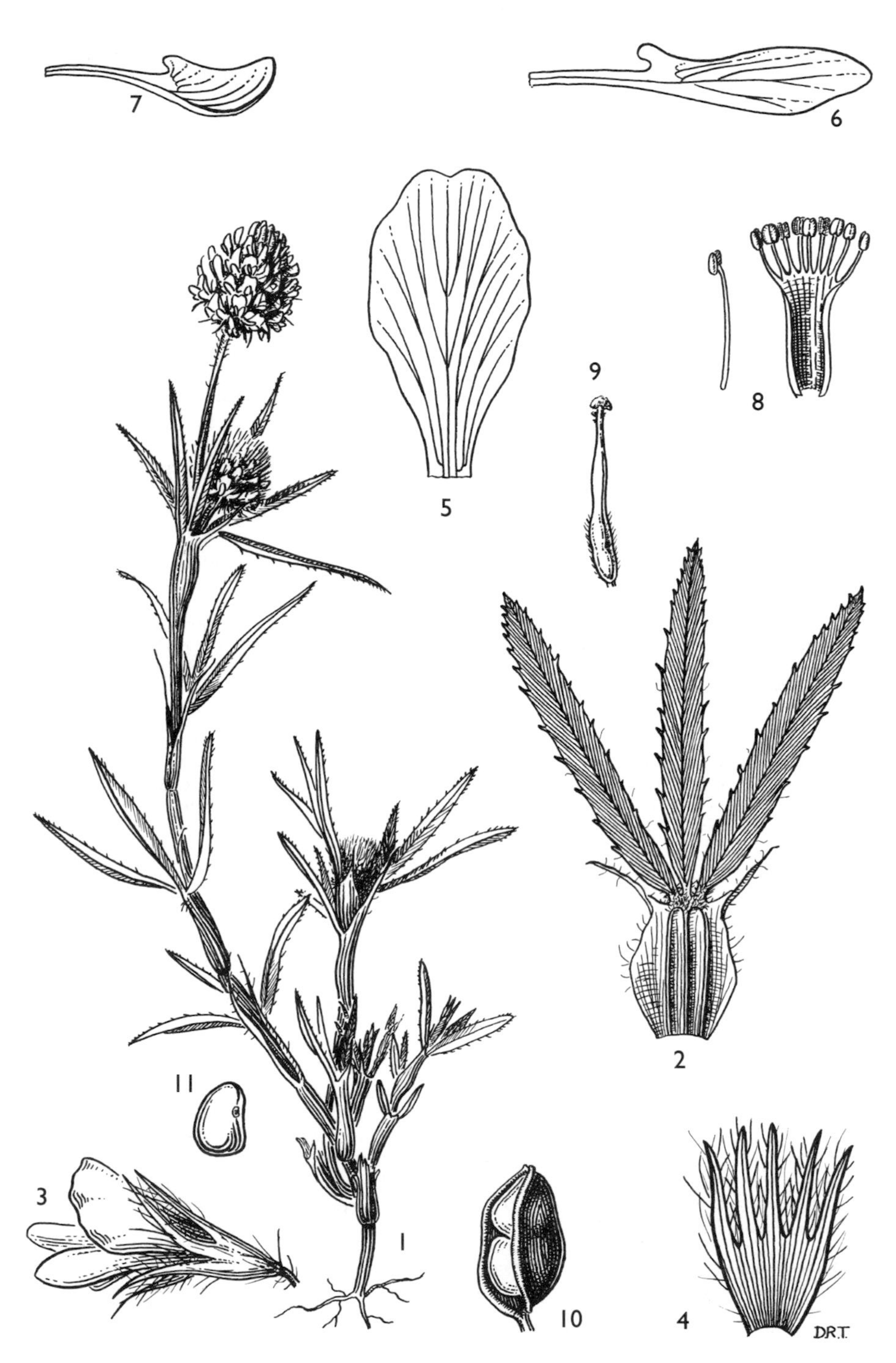

Tab. 3,7: **15**. TRIFOLIUM SIMENSE. 1, habit (× 1); 2, leaf, with sheath flattened out (× 4); 3, flower (× 3); 4, calyx, opened out (× 4); 5, standard (× 6); 6, wing (× 6); 7, keel (× 6); 8, androecium, spread out (× 6); 9, gynoecium (× 6); 10, fruit dehiscing (× 6); 11, seed (× 6), 1–11 from *Greenway* 7729. Drawn by D.R. Thompson. From F.T.E.A.

(1958). —White, F.F.N.R.: 166 (1962). —Gillett in F.T.E.A., Leguminosae, Pap.: 1024 (1971). —M. Zohary & Heller, Genus *Trifolium*: 104, t. 20 (1984). —Lock, Leg. Afr. Check-list: 498 (1989). —Agnew, Upland Kenya Wild Fl., ed. 2: 147 (1994). Syntypes from Tanzania.

Trifolium polystachyum sensu J.G. Baker in F.T.A. **2**: 58 (1871). —sensu Oliver in Trans. Linn. Soc., London **29**: 53 (1872), pro parte, non Fresen. (1839).

Trifolium pseudocryptopodium Chiov. ex Fiori in Nuovo Giorn. Bot. Ital., n.s. **55**: 343, fig. 20 (1949), nom. non rite public. Types from southern Ethiopia.

Annual herb. Stems suberect, weak, few-branched, often rooting at the lower nodes, glabrous below, ± pilose above. Leaves petiolate; leaflets 6–13 × 3–7 mm, cuneate-oblanceolate, truncate or rounded at the apex, glabrous or weakly pilose at the base and on the lower surface, the lateral nerves c. 15 on each side, at c. 20° to the midrib; petioles 5–7 mm long, adnate to the stem-sheathing stipules for all their length; stipules with free parts triangular, ending in a fine subulate point. Inflorescence ± oblong, many-flowered; peduncle 3–6 cm long, pilose towards the top; bracts c. 0.5 mm long, narrow. Flowers with pedicels 1–2 mm long, pilose, not reflexed in fruit. Calyx pilose; tube c. 1.8 mm long, 11-nerved; teeth subulate. Corolla purple; standard 4–5 mm long, oblong-elliptic. Stamens c. 3.5 mm long. Ovary 2-ovulate; style 2 mm long, persistent. Pod oblong, glabrous, usually 2-seeded. Seeds 1.3 × 1 mm, brown.

Zambia. W: Chingola, c. 1400 m, fl. 12.x.1955, *Fanshawe* 2497 (K). C: Serenje Distr., s. loc., fl. 1957, coll. unknown in *GHS* 100013 (SRGH). E: Nyika, fl. 30.xii.1962, *Fanshawe* 7336 (K; SRGH). **Zimbabwe**. C: Marondera (Marandellas), Grasslands Research Station, fl. 12.iv.1961, *Corby* 994 (K; LISC). **Malawi**. N: Rumphi Distr., Nyika National Park, fl. 8.iv.1981, *Salubeni & Tawakali* 3041 (MAL; SRGH). C: Dedza Distr., 11 km west of Dedza, fl. 14.vi.1989, *Brummitt* 18516 (K; MAL; MO). **Mozambique**. T: Tsangano Distr., 2 km north of Mlangeni near Ntcheu (Ncheu)–Dedza road, 1385 m, fl. 25.v.1970, *Brummitt* 11113 (K; SRGH).

Also in Nigeria, Cameroon, Equatorial Guinea, Dem. Rep. Congo, Rwanda, Ethiopia, Uganda, Tanzania and South Africa. Montane grassland along streams, dambos and seasonally waterlogged grassland, marshy ground by streams; c. 1380–2200 m.

G.A. Moore s.n. (K), Malawi, Dedza Distr., Bembeke Dambo, the specimen label data records the following information: "local Chichewa name *Therere*, reputed to be very palatable to stock".

4. **Trifolium polystachyum** Fresen. in Flora **22**: 50 (1839). —J.G. Baker in F.T.A. **2**: 58 (1871) pro parte. —E.G. Baker, Legum. Trop. Africa: 78 (1926). —Gillett in Kew Bull. **7**: 371, fig. 1/1–9 (1952); in F.T.E.A., Leguminosae, Pap.: 1024 (1971). —M. Zohary & Heller, Genus *Trifolium*: 96, t. 16 (1984). —Lock, Leg. Afr. Check-list: 495 (1989). Type from Ethiopia.

Var. **psoraleoides** Welw. ex Hiern, Cat. Afr. Pl. Welw. **1**: 205 (1896). —E.G. Baker, Legum. Trop. Africa: 78 (1926). —Gillett in Kew Bull. **7**: 371, fig. 1/4–6 (1952); in F.C.B. **4**: 293 (1953). —Torre in C.F.A. **3**: 80 (1962). Type from Angola.

Perennial herb, rhizomatous. Stems prostrate-ascending, often rooting at the nodes near the base. Leaves with petioles adnate to the stem-sheathing stipules for all their length; leaflets up to 30 mm long and 7 mm wide, oblanceolate, acute at the apex, cuneate to the narrowly rounded base, sharply spinulose-serrate, glabrous; stipules with free parts ending in a fine point, subulate. Inflorescence oblong, many-flowered; peduncle glabrous or slightly pilose; bracts 0.5–1 mm long, narrow. Flowers with pedicels pilose, erect in fruit. Calyx pilose; tube c. 2 mm long, 11-nerved; teeth rather longer, subulate. Corolla purple-mauve; standard c. 7 mm long, oblong-oblanceolate. Stamens 5–6 mm long. Pod glabrous, 1–2-seeded. Seeds 1.6 × 1 mm, brown.

Zambia. W: Kitwe, fl. 4.i.1957, *Fanshawe* 2918 (K). **Malawi**. C: Dedza Distr., Dedza Mountain Forest, fl. 21.iv.1969, *Salubeni* 1331 (SRGH).

Also in southern Dem. Rep. Congo and Angola. Montane grassland, moist dambo, and stream sides.

Herb similar to *T. usambarense* but larger in all its parts.

Var. *polystachyum* occurs in Sudan, Ethiopia, Uganda and Kenya. It differs in the more lax inflorescence, twice as long as broad, and the corolla longer than the calyx. Zohary & Heller (loc. cit.) do not maintain the varieties.

5. **Trifolium semipilosum** Fresen. in Flora **22**: 52 (1839). —J.G. Baker in F.T.A. **2**: 56 (1871). —E.G. Baker, Legum. Trop. Africa: 80 (1926). —Gillett in Kew Bull. **7**: 385, fig. 4/8 (1952). —J.M. Watt & Breyer-Brandwijk, Med. Pois. Pl. S. & E. Africa, ed. 2: 666 (1962). —Gillett in Kew Bull. **24**: 218 (1970); in F.T.E.A., Leguminosae, Pap.: 1027, fig. 142/1–4 (1971). —

M. Zohary & Heller, Genus *Trifolium*: 176, t. 54 (1984). —Lock, Leg. Afr. Check-list: 496 (1989). —Agnew, Upland Kenya Wild Fl., ed. 2: 148, t. 51 (1994). Type from northern Ethiopia.

Perennial herb from a strong rootstock. Stems prostrate, pilose, often rooting at the nodes. Leaves long-petioled; leaflets orbicular-obovate, rounded, truncate or emarginate at the tip, glabrous above but silky pilose at the margins, on the midrib beneath and also (except in var. *glabrescens*) on the underside of the lower half of the 2 lateral leaflets, faintly toothed, strongly veined; petiole glabrous or silky pilose; stipules whitish, with falcate-triangular tips usually rather longer than the base. Inflorescence ± globose; peduncle longer than the leaves, pilose; bracts up to 0.5 mm long or absent. Flowers with pedicels pilose, reflexed in fruit. Calyx glabrous or thinly silky; tube 11-nerved; teeth equal to the tube. Corolla white, pink or purplish-red. Stamens c. 5 mm long. Ovary stipitate; ovules 3–6. Pod oblong, 2–6-seeded. Seeds light brown, irregularly discoid.

Var. **semipilosum**

Trifolium semipilosum var. *microphyllum* Chiov. in Ann. Ist. Bot. Roma **8**: 406 (1908). —E.G. Baker, Legum. Trop. Africa: 81 (1926). —Gillett in Kew Bull. **7**: 385 (1952). Type from Eritrea.

Trifolium semipilosum var. *kilimanjaricum* Baker f., Legum. Trop. Africa: 81 (1926). — Gillett in Kew Bull. **7**: 385 (1952). Type from Tanzania.

Trifolium africanum sensu E.G. Baker, Legum. Trop. Africa: 80 (1926) pro parte, non Ser. (1825).

Trifolium semipilosum var. *sennii* Chiov. ex Fiori in Nuovo Giorn. Bot. Ital., n.s. **55**: 342 (1949). —Gillett in Kew Bull. **7**: 385 (1952) *nom. inval.* Type from Ethiopia.

Leaflets orbicular, elliptic or oblong-elliptic or rarely obovate, rounded, truncate or slightly emarginate at the apex, silky pilose beneath on the lower halves of the 2 lateral leaflets and on the midribs, with short and rounded teeth. Ovary pilose at least near the tip.

Malawi. N: Rumphi Distr., Nyika Plateau, Nyika National Park, Chelinda Area, 2250 m, fl. 19.xi.1967, *Richards* 22635 (K).

Also in Yemen, Eritrea, Ethiopia, Uganda, Kenya and Tanzania. Moist places in montane grassland; 2250 m.

Var. **glabrescens** J.B. Gillett in Kew Bull. **7**: 385 (1952); in F.T.E.A., Leguminosae, Pap.: 1027 (1971). —M. Zohary & Heller, Genus *Trifolium*: 177 (1984). Type from Kenya.

Trifolium repens sensu E.G. Baker, Legum. Trop. Africa: 81 (1926), non L. (1753).

Trifolium johnstonii sensu Edwards in Empire J. Exp. Agric. **3**: 153–159 (1935).

Leaflets cuneate-obovate, emarginate, glabrescent beneath, though always with a few hairs at least on the midrib, with larger and sharper teeth than those of var. *semipilosum*. Ovary glabrous.

Zimbabwe. C: Harare Distr., Harare (Salisbury) Research Station (in cultivation), fl. 23.viii.1974, *Biegel* 4555 (K; PRE). **Malawi**. N: Rumphi Distr., Nyika Plateau, Chelinda (Chelunduo) Stream at Chelinda Camp (Chilinda), 2200 m, fl. & fr. 26.xi.1958, *Robson & Angus* 379 (BM; K; LISC; PRE).

Also in southern Ethiopia, Uganda, Kenya and Tanzania. Montane grassland, in ± permanently moist areas, often competing successfully with *Pennisetum clandestinum*; 1200–2300 m.

6. **Trifolium repens** L., Sp. Pl.: 767 (1753). —Burtt Davy, Fl. Pl. Ferns Transvaal, pt. 2: 404 (1932). —Blakelock in Kew Bull. **3**: 421 (1948). —Pedro in Bol. Soc. Est. Moçambique **92**: 21 (1955). —J.M. Watt & Breyer-Brandwijk, Med. Pois. Pl. S. & E. Africa, ed. 2: 665 (1962). —Drummond in Kirkia **8**: 228 (1972). —Ross, Fl. Natal: 200 (1972). —M. Zohary & Heller, Genus *Trifolium*: 167, t. 52 (1984). —Lock, Leg. Afr. Check-list: 495 (1989). —Retief & Herman in Strelitzia **6**: 472 (1997). Type from Europe.

Perennial herb, glabrous. Stems creeping and rooting at the nodes. Leaves long-petioled; leaflets cuneate-obovate, the lateral nerves straight, pale; stipules small, obovate-lanceolate, with short, subulate points. Inflorescence globose; peduncle very long, erect; bracts small, ovate-oblong, acuminate. Flowers with pedicels slender, reflexed after flowering. Calyx glabrous; tube campanulate, 10-nerved; teeth triangular-lanceolate, subulate, pale-margined, the two upper longer. Corolla white or whitish or pinkish, ± twice as long as the calyx; standard oblong; wings subequalling the claw, the auricles short and broad; keel slightly shorter than the wings. Pod linear-oblong, 3–4 seeded. Seeds ovoid, pale brown.

Zimbabwe. C: Harare, Salisbury Research Station, 1500 m, (in cultivation) fl. & fr. 28.x.1977, *Biegel* 5539 (PRE). E: Chimanimani Distr., Skyline Junction (Halt) on main road c. 16 km from Chimanimani (Melsetter), fl. & fr. viii.1969, *Goldsmith* 68/69 (LISC; PRE; SRGH).

Also in Morocco, Algeria, Tunisia, Egypt, Ethiopia, Kenya, Tanzania and South Africa; and in Europe and the Middle East to Pakistan. In sub-montane grassland.

There are numerous cultivars of this species. The gathering *Biegel* 5539 (PRE), "Tamar white clover", from the Salisbury Research Station in Harare, Zimbabwe, is a cultivated plant and has a robust habit with longer petioles, larger leaflets, longer peduncles and larger heads, although in other respects it agrees well. The differences seem related to growing conditions.

7. **Trifolium masaiense** J.B. Gillett in Kew Bull. **7**: 387, fig. 4/7 (1952); in F.T.E.A., Leguminosae, Pap.: 1028 (1971). —M. Zohary & Heller, Genus *Trifolium*: 151, t. 44 (1984). —Lock, Leg. Afr. Check-list: 493 (1989). Type from Tanzania.

Annual herb. Stems erect or ascending, sometimes rooting at the nodes. Leaves with petioles free for the greater part of their length; leaflets ± elliptic, rounded at the apex, cuneate at the base, strongly toothed, pilose only on the midrib beneath, the lateral nerves c. 14 on each side; petiole often more than twice as long as the leaflets; stipules with the adnate part c. 7 mm long, the free part c. 7 mm long, acuminate. Inflorescence ± globose; peduncles longer than the leaves, pilose. Flowers with pedicels pilose, reflexed in fruit. Calyx sparsely pilose, green or straw-coloured; tube 10–11-nerved; teeth c. 5 mm long, 2–3 times as long as the tube. Corolla pinkish; standard c. 10 mm long. Ovary glabrous, 4-ovulate. Pod papyraceous, 4-seeded. Seeds not seen.

Zimbabwe. C: Marondera (Marandellas), Grasslands Research Station, fl. v.1953, *West* in *GHS* 42777 (K; LISC).

Also in Tanzania. Introduced for pasture trials on this agricultural research station.

Subsp. *morotoense* J.B. Gillett, rather more slender than subsp. *masaiense*, with slightly smaller flowers and more subulate calyx teeth, occurs in Uganda.

8. **Trifolium pseudostriatum** Baker f., Legum. Trop. Africa: 83 (1926). —Staner in Rev. Zool. Bot. Africaines **23**: 216 (1933). —Robyns, Fl. Sperm. Parc Nat. Alb. **1**: 290 (1948). —Gillett in Kew Bull. **7**: 387 (1952); in F.C.B. **4**: 294 (1953). —Binns, First Check List Herb. Fl. Malawi: 85 (1968). —Gillett in F.T.E.A., Leguminosae, Pap.: 1030 (1971). —M. Zohary & Heller, Genus *Trifolium*: 256, t. 92 (1984). —Lock, Leg. Afr. Check-list: 495 (1989). Type from Uganda.

Annual herb, almost glabrous. Stems prostrate or ascending. Leaves petiolate; leaflets c. 13(28) × 11 mm, ovate or oblong, rounded or truncate at the apex, cuneate at the base, the lateral nerves 20–30 on each side; petioles mostly 3 cm long, reducing in the upper leaves; stipules with the free part longer than the base, tapering gradually. Inflorescence terminal, ± globose, usually sessile in the axil of the uppermost pair of leaves; bracts up to 5 mm long, narrowly lanceolate, subulate at the tip or wholly subulate. Flowers with pedicels c. 1 mm long, not reflexed in fruit. Calyx sparsely pilose only at the margin; tube c. 2.5 mm long, strongly 10–11-nerved; teeth triangular with scarious margins at the base, subulate for three-quarters of their length, recurved in fruit. Corolla purple; standard oblong, rounded at the apex; wings with blade shorter than the claw, the auricle very short; keel with a white ventral nerve. Stamens c. 5 mm long. Ovary 1.5 mm long, 2-ovulate. Pod oblong, with papery walls and thickened sutures, dehiscing dorsally, glabrous, 2-seeded. Seeds yellowish-brown.

Malawi. N: Chitipa Distr., Misuku Hills, Mugesse (Mughesse), c. 1600 m, fl. 8.vii.1973, *Pawek* 7101 (K; MAL). S: Zomba, fl. 19.v.1933, *Agric. Dept.* 424 (K).

Also in Dem. Rep. Congo, Rwanda, Burundi, Uganda and Tanzania. Open submontane grassland in disturbed areas, also roadsides at forest edges; 1600–1800 m.

9. **Trifolium rueppellianum** Fresen. in Flora **22**: 51 (1839). —J.G. Baker in F.T.A. **2**: 58 (1871). —E.G. Baker, Legum. Trop. Africa: 81 (1926). —Gillett in Kew Bull. **7**: 389, fig. 4/2-6 (1952). —F.W. Andrews, Fl. Pl. Anglo-Egypt. Sudan **2**: 243 (1952). —Gillett in F.C.B. **4**: 296 (1953); in F.T.E.A., Leguminosae, Pap.: 1030 (1971). —M. Zohary & Heller, Genus *Trifolium*: 231, t. 80 (1984). —Lock, Leg. Afr. Check-list: 495 (1989). —Agnew, Upland Kenya Wild Fl., ed. 2: 148, t. 51 (1994). Type from Ethiopia.

Trifolium subrotundum A. Rich., Tent. Fl. Abyss. **1**: 172 (1847), pro majore parte. —Hutchinson & Dalziel, F.W.T.A. **1**: 400 (1928); App.: 265 (1948). Types from northern Ethiopia.

Trifolium preussii Baker f., Legum. Trop. Africa: 82 (1926). Type from Cameroon.

Trifolium rueppellianum var. *preussii* (Baker f.) J.B. Gillett in Kew Bull. **7**: 389 (1952); in F.W.T.A., ed. 2, **1**: 553 (1958).

Annual herb. Stems erect, sometimes prostrate, glabrous, not rooting at the nodes. Leaves petiolate; leaflets elliptic or obovate, rounded, truncate or less often retuse at the apex, cuneate at the base, glabrous, the lateral nerves c. 10 on each side, ending in strongly developed, subulate teeth up to 0.5 mm long; petioles up to 5 cm long in the lower leaves, c. 3 mm long and ± wholly adnate to the stipules in the upper leaves; stipules up to 15 mm long, acuminate at the apex, 3–4-nerved. Inflorescence ± globose, 15–30-flowered; peduncle many times longer than the subtending petiole, pilose near the top; bracts up to 3 mm long, linear-lanceolate. Flowers with pedicels up to 2 mm long, shorter than the calyx tube, glabrous or pilose. Calyx glabrous, except sometimes for a few hairs on the margin; tube pale, strongly 11-nerved, the commissural nerves always undivided to the top; teeth 2–5 mm long, abruptly narrowed near the base, subulate for most of their length, exceeding the corolla. Corolla pink or purple; standard oblong, rounded at the apex; wings with blade longer than the claw, with a well developed auricle. Ovary glabrous. Pod 3–4 × 2 mm, 2–3 seeded. Seeds ovoid, brown.

Zambia. S: Mazabuka Distr., Nanga, Kafue Pilot Polder near Mazabuka, fl. & fr. 5.ii.1960, *Angus* 2137 (PRE). **Zimbabwe**. E: Chimanimani (Melsetter), in nursery, fl. 7.v.1957, *West* 3554 (K; LISC; PRE). **Malawi**. N: Chitipa Distr., Mugesse (Mughesse), Misuku Hills, irrigation ditch, St. Mathias Parish, 1560 m, fl. & fr. 30.xii.1970, *Pawek* 4244 (K). C: Salima Distr., Chipoka (Cipoka), along Lake shore, fl. 15.iii.1972, *Salubeni* 1789 (PRE; SRGH).

Also in Nigeria, Cameroon, Equatorial Guinea, Dem. Rep. Congo, Sudan, Ethiopia, Uganda, Kenya and Tanzania. Along lake shores, by irrigation ditches and in heavily grazed roadside pasturage; 475–1560 m.

Tribe 17. **PODALYRIEAE**

Podalyrieae Benth., Comm. Legum. Gen.: 1 (1837). —van Wyk & Schutte in Adv. Leg. Syst. **7**: 304 (1995).

Shrubs or small trees. Leaves imparipinnate, digitately 3-foliolate on a prominent leaf-base or simple; stipels absent. Flowers in terminal and/or axillary racemes or 1–several from the axils; bracteoles small or lacking. Calyx-tube short with the hypanthium intruding at the base; upper lobes joined higher, lowermost lobe relatively well developed. Standard broad, shortly clawed; wings usually sculptured; keel obtuse to beaked. Stamens free or nearly so; anthers uniform to slightly dimorphic, with the alternate ones a little smaller and attached higher. Ovary with several–numerous ovules; style tapered to a small stigma, glabrous. Pods compressed or turgid, dehiscent, or flattened, narrowly winged along the upper suture and ± indehiscent. Seeds oblong-reniform, with a small hilum surrounded by a collar-like aril if the pods are dehiscent.

Five genera in South Africa, almost restricted to the Cape region, only *Calpurnia* extending into tropical Africa and India. *Calpurnia* comes very close to genera of the *Sophora* group of *Sophoreae*, and is included in the keys to both tribes; the intrusive calyx and chemical constituents are the principal features cited for its inclusion in *Podalyrieae* by van Wyk & Schutte, loc. cit. (1995).

Podalyria calyptrata (Retz.) Willd., native of South Africa, is mentioned as cultivated in Zimbabwe by Biegel, Check List Ornam. Pl. Rhod. Parks & Gard.: 87 (1977). Shrub or small tree

with simple leaves, short axillary racemes of large pink flowers and furry oblong-obovoid pods.
Virgilia oroboides (Berg.) Salter, native of South Africa, has been cultivated at la Rochelle Estate near Penhalonga in the Mutasa Distr., fl. 17.ix.1966, *Müller* 416 (K; SRGH). Tree to 12 m with pinnate leaves and racemes of mauve flowers. Biegel, Check List Ornam. Pl. Rhod. Parks & Gard.: 107 (1977) indicates that *V. divaricata* Adamson, with pink flowers, is also cultivated in Zimbabwe.

98. CALPURNIA E. Mey.

By R.K. Brummitt

Calpurnia E. Mey., Comment. Pl. Afr. Austr.: 2 (1836). —Beaumont, Beckett, Edwards & Stirton in Bothalia **29**: 5–23 (1999).

Shrubs or small trees. Leaves imparipinnate, 3–15-jugate, each leaflet up to 4(5) cm long, ovate, elliptic or obovate, obtuse or retuse and often mucronate at the apex, glabrous or usually sparsely to densely hairy. Flowers in axillary, or rarely terminal, few–many-flowered racemes; bracteoles very small and caducous or apparently absent. Calyx ± campanulate, the hypanthium intrusive at the base, the two upper teeth largely fused. Petals yellow; standard with a strongly channelled claw and ± reflexed limb; wings with slightly developed sculpturing between the veins proximally; keel obtuse. Ovary stalked, with several to many ovules. Pod stalked, ± membranous, flattened, sometimes with a narrow wing on the upper side, with the remains of the style usually persistent, usually indehiscent.

A genus of 7 species, all but one confined to South Africa.

Calpurnia aurea (Aiton) Benth., Comm. Legum. Gen.: 26 (1837). —J.G. Baker in F.T.A. **2**: 525 (1871). —Harms in Engler, Pflanzenw. Afrikas [Veg. Erde 9] **3**: 530 (1915). —Eyles in Trans. Roy. Soc. South Africa **5**: 369 (1916). —E.G. Baker, Legum. Trop. Africa: 593 (1929). —Gillett in Kew Bull. **19**: 421 (1965). —Hillcoat in C.F.A. **3**: 382 (1966). —Brummitt in Kirkia **6**: 123 (1967). —Polhill in F.T.E.A., Leguminosae, Pap.: 47, fig. 9 (1971). —Drummond in Kirkia **8**: 217 (1972). —Biegel, Check-list Ornam. Pl. Rhod. Parks & Gard.: 33 (1977). —K. Coates Palgrave, Trees South. Africa: 299 (1977). —Lock, Leg. Afr. Check-list: 473 (1989). —Beaumont et al. in Bothalia **29**: 19, fig. 11 (1999). Type from cultivation at Kew, originally from Ethiopia (BM).
Sophora aurea Aiton, Hort. Kew. **2**: 44 (1789).
Robinia subdecandra L'Hér., Stirp. Nov.: 157, t. 75 (1791). Type from Ethiopia.
Podalyria aurea (Aiton) Willd., Sp. Pl., ed. 4, **2**: 502 (1800).
Virgilia aurea (Aiton) Lam., Tabl. Encycl. [Ill. Gen.] **5**, pt. 2: 470 (1819).
Calpurnia lasiogyne E. Mey., Comment. Pl. Afr. Austr.: 3 (1836). —Harvey in F.C. **2**: 267 (1862). —E.G. Baker in J. Linn. Soc., Bot. **40**: 61 (1911). —Phillips in Ann. S. African Mus. **9**: 477 (1917). —Burtt Davy, Fl. Pl. Ferns Transvaal, pt. 2: 354 (1932). Syntypes from South Africa (KwaZulu-Natal and Eastern Cape Province).
Calpurnia subdecandra (L'Hér.) Schweick. in Bothalia **3**: 237 (1937). —Brenan, Check-list For. Trees Shrubs Tang. Terr.: 410 (1949). —Toussaint in F.C.B. **4**: 45 (1953). —Goodier & Phipps in Kirkia **1**: 56 (1961). —White, F.F.N.R.: 146 (1962).

Subsp. **aurea** —Brummitt in Kirkia **6**: 128 (1967). TAB. 3,7: **16**.
Calpurnia aurea var. *major* Oliv. & Baker ex Baker f., Legum. Trop. Africa: 594 (1929). Type from Tanzania (Kilimanjaro).

Bush or small tree up to 7 m high; young branches and inflorescences densely pubescent. Leaves (4)6–24 cm long, (4)5–15-jugate; leaflets mostly 2.5–4(5) × 1.2–1.6(2) cm, ovate or oblong to obovate, cuneate to rounded and slightly asymmetrical at the base, rounded or retuse and sometimes mucronate at the apex, pubescent on both surfaces. Racemes (5)7–21(24) cm long, 8–26-flowered; pedicels 6–16 mm long, slightly swollen at the top. Hypanthium 3–4.5 mm long. Calyx (5)6–9(10) mm long including teeth. Standard 12–20 mm long, the limb strongly reflexed; wings often slightly exceeding the standard, falcate; keel petals resembling the wings. Stamens 14–16 mm long. Pod on a stipe 5–6 mm long, 5–12 × 8–1.6 cm, straw-coloured or brownish, the suture with a wing 1–2.5 mm broad. Seeds 5–6 × 3–4 mm, oblong-ellipsoid, brownish.

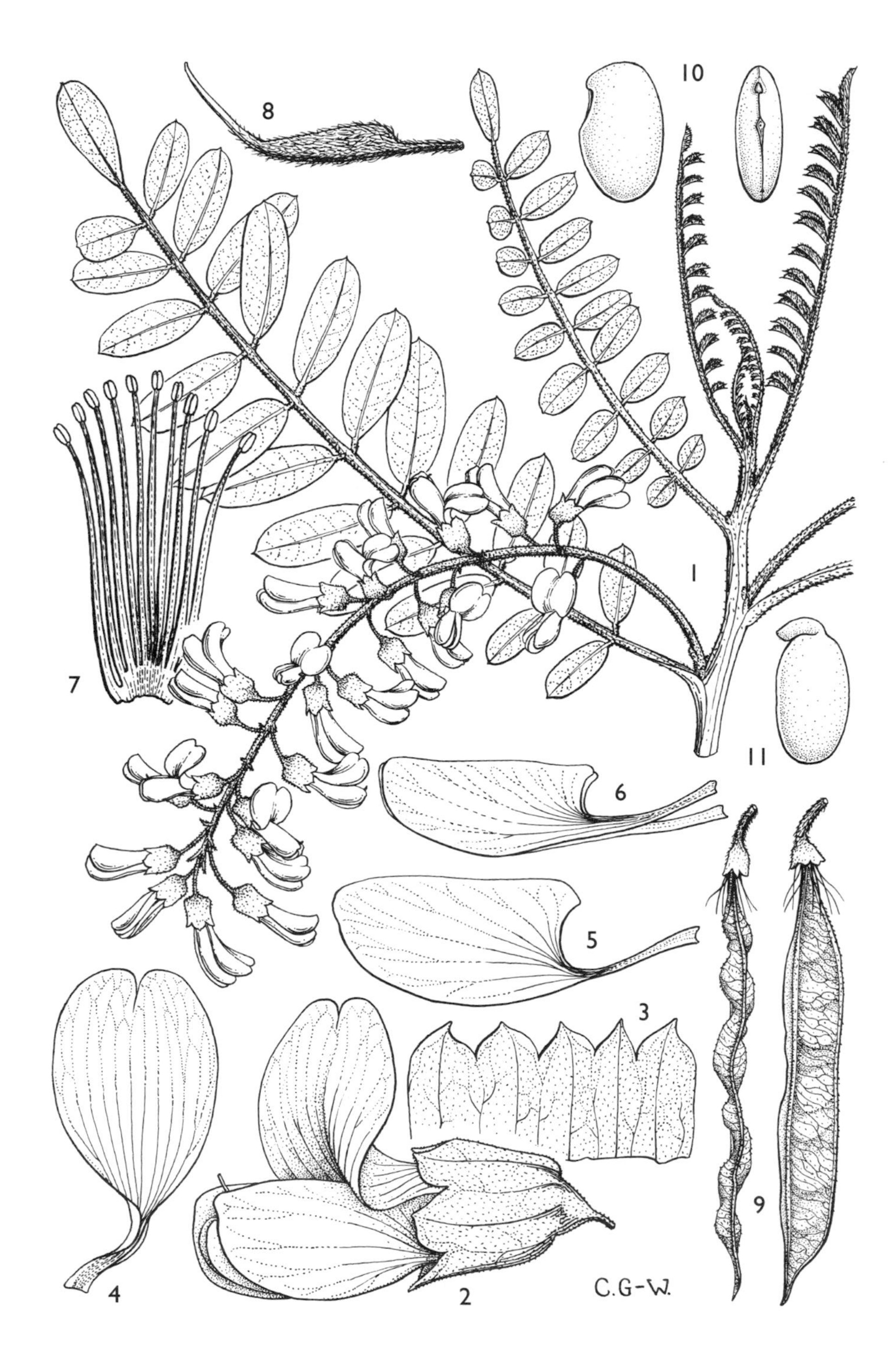

Tab. 3,7: **16**. CALPURNIA AUREA subsp. AUREA. 1, part of flowering branch ($\times \frac{2}{3}$); 2, flower ($\times 2\frac{1}{2}$); 3, calyx, opened out ($\times 2\frac{1}{2}$); 4, standard ($\times 2\frac{1}{2}$); 5, wing ($\times 2\frac{1}{2}$); 6, keel ($\times 2\frac{1}{2}$); 7, androecium, spread out ($\times 2\frac{1}{2}$); 8, gynoecium ($\times$ 2); 9, fruit, side and edge views ($\times \frac{2}{3}$), 1–9 from *Tweedie* 1584; 10, seed, side and hilar views ($\times$ 3); 11, embryo ($\times$ 3), 10 & 11 from *A.S. Thomas* 3216. Drawn by Christopher Grey-Wilson. From F.T.E.A.

Zimbabwe. C: Shurugwi Distr., Shurugwi (Selukwe), Ferny Creek, fl. 8.xii.1953, *Wild* 4286 (K; LISC; SRGH). E: Chipinge Distr., Chirinda outskirts, fl. & fr. 4.x.1906, *Swynnerton* 79 (BM; K; SRGH). S: Mberengwa Distr., Mt. Buhwa, c. 915 m, fl. 10.xii.1953, *Wild* 4335 (LISC; PRE; SRGH). **Mozambique**. M: Sábiè–Namaacha (Namahacha), fr. 21.xii.1947, *Pedro* 3560 (LMA).

Also in Eritrea, Ethiopia, Somalia, Sudan (Didinga Hills), Central African Republic, Uganda, Kenya, Tanzania, eastern Dem. Rep. Congo, Angola and eastern South Africa. Montane forest margins, apparently rare or absent below 920 m in the Flora Zambesiaca area, but occurring at lower altitudes (below 300 m) in South Africa. Also grown in parks and gardens, see Biegel, Check-list Ornam. Pl. Rhod. Parks & Gard.: 33 (1977).

The distribution of the subspecies appears to be somewhat disjunct and is discussed by Brummitt in Kirkia **6**: 123–132 (1967). Although it is found in eastern South Africa, in western Angola and in mountain districts from Tanzania to Eritrea, in the Flora Zambesiaca area it is apparently restricted to the higher parts of central and eastern Zimbabwe and probably adjacent Mozambique, and southern Mozambique close to the South African border. No certain records from central Mozambique are known, but one specimen seen was collected on the Zimbabwe/Mozambique border in the Chipinge Distr. near Espungabera, fl. 20.x.1947, *Rattray* 1153 (PRE; SRGH). The inclusion of Zambia in the distribution given in F.C.B. **4**: 45 (1953) is an error. The subspecies is itself variable and shows a marked tendency to shorter racemes, smaller flowers and leaves, and fewer leaflets towards the south in Africa.

Subsp. *sylvatica* (Burch.) Brummitt, from the south eastern Cape Province, differs from subsp. *aurea* mainly in having ± glabrous leaflets and a glabrous ovary. Two collections from central Zimbabwe — Somabhula (Somabula), x.1922, *Eyles* 6159 (SRGH) and Gweru (Gwelo), 1924, *Eyles* (S.R. Dept. Agric.) 3325 (PRE; SRGH) — resemble subsp. *aurea* in their leaflets but have glabrous ovaries, and so may be regarded as intermediate between the two subspecies. Similar plants are known from the Northern Province of South Africa. Beaumont et al., loc. cit. (1999), do not retain this subspecies, but it is morphologically striking and geographically disjunct (see map in Kirkia **6**: 132 (1967). It seems well worth recognition at subspecific rank, despite the intermediates noted. Subsp. *indica* Brummitt occurs in southern India.

Tribe 18. **CROTALARIEAE***

Crotalarieae (Benth.) Hutch., Gen. Fl. Pl. **1**: 364 (1964). —van Wyk & Schutte in Adv. Leg. Syst. **7**: 305 (1995).
Genisteae subtribe *Crotalariinae* Benth. in Bentham & J.D. Hooker, Gen. Pl. **1**: 440 (1865), as "*Crotalarieae*".
Borbonieae Hutch., Gen. Fl. Pl. **1**: 345 (1964).
Lotononideae Hutch., Gen. Fl. Pl. **1**: 357 (1964).

Shrubs or herbs. Leaves generally digitately 3(7)-foliolate, sometimes simple or 1-foliolate, generally pulvinate, without stipels. Flowers in terminal ± leaf-opposed or less often axillary racemes or heads, sometimes solitary or aggregated into leafy compound racemes. Calyx generally with 5 subequal teeth, less often the lateral sinuses shallower or the upper sinus deeper, sometimes the calyx 2-lipped, but then with the lower lobes much longer than the joined part. Standard sometimes with basal callous-like appendages; wings generally sculptured; keel obtuse, pointed or beaked. Stamens joined in a sheath or tube open at the base, the vexillary stamen rarely free or lacking; anthers generally dimorphic, alternately basifixed and shorter, dorsifixed, less often all similar or partially aborted. Style glabrous or with 1–2 lines of hairs. Pods generally 2-valved, occasionally indehiscent, flat or turgid, sometimes impressed between the seeds and occasionally folded. Seeds usually oblique-cordiform or rounded, with a well developed radicular lobe and small hilum, sometimes arillate.

A tribe of 11 genera, mostly in southern Africa, but extending thinly to the Mediterranean region, Asia and Australia, with *Crotalaria* widespread in the tropics and southern latitudes.

In recent years the limits between *Genisteae* and *Crotalarieae* have been clarified, see van Wyk & Schutte, loc. cit. (1995), with *Argyrolobium* now included with *Adenocarpus, Dichilus* and *Melolobium* in *Genisteae.* All these genera have a 2-lipped calyx with the lower lip entire to 3-fid; 2-lipped calyces, though rare, do occur in *Crotalarieae,* at least in *Crotalaria,* but then with the lower lobes much longer than the united part.

---

* *Pearsonia, Rothia, Robynsiophyton, Crotalaria, Lotononis* and *Lebeckia* by R.M. Polhill. *Bolusia* by B.-E. van Wyk.

1. Style straight or almost so, projecting forwards or downwards (flowers sometimes resupinate in *Pearsonia*) 2
– Style curved upwards 4
2. Standard 8–30 mm long; anthers elongate, 4 basifixed, 6 attached a little higher; upper calyx lobes joined higher; perennial herbs or small shrubs .... 99. **Pearsonia**
– Standard c. 5 mm long; anthers all small, rounded, some sometimes aborted; calyx lobes subequal in length (upper ones sometimes broader); small annuals or short-lived perennials 3
3. Stamens and anthers 10; stipules single at each node .... 100. **Rothia**
– Stamens 9, 5 with small anthers, the others aborted; stipules paired ... 101. **Robynsiophyton**
4. Keel distinctly beaked or coiled; distal part of style with 1–2 lines of hairs or helically coiled; pods markedly inflated 5
– Keel obtuse to shortly beaked; style glabrous above; pods flat to somewhat turgid 6
5. Beak of keel not coiled .... 102. **Crotalaria**
– Beak of keel helically coiled through several turns .... 103. **Bolusia**
6. Calyx usually with the lateral sinuses shallowest, the upper and lateral lobes on each side thus joined in pairs, the lowermost lobes narrower; stipules 1–2, often well developed, rarely lacking; leaves 3-foliolate in Flora Zambesiaca area .... 104. **Lotononis**
– Calyx subequally and often shortly lobed; stipules rarely present, if so then small; leaves simple in Flora Zambesiaca area .... 105. **Lebeckia**

99. PEARSONIA Dummer

By R.M. Polhill

Pearsonia Dummer in J. Bot. **50**: 353 (1912). —Polhill in Kew Bull. **29**: 383–410 (1974). —Campbell-Young & Balkwill in Nordic J. Bot. **20**: 547–555 (2000).
Pleiospora Harv., Thes. Cap. **1**: 51, t. 81 (1859), *nom. rejic.*, non *Pleiospora* Rabenh. ex Ces. & De Not. (1863).
Phaenohoffmannia Kuntze, Revis. Gen. Pl. **2**: 902 (1891), *nom. illegit.*
Gamwellia Baker f. in J. Bot. **73**: 160 (1935).
Edbakeria R. Vig. in Notul. Syst. (Paris) **13**: 364 (1949).

Perennial herbs or small shrubs. Leaves digitately 3-foliolate; stipules linear to foliaceous or lacking. Inflorescences terminal or leaf-opposed, the flowers in racemes or solitary on long peduncles, sometimes resupinate; bract and bracteoles linear to elliptic-obovate. Calyx not or slightly inflated; lobes rarely subequal, usually with the upper sinuses slightly to much shallower than the lower sinuses, the lateral sinuses sometimes the shallowest, the lowest lobe narrower than the others. Standard yellow or white marked purplish, generally elliptic to oblong-obovate, concave in the lower part, usually hairy outside; wings broadened upwards, generally sculptured; keel relatively small, narrow, long-clawed, usually pointed with a small forwardly directed tip. Stamens in a sheath open on the upper side and not much longer than the free parts, the vexillary filament sometimes free; anthers narrowly oblong, subequal, but 6 (including the carinal one) attached higher up. Style nearly straight or pointed downwards. Pods sessile, linear-oblong to ellipsoid, pointed, laterally compressed, dehiscent, 1–many-seeded. Seeds ± oblique-cordiform, with a small hilum.

A genus of 13 species, occurring in southern Africa, one just extending into Tanzania and Dem. Rep. Congo (Katanga), one in Madagascar.

1. Stipules present 2
– Stipules absent 6
2. Pods (1.5)2–3 times as long as the calyx, freely produced; flowers usually 1–2 on wiry mostly leaf-opposed peduncles (occasionally up to 7 in racemes in *P. flava*); leaflets 1–12 mm wide, with inconspicuous venation 3
– Pods less than 1.5 times as long as the calyx (rarely matured in prostrate species); flowers (3)8–30(40) in heads or elongate racemes; leaflets 8–35 mm wide, with conspicuous lateral nerves 5

3. Standard pale yellow, 10–14 mm long; calyx 7–11 mm long · · · · · · · · · · · · · · 3. *uniflora*
– Standard bright yellow, 20–27 mm long; calyx 16–20 mm long · · · · · · · · · · · · · · · · · · · 4
4. Stipules resembling leaflets and only a little smaller; leaflets oblanceolate; hairs on stem all long and spreading · 1. *flava*
– Stipules much smaller than the leaflets, subulate to lanceolate-caudate; leaflets narrowly elliptic-lanceolate to elliptic; stems with long spreading and short appressed hairs intermixed · 2. *mesopontica*
5. Stems erect; standard yellow, 8–13 mm long; stipules linear-subulate · · · · · · 4. *cajanifolia*
– Stems prostrate; standard cream-coloured with a purple centre, 16–22 mm long; stipules lanceolate to ovate or obovate-circular · 5. *grandifolia*
6. Calyx 5–8 mm long, with the tube almost always more than half as long as the whole calyx; standard 8–13(15) × 3.5–8 mm; petioles up to 1(2) mm long in subsp. *sessilifolia* · 8. *sessilifolia*
– Calyx (9)10–17 mm long, with the tube less than half as long as the whole calyx; standard (12)14–20 × (6)9–14 mm; petioles as in next couplet · 7
7. Leaflets narrowed to a pungent tip, hairy at first, usually glabrescent; petioles 0–3(6) mm long; pedicels 2–3 mm long · 6. *aristata*
– Leaflets obtuse to rounded to a small apiculum, sericeous-tomentose; petioles mostly 4–12 mm long; pedicels 0.5–1 mm long · 7. *metallifera*

1. **Pearsonia flava** (Baker f.) Polhill in Kew Bull. **29**: 390, fig. 2A (1974). —Lock, Leg. Afr. Check-list: 228 (1989). Type: Zambia, Mbala Distr., without precise locality, *A.H. Gamwell* 198 (BM, holotype).
Gamwellia flava Baker f. in J. Bot. **73**: 161 (1935). —White, F.F.N.R.: 152 (1962). —Milne-Redhead in F.T.E.A., Leguminosae, Pap.: 809, fig. 116 (1971).

Stems ascending and spreading, (10)15–45(60) cm long, numerous from a branched and rhizomatous rootstock, hirsute. Leaflets (1)1.5–3.5 × 0.25–1 cm, oblanceolate or linear-oblanceolate, pilose; petiole 6–13 mm long; stipules resembling the leaflets. Flowers bright yellow, turning orange, usually 1–3 on a terminal or leaf-opposed peduncle, rarely up to 7 in short racemes; bracts 0.5–1.2 cm long, linear-lanceolate; pedicels 2–5 mm long; bracteoles at the top of the pedicel, similar to the bract. Calyx 1.6–1.9(2.2) cm long, pilose; lobes c. 1.8–2.5 times as long as the tube, attenuately triangular-lanceolate, upper ones a little broader and ± downcurved; lateral sinus nearly equidistant from shallowest upper and deepest lower ones. Standard 2.4–2.7 cm long, elliptic, medially pubescent outside. Vexillary stamen usually free. Ovary 28–38-ovulate. Pod 3–3.8 × 0.4–0.6 cm, narrowly oblong-pointed, densely pilose. Seeds 2–2.5 mm across.

Subsp. **flava**. TAB. 3,7: **17**, fig. A.

Leaflets with scattered irregularly arranged hairs.

Zambia. N: Mbala Distr., Lunzua Road, fl. & fr. 6.vi.1961, *Richards* 15211 (EA; K); Mbala Distr., Chiyanga Village, along Chiyanga stream, fl. 4.ii.1969, *Sanane* 443 (K).
Also in southern Tanzania. Grassy places in miombo woodland, often in damp and disturbed places; 1500–1750 m.
Subsp. *mitwabensis* (Timp.) Polhill occurs in Dem. Rep. Congo (Katanga). It has densely sericeous-pilose leaflets.

2. **Pearsonia mesopontica** Polhill in Kew Bull. **29**: 390, fig. 2B–J (1974). —Lock, Leg. Afr. Check-list: 229 (1989). TAB. 3,7: **17**, fig. B. Type: Zimbabwe, near Chimanimani (Melsetter), on slopes of Nyamzure Hill (Pork Pie), *Corby* 1124 (K, holotype; SRGH).

Stems spreading or decumbent, 15–35 cm long, from a strong taproot or narrow rootstock, covered with a mixture of short appressed and longer spreading hairs. Leaflets 2.5–4.7 × 0.6–1.2 cm, mostly narrowly elliptic-lanceolate, shorter and relatively broad on lower leaves, subappressed pilose on both surfaces; petiole 6–10 mm long; stipules 0.4–1.1 cm long, subulate to linear-caudate. Flowers mostly erect or resupinate, yellow, sometimes red-brown veined on the standard, solitary on leaf-opposed peduncles 1.2–3 cm long; bracts 0.5–1 cm long, linear or subulate; pedicels 2–4 mm long; bracteoles on the pedicel, similar to the bract. Calyx 1.6–2 cm long,

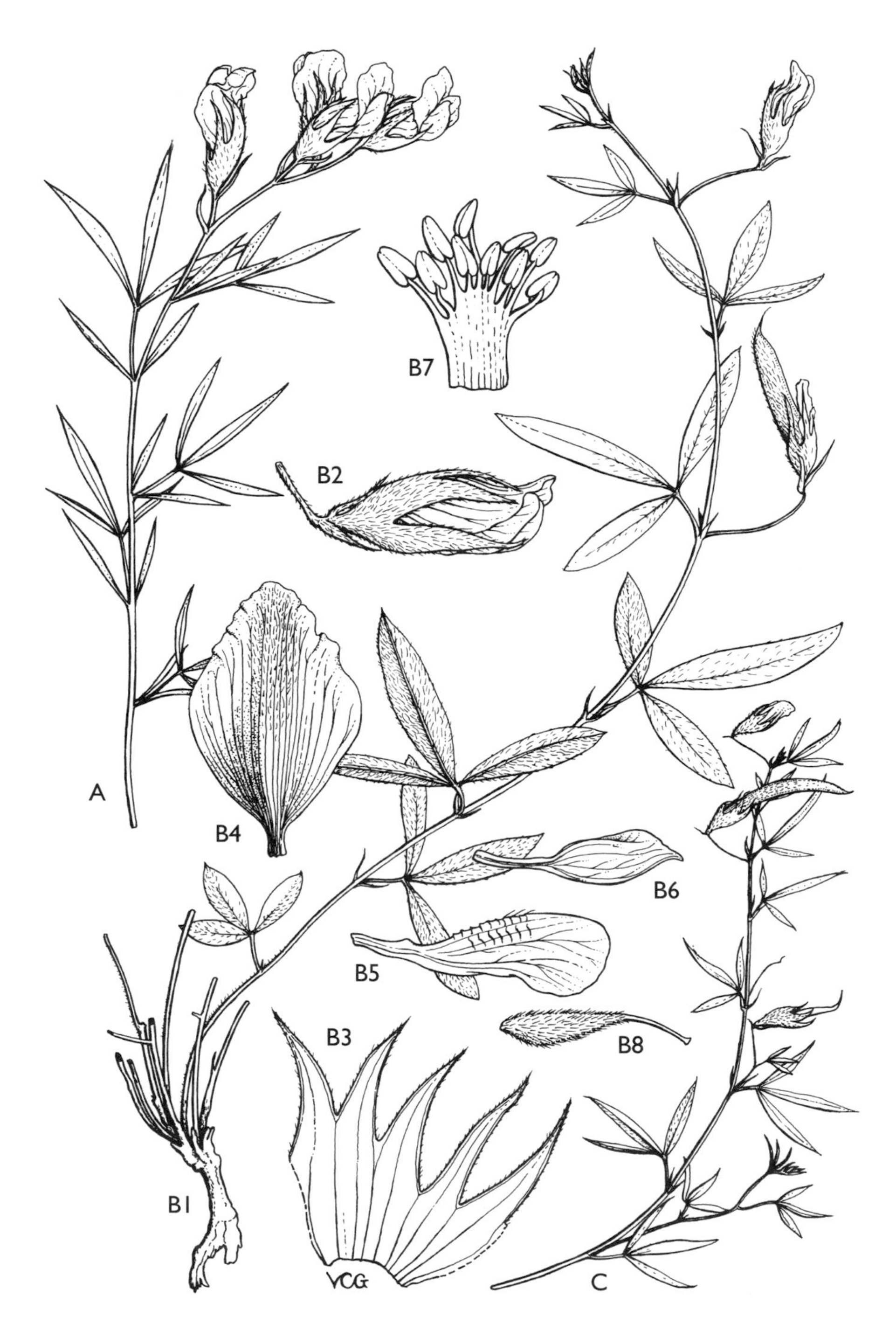

Tab. 3,7: **17**. A. —PEARSONIA FLAVA subsp. FLAVA, flowering branch (× $^2/_3$), from *Bullock* 2355 and *Sanane* 443. B. —PEARSONIA MESOPONTICA. B1, habit (× $^2/_3$), from *Corby* 1124; B2, flower (× $1^1/_2$); B3, calyx, opened out (× 2); B4, standard (× 2); B5, wing (× 2); B6, keel (× 2); B7, androecium from bud, spread out (× 3); B8, gynoecium (× 2), B2–B8 from *Corby* 1991. C. —PEARSONIA UNIFLORA, flowering branch (× $^2/_3$), from *Compton* 28995. Drawn by Victoria Friis. From Kew Bull.

covered with short and long hairs; lobes c. 1.8–2.5 times as long as the tube, attenuately triangular-lanceolate, the upper ones broadest and downcurved; lateral sinus nearly equidistant from shallower upper and deeper lower ones. Standard c. 2 cm long, elliptic-obovate, medially pilose outside. Stamens all joined. Ovary c. 30–32-ovulate. Pod c. 3.2 × 0.5 cm, oblong-pointed, densely pilose. Seeds 2–2.5 mm across.

Zimbabwe. E: Chimanimani Mts., where Bundi leaves Upper Bundi Plain, fl. 1.ii.1957, *Phipps* 334 (K; SRGH); Chimanimani Mts., 1.2 km from Mountain Hut in NE direction on Bundi Plain, fl. 6.iv.1969, *Kelly* 84 (K; PRE; SRGH).

Known only from the Chimanimani Mts. and hills just north of Chimanimani township. Montane grassland; c. 1550–1750 m.

3. **Pearsonia uniflora** (Kensit) Polhill in Kew Bull. **29**: 392, fig. 2K (1974). —Lock, Leg. Afr. Check-list: 230 (1989). TAB. 3,7: **17**, fig. C. Syntypes from South Africa (North-West and Northern Provinces).

Lotononis uniflora Kensit in Trans. Roy. Soc. South Africa **1**: 147 (1909). —Dummer in Trans. Roy. Soc. South Africa **3**: 316 (1913). —Burtt Davy, Fl. Pl. Ferns Transvaal, pt. 2: 388 (1932).

Stems slender, ascending or decumbent, 15–35(60) cm long, several to numerous from a taproot, appressed pubescent to pilose. Leaflets 15–30 × 2–6 mm, mostly narrowly elliptic-oblong, the lower ones and those on decumbent shoots relatively broad, thinly to densely pilose; petiole 4–16 mm long; stipules 2–7(11) mm long, linear or linear-lanceolate. Flowers mostly erect or resupinate, pale yellow, 1(2) on wiry leaf-opposed peduncles 0.7–2 cm long; bracts 2–5 mm long, linear; pedicels 1–3 mm long; bracteoles similar to the bract. Calyx (7)8–11 mm long, appressed pubescent or pilose; upper lobes 3.5–4.5 mm long, narrowly attenuately triangular; upper and lateral sinuses nearly at the same level and 1.5–2.5 mm higher than the lower sinuses. Standard 1–1.4 cm long, elliptic or ovate, medially pilose outside. Stamens all joined. Ovary 18–24-ovulate. Pod 2.5–3 × 0.3–0.5 cm, linear-oblong, appressed pubescent. Seeds not seen.

Mozambique. M: Moamba Distr., Ressano Garcia, fl. 24.xii.1897, *Schlechter* 11906 (BM; BOL; K; PRE); Ressano Garcia para Moamba, fl. 18.ii.1948, *Torre* 7359 (BM; K; LISC).

Also in South Africa (Northern and North-West Province east to Mpumalanga and KwaZulu-Natal) and Swaziland. Grassland; c. 300 m.

4. **Pearsonia cajanifolia** (Harv.) Polhill in Kew Bull. **29**: 393 (1974). —Lock, Leg. Afr. Check-list: 228 (1989). Syntypes from South Africa (Gauteng and Mpumalanga).

Pleiospora cajanifolia Harv., Thes. Cap. **1**: 51, t. 81 (1859); in F.C. **2**: 47 (1862). —Dummer in Trans. Roy. Soc. South Africa **3**: 331 (1913). —Burtt Davy, Fl. Pl. Ferns Transvaal, pt. 2: 382 (1932).

Phaenohoffmannia cajanifolia (Harv.) Kuntze, Revis. Gen. Pl. **2**: 940 (1891). —Gillett in Kew Bull. **20**: 298 (1966). —Drummond in Kirkia **8**: 225 (1972).

Pleiospora gracilior Dummer in Trans. Roy. Soc. South Africa **3**: 332 (1913). Type from South Africa.

Stems erect, well branched, (0.2)0.4–0.7(1.5) m tall, several from a deep taproot, pubescent to tomentose. Leaves mostly 3-foliolate, but upper ones 1-foliolate and intergrading to bracts; leaflets 3–6.5(9) × 0.8–2.5(3) cm, oblanceolate to elliptic or obovate, apiculate, thinly pubescent to silvery sericeous-tomentose; lower lateral nerves steeply ascending, reticulation rather prominent; petiole 0.5–2.8 cm long; stipules 0.4–1(1.4) cm long, linear-subulate. Flowers yellow, (3)8–30(40) in heads or short dense racemes often aggregated into panicles by suppression of upper leaves; bracts subtending flowers 4–8 mm long, linear or linear-lanceolate; bracteoles on the very short pedicels a little smaller. Calyx 6–10 mm long, tomentose; upper lip c. 1–1.5 times as long as the tube, with the lateral sinuses as high as or higher than the upper sinus; upper lobes 2–4 mm long, narrowly to broadly triangular-acuminate. Standard 8–13 × 3–7 mm, usually elliptic-oblong, pubescent outside. Stamens all joined. Ovary 6–12-ovulate. Pod little exserted from the calyx, 6–8 × 3–4 mm, oblong-ellipsoid, pubescent, usually only 2-seeded. Seeds c. 1 mm long, oblique-cordiform.

Subsp. **cryptantha** (Baker) Polhill in Kew Bull. **29**: 394 (1974). —Lock, Leg. Afr. Check-list: 228 (1989). Type: Malawi, Zomba Mt., *Whyte* s.n. (K, holotype).

Eriosema cryptanthum Baker in Bull. Misc. Inform., Kew **1897**: 264 (1897).

Pleiospora holosericea Schinz in Bull. Herb. Boissier **7**: 29 (1899). —Dummer in Trans. Roy. Soc. South Africa **3**: 333 (1913). —Burtt Davy, Fl. Pl. Ferns Transvaal, pt. 2: 382 (1932). Syntypes from South Africa (Northern Province and Mpumalanga).

Pleiospora paniculata Bolus ex Dummer in Trans. Roy. Soc. South Africa **3**: 331 (1913). —Burtt Davy, Fl. Pl. Ferns Transvaal, pt. 2: 382 (1932). Type from South Africa (Northern Province).

Pleiospora bolusii Dummer in Trans. Roy. Soc. South Africa **3**: 333 (1913). —Burtt Davy, Fl. Pl. Ferns Transvaal, pt. 2: 382 (1932). Type from South Africa (Northern Province).

Pleiospora macrophylla Dummer in Trans. Roy. Soc. South Africa **3**: 334 (1913). —Burtt Davy, Fl. Pl. Ferns Transvaal, pt. 2: 382 (1932). Type from South Africa (Northern Province).

Pleiospora buchananii Harms in Repert. Spec. Nov. Regni Veg. **14**: 145 (1915), *nom. illegit.* Types as for *Eriosema cryptanthum.*

Pleiospora cryptantha (Baker) Baker f., Legum. Trop. Africa: 14 (1926).

Phaenohoffmannia cajanifolia subsp. *cryptantha* (Baker) J.B. Gillett in Kew Bull. **20**: 298 (1966).

Undersurface of leaflets ± hidden by a dense silky indumentum. Calyx 8–10 mm long.

Zimbabwe. C: Shurugwi (Selukwe) Police Camp, fl. 27.xii.1968, *Corby* 2063 (K; PRE; SRGH). E: Nyanga Distr., Juliasdale, fl. 8.i.1965, *Wild* 6744 (K; PRE; SRGH). **Malawi**. S: Zomba Plateau, near Nowimbe Fire Tower, fl. 12.i.1979, *Blackmore & Patel* 141 (BM; K; MAL); Zomba Plateau, near Chingwe's Hole, fl. 7.ii.1985, *Salubeni & Nachamba* 4005 (K; MAL).

Also in South Africa (Northern Province, Mpumalanga and probably KwaZulu-Natal). Submontane grassland and often rocky hillsides in higher altitude miombo or *Uapaca* woodland; 1350–2100 m.

Subsp. *cajanifolia* has the surface of the leaflets clearly visible between the sparse covering of stiff hairs and slightly smaller flowers, the calyx 6–8 mm long. It occurs further south in the North-West, Gauteng and Mpumalanga Provinces of South Africa, with some intermediates in the Groblersdal and Lydenburg area. The Zimbabwean specimens from Shurugwi tend to have relatively sparse indumentum for subsp. *cryptantha,* but in general the differences have a distinct geographical bias.

5. **Pearsonia grandifolia** (Bolus) Polhill in Kew Bull. **29**: 396 (1974). —Lock, Leg. Afr. Check-list: 229 (1989). Type from South Africa (KwaZulu-Natal).

Lotononis grandifolia Bolus in J. Bot. **34**: 19 (1896). —Medley Wood, Natal Pl. **1**: 10, t. 9 (1898).

Pleiospora grandifolia (Bolus) Dummer in Trans. Roy. Soc. South Africa **3**: 334 (1913).

Phaenohoffmannia grandifolia (Bolus) J.B. Gillett in Kew Bull. **20**: 300 (1966).

Stems prostrate, c. 15–50 cm long, from a strong taproot or branched rootstock, somewhat flexuous, pubescent to tomentose with a mixture of short often crisped and long fine spreading hairs. Leaflets 3.5–7 × (1.5)2–3.5 cm, elliptic to obovate, apiculate, pilose to tomentose, with conspicuous lateral nerves (the lower ones steeply ascending); petiole 0.3–1.5(2.5) cm long; stipules 0.5–1 × 0.2–0.9 cm, usually broadly ovate-acuminate, less often lanceolate or obovate-circular. Flowers white or cream with a purple centre, (8)16–30 in subdense to lax pedunculate racemes 5–15(25) cm long; bract and bracteoles similar, at the top of the pedicel, linear-lanceolate to oblate. Calyx 1–1.8(2.2) cm long, densely hairy; upper lip c. 2–3 times as long as the tube, with the lateral sinuses as high as or higher than the upper sinus; upper lobes (3)5–7(9) mm long, narrowly to broadly triangular-lanceolate, acuminate. Standard 1.6–2.2 cm long, elliptic or oblong-elliptic, pubescent outside. Stamens all joined. Ovary 14–30-ovulate. Pod shortly exserted from the calyx, up to 17 × 3.5 mm, oblong-pointed, rather densely hairy. Seeds not seen.

Subsp. **latibracteolata** (Dummer) Polhill in Kew Bull. **29**: 396, fig. 3A (1974). —Lock, Leg. Afr. Check-list: 229 (1989). TAB. 3,7: **18**, fig. A. Syntypes from South Africa (Mpumalanga).

Pleiospora latibracteolata Dummer in Trans. Roy. Soc. South Africa **3**: 334 (1913) as "*latebracteolata*". —Burtt Davy, Fl. Pl. Ferns Transvaal, pt. 2: 382 (1932). —Goodier & Phipps in Kirkia **1**: 56 (1961).

Phaenohoffmannia latibracteolata (Dummer) J.B. Gillett in Kew Bull. **20**: 300 (1966). —Drummond in Kirkia **8**: 225 (1972).

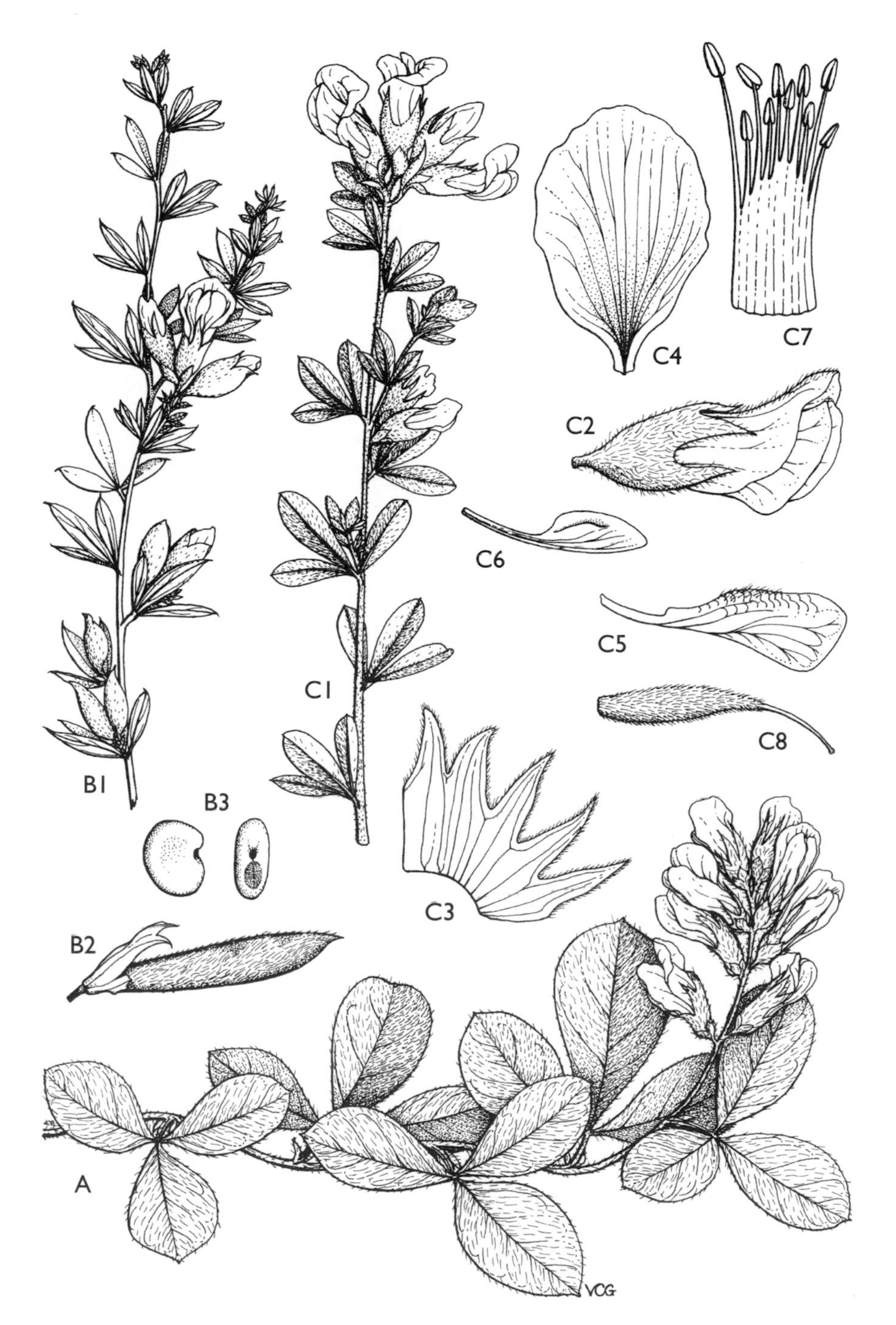

Tab. 3,7: **18**. A. —PEARSONIA GRANDIFOLIA subsp. LATIBRACTEOLATA, flowering branch (× 1), from *Chase* 664. B. —PEARSONIA ARISTATA. B1, flowering branch (× 1), from *Simon* 849; B2, pod (× 1); B3, seed, two aspects (× 4), B2 & B3 from *Codd* 9541. C. — PEARSONIA METALLIFERA. C1, flowering branch (× 1); C2, flower (× 2); C3, calyx, opened out (× 2); C4, standard (× 2); C5, wing (× 2); C6, keel (× 2); C7, androecium, spread out (× 3); C8, gynoecium (× 3), C1–C8 from *Wild* 6093. Drawn by Victoria Gordon. From Kew Bull.

Bracts 3–7(9) × (1)2.5–6(7) mm, obovate to oblate, rarely lanceolate, acuminate. Pedicels 2–4(6) mm long. Calyx (1.2)1.4–1.8(2.2) cm long.

Zimbabwe. E: Nyanga Distr., Nyamziwa Falls, fl. 11.i.1951, *Chase* 3676 (BM; K; SRGH); Chimanimani Mts., on lower SE slopes of Mt. Peza, fl. 30.xii.1957, *Goodier* 481 (K; SRGH).

Also in South Africa (Gauteng and Mpumalanga). Montane grassland, often on rocky hillsides or by streams; 1650–2300 m.

Subsp. *grandifolia*, with narrower bracts, shorter pedicels and mostly smaller flowers, occurs further south in South Africa in the Free State and KwaZulu-Natal. Specimens from the Chimanimani Mts., including *Goodier* 481 cited above, have unusually narrow bracteoles for the subspecies, but agree in other respects.

6. **Pearsonia aristata** (Schinz) Dummer in J. Bot. **50**: 356 (1912). —Burtt Davy, Fl. Pl. Ferns Transvaal, pt. 2: 383 (1932). —Polhill in Kew Bull. **29**: 398, fig. 3B–D (1974). —Lock, Leg. Afr. Check-list: 228 (1989). TAB. 3,7: **18**, fig. B. Types from South Africa (Mpumalanga).

Lotononis aristata Schinz in Bull. Herb. Boissier **7**: 32 (1899).

Lotononis aristata var. *gazensis* Baker f. in J. Linn. Soc., Bot. **40**: 51 (1911). —Eyles in Trans. Roy. Soc. South Africa **5**: 369 (1916). —Goodier & Phipps in Kirkia **1**: 56 (1961). —Drummond in Kirkia **8**: 225 (1972). Syntypes: Zimbabwe, Chimanimani Mts., *Swynnerton* 1418 (BM, syntype; K); Chimanimani (Melsetter), *Swynnerton* 6196 (BM, syntype).

Pearsonia aristata var. *gazensis* (Baker f.) Dummer in J. Bot. **50**: 356 (1912). —E.G. Baker, Legum. Trop. Africa: 19 (1926).

Pearsonia sp. 1 of Drummond in Kirkia **8**: 225 (1972).

Stems erect, spreading or prostrate, sometimes up to 1.5 m tall, numerous from a strong taproot or woody rootstock, much branched, sparsely to densely covered with rather short to long mostly appressed to spreading hairs. Leaflets often rather coriaceous, 8–28(38) × 2–6(14) mm, oblanceolate-elliptic, usually narrowly so, sometimes elliptic or obovate, narrowed to a pungent tip, hairy at first, but usually glabrescent; lower lateral nerves steeply ascending and running into the thickened margin; petiole 0–3(6) mm long; stipules lacking. Flowers bright yellow, turning orange in part, (1)3–8(24), laxly arranged at the end of main and short lateral branches; bracts 3–5(9) mm long, narrowly elliptic-oblanceolate to elliptic; pedicels 2–3 mm long; bracteoles on the pedicel, narrower and shorter than the bract. Calyx (0.9)1–1.4(1.7) cm long, usually appressed or subappressed pubescent, but hairs sometimes sparse or spreading; upper lip 2–3 times as long as the tube, with the lateral sinuses as high to much higher than the upper sinus; upper and lateral lobes narrowly to broadly triangular-falcate, ± acuminate, sometimes adherent or largely united. Standard (1.2)1.5–2 cm long, elliptic to oblong-obovate, hairy medially towards the apex outside. Stamens all joined. Ovary 18–32-ovulate. Pod 2–2.7(3.7) × 0.3–0.5(06) cm, narrowly oblong-pointed, straight or slightly curved, pubescent. Seeds 2.5–3 mm across, oblique-cordiform.

Zimbabwe. E: near Chimanimani (Melsetter) School, fl. 19.x.1950, *Chase* 2994 (BM; K; PRE; SRGH); Chimanimani Mts., between Bundi Plain and Southern Lakes, fl. 26.ix.1966, *Simon* 849 (EA; K; PRE; SRGH).

Also in mountainous parts of South Africa (Northern Province, Mpumalanga and KwaZulu-Natal) and Swaziland. Montane grassland and miombo woodland, often in steep rocky places; (1050)1450–2100 m.

A very distinctive species despite considerable variation in habit, indumentum and proportions of the leaves and flowers.

7. **Pearsonia metallifera** Wild in Kirkia **5**: 76 (1965). —Drummond in Kirkia **8**: 225 (1972). —Polhill in Kew Bull. **29**: 400, fig. 3E–M (1974). —Lock, Leg. Afr. Check-list: 229 (1989). TAB. 3,7: **18**, fig. C. Type: Zimbabwe, Charter Distr., Silverstream Ranch SW of Ngesi, *Wild* 5610 (K; PRE; SRGH, holotype).

Bushy perennial herb 0.35–1 m tall, with a strong taproot or woody rootstock, silvery tomentose throughout (except corolla). Leaflets 1.5–3 × 0.7–1.5 cm, elliptic-obovate to obovate or oblanceolate, obtuse or rounded to a small apiculum; petioles mostly 0.4–1.2(1.4) cm long; stipules lacking. Flowers yellow, (1)2–6(8), clustered at branch-ends, sometimes on very reduced lateral branches and appearing axillary; bracts 3–4 mm long, elliptic; pedicels 0.5–1 mm long; bracteoles 3–5 mm long, linear.

Calyx 1.1–1.4 cm long; lobes 1.2–2 times as long as the tube, attenuate-triangular, ± acuminate, with the lateral and upper sinuses progressively a little higher than the lower sinus. Standard 1.4–1.8 cm long, elliptic, pubescent outside. Stamens all joined. Ovary 16–20-ovulate. Pod 1.6–1.8 × 0.35–0.45 cm, narrowly oblong-pointed, densely sericeous-pubescent. Seeds not seen.

Zimbabwe. N: Zvimba Distr., Mvurwi (Umvukwes) Range, road between Mutorashanga (Mtoroshanga) and Muriel Mine, fl. 30.iii.1959, *Drummond* 6041 (K; PRE; SRGH). C: Chirumhanzu Distr., Lalapanzi (Lalapansi), fl., without date, *Walters* 2420 (K).

A distinctive and uniform species of grassland on serpentine soils of the Great Dyke in Zimbabwe; c. 1300–1450 m.

8. **Pearsonia sessilifolia** (Harv.) Dummer in J. Bot. **50**: 354 (1912). —Burtt Davy, Fl. Pl. Ferns Transvaal, pt. 2: 383 (1932). —Letty et al., Wild Fl. Transvaal: 164, t. 82/2 (1962). —Polhill in Kew Bull. **29**: 400 (1974). —Lock, Leg. Afr. Check-list: 229 (1989). Type from South Africa (Gauteng).

Lotononis sessilifolia Harv. in F.C. **2**: 66 (1862).

Pearsonia sessilifolia var. *conrathii* Dummer in J. Bot. **50**: 354 (1912) pro parte excl. specim. ex Zulu. —Burtt Davy, Fl. Pl. Ferns Transvaal, pt. 2: 383 (1932). Lectotype from South Africa (Gauteng/Mpumalanga), selected by Polhill in Kew Bull. **29** (1974).

Pearsonia mucronata Burtt Davy ex Baker f., Legum. Trop. Africa: 20 (1926). —Drummond in Kirkia **8**: 225 (1972). Type: Zimbabwe, Rusape, *Hislop* Z. 220 (K, holotype).

Stems erect, 0.1–1.2 m tall, usually several from a woody ± rhizomatous rootstock, often branching only towards the top, sparsely hairy to tomentose. Leaflets (0.8)1.5–5 × 0.1–1 cm, linear, oblanceolate or attenuately elliptic-oblong or obovate, apiculate, sparsely pilose to tomentellous or tomentose; petiole 0–1 cm long; stipules lacking. Flowers yellow, turning reddish-brown, sometimes few at branch-ends, sometimes in short to long, dense or lax racemes; bracts (2)4–7(10) mm long, obovate or subcircular; pedicels 1–2(3) mm long; bracteoles filiform to obovate or subcircular. Calyx 5–8 mm long, shortly appressed pubescent to tomentellous or coarsely tomentose; upper lip almost always shorter than the tube, with generally small upper and lateral lobes united to varying degrees. Standard 8–13(15) mm long, pubescent outside. Stamens all joined. Ovary 14–30-ovulate. Pod 1.5–2.5 × 0.4–0.5 cm, narrowly oblong-pointed. Seeds 2–2.5 mm across, oblique-cordiform.

Subsp. **sessilifolia**

Suffrutex or slender-stemmed shrublet, (5)15–40(75) cm tall, usually branching shortly only towards the top. Leaflets (2.5)3–10(15) times as long as broad, linear-oblanceolate or oblanceolate to narrowly oblong-elliptic or rarely obovate, narrowed or less often rounded to the apiculum, pilose to tomentose; petioles 0–1(2) mm long. Flowers 6–12 in frequently sessile heads or short spiciform racemes; bracteoles (3)3.5–5(7) mm long, linear. Upper calyx lobes 0.5–1.5(2) mm long, usually ± straight.

Zimbabwe. C: Makoni Distr., Rusape, fl. without precise date (received 3.ii.1921), *Hislop* Z 220 (K). E: Nyanga (Inyanga), fl. xii.1919, *Philomena* in *Eyles* 5183 (K); same locality, fl. 23.x.1935, *Eyles* 8504 (K). **Mozambique**. MS: Barué Distr., Serra de Chôa, 25 km de Catandica (Vila Gouveia), picada nova para a fronteira, fl. & fr. 10.xii.1965, *Torre & Correia* 13500 (LISC); same locality, 26 km de Catandica, fl. 26.iii.1966, *Torre & Correia* 15404 (LISC).

Also in South Africa (Northern, North-West, Gauteng, Mpumalanga and KwaZulu-Natal). Little information on habitat, but recorded from a dry hillside at Nyanga and in grassland with *Brachystegia spiciformis* and *Ericaceae* just over the Mozambique border; 1400–1750 m and perhaps rather lower at Rusape.

Only the above cited specimens seen, all those from Zimbabwe collected more than 60 years ago.

Pearsonia sessilifolia is an extremely variable aggregate. It includes ten of the eleven species included in the genus by Dummer in 1912, to which Burtt Davy added one more in 1932. Four subspecies are recognized (Polhill, 1974), with most of the variation and intermediates occurring in the northern part of the Drakensberg (in Swaziland and the former Transvaal).

The identity of *Pearsonia sp. 2* of Drummond in Kirkia **8**: 225 (1972), based on *Goodier* 192 (SRGH), has not been established, but is likely to be this species or *P. aristata*.

100. ROTHIA Pers.

By R.M. Polhill

Rothia Pers., Syn. Pl. **2**: 302, 638 (1807). —Polhill in Bot. Syst. **1**: 326 (1976) *nom. conserv.*
Xerocarpus Guill. & Perr. in Guillemin, Perrottet & Richard, Fl. Seneg. Tent.: 169 (1832).

Annual herbs. Leaves digitately 3-foliolate; stipules single at the nodes in Africa, elsewhere in pairs. Flowers small, few in leaf-opposed racemes; bracteoles absent. Calyx subequally lobed, the upper lobes largest and slightly curved. Standard not exceeding the calyx, yellow or cream-coloured, narrow; keel narrow, ± as long as the wings, bluntly pointed. Stamens all joined in a sheath open on the upper side; anthers all small, similar, rounded. Style straight. Pod membranous, sessile, flattened, narrowly pointed-oblong to linear, with the stylar beak projecting forwards, dehiscent, many-seeded. Seeds oblique-cordiform, small.

A genus of 2 species, one in the drier parts of Africa, the other extending from Pakistan to Australia.

Rothia hirsuta (Guill. & Perr.) Baker in F.T.A. **2**: 7 (1871). —Eyles in Trans. Roy. Soc. South Africa **5**: 369 (1916). —E.G. Baker, Legum. Trop. Africa: 21 (1926). —Hepper in F.W.T.A., ed. 2, **1**: 543 (1958). —White, F.F.N.R.: 164 (1962). —Torre in C.F.A. **3**: 5 (1962). —Schreiber in Merxmüller, Prodr. Fl. SW. Afrika, fam. 60: 107 (1970). —Milne-Redhead in F.T.E.A., Leguminosae, Pap.: 811, fig. 117 (1971). —Drummond in Kirkia **8**: 226 (1972). —Lock, Leg. Afr. Check-list: 232 (1989). TAB. 3,7: **19**. Type from Senegal.
Xerocarpus hirsutus Guill. & Perr. in Guillemin, Perrottet & Richard, Fl. Seneg. Tent.: 44 (1832).
Lotononis desertorum Dummer in Trans. Roy. Soc. South Africa **3**: 316 (1913). Type from Namibia.

Annual; stem shortly erect, developing longer slender procumbent to prostrate branches from the base; branches 10–40 cm long, pilose. Leaflets 10–25(30) × 4–10 mm, elliptic-oblanceolate to oblanceolate, obtuse to rounded to the apiculate tip, pilose on both surfaces; stipules single at the node, 2–8 mm long, shortly stalked, linear-lanceolate. Flowers 2–6, leaf-opposed in short racemes; bracts 1–2.5 mm long, filiform, usually slightly longer than the pedicel. Calyx 5–6(7 in fruit) mm long, thinly pilose; upper lobes 3–4.5 mm long, triangular-lanceolate, slightly downcurved. Standard white or cream-coloured lined purple, a little shorter than the calyx and slightly longer than the other petals, shortly clawed, narrowly elliptic, with a few hairs along the midvein outside. Pods 10–12 × 2.5–3 mm, narrow, tapered to the persistent forwardly directed style, ± densely pilose, 10–20-seeded. Seeds 1.2–1.5 mm long, cream-coloured to brown, sometimes mottled, slightly sculptured.

Zambia. B: near Senanga, fl. & fr. 4.viii.1952, *Codd* 7387 (K; PRE). C: Lusaka, fl. & fr. 6.iii.1971, *Fanshawe* 11179 (K; SRGH). S: Choma Distr., Mapanza Mission, fl. & fr. 29.iii.1953, *E.A. Robinson* 151 (K). **Zimbabwe**. N: Hurungwe Distr., Kariba Gorge, fl. & fr. 25.ii.1953, *Wild* 4036 (K; SRGH). W: Umguza Distr., Nyamandhlovu Pasture Research Station, fl. & fr. 25.iii.1954, *Plowes* 1691 (K; SRGH). C: Chegutu (Hartley), fl. & fr. 2.iv.1948, *R.M. Hornby* 2861 (K; SRGH). E: Nyanga North, at foot of Mica Hill, fr. 21.iv.1972, *Pope & Wild* 629 (K; SRGH). S: Chivi Distr., near Madzivire Dip, c. 6.5 km north of Runde (Lundi) R. Bridge, fr. 3.v.1962, *Drummond* 7888 (K; SRGH). **Malawi**. C: Salima Distr., Lifidzi Goat Breeding Centre, fl. & fr. 19.iv.1985, *Balaka & Tawakali* 1071 (K; MAL). S: Zomba Distr., L.

Tab. 3,7: **19**. ROTHIA HIRSUTA. 1, flowering and fruiting branch (× ²⁄₃); 2, portion of branch and leaf showing stipules (× 5); 3, flower (× 8); 4, calyx, opened out (× 8); 5, standard (× 8); 6, wing (× 8); 7, keel (× 8); 8, androecium and gynoecium (× 8); 9, androecium, spread out (× 8); 10, gynoecium, with ovary opened to show ovules (× 8), 1–10 from *Drummond & Hemsley* 3850; 11, pod (× 4); 12, seed (× 20), 11 & 12 from *Bogdan* 2205. Drawn by Margaret Stones. From F.T.E.A.

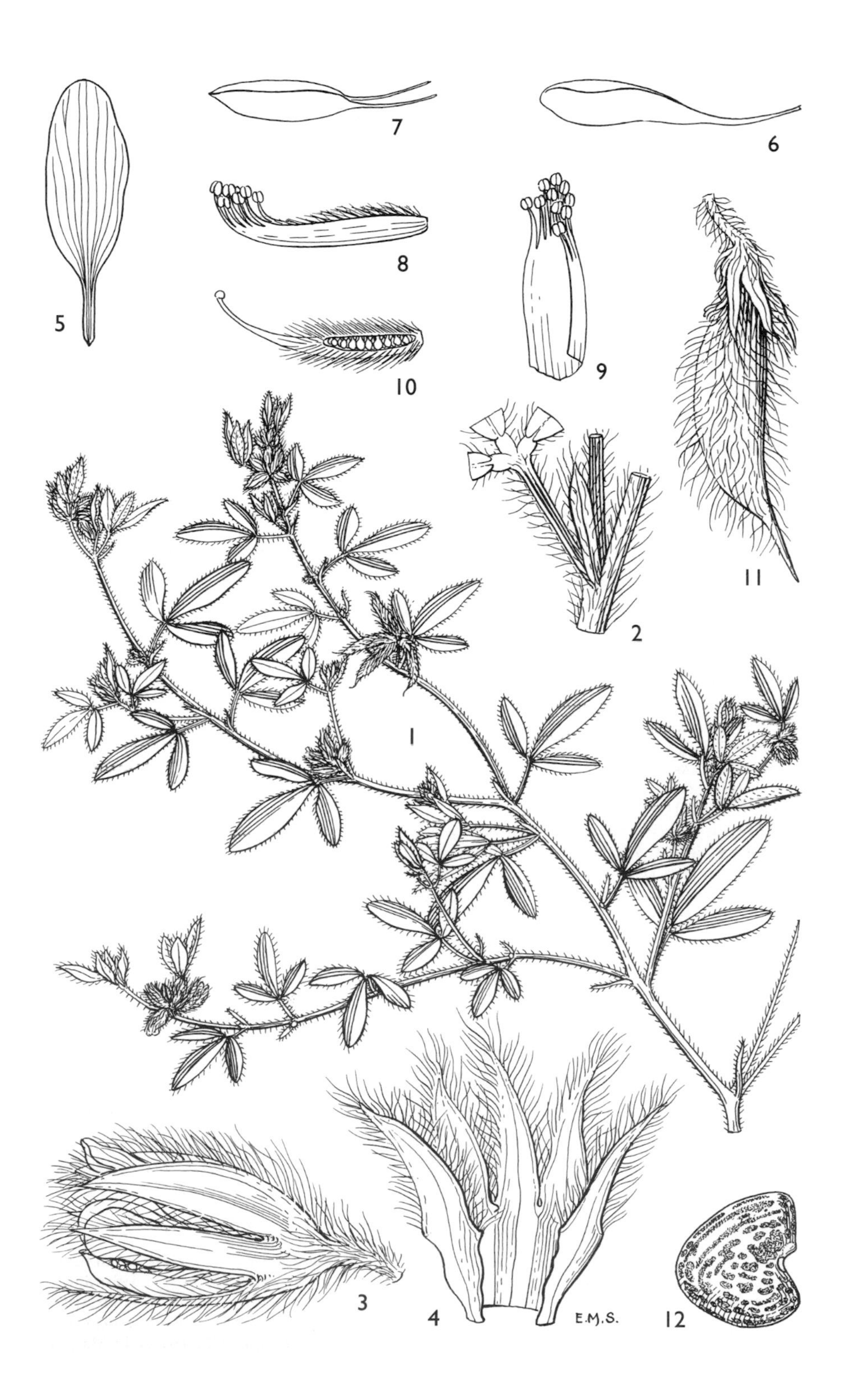
5
7
6
8
9
10
11
2
1
3
4
E.M.S.
12

Chilwa, Nchisi (Chisi) Island, fl. 10.xi.1986, *Salubeni & Kwatha* 4838 (K; MAL). **Mozambique**. N: Mossuril para Lumbo, 4.5 km na estrada de Nampula, fl. & fr. 5.v.1948, *Pedro & Pedrógão* 3131 (LMA; K); Mossuril Distr., Ilha de Moçambique (Mozambik I.), fr. 5.iv.1894, *Kuntze* s.n. (K).

Widespread in the drier parts of tropical Africa from Senegal to Eritrea and south to Angola and Namibia. Short grassland and disturbed places; 0–1350 m.

Superficially similar to *Lotononis*, but easily recognized by the broad downcurved upper calyx lobes and tapered fruit with ± straight style.

101. ROBYNSIOPHYTON R. Wilczek

By R.M. Polhill

Robynsiophyton R. Wilczek in Bull. Jard. Bot. État **23**: 128, t. 1 (1953); in F.C.B. **4**: 286, t. 17 (1953). —Polhill in Bot. Syst. **1**: 326 (1976).

Annual or short-lived perennial. Leaves mostly digitately 3-foliolate, some upper ones 1-foliolate; stipules paired, narrow. Flowers small, few in terminal and axillary heads; bracteoles absent. Calyx subequally lobed, the upper lobes narrow. Standard yellow, not exceeding the calyx, narrow; keel ± as long as the wings, narrow, obtuse. Stamens 9, all joined in a sheath open on the upper side, 4 sterile, the others with small rounded dorsifixed anthers. Style straight. Pod membranous, sessile, flattened oblong-ovate, narrowed to a straight stylar beak, dehiscent, rather few-seeded. Seeds oblique-cordiform, small.

A genus of one species, occurring along the Congo-Zambezi divide.

Robynsiophyton vanderystii R. Wilczek in Bull. Jard. Bot. État **23**: 128, t. 1 (1953); in F.C.B. **4**: 286, t. 17 (1953). —White, F.F.N.R.: 164 (1962). —Torre in C.F.A. **3**: 6 (1962). —Lock, Leg. Afr. Check-list: 232 (1989). TAB. 3,7: **20**. Type from Dem. Rep. Congo (Bas-Congo).

Annual or short-lived perennial, with erect pilose stem to 10–30 cm tall and often longer spreading to weakly ascending lower branches, sometimes perennating with extra basal branches but taproot undivided. Leaves mostly 3-foliolate, but uppermost mixed with 1-foliolate; leaflets mostly 15–25 × 4–10 mm, narrowly elliptic-oblanceolate to oblanceolate, obtuse to rounded to the apiculate tip, pilose on both surfaces; stipules 3–7 mm long, linear to linear-lanceolate. Flowers 2–9 in heads towards the tips of shoots, subsessile; bracts 2–3.5 mm long. Calyx 4–5 mm long, pilose; lobes 2.5–3 mm long, narrowly triangular. Standard pale yellow, a little shorter than the calyx and ± the same length as the wings and keel, spathulate, with a few hairs along the midvein outside. Pods 6–10 × 3–4.5 mm, ovate-oblong, tapered to the persistent forwardly directed style, densely pilose, 4–8-seeded. Seeds 1.2–1.5 mm long, cream-coloured to brown, slightly sculptured.

Zambia. N: Kawambwa Distr., Ntumbachushi (Timnatushi) Falls, fl. & fr. 18.iv.1957, *Richards* 9321 (K); Mbala (Abercorn) to Mpulungu by Chilongowelo turning, fl. & fr. 22.iv.1963, *Richards* 18144 (K). W: Mwinilunga, fr. 16.v.1969, *Mutimushi* 3416 (K; NDO).

Also in Dem. Rep. Congo and Angola. Short grassland, often in damp sandy places; 1150–1350 m.

Tab. 3,7: **20** ROBYNSIOPHYTON VANDERYSTII. 1, habit (× 1/2), from *Vanderyst* s.n.; 2, flower (× 10); 3, standard, external face (× 10); 4, wing, external face (× 10); 5, keel, profile (× 10); 6, androecium, spread out (× 10); 7, gynoecium, longitudinal section (× 10), 2–7 from *Vanderyst* 30528; 8, pod, open (× 5); 9, seed (× 20); 10, embryo (× 20), 8–10 from *Vanderyst* 21635. Drawn by M. Boutique. From Fl. Congo Belge. Reproduced with permission of Jardin Botanique National de Belgique.

1
2
3
4
5
6
7
8
9
10
M.BOUTIQUE DEL.

102. CROTALARIA L.

By R.M. Polhill

Crotalaria L., Sp. Pl.: 714 (1753); Gen. Pl., ed. 5: 320 (1754). —Polhill, Crotalaria Africa & Madagascar: 1–389 (1982).
Priotropis Wight & Arn., Prodr. Fl. Ind. Orient. **1**: 180 (1834).

Herbs or shrubs. Leaves simple, 1-foliolate or digitately 3(7)-foliolate, usually petiolate; stipules filiform to leaf-like or lacking. Flowers usually in terminal, leaf-opposed or less commonly axillary racemes, occasionally solitary or clustered; bract and paired bracteoles generally present. Calyx usually with 5 subequal lobes, sometimes the lower 3 on a short lip, sometimes 2-lipped with long lobes, occasionally with only 4 lobes or with the lateral and upper lobes largely joined on either side. Standard generally yellow, sometimes blue or white, often lined reddish, almost always with 2 appendages at the base inside, glabrous or hairy outside; keel rounded to angled, generally produced into a well developed beak. Stamens all joined into a sheath open at least at the base; anthers alternately long, basifixed, and smaller, dorsifixed. Style curved to geniculate, the basal portion thicker, almost always with 1–2 lines of hairs on the upper part; stigma small. Pod subsessile to long-stipitate, usually inflated, generally dehiscent, sometimes tardily so, 1–many-seeded. Seeds mostly oblique-cordiform to oblong-reniform, with a definite hilar sinus, sometimes with a conspicuous aril.

A genus of about 600 species throughout the tropics and extending into the subtropics particularly of the southern hemisphere, most numerous in eastern and southern tropical Africa.

Information on the classification of the genus, from which the order of species in this account is adopted, will be found in Polhill (loc. cit.), and a synopsis of the sections and subsections is given below:

1. Sect. GRANDIFLORAE (Baker f.) Polhill. Shrubs or perennial herbs, rarely annual. Leaves usually 3-foliolate with well-developed petioles, rarely 1-foliolate or simple; stipules narrow or rarely foliaceous, often caducous or absent. Bracteoles always inserted on the pedicel. Hypanthium prominent, often ridged. Calyx subequally 5-lobed or with the upper and lateral lobes united almost to the tips on either side, much shorter than the corolla. Standard yellow-green to bright yellow, often marked reddish-brown, with ridge-like or clavate appendages generally extending on to the claw; keel rounded, with an untwisted beak, (1.1)1.6–5.5 cm long, glabrous or ciliolate. Style curved, tapered, hairy on the inner side or rarely glabrous. Pod oblong-clavate or ellipsoid-clavate, conspicuously stipitate. *Species* 1–11.

2. Sect. CHRYSOCALYCINAE (Benth.) Baker f. Shrubs or herbs. Leaves (1)3(5)-foliolate, less commonly simple, sessile to long-petiolate; stipules filiform to compound or foliaceous, usually persistent, but often absent if leaves sessile or 1-foliolate. Bracteoles inserted on the pedicel or calyx. Hypanthium not very prominent. Calyx almost always subequally 5 and often deeply (4)5-lobed, sometimes slightly 2-lipped, usually at least 0.7 times as long as the keel (tending to be rather short in subsect. *Stipulosae* and parts of subsect. *Glaucae*) and often ± as long. Standard yellow, white or blue, sometimes flushed or speckled red or purplish, with ridge-like or clavate appendages extending on to the claw (except subsect. *Chrysocalycinae*); keel rounded, with a generally short untwisted (or rarely shortly twisted) beak or angled with a longer untwisted beak, sometimes lanate-pilose on the upper margin or more generally hairy. Style curved or geniculate, hairy on the inner side or both sides. Pod various, thin- to thick-walled and occasionally indehiscent, sometimes hairy inside. *Species* 12–61 and one cultivated species, *C. micans*, belonging to an American subsection.

2a. Subsect. INCANAE (Benth.) Bisby & Polhill. Leaves always 3(5)-foliolate; stipules always narrow. Bracteoles linear to ovate, usually curved and ascending from the pedicel. Calyx ± as long as the keel, deeply divided. Keel angled or rounded, generally lanate-pilose along the upper margin. Pod oblong-clavate or variously fusiform to subellipsoid but still broadest towards the apex and somewhat clavate, thin-walled. Seeds without a prominent rim aril. *Species* 12–17.

2b. Subsect. STIPULOSAE (Baker f.) Bisby & Polhill. Leaves 1–3-foliolate, petiolate; stipules expanded and ± foliaceous unless leaves 1-foliolate. Bracteoles narrow, usually caducous. Calyx as long as the keel or much shorter and then often rather shortly lobed. Keel usually abruptly rounded, almost angular, lanate towards the upper edge. Pod cylindrical or oblong-ellipsoid, occasionally somewhat compressed laterally, rounded at the base, sessile or stipitate. Seeds often with a small but prominent rim aril. *Species* 18–27.

2c. Subsect. GLAUCAE (Benth.) Bisby & Polhill. Leaves simple or (1)3-foliolate, sessile to petiolate; stipules, if present, usually narrow. Bracteoles narrow, usually persistent. Calyx generally as long as the keel, but sometimes much shorter. Keel either rounded, often half-round, frequently crested behind the short sometimes slightly twisted beak, or sometimes angled in the lower half with a practically straight beak, glabrous to ciliate or extensively hairy on the lateral surfaces away from the upper margin. Pod variously shaped, thin-walled to thick-walled. Seeds without a prominent aril. *Species* 28–59.

2d. Subsect. CHRYSOCALYCINAE. Leaves 1–3-foliolate, shortly petiolate; stipules simple to (2)3–5-fid. Bracteoles basally fused to the calyx tube, the free part linear-lanceolate and directed forwards. Calyx ± as long as the keel, deeply divided, sometimes somewhat 2-lipped. Keel rounded to angular, glabrous to hairy but not lanate on the upper margin. Pod subsessile, ovoid-ellipsoid to oblong-conoid or ellipsoid, shorter to a little longer than the calyx, glabrous inside. Seeds without a prominent rim aril. *Species* 60.

2e. Subsect. TETRALOBOCALYX (Harms) Bisby & Polhill. Similar to subsect. *Chrysocalycinae*, but the ovate bracteoles inserted below the calyx and the calyx 4-lobed. Pod sessile within the calyx, ellipsoid, sparsely hairy inside. *Species* 61.

3. Sect. HEDRIOCARPAE Wight & Arn. Shrubs or herbs. Leaves (1)3(5)-foliolate, less commonly simple; petioles usually well developed; stipules narrow or absent. Bracteoles inserted on the pedicel or calyx. Hypanthium scarcely developed. Calyx subequally 5-lobed, much shorter than the corolla. Standard yellow, sometimes flushed or conspicuously lined reddish-brown or purplish, with well-developed elongate appendages extending on to the claw; keel rounded, rarely subangular, with a straight or incurved untwisted beak, glabrous or ciliolate. Style curved, usually hairy only on the inner side. Pod sessile to stipitate, variously shaped, sometimes laterally compressed. *Species* 62–92 and two cultivated species, *C. brevidens* and *C. longithyrsa.*

3a. Subsect. HEDRIOCARPAE. Standard yellow, sometimes flushed or finely veined reddish-brown, but only in older flowers. Pod with the upper side evenly tapered to an asymmetrically inserted stipe (unless globose), globose, ellipsoid, fusiform, obovoid or oblong-clavate. *Species* 62–65.

3b. Subsect. MACROSTACHYAE (Benth.) Bisby & Polhill. Standard always conspicuously and heavily purplish lined or flushed from inception. Pod subsessile or with a short medially inserted stipe, usually cylindrical or ellipsoid, sometimes globose. *Species* 66–92 and two cultivated species, *C. brevidens* and *C. longithyrsa.*

4. Sect. GENICULATAE Polhill. Herbs. Leaves 1–3-foliolate and petiolate or simple and subsessile; stipules small and narrow or lacking. Bracteoles variously inserted, small. Hypanthium not prominent. Calyx subequally 5-lobed or with the tube slightly protracted on the lower side, usually much shorter than the keel. Standard yellow, often lined reddish or purplish, or sometimes bluish, with lamelliform, pyamidal or columnar appendages at the base of the blade inside; keel bent at right angles in the lower third, with a long narrow untwisted beak. Style geniculate, constricted at the bend, with hairs on the inner side or on both sides. Pod ssessile to shortly stipitate, oblong-obovoid to cylindrical, abruptly contracted at either end, less commonly ellipsoid-oblong or subglobose. *Species* 93–108.

5. Sect. CALYCINAE Wight & Arn. Herbs. Leaves usually simple and shortly petiolate; stipules usually present, sometimes decurrent. Bracteoles often linear-lanceolate and directed forwards. Hypanthium not prominent. Calyx deeply divided into 2 lips, usually ± as long as the calyx (a little less in *C. juncea*). Standard usually yellow and unlined, with lamelliform, cupular or irregularly obpyramidal appendages restricted to the blade inside; keel usually angular or subangular, with a nearly straight beak usually twisted through 90° or more. Style usually geniculate with a single spiral of hairs in Africa. Pod sessile or stipitate, variously shaped. *Species* 109–111 and one cultivated species, *C. paulina.*

6. Sect. CROTALARIA. Shrubs or herbs. Leaves simple or 3-foliolate; stipules narrow, rarely expanded or lacking. Bracteoles generally small. Hypanthium not very prominent except in largest flowered species. Calyx with the tube protracted as a lip on the lower side, shorter than the corolla. Standard usually yellow, sometimes finely lined reddish, or blue, with appendages restricted to the blade; keel rounded or angular, with a twisted beak in Africa, subglabrous to lanate-ciliolate especially along the upper margin. Style rounded or geniculate, usually pubescent along both sides. Pod obovoid to oblong-ellipsoid or oblong-clavate, rarely ± ellipsoid, generally distinctly stipitate. *Species* 112–139 and two cultivated species, *C. spectabilis* and *C. verrucosa.*

6a. Subsect. CROTALARIA. Keel rounded at or a little below the middle. Style curved, attenuate. Pod usually oblong-clavate and distinctly stipitate in native African species. *Species* 112–116 and two cultivated species, *C. spectabilis* and *C. verrucosa.*

6b. Subsect. LONGIROSTRES (Benth.) Polhill. Keel bent at right angles in the lower half, with a narrow practically straight beak. Style geniculate, constricted at the bend. Pod oblong-clavate, oblong-obovoid or oblong-ellipsoid to subglobose, subsessile to distinctly stipitate. *Species* 117–139.

7. Sect. DISPERMAE Wight & Arn. Herbs or small shrubs. Leaves 1–3-foliolate (often both on the same plant); stipules generally absent. Bracteoles narrow. Hypanthium negligible. Calyx with the tube slightly protracted on the lower side, much shorter than the corolla. Standard usually yellow, variously marked, with small appendages at the base of the blade; keel angular or subangular, with a narrow twisted beak. Pod subglobose to shortly oblong-ellipsoid, mostly 2–12-seeded. *Species* 140–208.

Some species are found in cultivation but have not yet become naturalized. These are described below and keyed out in the main key.

Crotalaria brevidens Benth. in Hooker, London J. Bot. **2**: 585 (1843). —Polhill, Crotalaria Africa & Madagascar: 189 (1982). —Lock, Leg. Afr. Check-list: 169 (1989). Type from Sudan.

Var. **intermedia** (Kotschy) Polhill in Kew Bull. **22**: 266 (1968); Crotalaria Africa & Madagascar: 190 (1982). Type from Sudan.

Crotalaria intermedia Kotschy in Sitz.-Ber. Math.-Nat. Akad. Wiss. Wien **50**, Abt. 1: 362, t. 3 (1865).

Similar to 68. *C. ochroleuca*, but calyx 5–8 mm long, puberulous, standard usually brighter yellow; keel 1.7–2.4 cm long; pod (3.5)4–5 × 0.5–0.7 cm, narrowly cylindrical.

Zimbabwe. N: Mazowe Distr., Archie Henderson Research Station, fl. & fr. 1.ii.1953, *Wild* 4001 (K; LISC; SRGH). W: Matobo Distr., Matopos Research Station, *Daniels* 6 (SRGH). C: Marondera, Grasslands Research Station, fl. 20.iii.1965, *Corby* 1295 (K; SRGH).

Native of tropical Africa from Nigeria to Ethiopia and south to Tanzania, introduced in trials for use as a green manure crop.

Crotalaria longithyrsa Baker f. in J. Linn. Soc., Bot. **42**: 349 (1914). —Wilczek in F.C.B. **4**: 166 (1953). —Polhill, Crotalaria Africa & Madagascar: 195 (1982). Type from Dem. Rep. Congo (Katanga).

Var. **latifolia** R. Wilczek in Bull. Jard. Bot. État **23**: 150 (1953); in F.C.B. **4**: 167 (1953). —Polhill, Crotalaria Africa & Madagascar: 196 (1982). Type from Dem. Rep. Congo (Yangambi).

Similar to 72. *C. subcapitata*, but with larger flowers and pods; the cultivated variant altogether more robust. Leaflets lanceolate to elliptic.

Zimbabwe. N: Mazowe Distr., Archie Henderson Research Station, fl. 1.ii.1953, *Wild* 4018 (K; SRGH).

Native to Dem. Rep. Congo and southern Sudan, introduced in trials for use as a green manure crop.

Crotalaria micans Link, Hort. Berol. **2**: 228 (1833). —Polhill, Crotalaria Africa & Madagascar: 371 (1982). —Lock, Leg. Afr. Check-list: 191 (1989). Type from Central America.

Crotalaria anagyroides Kunth, Nov. Gen. Sp. Pl. **6**: 404 (1824). Type from Venezuela.

Shrub 1–3 m tall. Leaves 3-foliolate; stipules linear or lacking; leaflets 3.5–10 × 1–4.5 cm, elliptic, appressed pubescent beneath. Racemes 15–30 cm long; bracts linear, usually caducous. Calyx 0.8–1.2 cm long, densely appressed pubescent. Standard ovate-circular, bright yellow, marked blackish and later reddish at centre; keel 1.3–1.5 cm long, shorter than the broad wings, abruptly rounded about the middle, with a fairly short slightly incurved obtuse beak, lanate towards the upper edge. Pod 3–4 × 1–1.4 cm, stipitate, cylindrical, pubescent.

Zimbabwe. C: Marondera Distr., Grasslands Research Station, fl. & fr. immat. 15.iv.1965, *Corby* 1304 (K; SRGH). E: tea estates near Nyamingura R., fl. & fr. 3.iv.1958, *Phipps* 1220 (BR; K; LISC; SRGH). **Mozambique**. M: Umbelúzi, fl. & fr. 19.viii.1949, *Myre* 755 (LISC; LMA).

Native of tropical America, grown as a garden ornamental and as shade for young tea plants.

Crotalaria paulina Schrank in Pl. Rar. Hort. Monac.: t. 88 (1822). —Polhill, Crotalaria Africa & Madagascar: 371 (1982). Type: grown in Europe from seed collected in Brazil.

Erect glaucous herb or shrub 1.5–3 m tall, glabrous or nearly so except on the underside of the leaflets. Leaves simple; blade 8–16 × 2.5–6.5 cm, elliptic to obovate, long cuneate to the base, appressed pubescent beneath; petiole 5–8 mm long; stipules variably developed as a broad decurrent wing or small or lacking. Racemes lax, many-flowered; bracts (6)10–20 × (1.5)3–4 mm, shortly stalked, lanceolate, acuminate; bracteoles at base of calyx, similar. Calyx 1.5–2 cm long, deeply divided into 2 lips. Standard clear yellow; keel 1.6–2 cm long, rounded below the middle with a slightly incurved beak, twisted at the tip, lanate on the upper margin. Pods 4–6 × 0.8–1 cm, shortly stipitate, oblong-clavate, glabrous, many-seeded.

Zimbabwe. N: Mazowe Distr., Archie Henderson Research Station, fl. & immat. fr. 7.v.1952, *Wild* 3818 (K; SRGH).

Native of Brazil, introduced in trials for use as a green manure crop.

Crotalaria spectabilis Roth, Nov. Pl. Sp.: 341 (1821). —Senn in Rhodora **41**: 326 (1939). —Polhill in F.T.E.A., Leguminosae, Pap.: 959 (1971); Crotalaria Africa & Madagascar: 373 (1982). Type from India.

Crotalaria sericea Retz., Observ. Bot. 5: 26 (1758). —Wilczek in F.C.B. **4**: 88 (1953) non Burm.f. (1768). Type presumably from India.

Similar to 116. *C. retusa*, but branches subglabrous, stipules obliquely ovate-oblong, bracts 8–20 × 5–10 mm, cordate, and pods a little fatter.

Zimbabwe. N: Mazowe Distr., Archie Henderson Research Station, fl. & fr. 1.ii.1953, *Wild* 4012 (K; SRGH). **Malawi**. S: Zomba, fl. 28.i.1982, *Berrie* 309 (K). **Mozambique**. Z: Lugela Distr., Namagoa Estate, fl. & fr. viii.1944, *Faulkner* Kew 33 (K). GI: Namaacha Distr., Umbelúzi, Estação Experimental, fl. & fr. 19.viii.1949, *Myre* 757 (LISC; LMA).

Native of tropical Asia, introduced in trials for use as a green manure crop.

Crotalaria verrucosa L., Sp. Pl.: 715 (1753). —Polhill in F.T.E.A., Leguminosae, Pap.: 959 (1971); Crotalaria Africa & Madagascar: 373 (1982). Type from Sri Lanka.

Erect annual to 1 m, branches 4-angled, sulcate, sometimes flexuous, pubescent. Leaves simple; blade 5–13 × 3.5–8 cm, ovate to elliptic, thinly pubescent on both surfaces; petiole 3–9 mm long; stipules 7–20 × 5–13 mm, ovate-falcate, acuminate. Racemes sublaxly 12–24-flowered; bracts 1–5 mm long, lanceolate-caudate. Calyx 7–11 mm long, thinly pubescent; lobes acuminately triangular, ± twice as long as the tube. Standard pale yellow, tinged blue; wings longer than the keel, blue or purple; keel 8–12 mm long, rounded, with a slightly incurved twisted beak. Pods 3–5 × 0.7–1.2 cm, very shortly stipitate, oblong-clavate, hairy, c. 12–20-seeded.

Zimbabwe. N: Mazowe Distr., Archie Henderson Research Station, fl. & fr. 1.ii.1953, *Wild* 4006 (K; LISC; SRGH).

Native of tropical Asia, introduced as a green manure crop, readily naturalized and widespread in the tropics.

Artificial key to species

To facilitate identification for the Flora Zambesiaca area, the species are keyed here by a series of artificial groups. It is generally necessary to determine whether the beak of the keel is spirally twisted or not. The twist may be seen quite easily by observing the course of the suture and veins with a ×10 lens. In species 109–208 the line of union between the keel-petals runs spirally up the beak from base to tip through 180–360° or more, see TAB. 3,7: **21**, fig. 7. This is generally, but not always, correlated with a slight extension of the calyx-tube on the lower side (TAB. 3,7: **21**, fig. 1) or with a ± 2-lipped calyx (TAB. 3,7: **21**, fig. 2), the development of hairs on both sides of the style (TAB. 3,7: **21**, fig. 12) and an upward shift of the standard appendages from the claw to the base of the blade (TAB. 3,7: **21**, fig. 9 & 10), modifications which presumably increase the economy of pollen transfer. Inevitably a few species show intermediate stages, as in TAB. 3,7: **21**, fig. 5, where the very short beak is circumflexed through no more than 90° or where the keel is very small, and these anomalies are allowed for in the keys. TAB. 3,7: **21** also shows other criteria used in the key and the conventions for measuring lengths of the various parts. Although this is

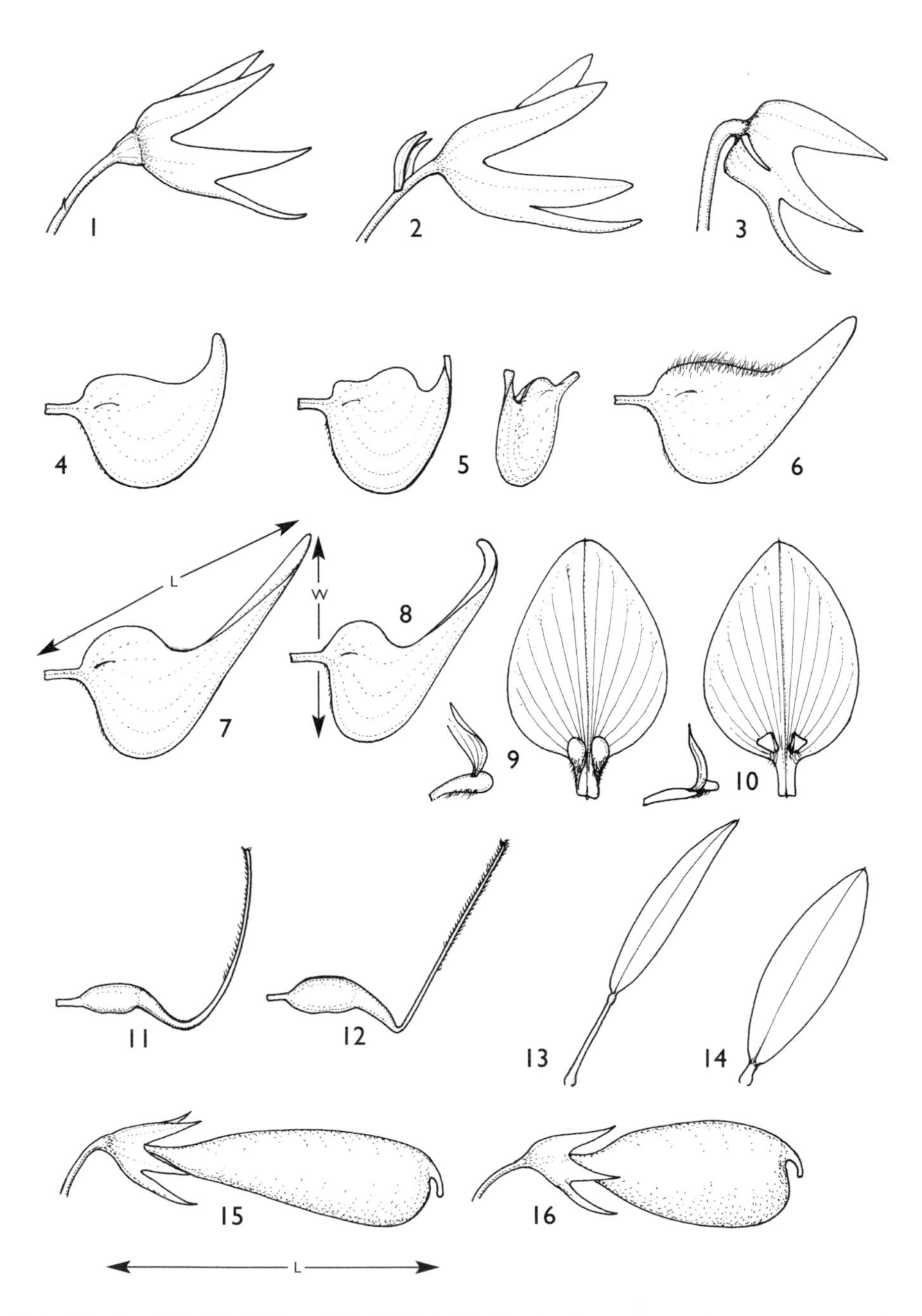

Tab. 3,7: **21**. CROTALARIA —KEY CHARACTERS. 1, calyx with prominent receptacle and tube protracted on the lower side; 2, bracteoles curved, ascending from the pedicel, and calyx 2-lipped; 3, calyx basally truncate and deflexed against the pedicel; 4, keel rounded; 5, keel crested behind the short circumflexed beak; 6, keel subangular, lanate towards the upper margin; 7, keel angular, with a narrow twisted beak; criteria for measuring length and breadth; 8, tip of beak strongly incurved, almost hooked; 9, standard with appendages extending on to the claw; 10, appendages restricted to the blade; 11, gynoecium, with curved tapered style hairy on the inner side; 12, style geniculate, hairy on both sides; 13, 1-foliolate leaf; 14, simple leaf; 15, pod oblong-clavate; criteria for measuring length (stipe included); 16, pod ellipsoid-clavate. Drawn by Roger Polhill. From Crotalaria Africa & Madagascar.

the largest genus of the family in the Flora Zambesiaca area, most of the species are fairly easily distinguished by readily visible features. Some difficulty will be found, however, with section *Dispermae* Wight & Arn., species 140–208, a section which is exceptionally well developed in the Flora Zambesiaca area. For these numerous, often diminutive, species it is important to collect rootstocks or at least note the habit carefully and try to collect pods as well as flowers.

Key to groups

1. Leaves all or almost all simple or 1-foliolate ········· **Group I** (page 87)
– Leaves all 3–5-foliolate or only the upper ones 1-foliolate ········· 2
2. Beak of the keel spirally twisted through 180° or more (use ×10 lens and observe course of suture and veins, see TAB. 3,7: **21**, fig. 7 & 8) ········· 3
– Beak of the keel untwisted or circumflexed through no more than 90° ········· 4
3. Stipules present ········· **Group G** (page 80)
– Stipules (not to be confused with bracts of axillary flowers) absent ·· **Group H** (page 82)
4. Leaves sessile or subsessile (petiole less than 1 mm long) ········· **Group A**
– Leaves petiolate ········· 5
5. Bracts fused to the pedicel, the free part spreading from just below the calyx, 3-fid (TAB. 3,7: **29**) ········· 60. *ononoides*
– Bracts not fused to the pedicel, simple ········· 6
6. Calyx lobes 4, ovate, ± enveloping the corolla and pod, drying blackish ···· 61. *nigricans*
– Calyx lobes 5 or sometimes seemingly 3 due to fusion of upper and lateral lobes on either side ········· 7
7. Stipules resembling small leaflets, very shortly stalked and expanded above into a narrow or broad blade-like portion, sometimes compound ········· **Group B** (page 74)
– Stipules not foliaceous, sometimes absent ········· 8
8. Keel abruptly bent ± at right angles in the lower half, with a practically straight beak (TAB. 3,7: **21**, fig. 6 & 7) ········· **Group C** (page 75)
– Keel rounded about the middle or a little below, the beak variously developed and orientated ········· 9
9. Keel ± half round, the upper margin crested behind the small beak (TAB. 3,7: **21**, fig. 5) ········· **Group D** (page 76)
– Keel not crested, without a marked sinus between the proximal part of the keel and the beak ········· 10
10. Calyx of mature flowers and fruits basally ± truncate and sharply deflexed against the pedicel (TAB. 3,7: **21**, fig. 3) ········· **Group E** (page 76)
– Calyx not becoming basally truncate and deflexed against the pedicel ·· **Group F** (page 78)

Group A

Leaves 3-foliolate, sessile or nearly so; beak of keel not twisted.

1. Stipules present; keel strongly rounded, crested behind the short beak (TAB. 3,7: **21**, fig. 5) ········· 2
– Stipules absent; keel not crested ········· 3
2. Leaflets flat or with the margins only slightly inrolled, 3–18 mm wide; standard glabrous outside except sometimes along the midvein ········· 38. *gazensis*
– Leaflets with inrolled margins, 1–3 mm wide; standard pubescent near the apex outside ········· 39. *phylicoides*
3. Flowers 1–2 on filiform leaf-opposed peduncles along the branches (not terminal); standard usually white becoming flushed red outside, rarely yellow; keel 3–4 mm long (in the Flora Zambesiaca area); pod cylindrical, 1–1.6 cm long ········· 53. *leptoclada*
– Flowers terminal on main branches, often also on reduced lateral branches and sometimes in the axils or leaf-opposed near branch-tips; peduncle, if present, not much more slender than the supporting branch; standard yellow; keel larger or else pod subglobose ····· 4
4. Pod oblong-ellipsoid to cylindrical, (2.5)3–5 times as long as broad, 4–20-seeded; keel 5–12 mm long ········· 5
– Pod subglobose-ellipsoid to shortly oblong-ellipsoid, up to 1.5 times as long as broad, 2–4-seeded; keel 3.5–6 mm long ········· 8
5. Leaflets beneath with prominent pale ascending lateral nerves, the basal pair marginal almost to the tip ········· 52. *sylvicola*

– Leaflets beneath with the venation obscure or hidden by hairs, never with conspicuous marginal nerves · 6
6. Keel falcate or more strongly rounded about the middle, with a narrow slightly incurved beak; standard orange-yellow; leaflets acute, with a short silvery (drying golden) silky tomentum when young · 51. *kerkvoordei*
– Keel subangular to shortly rounded in the lower half, with a ± straight beak; standard clear yellow, sometimes marked reddish-brown; leaflets mostly blunter or less hairy · · · · · · · 7
7. Pod 1.6–2.5 cm long, spreading pilose (hairs mostly 1 mm long); leaflets rounded to the apiculate tip, not coriaceous, not glossy, with the basal nerves not running into the upper part; calyx usually ± as long as the keel, densely silky pubescent · · · · · · · · · · 49. *caudata*
– Pod 0.7–1.2 cm long, pubescent with hairs up to 0.5 mm long; leaflets obtuse, slightly coriaceous, glossy, with the basal lateral nerves running close to the margin nearly to apex; calyx usually shorter than the keel, puberulous · 50. *kwengeensis*
8. Flowers 8–16 in short dense terminal racemes; keel 5–6 mm long; hairs on leaflets 1–2 mm long · 54. *abscondita*
– Flowers 1–few in clusters terminal on main and short lateral branches and in axils; keel 3.5–5 mm long; hairs on leaflets rarely exceeding 1 mm · · · · · · · · · · · · · · · · 55. *cistoides*

Group B

Leaves 3-foliolate, petiolate, stipules foliaceous; beak of keel not twisted (very rarely circumflexed).

1. Main nerve of stipules ± median · 2
– Main nerve of stipules (or segments thereof) markedly excentric · · · · · · · · · · · · · · · · 5
2. Keel ± half-round, crested behind the small beak (TAB. 3,7: **21**, fig. 5), not or not much longer than the calyx, 1–1.4 cm long* · 3
– Keel not crested, 1.5–2 times as long as the calyx · 4
3. Standard, apart from markings, yellow; pod 2.5–3.8 cm long; shrub · · · · · · · · · 37. *goetzei*
– Standard, apart from markings, violet-blue; pod 4–8 cm long; annual · · · · · 43. *glaucifolia*
4. Keel (1.5)2–2.4 cm long, rounded on the lower side; pod 4.5–7 cm long, pubescent · 5. *capensis*
– Keel 1–1.3 cm long, subangular; pod 1.6–1.8 cm long, glabrous · · · · · · · · · · · · 18. *stolzii*
5. Ovary and pod glabrous except sometimes (in *C. rhodesiae*) along the upper side · · · · · 6
– Ovary and pod uniformly hairy · 9
6. Perennial; standard subcircular; seeds smooth · 7
– Annual; standard ovate or ovate-elliptic; seeds rugose · 8
7. Stems erect; leaves mostly in tufts; pods 3.4–4.5 × 1.1–1.5 cm, glabrous · · · · · 19. *natalitia*
– Stems trailing; leaves mostly well spaced; pods 2–2.7 × 0.8–1 cm, with at least some hairs along upper side · 20. *rhodesiae*
8. Stipules entire · 23. *podocarpa*
– Stipules divided into (2)3–4 segments · 24. *damarensis*
9. Pod appressed puberulous to thinly pubescent; standard ovate or ovate-elliptic; wings oblong, fractionally shorter than the keel · 10
– Pod densely covered with spreading hairs; standard subcircular; wings oblong-obovate, slightly exceeding the keel · 11
10. Calyx 6–10 mm long; pod cylindrical, 2.3–3 cm long, c. 18–26-seeded; petioles mostly 1–3 cm long · 21. *cylindrocarpa*
– Calyx 4–5 mm long; pod oblong-ellipsoid, (1.2)1.5–2 cm long, c. 12–16 seeded; petioles mostly 3.5–6.5 cm long · 22. *goreensis*
11. Keel 1.4–1.6 cm long; pod c. 2–2.5 cm long (not seen mature); calyx hairs not completely covering the surface · 25. *prittwitzii*
– Keel 1.8–2.6 cm long; pod 2.8–4.5 cm long; calyx tomentum completely obscuring the surface · 12
12. Hairs of calyx spreading, all or some tawny; upper and lateral calyx lobes on either side rather broadly oblong-triangular, somewhat abruptly narrowed to the apex, cohering at the tips in mature flowers; bushy herb or shrub 1–3 m tall · · · · · · · · · · · · · · 26. *lachnophora*
– Hairs of calyx appressed, silvery; upper and lateral calyx lobes attenuate-triangular, separate in mature flowers; stem single and often simple, up to 1 m tall · · · · · · 27. *grandistipulata*

---

* If stipules not distinctly stalked, see **Group D**, p. 76.

Group C

Leaves 3-foliolate, petiolate; stipules narrow; keel angular, with an untwisted beak.

1. Bracteoles linear to oblong-falcate, curved-ascending from the pedicel (TAB. 3,7: **21**, fig. 2), usually long-ciliate; keel with white wool along the upper margin (TAB. 3,7: **21**, fig. 6); calyx lobes long, rather parallel-sided proximally · 2
– Bracteoles on the calyx, or if on the pedicel then linear to setaceous, neither curved-ascending nor pilose; keel glabrous to ciliolate on the upper margin, occasionally inconspicuously pubescent on the sides; calyx lobes triangular or subulate* · · · · · · · · · 6
2. Leaflets linear-lanceolate to elliptic (not drying blackish); bracteoles usually more than 1 mm broad; standard obovate or subcircular · 3
– Leaflets obovate-elliptic, obovate or subcircular (drying blackish); bracteoles linear, up to 0.5 mm broad; standard elliptic · 5
3. Pod (in the Flora Zambesiaca area) ellipsoid-clavate, 1.5–2.6 cm long, glabrous at least on underside; standard bright yellow; keel 0.9–1.4 cm long · · · · · · · · · · · · · · · · · 15. *barkae*
– Pod oblong-clavate, (3.5)4–5.5 cm long, pilose; standard usually blue or white; keel 1.5–1.8 cm long · 4
4. Leaves all 3-foliolate; leaflets narrowly to broadly elliptic, the mature ones mostly 1.5–3.5 cm wide; standard blue; erect rather bushy annual; racemes mostly 12–20-flowered · 13. *polysperma*
– Leaves 3–5-foliolate; leaflets linear-lanceolate to lanceolate, c. 0.4–1.6 cm wide; standard white, sometimes tinged mauve or yellow, darker veined; perennial developing long decumbent lower branches; racemes 5–12(16)-flowered · · · · · · · · · · · · · · · · 14. *burkeana*
5. Wings nearly as long as the keel; petioles c. 3–8 cm long; racemes mostly 8–30(60)-flowered · 16. *incana*
– Wings much shorter than the keel; petioles c. 1–2 cm long; racemes 1–4-flowered · 17. *lotoides*
6. Stems ribbed, 0.6–2 m tall; calyx 4.5–6 mm long, with lobes shorter than the tube; pods (1.2)1.6–2.2 × 0.7–0.8 cm · 93. *flavicarinata*
– Stems terete or finely striate; calyx lobes usually longer than the tube; pods 0.5–1.6 cm long · 7
7. Standard blue or white inside, flushed reddish outside; leaflets chartaceous, dark green and glossy above (unless tomentose), silvery silky tomentose beneath; stems trailing; racemes subumbelliform; pods oblong-ellipsoid, dehiscent · · · · · · · · · · · · · · · · · · 56. *variegata*
– Standard yellow apart from markings; leaflets membranous, neither glossy above nor silvery tomentose beneath; if racemes subumbelliform on long trailing branches then pods subglobose and tardily dehiscent · 8
8. Standard pubescent outside, at least medially · 9
– Standard glabrous outside · 11
9. Petioles 2–4 mm long; racemes short, usually dense; pods 2–3 mm in diameter; seeds 1–1.2 mm long · 101. *microcarpa*
– Petioles mostly 10–30 mm long; racemes lax; pods 3.5–6 mm in diameter; seeds 2–3 mm long · 10
10. Keel 9–13 mm long, glabrous on the sides; pod (8)10–12 × 4.5–6 mm, c. 8–12-seeded; racemes 4–6(12)-flowered · 99. *heidmannii*
– Keel (4)5–7(9) mm long, pubescent on the sides; pod 5–7 × 3.5–4.5 mm, 1–4(6)-seeded; racemes 12–many-flowered · 100. *sphaerocarpa*
11. Racemes subumbelliform, long-pedunculate, few-flowered; pods short, thick-walled, tardily dehiscent, 1–3-seeded · 12
– Racemes with an elongate rhachis; pods thin-walled, readily dehiscent, (10)14–24-seeded · 13
12. Pods laterally compressed, discoidal (more inflated with age), 7–12 × 3.5–6 × 6–9 mm, rugose; keel normally 7.5–12 mm long, but sometimes smaller in late season growth; calyx lobes 1.7–3 times as long as the tube · 58. *reptans*
– Pods subglobose, 5–6(7) × 4–6 × 4.5–6 mm, smooth or only slightly rugulose when old; keel 4.5–6(7.5) mm long; calyx lobes 1.3–2 times as long as the tube · · · · · · · · · · 59. *pisicarpa*

---

* If keel lanate near the upper margin, bracteoles caducous and stipules all expanded, re-check **Group B**

13. Calyx 0.7–1 times as long as the keel; racemes 2–5-flowered; pods usually ellipsoid-oblong, covered with hairs 0.5–1 mm long · 57. *lasiocarpa*
– Calyx ± half as long as the keel; racemes mostly more than 6-flowered; pods usually subglobose-ellipsoid (sometimes ellipsoid-oblong), covered with hairs 0.1–0.3 mm long · 85. *steudneri*

Group D

Leaves all or mostly all 3-foliolate, at least shortly petiolate; stipules narrow or lacking; keel ± half-round, crested behind the small beak; beak untwisted or circumflexed through no more than 90°.*

1. Standard blue, apart from markings · 2
– Standard yellow, apart from markings · 3
2. Inflorescence umbelliform; prostrate perennial; keel 7.5–10 mm long, without an ancillary spur · 36. *umbellifera*
– Inflorescence laxly racemose; erect annual; keel 11–14 mm long, with an oblique lateral spur from the leading edge · 43. *glaucifolia*
3. Annual; keel 4–5 mm long; pod 1.2–1.3 cm long, subcylindrical, 8–12-seeded · 42. *modesta*
– Perennial; keel 6–16 mm long · 4
4. Stems prostrate, keel 1.2–1.4 cm long; leaflets silky tomentose beneath · · · 35. *criniramea*
– Stems erect or if only weakly ascending (*C. gazensis* subsp. *herbacea*) then flowers smaller and leaves less hairy · 5
5. Standard tomentellous outside, the hairs obscuring the surface at least of the upper half · · 6
– Standard glabrous to pubescent outside, but hairs not wholly obscuring the surface · · · · 7
6. Calyx 6–8 mm long, ± half as long as the keel, with short oblong obliquely truncate upper lobes · 33. *tristis*
– Calyx 10–12 mm long, ± as long as the keel, with long narrowly attenuate-triangular lobes · 34. *amoena*
7. Leaflets elliptic, rarely obovate-elliptic, mostly more than 2 cm wide, discolorous, with a conspicuous golden closely appressed pubescence beneath; pods 1.8–2.2 cm long; keel 9–11 mm long · 32. *chrysotricha*
– Leaflets not as above; pods and keel various (see below) · 8
8. Pods 20–30 mm long, c. 8–14-seeded; calyx 7–13 mm long; racemes 1–6-flowered · · · · · 9
– Pods 7–11 mm long, 2–6-seeded; calyx 4–7 mm long; racemes often more than 6-flowered · 10
9. Keel 11–14 mm long; petioles mostly 15–45 mm long; standard pubescent outside, at least medially and apically · 37. *goetzei*
– Keel 7–10 mm long; petioles 0.5–16 mm long; standard glabrous outside except sometimes along the midvein · 38. *gazensis*
10. Bracts 2–6 mm long; calyx nearly as long as the keel · 40. *dura*
– Bracts 0.5–1.5 mm long (in the Flora Zambesiaca area); calyx ± two-thirds as long as the keel · 41. *huillensis*

Group E

Leaves 3–5-foliolate, petiolate; stipules small or lacking; bracteoles inserted on the calyx; calyx becoming truncate and deflexed against the pedicel, rarely much more than half as long as the corolla; petals usually strongly reddish-purple veined; keel rounded, with an untwisted beak.

1. Bracts and bracteoles caducous before the flowers fully open · 2
– Bracts and bracteoles persistent · 3
2. Pods 3.8–5 cm long; keel usually lined with purple, but without the colour diffused; annual or short-lived perennial · 66. *pallida*
– Pods 2.8–3.5 cm long; keel usually flushed purple, rarely wholly yellow-green; perennial, developing a woody rootstock (but also flowering in first year of growth) · · · · 67. *rogersii*

*Allied species with the keel-crest undeveloped come in the first part of the **Group F** key.

3. Bracteoles shorter than the calyx tube ································· 4
– Bracteoles longer than the calyx tube, mostly exceeding 2 mm ················ 15
4. Keel 17–24 mm long, shortly rounded with a rather long projecting beak; pods (4)5–7 cm long ································· 5
– Keel 4–14 mm long; pods up to 4.5 cm long ································· 6
5. Pods (12)15–20 mm in diameter; calyx glabrous; corolla generally cream-coloured or pale yellow ································· 68. *ochroleuca*
– Pods 5–7 mm in diameter; calyx puberulous (in the Flora Zambesiaca area); corolla generally bright yellow ································· *brevidens* (cultivated)
6. Beak of the keel attenuate, sharp, usually incurved (not in *C. lanceolata* subsp. *prognatha*), sometimes exceeding the wings ································· 7
– Beak of the keel short, broadly tapered, not or scarcely incurved, enveloped by the wings; keel 4–6(7) mm long ································· 12
7. Stipules absent ································· 8
– Stipules present (sometimes inconspicuous) ································· 10
8. Keel 5–9 mm long with an incurved beak or 9–11 mm long with a projecting beak; pods (1.6)1.8–3.8 × 0.4–0.6 cm ································· 71. *lanceolata*
– Keel 12–14 mm long, with a slightly incurved beak ································· 9
9. Pods (3.2)3.5–4.5 × 0.7–1.2 cm; bracts (1)2–4 mm long; leaflets very variable, but mostly lanceolate to elliptic-oblong, 1–4 cm wide ································· 69. *trichotoma*
– Pods 2.4–3 × 0.5–0.6 cm; bracts 0.5–2 mm long; leaflets narrow, mostly linear to linear-lanceolate, 0.4–1 cm wide ································· 70. *kirkii*
10. Pod ellipsoid, ± twice as long as broad, 2–4-seeded; buds reflexed almost from inception, not forming a tail above the flowers; petals bright yellow, unlined ·········· 63. *collina*
– Pod cylindrical, 4–6 times as long as broad, 20–30-seeded; buds ascending to spreading before the pedicel elongates, often forming a well developed tail above the open flowers; petals usually heavily purplish lined ································· 11
11. Pod 2.8–3.2 cm long; keel (10)11–12 mm long ·············· *longithyrsa* (cultivated)
– Pod 1.2–2.4 cm long; keel 5–11 mm long ································· 72. *subcapitata*
12. Pods subglobose, felty tomentose, 1–4-seeded; stipules 2–4.5 mm long, nearly as long as the petiole, leaflets pilose above ································· 80. *abbreviata*
– Pods oblong-ellipsoid to subcylindrical, pubescent, 10–40-seeded; stipules minute or lacking; leaflets glabrous or rarely thinly puberulous above ································· 13
13. Perennial, with a woody rootstock; leaflets all similar, elliptic to obovate ·· 75. *chrysochlora*
– Annual; leaflets, or at least the upper ones, mostly linear-lanceolate to oblong-lanceolate ································· 14
14. Pods 1–1.6 × 0.4–0.6 cm; lower branches often long, curved-ascending; rhachis longer than the peduncle (in the Flora Zambesiaca area) ································· 73. *onobrychis*
– Pods 2–2.5 × 0.6–0.8 cm; branches all ascending, the lower ones not longer nor curved; rhachis usually shorter than the peduncle ································· 74. *bernieri*
15. Keel 12–17 mm long; pods (3.5)3.8–5(6) cm long; leaves 3–5-foliolate ···· 76. *cleomifolia*
– Keel 4–10 mm long; pods 0.6–3 cm long; leaves 3-foliolate (uppermost rarely 1-foliolate) ································· 16
16. Petioles 1.5–11 cm long; pod-hairs less than 1 mm long; leaflets glabrous above ····· 17
– Petioles 0.2–0.7 cm long; pods very densely covered with hairs in two storeys, the longer ones c. 2 mm long; leaflets pilose on both surfaces ································· 19
17. Stipules absent; keel 6–8 mm long, ± as long as the wings; pod ellipsoid-oblong, 1.2–1.6 × 0.5–0.7 cm, 10–16-seeded ································· 78. *comosa*
– Stipules present; keel 8–11 mm long, exceeding the wings ································· 18
18. Pods ellipsoid, 1.2–1.8 × 0.6–0.8 cm, 2–4-seeded; bracteoles 1.5–2.5 mm long; petals bright yellow, unlined ································· 63. *collina*
– Pods cylindrical, 2.4–3 × 0.6–0.7 cm, c. 20–26-seeded; bracteoles 4–9 mm long; petals yellow lined purple or reddish-brown ································· 77. *chirindae*
19. Pods 1.6–1.8(2) cm long, subcylindrical; keel 6.5–8.5 mm long, exceeding the wings ································· 79. *vasculosa*
– Pods 0.6–0.8 cm long, subglobose; keel 4–5 mm long, shorter than the wings ································· 80. *abbreviata*

Group F

Leaves all or mostly 3-foliolate, petiolate; stipules narrow or lacking; calyx not deflexed against the pedicel; keel rounded with the beak untwisted or circumflexed through no more than 90°, rarely more twisted in a corkscrew manner (*C. orthoclada*).

1. Calyx a little shorter to a little longer than the keel, the lobes 2–4 times as long as the tube* 2
– Calyx ± half as long as keel or if up to two-thirds as long then lobes less than twice as long as the tube 9
2. Leaflets less than 2(3) times as long as the petiole 3
– Leaflets more than 4 times as long as the petiole 5
3. Keel 20–24 mm long; pod 4.5–7 cm long ..... 5. *capensis*
– Keel 9–16 mm long; pod 2–4 cm long 4
4. Pods less than twice as long as broad; seeds flat, conspicuously rugose; keel (12)14–16 mm long, with a projecting beak, annual ..... 10. *ulbrichiana*
– Pods more than 3 times as long as broad; seeds tumid, only slightly rugulose; keel 9–12 mm long, with a short incurved beak; shrubby ..... 12. *goodiiformis*
5. Stipules absent; leaflets 1.5–2.5(4) × 0.4–0.7(1.3) cm, oblanceolate, pubescent beneath; pods narrowly cylindrical, 7–9 mm across, covered in hairs c. 1 mm long; flowers up to 10 mm long in lax racemes ..... 48. *orthoclada*
– Stipules present (occasionally masked by indumentum); leaflets longer, pilose, often on both surfaces; pods shortly and broadly cylindrical to oblong-obovoid, 10–15 mm across, covered in hairs c. 2 mm long; flowers larger or crowded 6
6. Standard outside and sides of keel glabrous; keel 7.5–10 mm long; pods c. 4–6-seeded ..... 31. *densicephala*
– Standard outside and keel at least thinly hairy; keel 10–18 mm long; pods 12–30-seeded 7
7. Keel covered by a dense silky indumentum apart from the upper margin and beak ..... 28. *valida*
– Keel with hairs only in a narrow band near lower margin 8
8. Calyx with a tomentum obscuring the surface; beak of the keel bluntly pointed ..... 29. *lachnocarpoides*
– Calyx pilose with hairs not obscuring the surface; beak of the keel narrow, acute ..... 30. *polytricha*
9. Keel 3.5–5.5 cm long; upper and lateral calyx lobes joined on either side almost to their tips ..... 1. *agatiflora*
– Keel up to 3.2 cm long; upper and lateral calyx lobes free or, when young, lightly coherent, but not united 10
10. Keel strongly rounded, almost half-circular, the upper margin practically straight, abruptly produced into a well developed subtruncate upwardly directed beak, 2–3.2 cm long 11
– Keel less strongly rounded, more gradually narrowed to the beak, up to 2.4 cm long 13
11. Pods hairy; wings longer than the keel; stipules present ..... 3. *insignis*
– Pods glabrous; wings 0.5–0.75 times as long as the keel; stipules lacking 12
12. Leaflets acuminate, ± elliptic; petioles not basally constricted, mostly shorter than the terminal leaflet; pods (in the Flora Zambesiaca area) broadly ellipsoid-clavate, 5.7–6.7 × 1.6–1.8 cm ..... 2. *becquetii*
– Leaflets usually obtuse, rounded or retuse, rarely acute, variously shaped; petioles basally constricted, mostly longer than the terminal leaflet; pods oblong-clavate to subcylindrical, 5.5–10.5 × 1–1.5 cm ..... 4. *laburnifolia*
13. Sides of keel with white wool towards upper margin; bracts and bracteoles caducous before flowers fully open ..... *micans* (cultivated)
– Sides of keel glabrous; bracts and bracteoles more persistent 14
14. Calyx basally narrowed into a prominent obconical 2–7 mm long hypanthium (TAB. 3,7: **21**, fig. 1), altogether 6–18 mm long; pods 2–7.5 cm long, with a stipe 4–16 mm long 15
– Calyx without a conspicuous hypanthium, 3–8 mm long; pods 0.8–3 cm long, with a stipe up to 3 mm long 21
15. Plants perennial (sometimes with annual shoots from a woody rootstock); seeds smooth; seed-bearing part of pod 2.5 or more times as long as broad; keel 14–24 mm long 16

---

*Keel always rounded about the middle, with the small beak incurved (except *C. ulbrichiana*), so if keel at all angular see **Group C**.

- Plants annual; seeds flattened, rugose; seed-bearing part of the pod less than twice as long as broad; keel 11–16 mm long · 20

16. Plants perennial with several erect stems from a woody rootstock, generally somewhat glaucous, the bark not cracking and peeling, the branchlets subglabrous to puberulous · · 17
- Plants shrubby, not glaucous, the branchlets generally densely hairy at first, then glabrescent (variable in *C. capensis*) · 18

17. Petioles slender, terete; keel 14–15 mm long; pods oblong-clavate, 3.8–4.5 × c. 1.1 cm, pubescent · 8. *assurgens*
- Petioles broadly grooved, often slightly winged; keel 17–20 mm long; pods ellipsoid-clavate, 4–7.5 × c. 1.5 cm, soon glabrescent apart from sutures and apex · · · · · · · · · · 9. *unicaulis*

18. Leaflets mostly 3–7 × 1.5–3 cm, elliptic or obovate, usually acute or acuminate in atypical estipulate plants keying here, mostly a little longer than the petiole; leaves all well spaced; bark not cracking and peeling · 5. *capensis*
- Leaflets mostly smaller or more oblong, rounded or retuse, mostly shorter than the petiole unless very small; some leaves often clustered on short shoots; bark often cracking and peeling · 19

19. Pods (3.5)4–5.5 cm long, with a 8–12 mm long stipe; keel 20–23 mm long; petioles mostly 1.7–3.5 times as long as the terminal leaflet · 6. *pallidicaulis*
- Pods 2–3(3.5) cm long, with a 6–8 mm long stipe; keel 16–20 mm long; petioles mostly 0.5–1.7 times as long as the terminal leaflet · 7. *monteiroi*

20. Branchlets and undersurface of leaflets conspicuously pubescent; calyx 10–14 mm long; keel with a projecting obtusely pointed beak; stipe of the pubescent pod 4–5 mm long · 10. *ulbrichiana*
- Branchlets and undersurface of leaflets sparsely puberulous and glabrescent; calyx 6–8 mm long; keel with an incurved narrow acute beak; stipe of the glabrescent pod 9–13 mm long · 11. *barnabassii*

21. Plants shrubby or with strong erect stems from a woody rootstock; racemes subdense (pedicels longer than interval between their insertions); corolla clear yellow or with only fine darker veins on the standard · 22
- Plants annual, often with long decumbent slender lower branches, or perennials with trailing or weakly ascending stems from a small rootstock; racemes lax (flowers inserted at intervals greater than a pedicel length); corolla unusually heavily veined purplish or reddish-brown* · 25

22. Pods oblong-clavate, 2.5–3 cm long, 16–20-seeded, appressed pubescent; bracteoles inserted on the calyx, obliquely oblong-elliptic; branches appressed silvery tomentellous · 65. *schlechteri*
- Pods globose, fusiform or ellipsoid, 0.8–2 cm long, 2–8-seeded, densely spreading pilose or tomentose; bracteoles inserted at the base of the calyx or just below, linear to narrowly triangular; branches tomentose or densely pilose with longish subappressed to spreading hairs · 23

23. Calyx 7–8 mm long, with lobes ± twice as long as the tube; pods sessile, ovoid-fusiform, 6–8-seeded; petioles up to 10 mm long · 62. *torrei*
- Calyx 3.5–5 mm long, with lobes shorter to a little longer than the tube; pods shortly stipitate, globose to ellipsoid, 2–4-seeded; petioles mostly longer · · · · · · · · · · · · · · · · 24

24. Pods ellipsoid, 1.2–1.8 cm long; stipules 2–10 mm long; bracteoles inserted at the base of the calyx, narrowly triangular to linear-lanceolate, 1.5–2.5 mm long · · · · · · · · 63. *collina*
- Pods subglobose-ellipsoid, 0.8–1.1 cm long; stipules up to 1 mm long or lacking; bracteoles inserted on the upper part of the pedicel, smaller · · · · · · · · · · · · · · · · · · · 64. *uncinella*

25. Inflorescences subglabrous; pods narrowly subcylindrical (0.8)1.2–1.5 × 0.35–0.4 cm; bracteoles inserted on the pedicel near the middle · · · · · · · · · · · · · · · · · · · 87. *nudiflora*
- Inflorescences conspicuously hairy; pods broader; bracteoles inserted at the base of the calyx or just below · 26

26. Plants perennial, trailing or weakly ascending; pods oblong-fusiform to subcylindrical, slightly more than twice as long as broad (mature ones unknown in *C. sp. A*) · · · · · · · 27
- Plants annual, usually with a shortly erect stem and long decumbent lower branches; pods subglobose-ellipsoid to ellipsoid, rarely oblong-ellipsoid, less than twice as long as broad · 28

* Most species with such corolla markings come in **Group E**. If flowers blue, subumbelliform see *C. umbellifera*.

27. Stems spreading hirsute; stipules 3–6 mm long, linear-lanceolate; leaflets mostly oblong-elliptic to elliptic-obovate, 7–15 mm wide · · · 83. *schinzii*
– Stems appressed pubescent; stipules 0.5–1 mm long; leaflets narrowly oblong-elliptic, 3–6 mm wide · · · 84. *sp. A*
28. Branches rather long-hispid with spreading hairs · · · 85. *steudneri*
– Branches thinly appressed pilose · · · 86. *preladoi*

Group G

Leaves all or mostly 3-foliolate; stipules (not to be confused with bracts if flowers develop in the axils) present, usually conspicuous; racemes usually mostly leaf-opposed (except *C. prolongata* and *C. cuspidata*), sometimes dimorphic, short determinate ones sometimes pungent; keel rounded to angular with a twisted beak; pod subglobose to oblong-clavate, 2–many-seeded.

1. Keel rounded about the middle or a little below (TAB. 3,7: **21**, fig. 4), 1.3–2 cm long; pod oblong-clavate to fatly ellipsoid, 3–6 × 0.8–1.8 cm · · · 2
– Keel bent ± at right angles in the lower half (TAB. 3,7: **21**, fig. 7); pods smaller (except sometimes in *C. deserticola*) · · · 5
2. Flowers 2–6(12) in the axils, clustered or very shortly racemose · · · 114. *axillaris*
– Flowers in well developed terminal or leaf-opposed racemes · · · 3
3. Ovary and pod glabrous; branches hollow, ribbed, not woody; petioles flattened, broadly and shallowly grooved · · · 115. *recta*
– Ovary and pod hairy; branches woody; petioles slender, terete · · · 4
4. Pods 4.5–6 × 1–1.3 cm, oblong-clavate; bracts 3–9 mm long, caducous; calyx lobes lanate along margins inside · · · 112. *tabularis*
– Pods 3–4.5 × 1.4–1.8 cm, fatly ellipsoid; bracts 1–2.5 mm long, subpersistent; calyx lobes inconspicuously puberulous along margins inside · · · 113. *macrocarpa*
5. Flowers in elongate indeterminate racemes ending in a bud, sometimes with isolated flowers in the axils · · · 6
– Flowers all or partly (then inflorescences of two distinct types) 1–4 on short determinate axes ending in a spine or bristle (scarcely apparent in *C. misella*), often on abbreviated shoots; calyx lobes triangular* · · · 22
6. Upper calyx lobes oblong-triangular or oblong-lanceolate to spathulate, ± abruptly contracted to an excentric apiculate tip (TAB. 3,7: **37**, figs. B–E; examine preferably in late bud) · · · 7
– Upper calyx lobes symmetrically triangular or attenuate-triangular, sometimes curved in bud · · · 11
7. Perennial, with decumbent or ascending usually velutinous (occasionally glabrous) stems; leaflets all broadly elliptic to obovate, rounded or retuse, 8–17 × 5–14 mm, usually broader than petiole length · · · 124. *dinteri*
– Annual, rarely perennating, usually erect and spreading, but a decumbent short-lived perennial in *C. sp. C*; leaflets more variable, mostly more elongate, mostly narrower than petiole length · · · 8
8. Lateral calyx lobes oblong to oblong-spathulate, broadly obtuse, rounded or retuse to the slightly excentric tip, overlapped by upper lobes in bud; pod 0.8–1.2 × 0.5–0.6 cm; buds becoming reflexed then spreading · · · 127. *platysepala*
– Lateral calyx lobes oblong-triangular to oblong-lanceolate, valvate; pod 1–2.4 × 0.5–1.4 cm, 6–28-seeded; flower-buds ascending · · · 9
9. Calyx 4.5 mm long; keel 8.5 mm long; biennial, stem shortly erect with many branches from near the base · · · 128. *sp. C*
– Calyx 5–8 mm long; keel generally 9–15 mm long (rarely smaller in *C. senegalensis*); erect annuals, branched above · · · 10
10. Pod (1.4)1.6–2.4 × (0.8)1–1.4 cm, c. 16–28-seeded; upper calyx lobes becoming attenuate as flower matures, glabrous or glabrescent along margins inside · · · 125. *laburnoides*
– Pod 1–1.7 × 0.5–0.7(0.8) cm, 6–16-seeded; upper calyx lobes with a persistent excentric tip, persistently lanate along the inner margin · · · 126. *senegalensis*

* If oblong-lanceolate, take alternative lead.

11. Stems numerous from a woody rootstock or (*C. sp. B*) at least several from a taproot that thickens and produces secondary roots from the upper part, trailing to erect; leaves mostly well-spaced, not in clusters ········· 12
– Stems single from a taproot (sometimes branched from base), often annual, if perennial and somewhat shrubby then leaves usually clustered along the branches ········· 15
12. Racemes lax, flowers inserted 1–3 pedicel lengths apart; bracts ± half as long as the pedicel or less; wings oblong-obovate, abruptly expanded in upper half; stems finely pubescent (hairs 0.2–0.5 mm) ········· 13
– Racemes with flowers (except at base) mostly less than 1 pedicel length apart (occasionally laxer in *C. subcaespitosa*); bracts mostly 0.5–1 times as long as the pedicel; wings gradually broadened upwards; hairs on stem under high magnification rather coarse and irregularly shaped, appressed and spreading in varying proportions, sometimes sparse ········· 14
13. Keel 9–12 mm long; pods 10–14 mm long; stems from a rhizomatous rootstock ·· 121. *friesii*
– Keel 8–8.5 mm long; pods 8–9 mm long; stems from a taproot ········· 122. *sp. B*
14. Stems ascending, inconspicuously hairy to naked eye; pod 9–15 × 7–10 mm, 8–16-seeded; bracteoles on the pedicel ········· 129. *subcaespitosa*
– Stems trailing, upcurved at tips, coarsely hairy; pod 7–9 × 4–5 mm, 6–8-seeded; bracteoles often (not always) at calyx base ········· 130. *inyangensis*
15. Pod 2-seeded, 5–8 mm long; uppermost leaves 1-foliolate; racemes terminal, with flowers clustered in the axils below; leaflets markedly apiculate-mucronate ········· 16
– Pod 6–40-seeded; leaves all 3-foliolate; racemes mostly soon leaf-opposed, without numerous flowers in the axils below; leaflets minutely apiculate ········· 17
16. Standard uniformly pubescent outside; petioles c. 15–40 mm long; leaflets up to 14–25 mm wide ········· 140. *prolongata*
– Standard pubescent only towards the apex outside; petioles c. 7–10(15) mm long; leaflets c. 5–13 mm wide ········· 141. *cuspidata*
17. Racemes dense, the flowers and pods ± obscuring the axis, the rhachis mostly shorter than peduncle; bracts 4–6 mm long, caducous before the flowers open; standard pubescent only near the apex outside ········· 119. *onusta*
– Racemes laxer, the rhachis generally exceeding the peduncle; bracts smaller and more persistent ········· 18
18. Pods oblong-clavate, narrowed to the stipe, 4–6 times as long as broad; standard glabrous outside except sometimes for minute hairs along midvein ········· 117. *deserticola*
– Pods oblong-obovoid, ellipsoid or cylindrical, subsessile or rounded to the stipe, rarely 4 times as long as broad; standard more extensively hairy outside (occasionally and atypically sparsely hairy) ········· 19
19. Pods 1.2–2.8 cm long, 14–40-seeded; calyx lobes 0.8–1.5 times as long as the tube (occasionally longer in Zambia, but then flowers large); keel 1.1–1.8 cm long in the Flora Zambesiaca area; annual; racemes lax ········· 120. *distans**
– Pods 0.7–1.5 cm long, 6–10(12)-seeded; calyx lobes 1.5–2.5 times as long as the tube or if shorter (*C. dedzana*) then shrubby perennial with short subdense racemes; keel 0.7–1.2(1.4) cm long ········· 20
20. Leaves mostly well spaced; leaflets glabrous above, usually with small rather regularly arranged ± well spaced hairs beneath, usually elliptic, sometimes oblong-elliptic, 2–4.5 × 0.6–3 cm ········· 123. *mocubensis*
– Leaves in clusters along the branches; leaflets usually thinly hairy above when young, softly pubescent with ± irregularly overlapping hairs beneath, mostly smaller, narrower or more obovate; petioles often mostly shorter ········· 21
21. Racemes 3–6(8) cm long, subdense, the flowers mostly less than 1 pedicel length apart; calyx lobes 0.8–1.5 times as long as the tube, inconspicuously hairy along the margins inside ········· 131. *dedzana*
– Racemes mostly longer and laxer; calyx lobes 1.5–2.5 times as long as the tube, lanate along the margins inside (examine preferably lower lobes in late bud) ········· 134. *virgulata*
22. Rhachides of supplementary inflorescences ending in a bristle ········· 23
– Rhachides spine-tipped ········· 26
23. Pods globose, 4 mm in diameter; keel 4.5–5 mm long; prostrate annual; racemes 2-flowered, all leaf-opposed ········· 135. *misella*

---

* First year plants of *C. friesii* from the Kafue Basin will key here; the calyx lobes are short as *C. distans* but the pods are fewer-seeded and slightly smaller.

– Pods obovoid to oblong-obovoid, 8–15 mm long, 6–14-seeded, keel 6–14 mm long; stems ascending or erect, supplementary inflorescence from leaf-clusters 24
24. Bracts on long inflorescences 0.5–4 mm long, distinctly shorter than the flowering pedicels or if subequal then at least flowers mostly inserted much more than 1 bract length apart (except sometimes in subsp. *longistyla*); hairs on calyx up to 0.5 mm long, often whitish, often subappressed; pods often glabrous inside . . . . 134. *virgulata*
– Bracts on long inflorescences 3–7 mm long, subequal to longer than flowering pedicel; pedicels mostly inserted less or little more than 1 bract length apart; hairs on calyx 1 mm long, tangled, often yellowish; pods thinly pilose inside 25
25. Leaflets elliptic, acuminate to the apiculate tip, glabrous above . . . . 132. *pilosiflora*
– Leaflets oblanceolate to obovate, apically rounded, pilose on both surfaces . . 133. *oocarpa*
26. Keel 15–18 mm long; pod 20–30 × (10)12–14 mm . . . . 136. *kapiriensis*
– Keel 4–12 mm long; pod 7.5–23 × 3.5–9 mm 27
27. Keel 8–12 mm long; pod 10–23 × 5–9 mm . . . . 137. *aculeata*
– Keel 4–6 mm long; pod smaller 28
28. Pod 7.5–9 mm long, oblong-obovoid, c. 6–8-seeded; wiry annual or short-lived perennial, with pungent 1(3)-flowered inflorescences from leaf-tufts orientated as general branching pattern . . . . 138. *spinosa*
– Pod 4–7 mm long, subglobose, 2–4-seeded; perennial bearing reduced leaves and 1–2 flowers and with slender erect stems bearing numerous slender pale spreading spines . . . . 139. *eremicola* subsp. *parviflora*

Group H

Leaves all or partly 3-foliolate, without stipules; racemes terminal (branches sometimes largely suppressed), often with other flowers in the axils below; keel ± angular, with a twisted beak; pod usually ovoid-globose to ellipsoid, sometimes shortly cylindrical, mostly 2–10-seeded.

Note: this is the largest and most difficult group in the Flora Zambesiaca area. Characters of habit and standard-indumentum used in the first dichotomies are generally reliable, but if there is any doubt the alternative leads should be tried. A number of further species are likely to be found. In general the pyrophytes are much more variable than the annuals.

1. Stems several to numerous from a woody or rhizomatous rootstock, ascending to erect . . **H 1**
– Stems single from a taproot unless (*C. baumii*) a small prostrate mat-forming annual . . 2
2. Stems or long lower branches prostrate, radiating; leaflets obovate to obovate-circular, rarely oblanceolate; racemes often subumbelliform, sometimes with 1–few flowers below the main head, or short and dense 3
– Stems erect or ascending, sometimes lower branches curved-ascending; other features not combined 4
3. Pod 7–9 × 3–3.5 mm, subcylindrical, 12–16-seeded; stem very short with long procumbent branches; petioles 3–12 mm long . . . . 207. *sertulifera*
– Pod 3–4 × 2.5–3 mm, ovoid-globose, 2-seeded; stems numerous; petioles 1–4 mm long . . . . 208. *baumii*
4. Standard outside glabrous or pubescent only along the midvein and near the apex outside . . . . **H 2** (p. 83)
– Standard extensively pubescent to tomentose outside . . . . **H 3** (p. 84)

H 1. Stems several to numerous from a perennial rootstock.

1. Flowers in leafy pseudopanicles, the terminal racemes short, most of the flowers on many short lateral flowering branches progressively abbreviated upwards, and also clustered in the axils; dried plants often with a strong sweetish smell 2
– Flowers mostly in terminal racemes, often a few in the axils below, rarely 1–few upper branches largely suppressed; dried plants not sweet smelling 4
2. Stems velvety tomentose with short distinctly spreading hairs; rootstock bright yellow inside . . . . 149. *kambolensis*
– Stems strigulose with short closely appressed hairs; rootstock whitish inside 3
3. Leaflets slightly thickened, the hairs inconspicuous even when dry; standard glabrous or with a few hairs at the apex outside; beak of keel only slightly incurved at very tip; ovules 2 . . . . 148. *florida*
– Leaflets membranous, the white hairs closely arranged and conspicuous against the surface (black when dry); standard sparsely hairy on the upper part outside; beak of keel very

narrow and distinctly incurved at the tip; ovules 4–8 · · · · · · · · · · · · · · 150. *alemanniana*
4. Bracts linear-lanceolate to lanceolate, often exceeding the pedicel; standard ± densely hairy outside (except sometimes in *C. luondeensis*); stem hairs mostly 0.8–2 mm long, spreading · 5
– Bracts linear-subulate, generally shorter than the pedicel; standard glabrous to extensively but shortly hairy outside; stem hairs mostly shorter, appressed to spreading · · · · · · · · · 7
5. Petioles 3–15 mm long, channelled to narrowly winged; standard oblong-obovate, 9–11 mm long; pod 1–2-seeded · 164. *lepidissima*
– Petioles 1–5 mm long, terete; standard broadly obovate, 7–8 mm long; pod 4–10-seeded · · 6
6. Keel slightly exceeding wings, beak almost straight, 7–8 mm long; pod 4–5 mm long, ovoid-globose, 4–6-seeded; leaflets glabrous above; bracts 2–4 mm long, linear-lanceolate · 154. *luondeensis*
– Keel slightly shorter than wings, beak slightly incurved, 6–7.5 mm long; pod 7 × 4 mm, oblong-ellipsoid; leaflets thinly pilose above; bracts 5–7 mm long, lanceolate · · · · 155. *lukafuensis*
7. Pod obliquely inserted, shortly cylindrical when ripe, 7–8 mm long*; standard elliptic, apically pubescent outside; keel 9–12 mm long; stems well branched, terete · · · 144. *argyrolobioides*
– Pod ovoid-globular, 3.5–7 mm long; standard usually elliptic-obovate to subcircular, glabrous to hairy; keel 4–10 mm long; stems usually simple to sparingly branched at flowering time, sometimes finely ribbed · 8
8. Plants glabrous overall · 187. *nuda*
– Plants hairy · 9
9. Leaflets often little developed at flowering time, linear-oblanceolate to oblanceolate, often folded lengthwise, usually acute or obtuse; petioles 1–3(10) mm long, narrowly channelled; standard usually flushed crimson-purple, usually pubescent outside · · 153. *pseudodiloloensis*
– Leaflets linear-oblanceolate to obovate, flat, often rounded to emarginate at apex; petioles 2–25 mm long, broadly channelled to distinctly winged; standard sometimes flushed reddish-purple with age, mostly finely lined red or wholly yellow · · · · · · · · · · · · · · · · 10
10. Standard glabrous outside (very rarely with a few hairs); keel 6–10 mm long; hairs at base of stem almost always spreading, even when appressed above · · · · · · · · · 151. *quangensis*
– Standard almost always pubescent outside; keel 4–5(7) mm long; stem-hairs appressed to subappressed, rarely more spreading near base · 11
11. Standard 6–9 mm broad, ± sparsely pubescent outside; flowers inserted several bract lengths apart; ovules 2–14 · 152. *graminicola*
– Standard 4–5 mm across, densely strigulose-puberulous over upper half; flowers inserted less than 1 pedicel length apart; ovules 2 · 198. *pudica*

H 2. Stems single from a taproot, ascending to erect. Standard outside glabrous or hairy only along the midvein and towards the apex outside.

1. Plants glabrous overall · 2
– Plant puberulous to pilose at least on branches and leaves · 3
2. Upper calyx lobes attenuate-triangular, 2–3 times as long as the tube; keel 7–9 mm long, markedly exceeding the wings; standard elliptic; petioles 2–10 mm long · · · · 186. *laxiflora*
– Upper calyx lobes triangular-acuminate, 1.5–2 times as long as the tube; keel (4)5–7 mm long, ± as long as the wings; standard obovate-circular; petioles all short, 0.5–2(4) mm long · 187. *nuda*
3. Calyx lobes ovate-lanceolate to ovate or oblong-triangular and truncate · · · · · · · · · · · · 4
– Calyx lobes triangular, narrowed from the base · 6
4. Lobes of calyx irregularly oblong-triangular, truncate, involute and lanate inside, brownish strigulose outside · 175. *duboisii* subsp. *mutica*
– Lobes of calyx ovate-lanceolate to ovate, pointed, flat, thinly white pubescent to subglabrous · 5
5. Keel 11–13 mm long; ovules 8; upper calyx lobes c. 4 mm broad · · · · · · · · 180. *eurycalyx*
– Keel 7–9 mm long; ovules 2; upper calyx lobes 2.5–3 mm broad · · · · · 166. *streptorrhyncha*
6. Beak of the keel strongly incurved, almost hooked, at the tip (TAB. 3,7: **21**, fig. 8) · · · · 7
– Beak of the keel straight to only slightly incurved at the very tip (TAB. 3,7: **21**, fig. 7) · · · 10
7. Bracts leafy, lanceolate-caudate, 3–7 mm long; pedicels 2–4 times as long as the subglabrous calyx; pod glabrous · 179. *peregrina*

---

*One specimen seen from Mozambique, Lichinga with smaller pods, see note under *C. argyrolobioides.*

– Bracts linear-subulate to narrowly linear-lanceolate, mostly shorter; pedicels 1–2 times as long as the hairy calyx; pod hairy · · · 8
8. Stems strictly erect, undivided or later with some short flowering branches above, the flowers in short racemes and then in axils all down the stem, sometimes accompanied by reduced leaves; stem densely silvery-white silky-pubescent with hairs 0.4–0.7 mm long; leaflets pointed; keel 9–11 mm long · · · 182. *phyllostachys*
– Stems well branched, at least above, the flowers often mostly in terminal racemes (if mostly axillary then flowers smaller); stem strigose-pubescent to pilose; leaflets obtuse, rounded or emarginate · · · 9
9. Keel 11–13 mm long; pod (2)6–8-seeded; flowers mostly in terminal racemes · · 178. *arcuata*
– Keel 7–9.5 mm long; pod 2-seeded; flowers terminal at first, developing in axils below and ultimately predominantly there · · · 181. *elisabethae*
10. Calyx lobes inside lanate, outside brownish strigulose with finer white hairs interspersed · · · 175. *duboisii*
– Calyx lobes not woolly inside · · · 11
11. Pod 7–8 mm long, 6–16-seeded; seeds exarillate; keel 9–12 mm long; standard elliptic, apically pubescent outside; leaves all 3-foliolate · · · 12
– Pod 2–5 mm long, 1–2-seeded; seeds usually with a prominent aril; keel 3–11 mm long, but if 8–11 mm long uppermost leaves generally 1-foliolate; standard usually broadest above the middle · · · 13
12. Branches generally with short appressed hairs; leaflets almost always glabrous above; ripe pods obliquely inserted, shortly cylindrical; calyx 4–6 mm long, with lobes (1)1.5–2.5 times as long as the tube · · · 144. *argyrolobioides*
– Branches spreading pilose; leaflets pilose on both surfaces; pods obliquely but sub-basally inserted, ± ovoid-ellipsoid; calyx 6–7 mm long, with lobes 3–4 times as long as the tube · · · 145. *annua*
13. Branches spreading pilose; leaflets pilose on both surfaces · · · 185. *bredoi*
– Branches with short appressed hairs; leaflets glabrous or nearly so above* · · · 14
14. Petioles all 0.5–1.5 mm long; flowers subumbellate with a few more laxly inserted flowers below; most plants with a ± spongy stem base, many long thin lateral roots, and short sparse indumentum · · · 188. *limosa*
– Petioles mostly 2–20 mm long; flowers not subumbellate above a short raceme; not obviously marsh plants · · · 15
15. Flowers few in short terminal racemes, most on suppressed lateral branches to give clusters of leaves and flowers all along the main branches; keel 6–7 mm long · · · 197. *microthamnus*
– Flowers all or mostly in terminal racemes or heads · · · 16
16. Keel 8–11 mm long; upper leaves 1-foliolate · · · 183. *gamwelliae*
– Keel 3–5 mm long; leaves all 3-foliolate · · · 17
17. Racemes very short and dense, up to 1(3) cm long in flower (elongating in fruit); pod 2.5–3.5 mm long; keel 3–4 mm long · · · 202. *parvula*
– Racemes mostly more than 3 cm long in flower; pod 3–4.5 mm long; keel 3.5–5 mm long · · · 18
18. Branching often from base and irregularly above, lower branches curved ascending; flowers 1–5 bract lengths apart, often irregularly spaced · · · 200. *hyssopifolia*
– Branches rather stiffly spreading-ascending from the strictly erect stem, progressively shorter upwards; flowers 4–10 bract lengths apart, with regular incremental spacing · · · 201. *tenuirama*

H 3. Stems single from a taproot, erect. Standard extensively hairy outside.

1. Bracts usually leafy, linear-lanceolate to lanceolate, (3)5–12 mm long, often shortly stalked, exceeding the pedicel and sometimes ± as long as the flowers and pods, sometimes (*C. axilliflora*) linear; hairs mostly 0.5–2 mm long, often twisted, silvery-white when fresh; standard ± densely silvery hairy outside, and if bracts unusually small then standard broadly obovate; petioles 1–6 mm long · · · 2
– Bracts (except occasionally lowermost) linear to narrowly linear-lanceolate, 1–4 mm long, not stalked, often shorter than the pedicel; other features not combined · · · 8
2. Standard 6–12 mm wide; pod 6–10-seeded · · · 3

*If distinctly hairy above and from N Malawi, see also *C. johnstonii*.

– Standard 3–5 mm wide; pods 2–6-seeded · 6
3. Flowers in heads, not in axils below; beak of keel only slightly incurved; style hairy on both sides · 155. *lukafuensis*
– Flowers in heads or racemes, with other flowers developing in the axils below; beak of keel strongly incurved (TAB. 3,7: **21**, fig. 8) or if only slightly so then style hairy only on inner side · 4
4. Bracts as long as flowers and pods; pedicels 1–2(3 in fruit) mm long; racemes sessile, dense, 3–12 cm long · 157. *axillifloroides*
– Bracts distinct shorter than the flowers and pods; pedicels 2–6 mm long · · · · · · · · · · · 5
5. Flowers in short dense racemes, the majority developing in axils below; standard-indumentum obscuring the surface outside; keel 6–8 mm long · · · · · · · · · 156. *axilliflora*
– Flowers in laxer racemes, mostly inserted more than 1 pedicel length apart, ultimately some in the axils below; red veins of standard generally apparent between the hairs · 158. *argenteotomentosa*
6. Racemes lax; standard rounded to retuse; beak of keel slightly incurved at the tip, shorter than the wings · 160. *subtilis*
– Racemes dense; standard obtuse; beak of keel straight, usually slightly exceeding the wings · 7
7. Pods 4–6-seeded; seeds without a conspicuous aril; leaflets rounded · · · · · 159. *subspicata*
– Pods 2-seeded; seeds arillate; leaflets pointed · · · · · · · · · · · · · · · · · · 161. *vandenbrandii*
8. Unifoliolate leaves numerous above · 9
– Unifoliolate leaves 0–1(few) at apex of each shoot · 10
9. Keel 5.5–6 mm long; pod 4.5–5 mm long, 2-seeded · · · · · · · · · · · · · · · · · 174. *kuiririensis*
– Keel 10–12 mm long; pod 6–9 mm long, 16–20-seeded* · · · · · · · · · · · · · · · 167. *ionoptera*
10. Flowers mostly in axils far down stem and branches, either in clusters or on suppressed flowering branches 1–few mm long, terminal racemes usually short or laxly few-flowered; beak of keel straight · 11
– Flowers mostly in terminal racemes, often with some flowers developing in the axils below; beak of keel straight or incurved at the tip** · 16
11. Axils with suppressed flowering branches mostly comprising a small head of flowers and several reduced leaves on a very short axis · 12
– Axils with sessile clusters of 1-several flowers with or without 1(several) reduced leaves · · 14
12. Keel 8–10 mm long; pod 6 × 4 mm, 6–8-seeded · 168. *crebra*
– Keel 4–5 mm long; pod 3–4 mm long, 2-seeded · 13
13. Calyx thinly hairy on the lobe surfaces but densely so in a fringe around the margins · 205. *praetexta*
– Calyx uniformly puberulous · 206. *ephemera*
14. Keel 6–7 mm long; standard thinly puberulous on upper part · · · · · · · 197. *microthamnus*
– Keel 4.5–6 mm long; standard densely hairy outside · 15
15. Pod 7–12-seeded; bracts 2–4 mm long, exceeding the pedicel · · · · · · · · · · · 203. *alexandri*
– Pod 2-seeded; bracts 0.5–1.5 mm long, shorter than the pedicel · · · · · · · · 204. *bequaertii*
16. Flowers in heads surrounded by crowded leaves larger than those below (TAB. 3,7: **41**); pod hairy inside on the upper suture behind the placenta, 2-seeded · · · · · · · · 143. *cephalotes*
– Flowers not in heads surrounded by an involucre of relatively large leaves (sometimes capitate with subtending leaves not enlarged); pod glabrous inside · · · · · · · · · · · · · · 17
17. Ovules 4–20 (seeds more than 2) · 18
– Ovules 2 (seeds 1–2) · 23
18. Keel beak distinctly incurved towards the tip; standard 8–12 mm wide; petioles 0–4 mm long · 19
– Keel beak straight or almost so; standard 4–6 mm wide; lower petioles mostly more than 4 mm long · 20
19. Upper leaves sessile; keel 7–9 mm long · 176. *bemba*
– Upper leaves shortly petiolate; keel 10–12 mm long · 177. *blanda*
20. Keel 10–12 mm long; standard elliptic, pointed; small annual, with flowers in heads · 169. *pygmaea*

---

* If intermediate at this dichotomy and standard only thinly hairy, see *C. sparsifolia* and particularly the note after that species.

**If plants only just beginning to flower and not found under this group return to alternative lead.

– Keel 3.5–8 mm long; standard elliptic-obovate to obovate · 21
21. Standard 7–8 mm long; pod 6–8 mm long; seeds arillate; pedicels 4–8 mm long · 146. *congesta*
– Standard 4–5 mm long; pod 4–5 mm long; seeds without a conspicuous aril; pedicels 1–2.5 mm long · 22
22. Flowers in lax racemes; pods 3–6-seeded; leaflets linear-oblanceolate to oblanceolate · 170. *sparsifolia*
– Flowers in heads or short racemes; pods 12–16-seeded; leaflets elliptic-oblanceolate to obovate · 172. *quarrei*
23. Keel shorter than the wings, much shorter than the broad silvery tomentellous standard, 4–5 mm long, with beak little longer than basal part (TAB. 3,7: **45**, fig. B); racemes very lax, flowers mostly 5–10 mm apart; calyx 3–3.5 mm long, with broadly triangular upper lobes, silvery tomentellous · 173. *filicaulis*
– Keel as long as or usually slightly longer than wings, ± as long as standard; other features not combined · 24
24. Calyx 5–7 mm long, with narrow lobes 2–4 times as long as the tube; flowers mostly in dense sessile racemes with short pedicels exceeded by the bracts; beak of keel distinctly incurved at tip · 25
– Calyx 2–5 mm long; other features not combined · 26
25. Stems and calyx spreading pilose; keel 9–11 mm long · · · · · · · · · · · · · · · · · 164. *lepidissima*
– Stems and calyx appressed pubescent; keel 6–8 mm long · · · · · · · · · · · · · · · 165. *kipilaensis*
26. Keel 7–9 mm long; hairs on branches and/or calyx fine, irregularly arranged* · · · · · · · 27
– Keel 3–6.5 mm long; hairs mostly shorter · 28
27. Flowers almost always in heads, rarely elongating, dense; hairs often brownish in part; leaflets glabrous or almost so above · 147. *nyikensis*
– Flowers in elongate lax to subdense racemes; hairs white; leaflets pilose above · · 185. *bredoi*
28. Petioles up to 3(4) mm long; leaflets linear-oblanceolate (rarely oblanceolate), 0.5–2 mm wide · 29
– Petioles mostly more than 3 mm long; some leaflets generally broader · · · · · · · · · · · · · 30
29. Terminal racemes lax; standard pubescent outside, but hairs not obscuring the surface; stem branching freely from near the base · 195. *carsonii*
– Terminal racemes forming short heads; standard tomentellous outside; stem with 0–few weak branches from near the base · 196. *confertiflora*
30. Petiole narrowly winged, broadened upwards; seeds without a conspicuous aril; flowers in small terminal racemes or heads with pedicels 1.2–2 times as long as calyx; very small annual with several–numerous curved-ascending branches from near the base (TAB. 3,7: **49**)** · 206. *ephemera*
– Petiole grooved but not winged; seeds with a conspicuous aril; flowers in racemes or if in heads pedicel ± as long as calyx · 31
31. Keel with tip of beak hardened and incurved, (5)5.5–6.5 mm long (TAB. 3,7: **45**, fig. C); numerous flowers developing in the axils; leaflets oblanceolate to elliptic-oblanceolate, 2–7 mm wide, pubescent with fine irregularly arranged hairs above · · · · · · · · ·184. *johnstonii*
– Keel with straight beak; 0–few flowers in the axils; leaflets above glabrous or with subappressed forwardly directed hairs · 32
32. Racemes very short and dense, to 1(3) cm in flower (elongating in fruit); pod 2.5–3 mm long; keel 3–4 mm long · 202. *parvula*
– Racemes mostly more than 3 cm long in flower; pod more than 3 mm long; keel 3.5–6 mm long · 33
33. Calyx ± appressed white pubescent; hairs of standard white, fine, not covering the surface completely; pod 3–4(4.5) mm long · 200. *hyssopifolia*
– Calyx tomentose with ± spreading hairs; hairs of standard silvery or golden-brown, often dense · 34
34. Pedicels 1–2 mm long, shorter than the bract; calyx lobes ± twice as long as the tube, silvery tomentellous; pod 4–5 mm long; leaflets hairy above · · · · · · · · · · · · · · · · · 171. *schliebenii*
– Pedicels mostly 2–4 mm long, exceeding the bract; calyx lobes c. 1–1.5 times as long as the tube, brownish tomentellous; pod 5–6 mm long; leaflets glabrous above · 199. *pseudotenuirama*

---

* If calyx hairs appressed and beak of keel markedly incurved at tip, see also *C. elisabethae*.
** If calyx with dense fringe of hairs on lobe margins see also *C. praetexta*.

Group I

Leaves all or almost all simple or 1-foliolate.

1. Beak of the keel untwisted or circumflexed through 90°; calyx subequally lobed; style hairy on the inner side only (except *C. cornetii*) · · · 2
– Beak of the keel spirally twisted through at least 120° (TAB. 3,7: **21**, fig. 7), if less than 180° then calyx deeply divided and at least slightly 2-lipped (TAB. 3,7: **21**, fig. 2) and style with a spiral row of hairs · · · 25
2. Keel rounded about the middle or a little below (TAB. 3,7: **21**, fig. 4) · · · 3
– Keel bent sharply ± at right-angles in the lower third (TAB. 3,7: **21**, fig. 7) · · · 14
3. Keel half-round, crested behind the small usually circumflexed beak (TAB. 3,7: **21**, fig. 5) · 4
– Keel not crested · · · 5
4. Standard blue to violet; keel with an ancillary spur on the leading edge, (8)9–11 mm long; leaves 1-foliolate (TAB. 3,7: **21**, fig. 13) · · · 44. *anisophylla*
– Standard pale yellow; keel without a spur, 6–8 mm long; leaves simple (TAB. 3,7: **21**, fig. 14) · · · 47. *glauca*
5. Keel 17–22 mm long; pod 40–75 mm long, cylindrical · · · 9. *unicaulis*
– Keel 3–16 mm long; pod 6–40 mm long, variously shaped · · · 6
6. Stems several–numerous from a perennial rootstock, angular; calyx 0.7–1 times as long as the keel · · · 7
– Stems single from a taproot, terete; calyx usually ± half as long as the keel · · · 8
7. Standard pale yellow lined reddish-purple; stipules present; leaflets acute, puberulous beneath, not drying black · · · 45. *adamsonii*
– Standard mauve to purplish; stipules absent; leaflets pointed to rounded at the apex, glabrous, venose, drying blackish · · · 46. *cornetii*
8. Pods subglobose; branches hirsute; racemes dense · · · 9
– Pods subcylindrical or oblong-obovoid; branches with small appressed or crisped hairs; racemes lax · · · 10
9. Branches decumbent; racemes subsessile, rarely more than 10 cm long; pod (1)3–6-seeded · · · 81. *anthyllopsis*
– Branches ascending; racemes pedunculate, mostly more than 10 cm long; pod 1–2-seeded · · · 82. *cylindrostachys*
10. Keel much longer than the wings, 5–9 mm long · · · 11
– Keel not or little longer than the wings, 3–6.5 mm long · · · 13
11. Leaves simple (petiole very short, not jointed at the top, TAB. 3,7: **21**, fig. 14; 3,7: **33**) · · · 88. *spartea*
– Leaves 1-foliolate (petiole 3–12 mm long, jointed at the top, TAB. 3,7: **21**, fig. 13)) · · · 12
12. Pod subsessile; bracteoles at the base of the calyx; hairs crisped on the branches · · · 89. *paraspartea*
– Pod with a stipe 2–3 mm long; bracteoles on the pedicel; hairs appressed on the branches · · · 90. *incompta*
13. Leaves petiolate, flat, estipulate; calyx 1.5–2 mm long · · · 91. *germainii*
– Leaves sessile, often involute, sometimes minutely stipulate; calyx 2.5–3.5 mm long · · · 92. *involutifolia*
14. Stems distinctly ribbed, angled or winged · · · 15
– Stems terete · · · 19
15. Standard bluish to violet inside; stems 3–4-angled, not or scarcely winged, strongly ribbed · · · 16
– Standard yellow or white, generally lined reddish-brown or purplish outside; stems winged or ribbed · · · 17
16. Keel 18–21 mm long; perennial · · · 95. *trinervia*
– Keel 8–11 mm long; annual · · · 96. *angulicaulis*
17. Stems narrowly winged, the wings 0.5–2 mm wide at leaf insertions, with stipules on outer points, decurrent into angle of the stem lower down, glabrous · · · 94. *stenoptera*
– Stems ribbed, but without wings or stipules, hairy on youngest parts · · · 18
18. Standard usually whitish inside, yellow veined reddish-brown outside; petioles mostly 6–18 mm long; pedicels 6–9 mm long in flower · · · 97. *orientalis*
– Standard bright yellow; petioles 0.5–4(6) mm long; pedicels 3–6 mm long · · · 98. *spartioides*
19. Leaves 1-foliolate, with petioles up to 2–4 mm long, pilose · · · 102. *bongensis*
– Leaves simple, subsessile, glabrous to pubescent · · · 20

20. Standard pubescent outside; keel 8.5–14 mm long · · · · · · · · · · 21
– Standard glabrous outside; keel 3–8 mm long · · · · · · · · · · 22
21. Beak of keel slightly incurved; pod with a stipe 3 mm long; leaves 1–2 mm wide · · · · · · · · · · 103. *simoma*
– Beak of keel straight; pod sessile; leaves mostly 2–6 mm wide · · · · · · · · · · 104. *vanmeelii*
22. Racemes c. 3–12-flowered; keel 6–8 mm long; pod 5–8 mm long · · · · · · · · · · 105. *decora*
– Racemes mostly 1–3-flowered; keel 3–6 mm long; pods various · · · · · · · · · · 23
23. Leaves mostly terete, sometimes linear-oblanceolate, less than 1 mm wide, glabrous; pod 6–8 mm long; keel 4.5–5.5 mm long · · · · · · · · · · 107. *teretifolia*
– Leaves linear or elliptic, some at least more than 1 mm wide, puberulous to pubescent beneath; keel and pods as next couplet · · · · · · · · · · 24
24. Pods 8–13 mm long; keel 4–6 mm long; lower leaves mostly 2–10 mm wide · · 106. *shirensis*
– Pods 6–7.5(8) mm long; keel 3–4 mm long; lower leaves mostly less than 1.5 mm wide · · · · · · · · · · 108. *minutissima*
25. Leaves simple (petiole not jointed at the top, see TAB. 3,7: **21**, fig. 14); racemes becoming leaf-opposed; pods oblong-clavate to subcylindrical · · · · · · · · · · 26
– Leaves 1-foliolate (TAB. 3,7: **21**, fig. 13), with the petiole jointed at the top (except *C. debilis*); racemes terminal, sometimes developing flowers in the axils below; pods ovoid-globose to rounded-ellipsoid, 3–11 mm long · · · · · · · · · · 33
26. Keel 5 mm long; pods 12–16 mm long · · · · · · · · · · 110. *occidentalis*
– Keel 10–22 mm long; pods 20–60 mm long · · · · · · · · · · 27
27. Ovary and pod glabrous · · · · · · · · · · 28
– Ovary and pod hairy · · · · · · · · · · 31
28. Calyx as long or longer than the keel, 2-lipped · · · · · · · · · · 29
– Calyx much shorter than the keel, subequally lobed · · · · · · · · · · 30
29. Plants densely hirsute; stipules small, obscured by hairs · · · · · · · · · · 109. *calycina*
– Plants glabrous apart from puberulence on underside of leaflets; stipules variably developed, often extending as wings down the internodes · · · · · · · · *paulina* (cultivated)
30. Bracts subulate to lanceolate-caudate; stipules linear to subulate · · · · · · · · · · 116. *retusa*
– Bracts ovate-acuminate; stipules expanded · · · · · · · · · · *spectabilis* (cultivated)
31. Stipules ovate-falcate; wings petals blue · · · · · · · · · · *verrucosa* (cultivated)
– Stipules small; wing petals yellow · · · · · · · · · · 32
32. Calyx 16–20 mm long, brownish tomentellous · · · · · · · · · · 111. *juncea*
– Calyx 5–7 mm long, strigulose · · · · · · · · · · 118. *miranda*
33. Stipules setaceous, 2–4 mm long; pod pubescent along the lower suture inside · · · · · · · · · · 142. *morumbensis*
– Stipules absent; pod glabrous inside · · · · · · · · · · 34
34. Plants silky pilose · · · · · · · · · · 35
– Plants glabrous to shortly pubescent · · · · · · · · · · 36
35. Keel 10–11 mm long; racemes sublax · · · · · · · · · · 162. *egregia*
– Keel 6–7 mm long; racemes dense · · · · · · · · · · 163. *kipandensis*
36. Plant glabrous; keel 11–12 mm long; pod stipitate, 8–11 mm long, many-seeded · · · · · · · · · · 190. *campestris*
– Plant at least minutely hairy; keel 4–9 mm long; pod sessile, 3–5 mm long, 2-seeded · · 37
37. Keel (6)7–9 mm long, 2.5–5 times as long as the calyx; pod 4–5 mm long · · · · · · · · · · 38
– Keel 4–6 mm long, c. 1.5 times as long as the calyx; pod 3–4 mm long · · · · · · · · · · 39
38. Beak of the keel strongly incurved at the tip (TAB. 3,7: **21**, fig. 8); upper calyx lobes broadly triangular, slightly acuminate, 1.5 times as long as the tube; racemes lax but uppermost flowers subumbelliform, few developing in the axils below · · · · · · · · 189. *tenuipedicellata*
– Beak of the keel slightly incurved; upper calyx lobes (1.5)2–2.5 times as long as the tube; uppermost flowers not subumbelliform, many developing in the axils below · · 191. *basipeta*
39. Racemes very laxly 3–6-flowered; leaves 4–12 mm long, with a pulvinus but no petiole; stems wiry, glabrescent, 15–25 cm tall · · · · · · · · · · 194. *debilis*
– Racemes many-flowered; leaflets 10–45 mm long, mostly on distinct petioles; stems slender, puberulous to pubescent, 20–100 cm tall · · · · · · · · · · 40
40. Racemes lax, the flowers 3–10 bract lengths apart; bracts much shorter than the pedicel; standard inside with red-fringed basal patches · · · · · · · · · · 192. *ringoetii*
– Racemes subdense, the flowers c. 1–2 bract lengths apart; bracts often nearly as long as the pedicel; standard often marked red but without yellow red-fringed basal patches · · · · · · · · · · 193. *passerinoides*

1. **Crotalaria agatiflora** Schweinf. in von Hoehnel, Zum Rudolph-See und Stephanie-See, Anh.: 13 (1892). —N.E. Brown in Bot. Mag. **139**: t. 8505 (1913). —E.G. Baker in J. Linn. Soc., Bot. **42**: 315 (1914). —Verdoorn in Bothalia **2**: 389 (1928). —Brenan, Check-list For. Trees Shrubs Tang. Terr.: 414 (1949). —Polhill in Kew Bull. **22**: 202 (1968); in F.T.E.A., Leguminosae, Pap.: 850, fig. 10 (1971). —Biegel, Check-list Ornam. Pl. Rhod. Parks & Gard.: 43 (1977). —Polhill, Crotalaria Africa & Madagascar: 72, figs 5/3 & 6 (1982). —Lock, Leg. Afr. Check-list: 164 (1989). Type from Kenya.

Bushy woody herb or shrub, 1–10 m tall. Leaves 3-foliolate; leaflets 4–7 × 2–3.5 cm, elliptic-lanceolate to elliptic-ovate, glabrous to densely hairy; petioles mostly longer than leaflets; stipules linear and caducous or absent. Racemes to 40 cm long, many-flowered; bracts narrow or much expanded, caducous. Calyx plus hypanthium 1.8–3 cm long, with upper and lateral lobes on either side joined almost to their tips, ± twice as long as the tube. Standard ovate, lemon-yellow to greenish-yellow, sometimes medially pubescent outside; wings half to two thirds as long as keel; keel 3.5–4.5(5.5) cm long, obliquely ovate-elliptic, with a relatively short, forwardly directed, often greenish or purplish beak. Pod including the 1.5–2.5 cm long stipe 7.5–10 cm long, oblong-clavate, glabrous. Seeds 6–7(9) mm long, oblique-cordiform, smooth, pale brown.

A species occurring in the highlands of east and north-east Africa, just extending into the Flora Zambesiaca area. It is also grown in gardens as an ornamental shrub. Five subspecies are recognized, see Polhill, Crotalaria Africa & Madagascar: 72–74 (1982).

Subsp. **vaginifera** Polhill in Kew Bull. **22**: 205 (1968); in F.T.E.A., Leguminosae, Pap.: 853, fig. 119/3 (1971); Crotalaria Africa & Madagascar: 74, fig. 6/3 (1982). —Lock, Leg. Afr. Check-list: 164 (1989). Type from S Tanzania.

Woody herb or shrub, 2–3 m tall, glabrous to thinly pubescent only on the leaflets beneath or sometimes also on young twigs. Leaflets more than twice as long as broad. Bracts 3–3.6 × 2–3 cm, broadly elliptic, enveloping the subtended bud, abruptly acuminate, subglabrous; bracteoles less than 1 mm long, filiform.

Mozambique. N: mts. east of Lago Niassa (Lake Nyasa), *W.P. Johnson* s.n., comm. *H. Waller* iv.1884 (K).

Also in eastern and southern Tanzania. Montane forest, at edges and in clearings, but not recollected in Mozambique since the nineteenth century.

Subsp. **imperialis** (Taub.) Polhill in Kew Bull. **22**: 204 (1968); in F.T.E.A., Leguminosae, Pap.: 853 (1971); Crotalaria Africa & Madagascar: 73 (1982). —Lock, Leg. Afr. Check-list: 164 (1989). Type from western Tanzania.

Crotalaria imperialis Taub. in Engler, Pflanzenw. Ost-Afrikas **C**: 206 (1895). —E.G. Baker in J. Linn. Soc., Bot. **42**: 317, t. 14A (1914) pro parte.

Woody herb or shrub, 1–3(5) m tall, hairy on twigs and leaflets beneath, usually with some hairs on calyx and leaflets above. Leaflets mostly less than twice as long as broad. Bracts mostly 6–14 × 1–6 mm, subulate to narrowly lanceolate; bracteoles up to 3 mm long, filiform.

Zimbabwe. C: Harare, fl. 30.vii.1965, *Watmough* 647 (K; SRGH). E: Mutare, fl. 20.iv.1955, *Chase* 5555 (BM; K; SRGH). **Malawi**. S: Zomba Distr., Namadzi, Kapalasa Farm, Sven Gruner's garden, fl. 6.vi.1982, *Chapman* 6220 (K; MAL).

Introduced in the Flora Zambesiaca area, native to eastern Dem. Rep. Congo, Rwanda, Burundi, Uganda, Kenya and southern Ethiopia. The form generally grown in southern Africa, and of which specimens were received from the Cape at Kew as early as 1919, is somewhat intermediate between subsp. *imperialis* (Taub.) Polhill and *engleri* (Harms ex Baker f.) Polhill and probably originated from Elgon or the Mau Range in Kenya. The plant grown in Malawi, however, was collected much more recently by Mr Gruner from Lake Kivu in Rwanda.

2. **Crotalaria becquetii** R. Wilczek in Bull. Jard. Bot. État **23**: 188 (1953); in F.C.B. **4**: 229 (1953). —Polhill in F.T.E.A., Leguminosae, Pap.: 855 (1971); Crotalaria Africa & Madagascar: 75 (1982). —Lock, Leg. Afr. Check-list: 167 (1989). —F. White, Dowsett-Lemaire & Chapman, Evergreen For. Fl. Malawi: 324 (2001). Type from Rwanda.

Crotalaria ceciliae sensu Verdoorn in Bothalia **2**: 393 (1928) pro parte quoad specim. *Stolz* 641. —sensu Brenan, Check-list For. Trees Shrubs Tang. Terr.: 414 (1949).

Woody herb or shrub, 1.3–3 m tall; branches early glabrescent. Leaves estipulate, 3-foliolate; leaflets 5–10 × 1.5–4.8 cm, elliptic-lanceolate to broadly elliptic, acuminate, subglabrous to sparsely pubescent beneath; petiole usually a little shorter than leaflets. Racemes lax, many-flowered; bracts 2–5 mm long, subulate or filiform; bracteoles small. Calyx plus funnel-shaped hypanthium 1.4–1.8 cm long, glabrous to thinly puberulous; lobes narrowly attenuate-triangular. Standard ovate, yellow, finely veined purplish outside; wings a little over half to two-thirds as long as keel; keel 2.5–3.2 cm long, strongly rounded, ± straight along upper margin, with a subtruncate upwardly directed beak. Pods with 0.8–1.8 cm long stipe 5.5–9 cm long, oblong-clavate to ellipsoid-clavate, markedly inflated, glabrous. Seeds c. 4 mm long, broadly oblique-cordiform, smooth, brown (sometimes darker veined).

Subsp. **turgida** Polhill in Kew Bull. **22**: 208 (1968); in F.T.E.A., Leguminosae, Pap.: 856 (1971); Crotalaria Africa & Madagascar: 75, fig. 5/5 (1982). —Lock, Leg. Afr. Check-list: 168 (1989). Type: Malawi, Zomba Plateau, *Brass* 16306 (BM; K, holotype; MO; NY; SRGH).
Crotalaria laburnifolia sensu Brenan in Mem. New York Bot. Gard. **8**: 247 (1953).

Pods 5.7–6.7 × 1.6–1.8 cm, broadly ellipsoid-clavate, abruptly contracted to the 1.5–1.8 cm long stipe; keel subglabrous to shortly lanate along upper margin at base; keel beak 8–10 mm long; leaflets up to 2.4–4.8 cm wide, usually elliptic, membranous, with the tertiary venation often rather inconspicuous.

Zambia. E: Nyika, fr. 30.xii.1962, *Fanshawe* 7360 (K; NDO). **Malawi**. N: Viphya, fl. & fr. 10.vii.1952, *Jackson* 965 (K). S: Zomba Plateau, Mlunguzi R., fl. & fr. 31.v.1957, *Boughey* 1515 (K; SRGH).

Also in southern Tanzania. Montane forest edges, streamsides, disturbed places and secondary associations; 1500–2000 m.

Subsp. *becquetii*, with oblong-clavate pods 7.5–9 × 1–1.3 cm, with a more conspicuously ciliate longer beaked keel and narrower more conspicuously veined leaflets, occurs on the mountainous fringes of the Western Rift Valley, particularly Lake Kivu area and in Tanzania north of Lake Malawi.

3. **Crotalaria insignis** Polhill, Crotalaria Africa & Madagascar: 76, fig. 7 (1982). —Lock, Leg. Afr. Check-list: 183 (1989). TAB. 3,7: **22**. Type: Zimbabwe, Mt. Pene, SW slope, Mermaid's Grotto, *Chase* 6893 (K, holotype; LISC; P; SRGH).

Virgately branched shrub, 1.5–3 m tall, with shortly, densely and persistently pubescent branches. Leaves 3-foliolate; leaflets mostly 3.4–5 × 0.9–2.2 cm, narrowly elliptic to elliptic, acute to rounded at the apex, glabrous above, densely appressed puberulous beneath; petiole shorter than leaflets; stipules 2–3 mm long, subulate, persistent. Racemes lax, 6–12-flowered; bracts 3–4 mm long, linear-subulate; bracteoles setaceous. Calyx with funnel-shaped hypanthium 1.5–1.7 cm long, densely puberulous; upper lobes narrowly attenuate-triangular, ± as long as calyx tube plus hypanthium. Standard broadly elliptic, yellow turning bronze; wings longer than the keel; keel 2.4–2.8 cm long, strongly rounded, ± straight along upper margin, abruptly contracted into a well developed darker marked upwardly directed truncate beak. Pod including the 0.8–1.2 cm long stipe 5.5–6.5 cm long, oblong-clavate, densely pubescent. Seeds not seen.

Zimbabwe. E: Chimanimani Distr., Chisengu Forest Reserve, fl. v.1968, *Goldsmith* 73/68 (K; SRGH); Kasipiti, fl. 1.vi.1966, *Loveridge* 1565 (K; SRGH). **Mozambique**. MS: Chimanimani Mts., fl. 7.vi.1949, *Wild* 2931 (K; SRGH). Without locality, fl. 1948, *Pedro* 2324 (LISC).

Known only from the Zimbabwe-Mozambique border mountains. Forest margins, streamsides and secondary scrub; 1050–1800 m. Cultivated at Marondera, Grasslands Research Station, *Corby* 1869 (K; LISC; SRGH).

4. **Crotalaria laburnifolia** L., Sp. Pl. **2**: 715 (1753). —E.G. Baker in J. Linn. Soc., Bot. **42**: 318 (1914). —Eyles in Trans. Roy. Soc. South Africa **5**: 371 (1916) pro majore parte. —Verdoorn in Bothalia **2**: 390 (1928) pro parte. —Brenan, Check-list For. Trees Shrubs Tang. Terr.: 415 (1949). —Wilczek in F.C.B. **4**: 227 (1953). —Martineau, Rhod. Wild Fl.: 34, pl. 9 fig. 3 (1953). —Polhill in Kew Bull. **22**: 209 (1968); in F.T.E.A., Leguminosae, Pap.: 856 (1971). —Drummond in Kirkia **8**: 218 (1972). —Moriarty, Wild Fl. Malawi: 128, t. 64, fig. 5 (1975). —Tredgold & Biegel, Rhod. Wild Fl.: 27, fig. 20 (1979) excl. pod. —Polhill,

Tab. 3,7: **22**. CROTALARIA INSIGNIS. 1, flowering branches (× 1), from *Corby* 1869; 2, calyx, opened out (× 1); 3, standard (× 1); 4, wing (× 1); 5, keel (× 1); 6, anthers (× 2 1/2); 7, gynoecium (× 1), 2–7 from *Goldsmith* 74/68. Drawn by Victoria Gordon. From Crotalaria Africa & Madagascar.

Crotalaria Africa & Madagascar: 76 (1982). —Lock, Leg. Afr. Check-list: 185 (1989). Syntypes from Sri Lanka.

Erect scarcely woody bushy herb, 0.6–2 m tall, with glabrous or thinly pubescent branches. Leaves 3-foliolate, estipulate; leaflets variable, 1–10.5 × 0.5–5.5 cm, narrowly oblong to elliptic or elliptic-obovate, bluntly pointed to rounded or retuse at apex, glabrous or pubescent; petiole usually longer than leaflets. Racemes lax, usually many-flowered; bracts up to 4 mm long, linear or subulate; bracteoles small. Calyx plus funnel-shaped hypanthium 1.2–1.7 cm long, glabrous to sparsely puberulous; lobes attenuate-triangular, longer than the tube. Standard ovate, yellow, often marked red-brown outside; wings little over half to three quarters length of keel; keel 2–3 cm long, strongly rounded, ± straight along upper margin, abruptly contracted into a subtruncate upwardly directed beak, often flecked or flushed maroon. Pod including the 1.7–9.5 cm long stipe up to 4.5–15 cm long, oblong-clavate to subcylindrical, glabrous, often mottled when young. Seeds 4–5 mm long, oblique-cordiform, granulate, ochre or brown.

Subsp. **laburnifolia** —Polhill in Kew Bull: **22**: 209 (1968); in F.T.E.A., Leguminosae, Pap.: 857 (1971) —Gonçalves in Garcia de Orta, Sér. Bot. **5**: 66 (1982). —Moriarty, Wild Fl. Malawi: t. 64 (1975). —Polhill, Crotalaria Africa & Madagascar: 78 (1982). —Lock, Leg. Afr. Check-list: 185 (1989).

Crotalaria petiolaris ?sensu Steedman, Trees, Shrubs & Lianes S. Rhod.: 23 (1933).

Leaflets usually elliptic or obovate-elliptic, mostly bluntly pointed or rounded at the apex, usually pubescent in the Flora Zambesiaca area at least when young. Pods 5.5–7.5 cm long; seed-bearing part abruptly contracted to the stipe and less to little more than twice as long.

Caprivi Strip. Katima Mulilo, fl. 30.xii.1958, *Killick & Leistner* 3191 (BM; K; M; PRE; SRGH). **Botswana**. N: Ngamiland Distr., Maun, fl. & fr. ii.1967, *Lambrecht* 55 (K; SRGH). **Zambia**. N: Isoka Distr., R. Luija, fl. i.1897, *T.G. Nicholson* s.n. (K). E: locality and date not recorded, *Verboom* 484 (SRGH). S: 16 km Livingstone–Katombora, fl. & fr. 6.i.1953, *Angus* 1115 (BM; FHO; K). **Zimbabwe**. N: Hurungwe Distr., Zambezi Valley, Rifa R., fl. & fr. 24.ii.1953, *Wild* 4101 (K; LISC; SRGH). W: Umguza Distr., Nyamandhlovu Pasture Research Station, fl. & fr. ii.1954, *Plowes* 1679 (K; SRGH). C: KweKwe (Que Que), fl. & fr. immat. 8.ii.1967, *Corby* 1742 (K; SRGH). E: Nyanga Distr., Cheshire, fl. 15.i.1931, *Norlindh & Weimarck* 4358 (K; LD; M). S: Masvingo Distr., Makaholi Experimental Farm, fl. & fr. 23.iii.1948, *D.A. Robinson* 247 (K; LISC; SRGH). **Malawi**. C: 8 km north of Kasungu, fr. 7.v.1970, *Brummitt* 10435 (K; LISC; MAL; PRE; SRGH). S: Zomba, fl. & fr. 5.ii.1955, *Jackson* 1460 (K; LISC; SRGH). **Mozambique**. N: Montepuez, rio Lúrio, fl. & fr. immat. 2.ix.1948, *Barbosa* 1974 (K; LISC). T: 26 km Luenha (Changara)–Catandica (Vila Gouveia), fl. 23.ii.1968, *Torre & Correia* 17763 (LISC). MS: Gondola Distr., Matsinho (Chimoio), Tembe, fl. & fr. 1.iii.1948, *Garcia* 439 (K; LISC).

Widespread in drier parts of east and northeast Africa, Mascarene Islands, India, Malesia and NE Australia. Savanna woodland and tree savanna of various types (rarely in miombo), often by rivers, roads or other disturbed places; 50–1450 m.

Subsp. **australis** (Baker f.) Polhill in Kew Bull. **22**: 213 (1968); Crotalaria Africa & Madagascar: 79 (1982). —Lock, Leg. Afr. Check-list: 185 (1989). Lectotype from South Africa (Northern Province).

Crotalaria petiolaris var. *australis* Baker f. in J. Linn. Soc., Bot. **42**: 319 (1914). —Eyles in Trans. Roy. Soc. South Africa **5**: 371 (1916).

Crotalaria australis (Baker f.) Verdoorn in Bothalia **2**: 389 (1928). —Burtt Davy, Fl. Pl. Ferns Transvaal, pt. 2: 398 (1932) pro specim. Afr. austr.

Crotalaria laburnifolia sensu Verdoorn in Bothalia **2**: 390 (1928) pro minore parte. — sensu Burtt Davy, Fl. Pl. Ferns Transvaal, pt. 2: 398 (1932) pro specim. Afr. austr.

Leaflets narrowly oblong-elliptic to broadly elliptic, rounded or occasionally emarginate at the apex, glabrous. Pods (8)9–10.5 cm long; seed-bearing part tapered to the stipe and more than twice as long.

Botswana. SE: 6 km SSE of Mahalapye, fl. 4.xi.1978, *Hansen* 3520 (GAB; K; PRE; SRGH). **Zimbabwe**. E: Chipinge Distr., Chisumbanje, fl. 23.ii.1960, *Goodier* 957 (K; LISC; SRGH). S: 32 km Birchenough Bridge–Masvingo (Fort Victoria), fl. & fr. 17.iii.1965, *Corby* 1289 (K; SRGH); Beitbridge, fl. & fr. 16.ii.1955, *Exell, Mendonça & Wild* 433 (BM; LISC; SRGH). **Mozambique**. GI: Chibuto Distr., Mundiane, fl. & fr. 8.vii.1947, *Pedro & Pedrógão* 1362 (COI; K; LISC; LMA;

SRGH); Magude–Chókwe (Guijá), fl. & fr. 5.v.1944, *Torre* 6577 (BM; K; LISC). M: Manhica–R. Maputo, fl. & fr. 28.xi.1961, *Myre & Macêdo* 4483 (LMA).

Also in South Africa and Swaziland. Tree savanna and grassland; near sea-level to 1050 m.

No difficulty has been found in naming fruiting material to subspecies, but the two meet in south and southeast Zimbabwe and it would be interesting to have a detailed report on distribution and variation in this region. The LISC sheet of *Lemos & Balsinhas* 84, from Mozambique, Chibuto Distr., has two pods typical of subsp. *australis* (as sheets at COI, K and LMA), but the third has the lower part of the seed-bearing portion largely aborted so that it looks almost exactly like a pod of subsp. *laburnifolia*; the leaves are quite normal for subsp. *australis*.

Subsp. *australis* has been grown as a garden ornamental in Harare, e.g. *Nicholls* 233/142 (K; SRGH).

5. **Crotalaria capensis** Jacq., Hort. Vindob. **3**: 36, t. 64 (1776–7). —Harvey in F.C. **2**: 46 (1862). —Medley Wood, Natal Pl. **1**: 74, t. 92 (1899). —Hemsley in Bot. Mag. **130**: t. 7950 (1904). —E.G. Baker in J. Linn. Soc., Bot. **42**: 405 (1914). —Verdoorn in Bothalia **2**: 379 (1928). —Pole Evans in Fl. Pl. South Africa **10**: t. 386 (1930). —Burtt Davy, Fl. Pl. Ferns Transvaal, pt. 2: 397 (1932). —Mogg in Macnae & Kalk, Nat. Hist. Inhaca Isl., Moçamb.: 146 (1958). —Drummond in Kirkia **8**: 217 (1972). —Polhill, Crotalaria Africa & Madagascar: 81, fig. 8 (1982). —Lock, Leg. Afr. Check-list: 170 (1989). Type grown in Vienna from Cape seed.

Crotalaria ceciliae Verdoorn in Bothalia **2**: 392 (1928), excl. specim. *Stolz* 641. Type: Mozambique, railway between Beira and Manica (Massi Kessi), *Cecil* (K, holotype).

Bushy shrub, 1–3 m tall, usually with rather persistently tomentose branches, but sometimes glabrescent or even glabrous. Leaves 3-foliolate; leaflets (1.5)3–7 × (1)1.5–3 cm, elliptic to obovate, pointed to rounded at the apex, glabrous to densely pubescent (particularly beneath); petioles mostly shorter than leaflets; stipules usually leaf-like, 0.8–2(3) cm long, with 1–4 mm long stalk and obovate blade, sometimes caducous or lacking. Racemes with few to many fairly closely arranged flowers; bracts (1.5)3.5–7.5(10) mm long, linear-lanceolate; bracteoles filiform, smaller. Calyx with funnel-shaped hypanthium (8)10–16(18) mm long, glabrous to thinly pubescent; lobes attenuate-triangular, longer than the tube. Standard ovate or elliptic, yellow, marked and later flushed with red, glabrous or medially pubescent outside; wings three quarters to nearly as long as the keel; keel (1.5)2–2.4 cm long, rounded about the middle, with a rather short blunt beak. Pod including (5)12–16 mm long stipe c. 4.5–6.8 cm long, oblong-clavate, shortly appressed-pubescent. Seeds 4–4.5 mm long, oblique-cordiform, practically smooth, brown.

Zimbabwe. E: Mutare Distr., Vumba, Elephant Forest, fl. 20.i.1955, *Chase* 5447 (BM; COI; K; LISC; SRGH). **Malawi**. S: Shire Highlands, Blantyre, fl. 6.vii.1879, *Buchanan* 131 (K). **Mozambique**. Z: 25 km da costa para Maganja da Costa, fl. & fr. 27.ix.1949, *Barbosa & Carvalho* 4223 (K; LISC; LMA). MS: Báruè Distr., Serra de Chôa, fl. & fr. immat. 28.iii.1966, *Torre & Correia* 15490 (LISC). GI: Chokwe Distr., Chipenhe, fl. & fr. 9.vi.1960, *Lemos & Balsinhas* 51 (BM; COI; K; LISC; LMA; SRGH). M: Maputo (Lourenço Marques), Costa do Sol, fr. 30.iii.1960, *Balsinhas* 153 (COI; K; LISC; SRGH); Inhaca Island, fl. 10.vii.1956, *Mogg* 26731 (BM; K; SRGH); fl. & fr. 14.vii.1957, *Mogg* 27237 (LMA; SRGH).

Also in South Africa. Forest margins, woodland, less often in tree or shrub savanna, but persisting in disturbed places; 0–1500 m.

Rather variable throughout its range — in hairiness, leaflet shape, stipule development and flower size, but the most extreme forms occur in the Flora Zambesiaca area. The single gathering from Malawi is scarcely recognizable, being subglabrous, estipulate, with exceptionally small bracts and flowers. Specimens from the Zimbabwe-Mozambique border mountains have sparsely hairy, relatively large pointed leaflets and somewhat inconstant stipular development; this is not just a shade effect as the features persist in cultivation, e.g. *Corby* 1445 (K; SRGH). These features are present to a greater or lesser extent in the populations that occur in the region from these two areas to the Mozambique coast. The species is not recorded from the main part of Gaza and Inhambane, but is common south of Inhambane, where the plants are generally characterized by a well developed indumentum, large stipules, and smaller more rounded leaflets. Specimens of the northern populations are frequently confused with *C. becquetii* and *C. laburnifolia*, but are easily distinguished by the keel shape and hairy pods.

6. **Crotalaria pallidicaulis** Harms in Bot. Jahrb. Syst. **54**: 382 (1917). —E.G. Baker, Legum. Trop. Africa: 40 (1926). —Brenan, Check-list For. Trees Shrubs Tang. Terr.: 415 (1949). —Polhill in Kew Bull. **22**: 214 (1968); in F.T.E.A., Leguminosae, Pap.: 859 (1971). —Drummond in Kirkia **8**: 218 (1972). —Jacobsen in Kirkia **9**: 160 (1973). —Polhill, Crotalaria Africa & Madagascar: 82 (1982). —Lock, Leg. Afr. Check-list: 195 (1989). Type from Tanzania.

Crotalaria athroophylla Verdoorn in Bothalia **2**: 390 (1928). —Steedman, Trees, Shrubs & Lianes S. Rhod.: 23 (1933). —White, F.F.N.R.: 147 (1962). Type: Zimbabwe, Harare, *Eyles* 1330 (BM; PRE, holotype; SRGH).

Crotalaria multicolor Merxm. in Proc. & Trans. Rhodesia Sci. Assoc. **43**: 19 (1951). Type: Zimbabwe, Marondera (Marandellas), *Dehn* 164 (BR; M, holotype).

Shrub, 1–3(4) m tall, with hairy sometimes glabrescent branches. Leaves often with short axillary shoots bearing a cluster of smaller leaves, estipulate, 3-foliolate; leaflets 1.5–4 × 0.4–1.2 cm, very variable in shape from linear-oblanceolate to elliptic or oblong-obovate, puberulous at least beneath; petioles mostly 1.7–3.5 times as long as leaflets. Racemes with usually numerous laxly arranged flowers; bracts 2–3.5 mm long, subulate or linear-lanceolate; bracteoles small. Calyx (8)9–12 mm long, puberulous; lobes narrowly triangular, sometimes acuminate, ± as long as the tube. Standard elliptic-ovate, yellow, usually marked reddish or purple and glabrous outside; wings 0.75–0.9 times as long as the keel; keel 2–2.2(2.3) cm long, rounded about the middle with a rather short bluntly pointed beak. Pod including 0.8–1.2 cm long stipe up to 4–5.5 cm long, oblong-clavate, puberulous. Seeds c. 5 mm long, oblong-reniform, granulate, ochre or brown.

Botswana. N: Hwange Distr., Kazuma, fr. xi.1966, *Mutakela* 149 (SRGH). **Zambia**. W: Ndola, fl. & fr. 23.iv.1954, *Fanshawe* 1137 (BR; K; LISC; NDO; SRGH). C: 16 km south of Lusaka, fl. & fr. 6.iv.1955, *Exell, Mendonça & Wild* 1413 (BM; LISC; SRGH). S: Gwembe Distr., 61 km Mochipapa–Sinazongwe, 2.iii.1960, *White* 7557 (FHO; K; SRGH). **Zimbabwe**. N: Hurungwe Distr., Mwami (Miami), Experimental Farm, fl. 7.iii.1947, *Wild* 1802 (K; SRGH). W: Nkayi Distr., Gwampa Forest Land (Reserve), fl. & fr. immat. iii.1955, *Goldsmith* 99/55 (K; SRGH). C: Gweru Distr., Athlone, fl. 4.i.1963, *Loveridge* 518 (K; LISC; SRGH). E: Mutare Distr., Banti Forest Reserve, fl. 10.xi.1967, *Mavi* 582 (K; SRGH). S: 32 km north of Masvingo (Fort Victoria), fr. 4.v.1962, *Drummond* 7948 (K; LISC; SRGH). **Mozambique**. MS: Sussundenga Distr., Tsetserra, fl. & fr. immat. 6.iv.1966, *Torre & Correia* 15745 (LISC).

Also in Dem. Rep. Congo (Katanga) and Tanzania. Miombo woodland or sometimes tree savanna, quite commonly on rock outcrops, around termite mounds or in disturbed places; 750–1500 m.

In southern Zimbabwe *Crotalaria pallidicaulis* may be confused with *C. monteiroi* var. *galpinii* (see below).

7. **Crotalaria monteiroi** Taub. ex Baker f. in J. Linn. Soc., Bot. **42**: 387 (1914). —Verdoorn in Bothalia **2**: 391 (1928) excl. specim. *F.A. Rogers* 7712. —Mogg in Macnae & Kalk, Nat. Hist. Inhaca Isl., Moçamb.: 146 (1958). —Polhill, Crotalaria Africa & Madagascar: 82, fig. 5/9 (1982). —Lock, Leg. Afr. Check-list: 192 (1989). Lectotype, selected by Polhill, loc. cit.: Mozambique, Delagoa Bay, *Monteiro* 18 (K, lectotype; P).

Shrub or bushy subshrub, 1–2 m tall, with hairy ± glabrescent branches. Leaves often with short axillary shoots bearing clusters of smaller leaves, estipulate, 3-foliolate; leaflets 7–20(30) × 4–10 mm, usually small and obovate but in var. *galpinii* varying from oblong-oblanceolate to elliptic-obovate, puberulous beneath; petioles mostly 0.5–1.7 times as long as leaflets. Racemes mostly few-flowered; bracts 1–3 mm long, subulate or linear; bracteoles minute. Calyx 7–10 mm long, puberulous, rarely subglabrous; lobes triangular, sometimes acuminate, ± as long as tube. Standard ovate, yellow, usually marked orange or brown, glabrous or medially puberulous outside; wings 0.75–0.9 times as long as keel; keel 1.6–2 cm long, rounded about the middle with a rather short bluntly pointed beak. Pod including 6–8 mm long stipe 2–3(3.5) cm long, oblong-clavate, pubescent. Seeds c. 3.5 mm long, subcircular-cordiform, granulate or slightly rugulose, ochre or pale brown.

Var. **monteiroi** —Verdoorn in Bothalia **2**: 391 (1928). —Drummond in Kirkia **8**: 218 (1972). —Polhill, Crotalaria Africa & Madagascar: 83 (1982).

Leaflets c. 5–14 mm long; petioles 3–10(17) mm long. Bracts usually less than 2 mm long. Keel 16–18 mm long.

Zimbabwe. S: Chiredzi Distr., Gonakudzingwa, fr. 4.iv.1961, *R. Goodier* 1058 (K; LISC; M; SRGH). **Mozambique**. MS: 96 km do T.Z.R., fr. 25.vii.1946, *Simão* 817 (LISC; LMA). GI: Homoíne, fl. & fr. 25.ii.1955, *Exell, Mendonça & Wild* 582 (BM; BR; LISC; SRGH). M: Matutuíne (Bela Vista), Tinonganine, fl. & fr. 28.iii.1957, *Barbosa & Lemos* 7567 (COI; LMA).

Also in South Africa (Northern Province, Mpumalanga and KwaZulu-Natal). Woodland, tree savanna and grassland, most common in open disturbed places on sandy soils; 0–500 m.

Var. **galpinii** Burtt Davy ex Verdoorn in Bothalia **2**: 391 (1928) excl. specim. *F.A. Rogers* 11810. —Burtt Davy, Fl. Pl. Ferns Transvaal, pt. 2: 398 (1932). —Drummond in Kirkia **8**: 218 (1972). —Polhill, Crotalaria Africa & Madagascar: 83 (1982). Type from South Africa (Mpumalanga).

Crotalaria rigidula Baker f. in J. Bot. **58**: 75 (1920). —Verdoorn in Bothalia **2**: 391 (1928). —Burtt Davy, Fl. Pl. Ferns Transvaal, pt. 2: 398 (1932). Type from South Africa (Northern Province).

Crotalaria breyeri N.E. Br. in Bull. Misc. Inform., Kew **1922**: 27 (1922). Type from South Africa (Northern Province).

Crotalaria inhabilis Verdoorn in Bothalia **3**: 238 (1937). Type from South Africa (Northern Province).

Leaflets (10)15–30 mm long; petioles 10–30 mm long. Bracts (1.5)2–3 mm long. Keel (16)18–20 mm long.

Zimbabwe. S: Chiredzi Distr., west of Save (Sabi) R, Chionja, fl. 29.i.1957, *Phipps* 227 (BM; K; SRGH); Beitbridge Distr., Tshiturapadsi (Chiturupazi)–Chikwarakwara, fl. & fr. 24.ii.1961, *Wild* 5375 (K; SRGH). **Mozambique**. T: Magoe Distr., 46 km de Mphende (Mágoè) para Mukumbura (Mucumbura), fr. 9.iii.1970, *Torre & Correia* 18229 (LMU). M: Moamba–Ressano Garcia, fl. 3.xii.1940, *Torre* 2224 (BM; LISC); Namaacha Falls, fl. 7.xii.1942, *Torre* 1693 (LISC).

Also in South Africa (Northern Province, Mpumalanga, KwaZulu-Natal), mostly more inland and further within South Africa than var. *monteiroi*. Miombo and *Colophospermum* woodland, tree savanna; up to 550 m.

Var. *galpinii* forms a morphological and geographical link between var. *monteiroi* and *C. pallidicaulis*, which in their typical forms look quite different. Without pods some difficulty may be found in naming plants from southern Zimbabwe, and in the province of Maputo there is an intergradation of the two varieties. Although there is a case for treating these taxa as three subspecies, the changing of familiar names does not really seem to be justified, particularly as the two existing varieties seem to intergrade to the greater extent (see Polhill, loc. cit. (1982)).

8. **Crotalaria assurgens** Polhill in Kew Bull. **22**: 214, fig. 5 (1968); Crotalaria Africa & Madagascar: 83 (1982). —Lock, Leg. Afr. Check-list: 166 (1989). Type from SE Tanzania.

Perennial herb with several erect much branched stems to 60 cm tall, with sparsely puberulous branches. Leaves 3-foliolate, somewhat glaucous; leaflets 1.8–2.8(5.8) × 0.4–1.4 cm, linear-lanceolate to elliptic, glabrous or practically so above, thinly appressed pubescent beneath; stipules absent. Racemes lax, mostly 8–16-flowered; bracts 2–2.5 mm long, linear; bracteoles small. Calyx 7–8.5 mm long, subglabrous; lobes acuminately triangular, ± as long as the tube. Standard elliptic to subcircular, yellow, flecked and veined outside reddish-brown to purple, glabrous outside; wings 0.75–0.8 times as long as the keel; keel 1.4–1.5 cm long, rounded about the middle with a rather short obtuse beak. Pod with the 5–7 mm long stipe 3.8–4.5 cm long, oblong-clavate, puberulous. Mature seeds not seen.

Mozambique. N: c. 7 km de Marrupa para Lichinga, fl. 19.ii.1982, *Jansen & Boane* 7881 (LMU).

Also in SE Tanzania. Miombo woodland, sandy soil; c. 800 m.

9. **Crotalaria unicaulis** Bullock in Bull. Misc. Inform., Kew **1931**: 271 (1931). —Milne-Redhead in Kew Bull. **2**: 26, fig. (1947). —Torre in C.F.A. **3**: 33 (1962). —Polhill in Kew Bull. **22**: 213 (1968); Crotalaria Africa & Madagascar: 80 (1982) excl. specim. ex Kafue. —Lock, Leg. Afr. Check-list: 208 (1989). Type: Zambia, Mwinilunga, *Marks* 10 (K, holotype).

Crotalaria tabularis sensu Torre in C.F.A. **3**: 51 (1962) quoad specim. cit.

Suffrutex with annual or perennial scarcely woody glaucous subglabrous stems erect from a well developed rootstock. Leaves estipulate, simple or (1)3-foliolate; leaflets 3.5–9 × 1.5–5 cm, elliptic, subglabrous; petiole broadly grooved, often slightly winged, much shorter to a little longer than the leaflets. Racemes lax, many-flowered; bracts mostly 4–5 mm long, linear-subulate; bracteoles small. Calyx 1–1.3 cm long, glabrous or with few scattered hairs; lobes narrowly attenuate-triangular, ± twice as long as the tube. Standard elliptic to subcircular, yellow veined and speckled reddish-brown, glabrous or nearly so; wings ± three quarters as long as the keel; keel

1.7–2 cm long, rounded about the middle, narrowed to the rather short bluntly pointed beak. Pod 4–7.5 cm long, ellipsoid-clavate, rather abruptly narrowed to the mostly 0.6–0.7 cm long stipe, pubescent, glabrescent except along sutures and near apex. Mature seeds not seen.

Zambia. N: Mbala (Abercorn) to Mpulungu road, fl. & fr. 5.iv.1955, *Richards* 5304 (K); Kasama to Mungwi road, fl. & fr. 24.ii.1962, *Richards* 16158 (K). W: Mwinilunga, fl. xi.1929, *Marks* 10 (K).

Also in Angola. Miombo woodland and secondary associations, commonly in disturbed places; 1000–1500 m.

The type has mostly simple leaves with the uppermost 1-foliolate, *Milne-Redhead* 4133 from Moxico has all 1-foliolate leaves, whereas the other material has constantly 3-foliolate leaves. It is still uncertain whether these plants are rightly associated as one species, but there do not seem to be any other significant differences.

Trapnell CRS 534 (K), a rather poor specimen from Mazabuka in the S Province of Zambia, included under this species in Polhill (1982), seems on second thoughts to be *C. laburnifolia* subsp. *laburnifolia*, which is well known from this locality.

10. **Crotalaria ulbrichiana** Harms in Bot. Jahrb. Syst. **54**: 384 (1917). —E.G. Baker, Legum. Trop. Africa: 46 (1926). —Polhill in Kew Bull. **22**: 216 (1968). —Schreiber in Merxmüller, Prodr. Fl. SW. Afrika, fam. 60: 28 (1970). —Drummond in Kirkia **8**: 219 (1972). —Polhill, Crotalaria Africa & Madagascar: 84 (1982). —Lock, Leg. Afr. Check-list: 208 (1989). Syntypes from Namibia.

Crotalaria barnabassii var. *cunenensis* Torre in Mem. Junta Invest. Ultramar, sér. 2, **19**: 41, t. 20 (1960); in C.F.A. **3**: 59 (1962). Type from Angola.

Erect laxly branched annual, mostly 40–120 cm tall, with pubescent branches. Leaves 3-foliolate; leaflets 3.6–6.2 × 1.8–3.6(4) cm, elliptic, glabrous above, pubescent beneath; petiole a little shorter to a little longer than the leaflets; stipules 3–7 mm long, linear. Racemes short and few-flowered at first, elongating and ultimately c. 20 cm long and many-flowered; bracts 4–5 mm long, linear-lanceolate; bracteoles up to 2 mm long, filiform. Calyx 1–1.4 cm long, sparsely to conspicuously puberulous; lobes narrowly attenuate-triangular, 2–3 times as long as the tube. Standard elliptic-oblong, yellow, subglabrous outside; wings 0.75–0.9 times as long as keel; keel (1.2)1.4–1.6 cm long, rounded about the middle with a projecting obtusely pointed beak. Pod up to 2.4 cm long, broadly ellipsoid-clavate, abruptly contracted to the 4–5 mm long stipe, pubescent. Seeds c. 5 mm in diameter, subcircular, flattened, rugose, brown.

Zambia. S: Mazabuka, fl. & fr. 26.iii.1963, *van Rensburg* 1820 (K; SRGH). **Zimbabwe**. N: Tete–Harare, fl. & fr. immat. 2.iii.1961, *Richards* 14512 (K; SRGH); Mount Darwin Distr., Mavuradonha (Mvuradona) Mission, fl. & fr. 31.iv.1964, *Wild* 6550 (K; LISC; SRGH).

Also in Angola and Namibia, in the Zambezi–Cunene drainage area. Tree savanna, streamsides; c. 900–1000 m.

11. **Crotalaria barnabassii** Dinter ex Baker f. in J. Linn. Soc., Bot. **42**: 356 (1914). —Eyles in Trans. Roy. Soc. South Africa **5**: 370 (1916). —Verdoorn in Bothalia **2**: 393 (1928). —Torre in C.F.A. **3**: 58 (1962) excl. var. *cunenensis*. —Schreiber in Merxmüller, Prodr. Fl. SW. Afrika, fam. 60: 22 (1970). —Polhill in F.T.E.A., Leguminosae, Pap.: 860 (1971). —Drummond in Kirkia **8**: 217 (1972). —Gonçalves in Garcia de Orta, Sér. Bot. **5**: 66 (1982). —Polhill, Crotalaria Africa & Madagascar: 84, fig. 9 (1982). —Lock, Leg. Afr. Check-list: 167 (1989). Lectotype, selected by Polhill, loc. cit. (1971), from Namibia.

Crotalaria mossamedesiana Baker f. in Bol. Soc. Brot., Sér. 2, **8**: 103 (1933). —Torre in C.F.A. **3**: 59 (1962). Type from Angola.

Erect laxly branched annual, mostly 1–2 m tall, with sparsely appressed puberulous and glabrescent branches. Leaves 3-foliolate; leaflets mostly 3–6 × 0.8–2.8 cm, lanceolate to elliptic, glabrous above, sparsely puberulous or subglabrous beneath; petiole usually a little shorter than leaflets; stipules 1–4 mm long, linear, caducous. Racemes lax, short and few-flowered at first, but elongating and ultimately 10–35 cm long and many-flowered; bracts 1.5–2.5 mm long, linear; bracteoles small. Calyx 6–8 mm long, sparsely puberulous; lobes attenuate-triangular, longer than tube. Standard elliptic-obovate to subcircular, pale yellow, fading orange, glabrous outside; wings 0.7–0.8 times as long as keel; keel 1.1–1.4 cm long, rounded a little below the middle, with a slightly incurved acute beak. Pod (2.5)2.8–3.6 cm long, ellipsoid-clavate, abruptly contracted to the 9–13 mm long stipe, puberulous, glabrescent. Seeds mostly 4.5–6 mm across, subcircular, thin, markedly rugose towards the margin, brown with darker speckling.

Botswana. SE: Ngamiland, fl. & fr. 30.iii.1987, *Long & Rae* 503 (SRGH). **Zambia**. E: Petauke Distr., R. Nyamadzi, fl. & fr. 25.iii.1955, *Exell, Mendonça & Wild* 1179 (BM; LISC; SRGH). S: Choma Distr., Mapanza, fr. 8.v.1956, *E.A. Robinson* 1499 (K; LISC; SRGH). **Zimbabwe**. N: Hurungwe Distr., Rifa R., fl. & fr. 25.ii.1953, *Wild* 4106 (K; LISC; SRGH). W: Hwange Distr., Kazungula, fl. & fr. iv.1955, *R.M. Davies* 1098 (K; SRGH). E: Lower Save (Sabi) R., fl. & fr. 28.i.1948, *Wild* 2314 (BR; K; SRGH). S: Mwenezi Distr., Malangwe R., SW of Mateke Hills, fl. & fr. 5.v.1958, *Drummond* 5580 (K; SRGH). **Malawi**. S: Mangochi Distr., Lake Malawi, Boadzulu Is., fl. & fr. 14.iii.1955, *Exell, Mendonça & Wild* 877 (BM; LISC; SRGH). **Mozambique**. N: Malema Distr., Mutuáli, fl. & fr. 6.vi.1948, *Pedro & Pedrógão* 4170 (EA; LMA). T: Magoe Distr., Zambeze R., Msusa, fl. & fr. 25.vii.1950, *Chase* 2794 (BM; COI; K; LISC; SRGH).

Also in southern Tanzania, Angola, Namibia and South Africa, in the drainage basins of the major rivers, notably the Ruaha, Cunene, Zambezi, Orange and Limpopo. Near lakes and rivers and on newly disturbed ground by roads and fields; 200–1050 m.

12. **Crotalaria goodiiformis** Vatke in Oesterr. Bot. Z. **29**: 220 (1879). —Polhill in F.T.E.A., Leguminosae, Pap.: 861 (1971); Crotalaria Africa & Madagascar: 88, fig. 10/4 (1982). —Lock, Leg. Afr. Check-list: 180 (1989). Type from Kenya.

Crotalaria saxatilis Vatke in Oesterr. Bot. Z. **29**: 219 (1879) non Zoll. (1847), *nom. illegit.* —E.G. Baker in J. Linn. Soc., Bot. **42**: 392 (1914). —Brenan, Check-list For. Trees Shrubs Tang. Terr.: 414 (1949). —Wilczek in F.C.B. **4**: 163 (1953). Type from Kenya.

Bushy shrub or subshrub 1–3 m tall, with slender pubescent branches, often glabrescent. Leaves 3-foliolate; leaflets 1–5 × 0.5–3 cm, broadly oblanceolate, elliptic or obovate, sparsely pilose on both surfaces; petiole slender, shorter to a little longer than leaflets; stipules 1–4 mm long, filiform, caducous. Racemes with few flowers laxly inserted on a very slender rhachis; bracts 1–3 mm long, subulate or filiform; bracteoles inserted just below calyx, 3–4(6) mm long, linear, ascending and sometimes curved. Calyx 8–10 mm long, sparsely puberulous; lobes narrowly attenuate-triangular, more than twice as long as the tube. Standard subcircular, yellow or orange, marked darker, usually puberulous along midvein outside; wings exceeding keel; keel 9–12 mm long, strongly rounded about the middle with a short slightly incurved beak, lanate-pubescent towards base of upper margin, rarely sparsely so. Pod (2)3–4 cm long, oblong-clavate, shortly stipitate, thinly spreading pubescent. Seeds 4–5 mm long, oblique-cordiform, slightly rugulose, dark brown, sometimes mottled greenish-brown.

Mozambique. N: Nampula Distr., mata Nampaha, a c. 27 km de Nampula, prox. da estrada velha de Nampula para Muecate, c. 380 m, fl. 2.iv.1964, *Torre & Paiva* 11585 (LISC). Z: Lugela Distr., Namagoa Estate, fl. & fr. vii.1946, *Faulkner* Kew 15 (K). MS: Cheringoma Distr., Inhaminga–Mupa, fl. & fr. 11.vii.1946, *Simão* 786 (LISC; LMA). GI: Mandlakazi (Manjacaze), fl. 18.iii.1948, *Torre* 7514 (BM; K; LISC).

Also in eastern Dem. Rep. Congo, Kenya and Tanzania. Woodland and mixed dry forest-woodland associations, streamsides; below 500 m.

13. **Crotalaria polysperma** Kotschy ex Schweinf., Reliq. Kotschy.: 18, t. 14 (1868). —J.G. Baker in F.T.A. **2**: 42 (1871). —E.G. Baker in J. Linn. Soc., Bot. **42**: 321 (1914) pro majore parte. —Polhill in F.T.E.A., Leguminosae, Pap.: 864 (1971). —Drummond in Kirkia **8**: 218 (1972). —Gonçalves in Garcia de Orta, Sér. Bot. **5**: 70 (1982). —Polhill, Crotalaria Africa & Madagascar: 91, fig. 10/2, 7 (1982). —Lock, Leg. Afr. Check-list: 198 (1989). Type from the Sudan.

Crotalaria stewartii Baker in F.T.A. **2**: 32 (1871). —Verdoorn in Bothalia **2**: 406 (1928) pro majore parte excl. specim. *Rogers*. Type: Malawi/Mozambique, Shire–Chupanga area, *Stewart* (K, holotype).

Crotalaria polysperma subsp. *stewartii* (Baker) Baker f. in J. Linn. Soc., Bot. **42**: 321 (1914).

Erect well-branched annual, 30–120 cm tall, with branches densely covered usually by both appressed and longer spreading hairs (sometimes all spreading). Leaves all 3-foliolate; leaflets 3–7 × 1.5–3.5 cm, elliptic, pointed, pilose at least beneath; petioles shorter to longer than leaflets; stipules 8–16 mm long, linear-caudate or filiform.

Racemes lax, mostly 12–20-flowered; bracts filiform, 4–10 mm long; bracteoles curved-ascending from just below calyx, 3.5–10 mm long, linear-lanceolate. Calyx accrescent, 1.4–2(2.4) cm long, hairy; upper lobes lanceolate, c. 3 times as long as the tube. Standard subcircular, blue, pubescent outside; wings oblong, as long as the keel; keel 1.6–1.8 cm long, bent at right-angles in lower half, with a well developed practically straight beak, densely lanate-pilose on the upper edge. Pod (3.5)4–5 cm long, broadly oblong-clavate, shortly stipitate, pilose. Seeds c. 3.5 mm long, oblique-cordiform, slightly rugulose, pale brown, mottled darker.

Zambia. S: Mazabuka Distr., Nega Nega, fl. 19.ii.1995, *Bingham* 10394 (K). **Zimbabwe**. N: Mutoko (Mtoko), fl. & fr. 16.ii.1966, *Corby* 1452 (K; SRGH). **Malawi**. S: Zomba Distr., Nchisi (Chilwa) Island, fl. & fr. 11.iii.1955, *Jackson* 1657 (K; SRGH). **Mozambique**. Z: Morrumbala Distr., entre Morrumbala e Megaza, a 37 km do Morrumbala, fl. 13.vi.1949, *Barbosa & Carvalho* 3070 (K; LISC; LMA). T: de Zóbuè para Tete, fr. 16.vi.1941, *Torre* 2855 (LISC). MS: Chupanga–Mopeia, fl. & fr. 11.ix.1944, *Mendonça* 2035 (K; LISC).

Also in Sudan, Ethiopia, Uganda, Kenya and Tanzania; in the Flora Zambesiaca area virtually restricted to valleys of R. Zambezi and its tributaries (plants from Harare are cultivated). Tree or shrub savanna or grassland, sometimes in woodland, often in disturbed places; up to 1050 m.

14. **Crotalaria burkeana** Benth. in Hooker, London J. Bot. **2**: 593 (1843). —Harvey in F.C. **2**: 43 (1862). —E.G. Baker in J. Linn. Soc., Bot. **42**: 322 (1914). —Verdoorn in Bothalia **2**: 398 (1928). —Burtt Davy, Fl. Pl. Ferns Transvaal, pt. 2: 399 (1932). —Polhill, Crotalaria Africa & Madagascar: 91 (1982). —Lock, Leg. Afr. Check-list: 169 (1989). Syntypes from South Africa (Gauteng).

Perennial herb (sometimes flowering in first year), erect or decumbent, up to 70 cm tall, with 1–several patently pilose stems from an ultimately woody tuberous stock. Leaves 3–5-foliolate; leaflets 2.5–5.5 × 0.4–1.6 cm, linear-lanceolate to lanceolate, rarely elliptic, glabrescent above, pilose beneath; petioles shorter to ± as long as leaflets; stipules 8–17 mm long, linear-caudate. Racemes lax, 5–12(16)-flowered; bracts 3–8 mm long, linear-lanceolate, sometimes caudate; bracteoles curved-ascending from just below the calyx, 5–8 mm long, lanceolate, ± subfalcate. Calyx accrescent, 1.5–2.2 cm long, pilose; upper lobes oblong-lanceolate, 2–3 times as long as the tube. Standard subcircular or obovate, white, sometimes tinged yellow or mauve, veined brown or dark blue, subglabrous to pubescent outside; wings as long as keel; keel 1.5–1.8 cm long, abruptly rounded ± at right-angles in the lower third, with a well developed straight beak, lanate-pubescent along the upper edge. Pod (3.5)4–5.5 cm long, broadly oblong-clavate, subsessile, pilose. Seeds c. 4 mm long, oblique-cordiform, rugulose, brown.

Botswana. SE: 8 km NW of Derdepoort, fl. & fr. 30.xi.1954, *Codd* 8879 (BM; K; PRE; SRGH). **Mozambique**. GI: de Chibuto para Guijá (Caniçado), Macalane (Macalanane), fl. & fr. 2.vi.1959, *Barbosa & Lemos* 8560 (BR; COI; K; LMA). M: Boane, fl. & fr. 18.ix.1963, *Balsinhas* 609 (K; LISC).

Also in South Africa (Northern, North-West, Gauteng, Mpumalanga, Free State and northern KwaZulu-Natal), largely in the catchment area of the Limpopo R. Grassland, *Acacia* savanna, disturbed places, usually on sandy soils; up to 900 m.

15. **Crotalaria barkae** Schweinf. in Bull. Herb. Boissier **4**, Appendix 2: 226 (1896). —E.G. Baker in J. Linn. Soc., Bot. **42**: 390 (1914). —Hepper in F.W.T.A., ed. 2, **1**: 550 (1958). —Schreiber in Merxmüller, Prodr. Fl. SW. Afrika, fam. 60: 22 (1970). —Polhill in Kew Bull. **25**: 276 (1971); in F.T.E.A., Leguminosae, Pap.: 865 (1971). —Drummond in Kirkia **8**: 217 (1972). —Polhill, Crotalaria Africa & Madagascar: 93, fig. 12 (1982). —Lock, Leg. Afr. Check-list: 195 (1989). —Lock, Leg. Afr. Check-list: 167 (1989). Type from Eritrea.

Annual or short-lived perennial with erect leading shoot to 0.5(1) m tall and long decumbent lower branches, hirsute with mostly spreading hairs in the Flora Zambesiaca area. Leaves 3-foliolate; leaflets 1.5–7 × 0.5–2 cm, lanceolate to elliptic, sparsely pilose at least beneath; petioles shorter to longer than leaflets; stipules 2–11 mm long, filiform to linear-lanceolate. Racemes long-pedunculate, laxly 2–6-flowered; bracts 2–10 mm long, subulate to lanceolate; bracteoles curved-ascending from upper part of the pedicel, 2.5–9 mm long, obliquely lanceolate. Calyx accrescent, (9)10–20 mm long; upper lobes attenuately triangular-lanceolate to oblong-cordate, mostly 3–4 times as long as the tube. Standard obovate to subcircular, medially and apically pubescent outside; wings a little shorter than keel; keel (9)10–17 mm long, angled in lower half, with a long slightly incurved beak, lanate on upper margin proximal to the beak. Pod (1.5)1.8–3.5 cm long, clavate-ellipsoid, shortly stipitate, glabrous to thinly pilose (often only on upper side). Seeds 2–2.5 mm long, oblique-cordiform, usually obscurely rugulose, brown.

Widespread in the drier savanna regions of the Sudanian and Zambezian Domains of tropical Africa (generally absent from miombo woodland), with three subspecies in East Africa.

Subsp. **barkae** —Polhill in Kew Bull. **25**: 278 (1971); in F.T.E.A., Leguminosae, Pap.: 866, fig. 120 (1971); Crotalaria Africa & Madagascar: 94 (1982). —Lock, Leg. Afr. Check-list: 167 (1989). TAB. 3,7: **23**.

Crotalaria geminiflora Baker f. in J. Linn. Soc., Bot. **42**: 391 (1914). —Verdoorn in Bothalia **2**: 403 (1928). —Burtt Davy, Fl. Pl. Ferns Transvaal, pt. 2: 399 (1932). —Torre in C.F.A. **3**: 65 (1962). Type from Namibia.

Crotalaria homalocarpa Baker f. in J. Bot. **58**: 75 (1920). Type: Zimbabwe, Hwange (Wankie), *F.A. Rogers* 13302 (BM, holotype; K).

Bracts 0.5–1.5 mm wide, subulate to lanceolate. Upper calyx lobes attenuate-lanceolate to ovate-lanceolate, 2–4.5 times as long as wide, usually ciliate, but glabrous or nearly so on the surface. Standard usually bright yellow, rarely white; keel (9)10–13(14) mm long. Pod (1.5)1.8–2.6 cm long, usually hairy on the upper side, sometimes glabrous overall.

Botswana. N: Ngamiland Distr., Moremi Game (Wildlife) Reserve, on Maxwee–Khwai Gate road, fl. 3.iii.1976, *P.A. Smith* 1583 (K; SRGH). SE: Kgatleng Distr., Mochudi R., fl. 24.iii.1967, *Mitchison* A7 (K). **Zambia**. B: Sesheke Distr., Masese, fl. & fr. 14.iii.1961, *Fanshawe* 6430 (K; SRGH). C: Lusaka Distr., R. Kafue near Kafue town, Iolanda, fl. & fr. 14.iii.1965, *E.A. Robinson* 6431 (BR; K; M; SRGH). E: Chipata Distr., Jumbe, fl. & fr. 24.iii.1963, *Verboom* 805 (K; LISC; SRGH). **Zimbabwe**. W: Umguza Distr., Redbank, Whinburn Farm, fl. & fr. 20.i.1950, *Plowes* 1484 (K; LISC; SRGH). C: Chegutu Distr., Mupfure (Umfuli) R., Poole Farm, fl. & fr. 24.ii.1953, *R.M. Hornby* 3308 (K; SRGH). **Malawi**. N: 16 km NW of Rumphi, fl. & fr. 11.iii.1978, *Pawek* 14034 (MAL; MO; K; SRGH). S: Mangochi Distr., Monkey Bay, fl. & fr. 25.iii.1974, *Patel* 96 (K; SRGH). **Mozambique**. N: Cuamba Distr., andados 3 km de Cuamba (Nova Freixo) para Mutuáli, c. 600 m, fl. & fr. 21.ii.1964, *Torre & Paiva* 10711 (LISC).

Distribution as for species. Grassland, tree savanna and mopane woodland, streamsides, disturbed places, usually on sandy soils; 550–1350 m.

16. **Crotalaria incana** L., Sp. Pl. **2**: 716 (1753). —Milne-Redhead in Kew Bull. **15**: 159 (1961). —Polhill in F.T.E.A., Leguminosae, Pap.: 869 (1971); Crotalaria Africa & Madagascar: 95, fig. 13 (1982). —Lock, Leg. Afr. Check-list: 182 (1989). Syntypes from Jamaica.

Erect or spreading usually rather bushy herb, up to 1.5(3.5) m tall, variously hairy. Leaves 3-foliolate; leaflets up to 2.5–5 × 1.7–4.5 cm, elliptic-obovate to obovate or subcircular, rounded to retuse at apex, usually thinly pilose beneath; petioles mostly longer than leaflets; stipules 2.5–12(20) mm long, filiform. Racemes usually lax, few–many-flowered; bracts 1–10 mm long, subulate or linear; bracteoles curved-ascending from just below calyx, 2–6 mm long, linear. Calyx 7–11 mm long, subglabrous to spreading pilose; lobes narrow, ± twice as long as tube. Standard elliptic, yellow, veined reddish-brown or purple, glabrous or medially pubescent outside; wings nearly as long as keel; keel 8–11.5 mm long, bent at right-angles in lower half, lanate-pilose on upper margin. Pod 3–4.5 cm long, fusiform-clavate, subsessile, somewhat pointed, pilose. Seeds 2.5–3 mm long, oblique-cordiform, smooth or obscurely papillose, pale brown, mottled darker, or olive-green.

Subsp. **incana** —Milne-Redhead in Kew Bull. **15**: 159 (1961). —Polhill in F.T.E.A., Leguminosae, Pap.: 870 (1971); Crotalaria Africa & Madagascar: 97 (1982). —Lock, Leg. Afr. Check-list: 182 (1989).

Crotalaria incana forma *glabrescens* R. Wilczek in Bull. Jard. Bot. État **23**: 147 (1953); in F.C.B. **4**: 150 (1953). Type from Dem. Rep. Congo.

Stem shortly pubescent, without long spreading hairs. Bracts 1–3 mm long. Calyx lobes subglabrous. Seeds brown, mottled darker, obscurely papillose.

Zimbabwe. E: Mutare Distr., Vumba, fl. & fr., ii.1959, *Head* s.n. (BM).

Tab. 3,7: **23**. CROTALARIA BARKAE subsp. BARKAE. 1, flowering and fruiting branch (× 1); 2, flower (× 2); 3, calyx (× 2); 4, standard (× 2); 5, wing (× 2); 6, keel (× 2); 7, gynoecium (× 2), 1–7 from *Polhill & Paulo* 1282; 8, seed (× 3), from *Polhill & Paulo* 1953. Drawn by Roger Polhill. From F.T.E.A.

Pantropical, almost certainly originating in South America but so readily naturalized that native distribution now obscure. Weed.

Subsp. **purpurascens** (Lam.) Milne-Redh. in Kew Bull. **15**: 159 (1961). —Polhill in F.T.E.A., Leguminosae, Pap.: 870 (1971). —Drummond in Kirkia **8**: 218 (1972). —Gonçalves in Garcia de Orta, Sér. Bot. **5**: 66 (1982). —Polhill, Crotalaria Africa & Madagascar: 97, fig. 13 (1982). —Lock, Leg. Afr. Check-list: 182 (1989). Type from Madagascar, cultivated in Paris.
Crotalaria purpurascens Lam., Encycl. Méth. Bot. **2**: 200 (1786).
Crotalaria incana sensu J.G. Baker in F.T.A. **2**: 31 (1871). —sensu E.G. Baker in J. Linn. Soc., Bot. **42**: 357 (1914), pro parte. —sensu Verdoorn in Bothalia **2**: 398 (1928) pro majore parte excl. specim. *Eyles*. —sensu Brenan, Check-list For. Trees Shrubs Tang. Terr.: 416 (1949). —sensu Wilczek in F.C.B. **4**: 148 (1953). —sensu Hepper in F.W.T.A., ed. 2, **1**: 550 (1958) pro parte.
Crotalaria incana forma *lanata* R. Wilczek in Bull. Jard. Bot. État **23**: 147 (1953); in F.C.B. **4**: 150 (1953). Type from Dem. Rep. Congo.

Stem with long yellow-brown (or rarely white) spreading hairs. Bracts mostly 4–10 mm long. Calyx lobes spreading pilose. Seeds smooth, dark olive-green.

Zambia. N: Mbala Distr., Kapata, fl. & fr. 11.ix.1958, *Fanshawe* 4795 (K; LISC; NDO; SRGH). **Zimbabwe**. N: locality and date not recorded, *Gebbie* in *GHS* 3994 (SRGH). C: Marondera (Marandellas), Grasslands Research Station, fl. & fr. 17.ii.1967, *Corby* 1766 (K; LISC; SRGH). E: Mutare Distr., Engwa, fl. 1.ii.1955, *Exell, Mendonça & Wild* 24 (BM; SRGH). **Malawi**. N: Viphya, fl. & fr. 8.vii.1952, *Jackson* 934 (K). C: Dedza Distr., south of Dedza, Chincherere Hill, fr. 24.iv.1971, *Pawek* 4698 (K). S: Zomba Distr., Likangala, fl. & fr. 15.vi.1956, *Jackson* 1863 (K; SRGH). **Mozambique**. N: Lichinga (Vila Cabral), fl. & fr. 8.viii.1934, *Torre* 275 (BM; COI; LISC). T: Angónia Distr., andados 4 km de Calóbuè para Vila Coutinho, c. 1550 m, fl. & fr. 7.iii.1964, *Torre & Paiva* 11050 (LISC). MS: Messambize Valley, fl. & fr. 5.iii.1948, *Pedro* 3646 (LMA).

Principally in the highlands of eastern Africa and Madagascar, but with scattered records in western Africa and doubtfully in South America. Montane forest margins and secondary associations, grassland, disturbed places, readily becomes a weed of cultivation; 1000–1800 m.

17. **Crotalaria lotoides** Benth. in Hooker, London J. Bot. **2**: 575 (1843). —Harvey in F.C. **2**: 42 (1862). —E.G. Baker in J. Linn. Soc., Bot. **42**: 399 (1914). —Verdoorn in Bothalia **2**: 387 (1928). —Burtt Davy, Fl. Pl. Ferns Transvaal, pt. 2: 398 (1932). —Miller in J. S. African Bot. **18**: 30 (1952). —Drummond in Kirkia **8**: 218 (1972). —Polhill, Crotalaria Africa & Madagascar: 97 (1982). —Lock, Leg. Afr. Check-list: 189 (1989). Syntypes from South Africa (Gauteng).

Spreading or semidecumbent perennial, with a number of densely subappressed pubescent to patently pilose stems to 15–50 cm long. Leaves 3-foliolate; leaflets 8–20 × 8–18 mm, obovate to subcircular, rounded to emarginate at apex, glabrous to thinly pilose above, pilose beneath; petioles shorter to longer than leaflets; stipules 2–5 mm long, subulate or linear. Racemes laxly 1–3(4)-flowered, with peduncle much longer than the rhachis; bracts 2–4 mm long, linear-lanceolate; bracteoles curved-ascending from upper part of pedicel, similar to bract. Calyx 7–9 mm long, thinly pilose; lobes attenuately lanceolate-triangular, 4–5 times as long as the tube. Standard elliptic, yellow, ± flushed reddish, medially puberulous outside; wings 0.6–0.7(0.8) times as long as the keel; keel 8–11 mm long, abruptly rounded in lower half, with a practically straight beak, lanate-pubescent along upper margin. Pod 2.4–3.6 cm long, subcylindrical, practically sessile, somewhat pointed, pilose. Seeds 3 mm long, oblique-cordiform, rugulose, pale yellow to brown.

Botswana. SE: Mahalapye, fl. & fr. 17.xii.1958, *de Beer* 796 (K; SRGH); 175 km west of Gaborone on road to Molepolole, fl. & fr. 31.iii.1977, *Hansen* 3100 (GAB; K; PRE; SRGH). **Zimbabwe**. W: Bulawayo Distr., fr. *Eyles* 84 (K; SRGH); Matobo Distr., fr. 3.ii.1948, *West* 2671 (K; SRGH).

Also in South Africa (Northern Cape, North-West, Gauteng and Northern Province). Little ecological information for the Flora Zambesiaca area, but probably *Acacia* or *Terminalia* tree savanna, grassland and disturbed places, on sandy soils; 950–1350 m.

18. **Crotalaria stolzii** (Baker f.) Polhill in Kew Bull. **22**: 221 (1968); in F.T.E.A., Leguminosae, Pap.: 871 (1971). —Drummond in Kirkia **8**: 219 (1972). —Polhill, Crotalaria Africa & Madagascar: 99, fig. 14/1, 7 (1982). —Lock, Leg. Afr. Check-list: 205 (1989). Type from southern Tanzania.

Crotalaria xanthoclada var. *stolzii* Baker f. in J. Linn. Soc., Bot. **42**: 412, t. 14B (1914). —Verdoorn in Bothalia **2**: 380 (1928). —Wilczek in F.C.B. **4**: 121 (1953). —Binns, First Check List Herb. Fl. Malawi: 80 (1968).

Straggling slightly woody herb ascending to 1–3 m, with lax spreading glabrescent branches. Leaves 3-foliolate; leaflets mostly 1.5–3 × 0.4–1 cm, oblanceolate, sparsely puberulous beneath; petiole shorter than leaflets; stipules leaf-like, (10)15–25 × 6–15 mm, ovate- or lanceolate-cordate, subequal-sided. Racemes sublaxly 12–20-flowered; bracts up to 7 mm long, spathulate, acuminate, early caducous. Calyx 7–8 mm long, appressed puberulous; lobes attenuately triangular, as long as to longer than the tube. Standard subcircular, bright yellow, glabrous outside; wings as long as keel; keel 10–13 cm long, somewhat angled about the middle with a pointed slightly incurved beak, ciliate-lanate along upper edge. Pod mostly 16–18 × 5–8 mm, shortly stipitate, ± oblong, slightly compressed laterally, narrowed at either end, glabrous, mostly 6–8-seeded. Seeds c. 2.5 mm long, oblique-cordiform, minutely punctate, greenish-brown.

Zimbabwe. E: Nyanga Distr., Juliasdale, Cotswold Farm, fl. 1.iv.1961, *Leach* 10770 (K; LISC; M; SRGH); Mutare Distr., SW Vumba, Zimunya's C.L. (Reserve), fl. 7.iv.1957, *Chase* 6394 (K; LISC; P; SRGH). **Malawi**. N: Rumphi Distr., Nyika Plateau, Chelinda Bridge, fl. 29.iii.1970, *Pawek* 3418 (K). C: Dedza Distr., Chongoni Forest, Namungu Stream, fr. 10.vi.1972, *Salubeni* 1818 (MO).

Also in Kenya, Tanzania and Dem. Rep. Congo (Katanga). Montane forest, usually by streams, sometimes extending into open secondary grassland; 1200–2250 m.

19. **Crotalaria natalitia** Meisn. in Hooker, London J. Bot. **2**: 67 (1843). —Harvey in F.C. **2**: 46 (1862). —E.G. Baker in J. Linn. Soc., Bot. **42**: 410 (1914) excl. var. *sengensis.* —Eyles in Trans. Roy. Soc. South Africa **5**: 371 (1916). —Verdoorn in Bothalia **2**: 381 (1928). —Burtt Davy, Fl. Pl. Ferns Transvaal, pt. 2: 397 (1932). —Hutchinson, Botanist South. Africa: 485, 496 (1946). —Merxmüller in Proc. & Trans. Rhodesia Sci. Assoc. **43**: 17 (1951). —Brenan, Check-list For. Trees Shrubs Tang. Terr.: 413 (1949); in Mem. New York Bot. Gard. **8**: 248 (1953). —Wilczek in F.C.B. **4**: 122 (1953). —Topham, Check List For. Trees Shrubs Nyasaland Prot.: 74 (1958). —White, F.F.N.R.: 146, fig. 30A (1962). —Binns, First Check List Herb. Fl. Malawi: 79 (1968). —Polhill in F.T.E.A., Leguminosae, Pap.: 871, fig. 121 (1971); Crotalaria Africa & Madagascar: 100, fig. 15 (1982). —Lock, Leg. Afr. Check-list: 193 (1989). Type from South Africa (KwaZulu-Natal).

Woody herb or small shrub, 1–2.6 m tall, with 1–several thin stems and rather strictly ascending pubescent branches mostly towards the top, rarely somewhat procumbent in burned and heavily grazed places. Leaves mostly in tufts, 3-foliolate; leaflets mostly (1.5)2–4.5 × 0.4–1.5 cm, linear-oblanceolate to oblanceolate-elliptic, appressed pubescent beneath; petioles mostly a little shorter than leaflets; stipules (3)5–15 × 1–4 mm, shortly stalked, linear-lanceolate to oblong-falcate, very unequal-sided. Racemes terminal, up to 10–22 cm long, with few–many fairly closely to rather laxly arranged flowers; bracts 2.5–6 mm long, linear to elliptic-caudate, caducous; pedicels rather stiffly erect. Calyx 6–9(10) mm long; upper lobes narrowly attenuate-triangular, as long as to a little longer than the tube. Standard subcircular, yellow, sometimes reddish-brown tinged outside, fading orange-red, glabrous outside; wings broad, exceeding the keel; keel 11–14 mm long, abruptly rounded about the middle, with a rather short slightly incurved beak, lanate towards the upper margin. Pod 3.4–4(4.5) × 1.1–1.5 cm, broadly cylindrical, abruptly contracted to a 4–5 mm long stipe, glabrous, often mottled dark green to blue-black before ripening, 25–40-seeded. Seeds 3.5–4 mm long, oblong-reniform, with a small aril, granulate or smooth, dark brown.

Var. **natalitia** —Drummond in Kirkia **8**: 218 (1972). —Gonçalves in Garcia de Orta, Sér. Bot. **5**: 68 (1982).

Crotalaria gymnocalyx Baker in Bull. Misc. Inform., Kew **1897**: 252 (1897). Syntypes: Malawi, Chitipa (Fort Hill), *Whyte* s.n.; Mpata–Tanganyika Plateau, *Whyte* s.n. (K, syntypes).

Calyx and upper part of the pedicel glabrous.

Zambia. N: Kasama–Senga Hill, fl. & fr. 24.ii.1962, *Richards* 16162 (K; LISC; SRGH). W: Kitwe, fl. & fr. 20.v.1955, *Fanshawe* 2294 (BR; K; NDO; SRGH). C: Mkushi Distr., 8 km east of

Chiwefwe, fl. & fr. 15.vii.1930, *Hutchinson & Gillett* 3670 (BM; BR; COI; K; LISC; SRGH). E: Katete, fl. 24.iii.1955, *Exell, Mendonça & Wild* 1158 (BM; LISC; SRGH). S: Choma Distr., 3 km Kanchomba–Sikalongo, fl. & fr. 19.ii.1960, *White* 7253 (FHO; K; SRGH). **Zimbabwe**. N: Mutoko Reserve, fl. iv.1956, *R.M. Davies* 1925 (K; LISC; SRGH). C: Rusape Distr., fl. & fr. iv.1953, *Dehn* 366/53 (K; M; SRGH). E: Nyanga Distr., Juliasdale, Pienaar's Farm, fl. & fr. 11.ii.1961, *Rutherford-Smith* 508 (K; SRGH). S: 6 km north of Runde (Lundi) R., near Madzivire Dip, fr. 4.v.1962, *Drummond* 7926 (K; SRGH). **Malawi**. N: Nkhata Bay Distr., Mazamba Forest, fl. 27.iv.1967, *Salubeni* 684 (K; SRGH). C: Chongoni–Lilongwe, fl. & fr. 20.vi.1962, *Richards* 16778 (K). S: Zomba Plateau, fl. & fr. 5.vi.1946, *Brass* 16257 (BM; K; L; NY; SRGH). **Mozambique**. N: Lichinga (Vila Cabral), fl. & fr. v.1934, *Torre* 18 (BM; COI; LISC). Z: Milange, fr. 9.ix.1941, *Torre* 3381 (K; LISC). T: Angónia Distr., base do monte Dómuè, c. 1400 m, fl. & fr. 9.iii.1964, *Torre & Paiva* 11063 (LISC). MS: Gorongosa, serra da Gorongosa, monte Unora, a caminho das quedas de agua, na propriedade do Sr. M. Ferrao, c. 800 m, fl. & fr. 6.v.1964, *Torre & Paiva* 12256 (LISC). GI: Jangamo Distr., Cumbana (Cumbane), fl. iv.1936, *Gomes e Sousa* 1747 (BR; COI; K).

Plateau regions of eastern Africa from Ethiopia to South Africa (KwaZulu-Natal), descending to the coast in places. Montane forest edges, miombo woodland, tree savanna, montane and valley grassland, near lakes and streams, disturbed places and sometimes a weed of cultivation; near sea-level to 1700 m.

Var. **rutshuruensis** De Wild. in Bull. Jard. Bot. État **8**: 137 (1923). —Wilczek in F.C.B. **4**: 123 (1953). —Drummond in Kirkia **8**: 218 (1972). —Polhill in F.T.E.A., Leguminosae, Pap.: 872 (1971); Crotalaria Africa & Madagascar: 101 (1982). Type from Dem. Rep. Congo (Kivu).

Crotalaria natalitia sensu J.G. Baker in F.T.A. **2**: 34 (1871). —sensu Verdoorn in Bothalia **2**: 381 (1928) quoad specim. *Teague*. —sensu Brenan, Check-list For. Trees Shrubs Tang. Terr.: 413 (1949) pro parte. —sensu Torre in C.F.A. **3**: 73 (1962).

Calyx and upper part of the pedicel appressed-pubescent.

Zambia. E: Isoka Distr., Nyika Plateau, Kangampande Mt., fl. 2.v.1952, *White* 2558 (FHO; K). **Zimbabwe**. E: Nyanga Distr., Punch Rock, fl. & fr. ix.1959, *R.L. Williams* 175A (K; SRGH). **Malawi**. N: Rumphi Distr., Nyika Plateau, Lake Kaulime, fl. 3.ix.1962, *Tyrer* 787 (BM). **Mozambique**. N: Lichinga (Vila Cabral), fl. & fr. 15.ix.1958, *Rui Monteiro* 66 (LISC). Z: Gurué Distr., near Namuli Peak, fl. & fr. 8.iv.1943, *Torre* 5122 (LISC).

Relatively uncommon and only in mountainous parts of the Flora Zambesiaca area, becoming more common north of the Flora Zambesiaca area and also extending the range of the species to Angola. Montane forest edges, tree savanna, streamsides; c. 1200–2200 m.

20. **Crotalaria rhodesiae** Baker f. in J. Linn. Soc., Bot. **42**: 401 (1914). —Verdoorn in Bothalia **2**: 381 (1928). —Merxmüller in Proc. & Trans. Rhodesia Sci. Assoc. **43**: 94 (1951). —Brenan in Mem. New York Bot. Gard. **8**: 248 (1953). —Binns, First Check List Herb. Fl. Malawi: 79 (1968). —Polhill in F.T.E.A., Leguminosae, Pap.: 874 (1971). —Drummond in Kirkia **8**: 218 (1972). —Jacobsen in Kirkia **9**: 160 (1973). —Polhill, Crotalaria Africa & Madagascar: 101, fig. 14/8 (1982). —Lock, Leg. Afr. Check-list: 201 (1989). Type: Zambia, Kalomo, *F.A. Rogers* 8214 (K, holotype; SRGH).

Crotalaria natalitia var. *procumbens* Baker f., Legum. Trop. Africa: 58 (1926). Type: Zimbabwe, Harare, *Eyles* 2161 (BR; K, holotype; P; PRE; SRGH).

Crotalaria natalitia var. *pseudo-rhodesiae* Merxm. in Proc. & Trans. Rhodesia Sci. Assoc. **43**: 94 (1951). Type: Zimbabwe, Marondera (Marandellas), *Dehn* 164a (M, holotype).

Prostrate perennial, with many well branched puberulous radiating stems. Leaves mostly well spaced, 3-foliolate; leaflets (1)1.5–3.2 × (0.4)0.6–1.5 cm, oblanceolate, oblanceolate-elliptic or obovate, appressed puberulous beneath; petiole shorter than leaflets; stipules (3)5–10 mm long, shortly stalked, linear to lanceolate-falcate, acuminate, markedly unequal-sided. Racemes up to 7 cm long, laxly 3–8-flowered; bracts up to 3 mm long, subulate or filiform, caducous. Calyx 8–11(13) mm long, appressed puberulous; lobes narrowly attenuate-triangular, longer than the tube. Standard subcircular, yellow, often reddish tinged outside, fading orange, glabrous outside; wings broadly oblong, exceeding the keel; keel 11–13 mm long, abruptly rounded in the lower half, tapered gradually into a short straight blunt beak, lanate towards upper edge. Pod 20–27 × 8–10 mm, subsessile, cylindrical, with scattered hairs or pubescent only along upper side, 16–22-seeded. Seeds 3–3.5 mm long, oblong-reniform, with a small aril, finely granulate or smooth, orange, red or rarely pale yellow.

Zambia. N: Samfya, fl. & fr. 10.vii.1955, *Fanshawe* 2372 (BR; K; NDO; SRGH). C: Lusaka, fl. & fr. 29.v.1958, *E.A. Robinson* 2871 (K; M; SRGH). S: Choma Distr., Choma National Forest (Siamambo Forest Reserve), fl. 13.iii.1960, *White* 7748 (FHO; K). **Zimbabwe**. N: Zvimba Distr., Trelawney, fl. & fr. 27.iii.1943, *Jack* 148 (K; LISC; PRE; SRGH). C: Chegutu Distr., Poole Farm, fl. & fr. 26.iv.1953, *R.M. Hornby* 3306 (K; LISC; SRGH). E: Mutare Distr., Umtali Sports Ground, fl. & fr. 15.iii.1960, *Chase* 7303 (BM; K; M; SRGH). S: Masvingo Distr., Makaholi Experimental Farm, fl. 23.iii.1948, *D.A. Robinson* 315 (K; SRGH). **Malawi**. C: Dedza, fl. 13.ix.1946, *Brass* 17636 (K; NY; SRGH).

Also in southern Tanzania and South Africa (Northern Province). Miombo woodland, grassland (sometimes in seasonal seepage areas) and disturbed places, usually on sandy soils; 900–1900 m.

21. **Crotalaria cylindrocarpa** DC., Prodr. **2**: 133 (1825). —J.G. Baker in F.T.A. **2**: 40 (1871). —E.G. Baker in J. Linn. Soc., Bot. **42**: 414 (1914). —Verdoorn in Bothalia **2**: 383 (1928). —Wilczek in F.C.B. **4**: 126 (1953). —Hepper in F.W.T.A., ed. 2, **1**: 548 (1958). —Torre in C.F.A. **3**: 75 (1962). —Polhill in Kew Bull. **22**: 224 (1968); in F.T.E.A., Leguminosae, Pap.: 874 (1971); Crotalaria Africa & Madagascar: 102 (1982). —Lock, Leg. Afr. Check-list: 173 (1989). Type from Senegal.

Crotalaria pseudopodocarpa R.E. Fr., Wiss. Ergebn. Schwed. Rhod.-Kongo-Exped. **1**: 75 (1914). —Verdoorn in Bothalia **2**: 383 (1928). Type: Zambia, Kabendemushi (Kawendi Musi), *R.E. Fries* 801 (K, fragment; UPS, holotype).

Perennial up to 60 cm tall, with generally several pubescent ascending or spreading stems. Leaves 3-foliolate; leaflets variable, (1.5)2–7 × 0.4–2.3 cm, oblong-oblanceolate to elliptic, subrhombic or obovate-oblong, subglabrous to appressed pubescent beneath; petioles mostly 1–3 cm long; stipules (4)7–14 mm long, obliquely linear-lanceolate to falcate, acuminate, markedly unequal-sided. Racemes 3.5–23 cm long, with few to many flowers rather closely arranged at least towards the top; bracts 2–5(8) mm long, filiform, early caducous. Calyx 6–8(10) mm long, appressed puberulous; lobes triangular-subulate, longer than the tube. Standard ovate, yellow, marked reddish-brown or orange, glabrous or practically so outside; wings as long as the keel; keel (1)1.15–1.5 cm long, abruptly rounded in the lower half, with a straight tapered beak, lanate towards the upper edge. Pod 2.3–3 × 0.8–1 cm, cylindrical, abruptly contracted to the 1–4 mm long stipe, thinly appressed puberulous, c. 18–26-seeded. Seeds c. 4 mm long, oblong-reniform, with a very small aril, minutely granulate, pale yellow to brown.

Zambia. B: Mongu, fl. & fr. 30.iv.1964, *Verboom* 1052 (K). N: Samfya Distr., Kabendemushi (Kawendi Musi), fl. & fr. 26.ix.1911, *R.E. Fries* 801 (K; UPS). W: Kitwe Distr., Mindolo, fl. & fr. 26.ii.1966, *Mutimushi* 1259 (K; NDO; SRGH). S: near Mumbwa, 1911, *Macaulay* 608 (K).

Also in West Africa, Gabon, Dem. Rep. Congo, Uganda, Kenya and Angola. Miombo woodland, grassland, edges of streams and dambos, disturbed places; c. 900–1300 m.

Plants from the southern part of the range, including those from the Flora Zambesiaca area, tend to have relatively long stipes to the pod and a more condensed shoot arrangement (Polhill, loc. cit. (1968)).

22. **Crotalaria goreensis** Guill. & Perr. in Guillemin, Perrottet & Richard, Fl. Seneg. Tent.: 165 (1832). —J.G. Baker in F.T.A. **2**: 28 (1871). —E.G. Baker in J. Linn. Soc., Bot. **42**: 413 (1914). —Verdoorn in Bothalia **2**: 383 (1928). —Wilczek in F.C.B. **4**: 124, fig. 10 (1953). —Hepper in F.W.T.A., ed. 2, **1**: 548 (1958). —Torre in C.F.A. **3**: 74 (1962). —Polhill in F.T.E.A., Leguminosae, Pap.: 875 (1971). —Drummond in Kirkia **8**: 218 (1972). —Jacobsen in Kirkia **9**: 160 (1973). —Polhill, Crotalaria Africa & Madagascar: 104, fig. 14/3, 6, 9 (1982). —Lock, Leg. Afr. Check-list: 180 (1989). Syntypes from Senegal.

Annual or short-lived perennial of rather varied habit, 0.3–2.6 m tall; branches densely pubescent with appressed to slightly spreading hairs. Leaves 3-foliolate; leaflets variable, 2.5–8.5 × 0.8–2.5 cm, linear-oblanceolate to lanceolate, oblong-elliptic or obovate, appressed pubescent at least beneath; petioles mostly 3.5–6.5 cm long; stipules 9–25 × 3–7 mm, oblong-falcate, acuminate, very unequal-sided. Racemes 4–24 cm long, with many flowers crowded at least towards the top; bracts 3–4 mm long, subulate or filiform, early caducous. Calyx 4–5 mm long, pubescent; lobes subulate-triangular, ± as long as the tube. Standard ovate, yellow, veined or suffused with orange, brown or purple, glabrous outside; wings a little shorter than the keel; keel 8.5–10(11.5) mm long, angled in the lower half, with a straight beak, lanate towards the upper edge. Pod (1.2)1.5–2 × 0.7–0.9 cm, subsessile, oblong-ellipsoid, pubescent, 12–16-seeded. Seeds c. 4 mm long, oblong-reniform, with a small aril, minutely granulate, sometimes radially veined, yellow, usually tinged orange or red.

Zambia. B: Mongu, fl. & fr. 28.iii.1964, *Verboom* 1018 (K; SRGH). N: Kawambwa Distr., Mbereshi, fl. & fr. 24.vi.1957, *E.A. Robinson* 2399 (K; M; SRGH). W: Kitwe, fl. & fr. 20.v.1955, *Fanshawe* 2292 (K; LISC; NDO; SRGH). C: Lusaka Distr., Mt. Makulu Research Station, fl. 2.iv.1957, *Angus* 1537 (K; LISC; NDO; SRGH). E: Chipata Distr., Nsefu, fr. 12.iv.1968, *R. Phiri* 154 (K). S: Mazabuka Distr., Magoye, fl. & fr. 14.iv.1952, *White* 2658 (FHO; K). **Zimbabwe**. N: Makonde Distr., Chinhoyi (Sinoia), near Umboe Club, fl. & fr. 5.iv.1966, *Corby* 1605 (K; LISC; SRGH). **Malawi**. N: Karonga, fl. & fr. 20.vii.1952, *J. Williamson* 42 (BM). **Mozambique**. N: Mocímboa da Praia Distr., andados 5 km de Mocímboa da Praia para Diaca, c. 20 m, fl. 14.iv.1964, *Torre & Paiva* 11922 (LISC).

Throughout most of tropical Africa except the NE Horn and the extreme south. Essentially a species of grassland and disturbed places, particularly in seasonally damp spots by rivers, dambos, roadsides, etc., but also recorded from various woodland and tree savanna associations; 20–1950 m.

23. **Crotalaria podocarpa** DC., Prodr. **2**: 133 (1825). —J.G. Baker in F.T.A. **2**: 17 (1871). —E.G. Baker in J. Linn. Soc., Bot. **42**: 406 (1914). —Eyles in Trans. Roy. Soc. South Africa **5**: 371 (1916). —Verdoorn in Bothalia **2**: 381 (1928). —Hepper in F.W.T.A., ed. 2, **1**: 548 (1958). —Torre in C.F.A. **3**: 72 (1962). —Polhill in Kew Bull. **22**: 224 (1968). —Schreiber in Merxmüller, Prodr. Fl. SW. Afrika, fam. 60: 26 (1970). —Polhill in F.T.E.A., Leguminosae, Pap.: 876 (1971). —Drummond in Kirkia **8**: 218 (1972). —Gonçalves in Garcia de Orta, Sér. Bot. **5**: 70 (1982). —Polhill, Crotalaria Africa & Madagascar: 104, fig. 17/1–10 (1982). —Lock, Leg. Afr. Check-list: 197 (1989). TAB. 3,7: **24**, fig. A. Type from Senegal.

Crotalaria arvensis Klotzsch in Peters, Naturw. Reise Mossambique **6**, pt. 1: 53 (1861). Type: Mozambique, Sena, *Peters* (B†, holotype).

Crotalaria pilifera Klotzsch in Peters, Naturw. Reise Mossambique **6**, pt. 1: 54 (1861). Syntypes: Mozambique, Mossuril (Mesuril) and Cabaceira, *Peters* (B†, syntypes).

Crotalaria flexuosa Baker in F.T.A. **2**: 29 (1871). Syntypes: "South tropical Africa", *Baines*; Mozambique, Tete, *Kirk*; and near Tete, *Kirk* (all K, syntypes).

Crotalaria fulgida Baker in F.T.A. **2**: 32 (1871). —E.G. Baker in J. Linn. Soc., Bot. **42**: 408 (1914). Type from Angola.

Crotalaria belckii Schinz in Verh. Bot. Vereins Prov. Brandenburg **30**: 159 (1888). Type from Namibia.

Crotalaria podocarpa var. *villosa* Schinz in Verh. Bot. Vereins Prov. Brandenburg **30**: 160 (1888). —E.G. Baker in J. Linn. Soc., Bot. **42**: 407 (1914). Type from Namibia.

Crotalaria hirsutissima Schinz in Bull. Herb. Boissier, sér. 2, **3**: 819 (1903). —E.G. Baker in J. Linn. Soc., Bot. **42**: 409 (1914). —Verdoorn in Bothalia **2**: 383 (1928). Type from Namibia.

Crotalaria mutabilis Schinz in Bull. Herb. Boissier, sér. 2, **3**: 820 (1903). —E.G. Baker in J. Linn. Soc., Bot. **42**: 408 (1914). Syntypes from Namibia.

Crotalaria podocarpa subsp. *flexuosa* (Baker) Baker f. in J. Linn. Soc., Bot. **42**: 407 (1914).

Crotalaria podocarpa subsp. *belckii* (Schinz) Baker f. in J. Linn. Soc., Bot. **42**: 407 (1914).

Crotalaria damarensis var. *maraisiana* Torre in Mem. Junta Invest. Ultramar, sér. 2, **19**: 48, t. 27 (1960); in C.F.A. **3**: 73 (1962). Type from Angola.

Annual with an erect stem to 70 cm tall and long spreading or decumbent lower branches, with generally spreading hears. Leaves 3-foliolate; leaflets variable, mostly 1.5–5(8) × 0.3–1.6(2.5) cm, linear or oblong-lanceolate to obovate-elliptic, usually thinly hairy at least beneath; petioles mostly shorter than the leaflets; stipules 5–17 × 3–7 mm, shortly stalked, linear-lanceolate to lanceolate-falcate, acuminate, usually truncate or rounded at the base, very unequal-sided. Racemes (3)6–14 cm long, long-pedunculate, laxly 2–5(8)-flowered; bracts up to 3.5 mm long, filiform and caducous, or lacking; pedicels 2.5–6 mm long. Calyx 7–9 mm long, hairy or subglabrous; lobes triangular-subulate, longer than the tube, ultimately reflexed. Standard ovate, clear yellow, sometimes slightly reddish tinged, glabrous outside; wings a little shorter than the keel; keel 1.2–1.7 cm long, angled in the lower half, with a straight beak, lanate towards the upper edge (particularly on the beak). Pod 2–3 × 0.9–1.4 cm, cylindrical, often rather broadly so and somewhat inflated when young, abruptly contracted to the slender 2.5–5 mm long stipe, glabrous, 6–12-seeded. Seeds 4.5–6 mm in diameter, subcircular-reniform, compressed with a broad raised rim, irregularly tuberculate, orange or pale yellow.

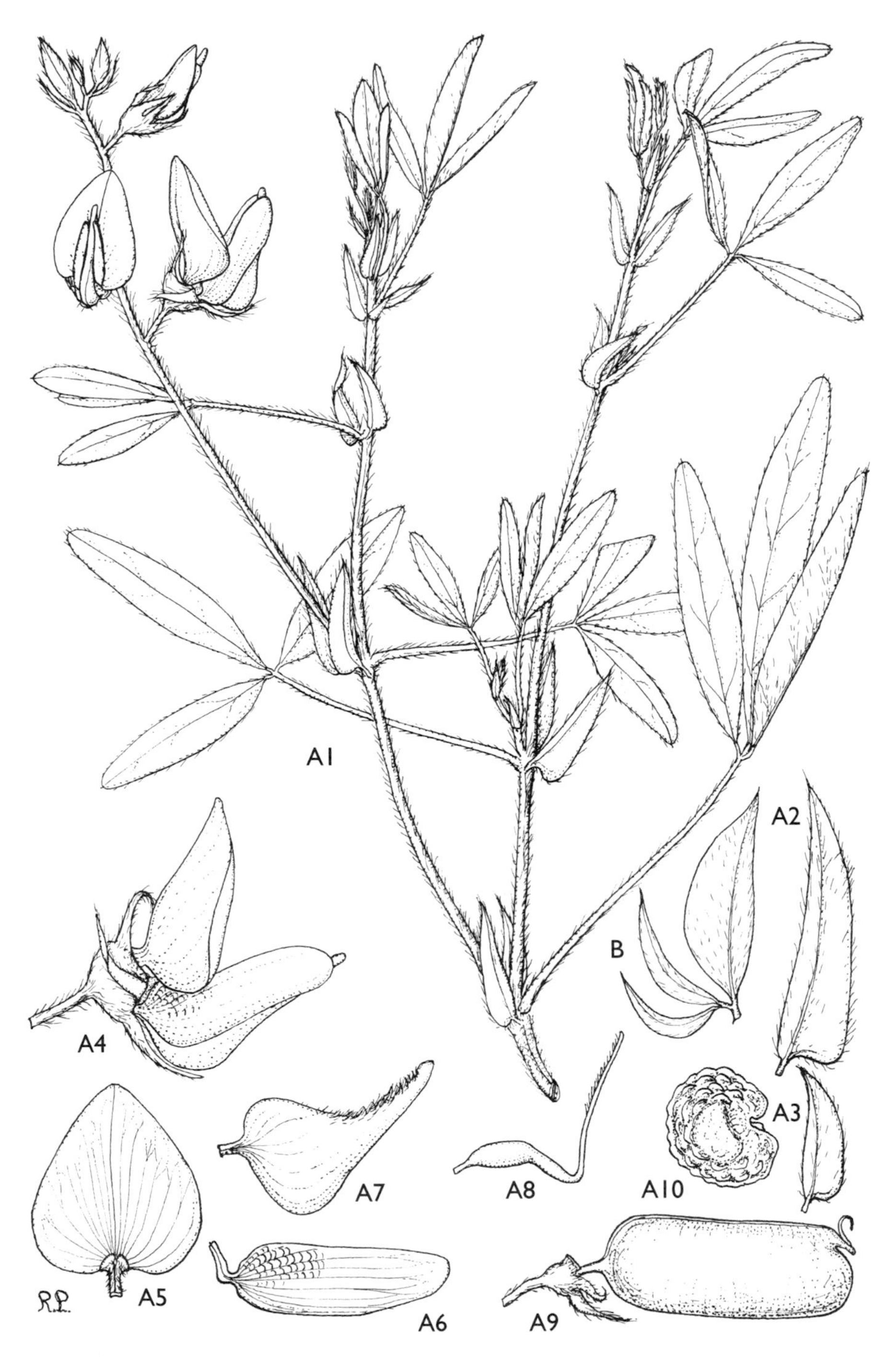

Tab. 3,7: **24**. A. —CROTALARIA PODOCARPA. A1, flowering branch (× 1); A2 & A3, stipules (× 2); A4, flower (× 2); A5, standard (× 2); A6, wing (× 2); A7, keel (× 2); A8, gynoecium (× 2), A1–A8 from *Richards* 14424; A9, pod (× $1^1/_3$), from *de Winter* 2673; A10, seed (× 3), from *Bogdan* 5173. B. —CROTALARIA DAMARENSIS, stipule (× 2), from *Corby* 1072. Drawn by Roger Polhill. From Crotalaria Africa & Madagascar.

Botswana. N: Ngamiland Distr., Khwebe (Kwebe) Hills, fl. & fr. immat. 17.ii.1898, *Mrs. E.J. Lugard* 179 (K). SW: Kgalagadi Distr., 27 km north of Kang, fl. 18.ii.1960, *Wild* 5034 (K; SRGH). SE: Central Distr., Mosetse R., fl. & fr. 8.iii.1961, *Richards* 14599 (K; SRGH). **Zambia**. B: Sesheke Distr., Masese, fl. & fr. 14.iii.1961, *Fanshawe* 6441 (BR; K; SRGH). S: Mazabuka, fl. & fr. 27.ii.1963, *van Rensburg* 1464 (BR; K; SRGH). **Zimbabwe**. N: 18 km ESE of Chirundu Bridge, Mhenza (Mensa) Pan, fl. 29.i.1958, *Drummond* 5334 (COI; K; LISC; SRGH). W: Hwange Distr., Dete (Dett), fl. & fr. 17.ii.1956, *Wild* 4754 (K; LISC; SRGH). C: Hwedza Distr., Ziyambe, fl. & fr. ii.1963, *R.M. Davies* 2898 (K; SRGH). E: Chimanimani Distr., Mutare (Umtali) to Chipinge (Chipinga), near Hot Springs, fl. & fr. 18.iii.1964, *Corby* 1109 (K; SRGH). S: Save (Sabi) R. to Masvingo (Fort Victoria), fl. i.1969, *Goldsmith* 8/69 (K; SRGH). **Malawi**. N: Chisenga–Rumphi, fl. 19.ii.1961, *Richards* 14424 (K). C: Lilongwe, fl. & fr. 23.iii.1970, *Brummitt & Little* 9334 (K; MAL; SRGH). S: Blantyre Distr., near Shire R., Mpatamanga Gorge, fl. & fr. immat. 28.ii.1961, *Richards* 14501A (K; SRGH). **Mozambique**. N: Mossuril (Mesuril) and Cabaceira Pequena (Cabaceira), *Peters* s.n. (B†). T: Tete–Mandié, fl. i.1932, *Sofia Pomba* 80 (COI). MS: Caia Distr., Sena, *Peters* s.n. (B†). GI: Massangena Distr., Kapateni, fl. & fr. immat. 25.iv.1962, *Drummond* 7729 (K; SRGH). M: entre Chobela e Magude, fl. 6.i.1948, *Torre* 7051 (BM; K; LISC).

Widespread in the drier parts of the Sudanian and Zambezian Domains of tropical Africa. Mopane woodland (rarely in miombo), *Acacia*, *Combretum* tree savanna and grassland, usually on sandy soils, often in temporary water catchment areas by roads, pans and rock outcrops; up to 1100 m.

Introgression seems to have occurred in certain places where the ranges of *C. podocarpa* and *C. damarensis* approximate or overlap (Polhill, loc. cit. (1968)). In the Flora Zambesiaca area this is most apparent along the upper reaches of the Zambezi R. Around Livingstone plants collected over a period of fifty years have appressed rather than spreading hairs but otherwise nearly resemble other material of *C. podocarpa*. In Barotseland most plants tend to have appressed hairs and in addition have shorter pedicels and smaller flowers, which are further features of *C. damarensis*, but the short bracts and the small number of flowers per raceme are more characteristic of *C. podocarpa*. If a name is wanted for these plants with appressed hairs and smaller flowers, var. *villosa* Schinz is applicable.

24. **Crotalaria damarensis** Engl., Bot. Jahrb. Syst. **10**: 27 (1888). —E.G. Baker in J. Linn. Soc., Bot. **42**: 409 (1914). —Verdoorn in Bothalia **2**: 382 (1928). —Burtt Davy, Fl. Pl. Ferns Transvaal, pt. 2: 397 (1932). —Torre in C.F.A. **3**: 72 (1962) excl. var. *maraisiana*. —Schreiber in Merxmüller, Prodr. Fl. SW. Afrika, fam. 60: 22 (1970). —Drummond in Kirkia **8**: 217 (1972). —Polhill, Crotalaria Africa & Madagascar: 105, fig. 17/11 (1982). —Lock, Leg. Afr. Check-list: 173 (1989). TAB. 3,7: **24**, fig. B. Type from Namibia.

Crotalaria marlothii Engl., Bot. Jahrb. Syst. **10**: 27 (1888). —E.G. Baker in J. Linn. Soc., Bot. **42**: 409 (1914). —Verdoorn in Bothalia **2**: 382 (1928). Type from Namibia.

Crotalaria fleckii Schinz in Mém. Herb. Boissier No. 1: 128 (1900). —E.G. Baker in J. Linn. Soc., Bot. **42**: 408 (1914). Syntypes from Namibia.

Similar to *C. podocarpa*, distinguished certainly only by the divided stipules. Branches usually appressed pubescent. Stipules shortly stalked, (2)3–4-partite; segments very shortly stalked, unequal, the proximal one 4–18 mm long, linear-lanceolate to falcate, acuminate, very unequal-sided, the others smaller. Racemes often 6–12-flowered; bracts up to 4–8 mm long; pedicels 1–2.5 mm long. Keel 10.5–14 mm long.

Botswana. N: c. 6 km upstream from Nata R. delta, fl. & fr. 14.iv.1976, *Ngoni* 499 (K; SRGH). SE: Kgatleng Distr., Mochudi, fl. & fr. 14.iv.1967, *Mitchison* A37 (K). **Zimbabwe**. W: Bulawayo–Nyamandhlovu road, fl. & fr. 1.iii.1966, *Corby* 1507 (K; SRGH). C: Gutu Distr., Chirumhanzu (Chilimanzi), Serima C.L (Reserve), fl. & fr. 13.i.1954, *D.A. Robinson* 416 (K; SRGH). E: Chipinge Distr., Lower Save (Sabi), Mutema (Mtema), fl. & fr. 28.i.1948, *Wild* 2396 (BR; K; SRGH). S: Chiredzi Distr., Runde (Lundi) R., Chipinda Pools, fl. 12.i.1960, *Goodier* 804 (K; LISC; SRGH). **Mozambique**. M: Maputo, *Borle* 406 (P; SRGH).

Also in SW Angola, Namibia, South Africa (Northern Cape, Northern Province and Mpumalanga). Mopane woodland, river banks and disturbed places on sandy soils; 300–900 m.

See note above under *C. podocarpa*.

25. **Crotalaria prittwitzii** Baker f. in J. Linn. Soc., Bot. **42**: 326 (1914). —Polhill in Kew Bull. **22**: 227 (1968); in F.T.E.A., Leguminosae, Pap.: 877 (1971); Crotalaria Africa & Madagascar: 106 (1982). Type from Tanzania.

Crotalaria natalitia var. *sengensis* Baker f. in J. Linn. Soc., Bot. **42**: 411 (1914). —Lock, Leg. Afr. Check-list: 198 (1989). Type from Dem. Rep. Congo (Katanga).

Crotalaria pseudonatalitia R. Wilczek in F.C.B. **4**: 120 (1953). Type as for *C. natalitia* var. *sengensis*.

Erect bushy and slightly woody perennial, up to 1 m tall, densely covered with both appressed and spreading hairs. Leaves 3-foliolate; leaflets 2–5.5 × 0.5–1.5 cm, oblanceolate to elliptic-oblanceolate, silky pubescent beneath; petiole mostly shorter than leaflets; stipules 7–15(20) mm long, oblong-falcate, acuminate, very unequal-sided. Racemes terminal, becoming lax, mostly 6–15-flowered; bracts 3–9 mm long, narrow, evanescent. Calyx (5.5)8–10 mm long, rather densely pubescent; lobes triangular, somewhat acuminate, a little longer than the tube. Standard subcircular, bright yellow, sometimes faintly lined brown, glabrous outside; wings oblanceolate to broadly obovate-oblong, exceeding the keel; keel 1.4–1.6 cm long, angled in the lower half, with a straight blunt beak, lanate towards the upper edge. Pod at least 2.2 cm long, sessile, shortly and broadly cylindrical, shaggy tomentose, mostly 12–16-seeded. Mature seeds not seen.

Zambia. N: Mbala Distr., Mpulungu–Mbala road, fl. 20.iii.1957, *Richards* 8836 (K); Mbala Distr., Lake Chila, fl. 15.iii.1955, *Richards* 4952 (K).

Also in Angola, Dem. Rep. Congo (Katanga) and Tanzania. Miombo woodland, usually in disturbed places, sometimes in seasonally wet places, often on sandy soils; 1500–1650 m.

26. **Crotalaria lachnophora** A. Rich., Tent. Fl. Abyss. **1**: 151 (1847). —Milne-Redhead in Kew Bull. **5**: 351, fig. 3/1, 3, 5, 7, 9 (1951). —Wilczek in F.C.B. **4**: 129, fig. 6C (1953). —Martineau, Rhod. Wild Fl.: 35 (1953). —Hepper in F.W.T.A., ed. 2, **1**: 548 (1958). —Torre in C.F.A. **3**: 70 (1962). —Binns, First Check List Herb. Fl. Malawi: 79 (1968). —Polhill in F.T.E.A., Leguminosae, Pap.: 877 (1971). —Drummond in Kirkia **8**: 218 (1972). —Brummitt in Wye Coll. Malawi Proj. Rep.: 65 (1973). —Gonçalves in Garcia de Orta, Sér. Bot. **5**: 67 (1982). —Polhill, Crotalaria Africa & Madagascar: 108 (1982). —Lock, Leg. Afr. Check-list: 186 (1989). Syntypes from Ethiopia.

Crotalaria lachnocarpa Hochst. ex Baker in F.T.A. **2**: 33 (1871) pro parte, *nom. illegit.* —E.G. Baker in J. Linn. Soc., Bot. **42**: 404 (1914) pro parte. —Eyles in Trans. Roy. Soc. South Africa **5**: 371 (1916). —Verdoorn in Bothalia **2**: 380 (1928) pro parte. —Hutchinson, Botanist South. Africa: 512 (1946). —Brenan, Check-list For. Trees Shrubs Tang. Terr.: 413 (1949). —Merxmüller in Proc. & Trans. Rhodesia Sci. Assoc. **43**: 19 (1951). Types as for species.

Bushy herb or shrub, 1–3 m tall, with numerous densely hairy branches above. Leaves 3-foliolate; leaflets mostly 3–7.5 × 0.8–3.5 cm, oblanceolate-oblong, oblong-elliptic or obovate, silky pubescent at least beneath; petiole usually shorter than leaflets; stipules mostly 8–25 mm long, oblong-falcate, caudate, very unequal-sided. Racemes terminal, laxly few–many-flowered; bracts 6–12 mm long, ovate to obovate, acuminate, caducous. Calyx 1.1–1.5(1.8) cm long, brownish spreading tomentose; upper and lateral lobes coherent, narrowly triangular, ± twice as long as the tube. Standard subcircular, clear yellow, fading orange-red, glabrous outside; wings broad, exceeding the keel; keel (1.8)2.1–2.4(2.6) cm long, abruptly rounded in the lower half, with a relatively short straight blunt beak, lanate towards the upper edge. Pod 2.8–4.5 × 1.2–2 cm, subsessile, broadly cylindrical, with a dense shaggy tomentum, mostly 16–18-seeded. Seeds 4.5–5 mm long, oblong-reniform, granulate, orange-yellow.

Zambia. N: 11 km NW of Mbala, fl. & fr. 19.vii.1930, *Hutchinson & Gillett* 3909 (BM; BR; K; LISC; SRGH). W: Kitwe, fl. 27.ii.1968, *Mutimushi* 2499 (K; NDO; SRGH). E: Katete, fl. & fr. 22.iv.1952, *White* 2439 (FHO; K). S: Choma–Mochipapa, fl. 21.ii.1963, *van Rensburg* 1415 (K; SRGH). **Zimbabwe**. N: Centenary Distr., Mavuradonha (Mvuradona) Mts., Musingwa (Msingwa) R. Falls, fr. 1.iv.1965, *Bingham* 1432 (K; LISC; SRGH). C: Harare (Salisbury), fl. & fr. ii.1917, *Eyles* 665 (BM; K; SRGH). E: Mutare Distr., SW Vumba Mts., Hawkdale, fl. 24.iii.1961, *Chase* 7449 (BM; K; M; SRGH). S: Chivi Distr., 6 km north of Runde (Lundi) R., Madzivire Dip, fl. & fr. 4.v.1962, *Drummond* 7925 (K; LISC; SRGH). **Malawi**. N: Mzimba, fl. 30.vii.1960, *Leach & Brunton* 10368 (K; SRGH). C: Dedza Distr., Chongoni, fl. 8.iii.1961, *Chapman* 1165 (K; SRGH). S: Ntcheu Distr., Lower Kirk Range, Chipusiri, fl. 17.iii.1955, *Exell, Mendonça & Wild* 956 (BM; LISC; SRGH). **Mozambique**. N: near Lake Malawi, fl. 1902, *W.P. Johnson* 411 (K). T: Angónia Distr., Missão de Lifidzi (Lipidzi), fl. & fr. 8.iii.1964, *Correia* 179 (LISC). MS: Mavita–Chimoio (Vila Pery), fl. & fr. immat. 18.vi.1942, *Torre* 4330 (BM; K; LISC).

Plateau regions of tropical Africa, with a wide discontinuous distribution from Senegal to Ethiopia and south to Angola and southern Zimbabwe. Miombo and related woodlands, grassland, sometimes beside lakes, streams and dambos; 750–1650 m.

27. **Crotalaria grandistipulata** Harms in Bot. Jahrb. Syst. **28**: 399 (1900). —Milne-Redhead in Kew Bull. **5**: 349, fig. 3/2, 4, 6, 8, 10 (1951). —Wilczek in F.C.B. **4**: 130 (1953). —Torre in

C.F.A. **3**: 71 (1962). —Binns, First Check List Herb. Fl. Malawi: 79 (1968). —Polhill in F.T.E.A., Leguminosae, Pap.: 878 (1971). —Drummond in Kirkia **8**: 218 (1972). —Polhill, Crotalaria Africa & Madagascar: 108 (1982). —Lock, Leg. Afr. Check-list: 180 (1989). Type from S Tanzania.

Crotalaria lachnocarpa sensu J.G. Baker in F.T.A. **2**: 33 (1871) pro parte. —sensu E.G. Baker in J. Linn. Soc., Bot. **42**: 404 (1914) pro parte. —sensu Verdoorn in Bothalia **2**: 380 (1928) pro parte.

Erect perennial, with 1 (several) simple or sparsely branched ± sericeous-tomentose stems up to 45–70(100) cm tall, from a well developed swollen taproot. Leaves 3-foliolate; leaflets (4)6–9.5(12) × (1.5)2.5–5 cm, oblong-obovate, sparsely to conspicuously pubescent above, silky pubescent beneath; petiole shorter than leaflets; stipules 1.5–3.5 cm long, oblong-falcate, acuminate, very unequal-sided. Racemes terminal, subdense, becoming lax, mostly 8–15-flowered; bracts up to 11 mm long, ovate, caudate, caducous. Calyx 1.2–1.6 cm long, densely silvery pubescent; lobes free, attenuate-triangular, 2–3 times as long as the tube. Standard subcircular, yellow, tinged and fading reddish-orange, glabrous outside; wings broadly oblong, exceeding the keel; keel 1.9–2.1 cm long, abruptly rounded near the middle, with a rather short practically straight beak, lanate towards the upper edge. Pod 3–4.5 × 1.2–1.5 cm, shortly stipitate, broadly cylindrical, velvety tomentose, mostly 14–18-seeded. Seeds 5–5.5 mm long, oblong-reniform, smooth or granulate, brownish, tinged orange-red.

Zambia. B: Kaoma Distr., Luampa Mission, fl. & fr. 21.ii.1952, *White* 2113 (FHO; K). N: Mpika, fl. & fr. immat. 6.ii.1956, *Fanshawe* 2005 (BR; K; NDO; SRGH). W: Solwezi Distr., Mutanda Bridge, fr. 26.vi.1930, *Milne-Redhead* 601 (K). C: near Kabwe (Broken Hill), Mulungushi R., fl. ii.1907, *C.E.F. Allen* 487 (K). **Zimbabwe**. N: Gokwe, fl. & fr. 17.iii.1962, *Bingham* 173 (K; LISC; SRGH). **Malawi**. N: Mzimba Distr., Njakwa, fl. 16.ii.1947, *Benson* 1177 (K). C: Dzalanyama Forest Reserve, near Chionjeza (Chiungiza), fl. & fr. immat. 9.ii.1959, *Robson* 1515 (BM; K; LISC; SRGH).

Also in Angola, Dem. Rep. Congo (Katanga) and southern Tanzania. Miombo woodland, sometimes at grassy edges or near dambos; 1100–1750 m.

28. **Crotalaria valida** Baker in Bull. Misc. Inform., Kew **1897**: 253 (1897) excl. specim. *Whyte* 373. —Wilczek in F.C.B. **4**: 152 (1953). —Binns, First Check List Herb. Fl. Malawi: 80 (1968). —Polhill in Kew Bull. **22**: 231 (1968); in F.T.E.A., Leguminosae, Pap.: 879 (1971). —Drummond in Kirkia **8**: 219 (1972). —Brummitt in Wye Coll. Malawi Proj. Rep.: 65 (1973). —Polhill, Crotalaria Africa & Madagascar: 117 (1982). —Lock, Leg. Afr. Check-list: 208 (1989). Type: Malawi, Nyika Plateau, *Whyte* 94 (K, lectotype).

Crotalaria lachnocarpoides subsp. *valida* (Baker) Baker f. in J. Linn. Soc., Bot. **42**: 323 (1914) pro parte.

Crotalaria lachnocarpoides var. *valida* (Baker) Verdoorn in Bothalia **2**: 395 (1928) pro parte.

Very similar to *C. lachnocarpoides*, but easily distinguished by the keel which is covered by a dense silky indumentum apart from the upper margin and the short beak, it is also more commonly suffrutescent, its uppermost leaves often 1-foliolate, the leaflets pointed to rounded at the apex and sometimes not densely hairy above, the racemes 4–20-flowered and the keel 1.4–1.8 cm long.

Zambia. N: Mbala Distr., Chilongowelo Farm, fl. 15.iii.1952, *Richards* 1099 (BR; K). E: Katete, fl. & fr. 22.iv.1952, *White* 2438 (BR; FHO; K). **Zimbabwe**. C: locality and date not recorded, *Eyles* 5743 (SRGH). E: Nyanga Distr., Mt. Nyangani (Inyangani), fl. 9.xii.1959, *Wild* 4885 (K; SRGH). **Malawi**. N: Rumphi Distr., Chelinda Bridge, fl. 21.xii.1969, *Pawek* 3281 (K). C: Lilongwe Distr., Dzalanyama Forest Reserve, fl. & fr. 27.iv.1958, *Jackson* 2216 (K). **Mozambique**. N: 16 km Marrupa–Nungo, fr. 6.viii.1981, *Jansen et al.* 80 (K; WAG).

Also in Dem. Rep. Congo (Katanga) and Tanzania. Miombo woodland (particularly *Julbernardia paniculata* associations) and montane grassland; 750–2250 m.

29. **Crotalaria lachnocarpoides** Engl., Hochgebirgsfl. Afrika: 246 (1892). —E.G. Baker in J. Linn. Soc., Bot. **42**: 323 (1914) pro parte. —Hutchinson, Botanist South. Africa: 470 (1946). —Brenan, Check-list For. Trees Shrubs Tang. Terr.: 415 (1949); in Mem. New York Bot. Gard. **8**: 247 (1953) excl. syn. —Wilczek in F.C.B. **4**: 153 (1953). —Binns, First Check List Herb. Fl. Malawi: 79 (1968). —Polhill in F.T.E.A., Leguminosae, Pap.: 879 (1971). —Drummond in Kirkia **8**: 218 (1972). —Polhill, Crotalaria Africa & Madagascar: 117 (1982). —Lock, Leg. Afr. Check-list: 186 (1989). Type from Ethiopia.

Crotalaria lachnocarpoides subsp. *valida* sensu E.G. Baker in J. Linn. Soc., Bot. **42**: 323 (1914) pro majore parte.

Crotalaria lachnocarpoides var. *valida* sensu Verdoorn in Bothalia **2**: 395 (1928) pro majore parte.

Bushy slightly woody herb or small shrub, up to 0.5–2(2.7) m tall, very densely covered with mostly rather long fine spreading hairs. Leaves shortly petiolate, 3-foliolate; leaflets mostly 3.5–7 × 1–2.5 cm, oblanceolate-oblong or narrowly oblong-elliptic, rounded to truncate and apiculate, densely pilose on both surfaces, rather felty; stipules 5–8 mm long, linear. Racemes lax to dense, mostly 4–12-flowered; bracts up to 8–11 mm long, linear to attenuate-lanceolate; bracteoles inserted on the pedicel, similar. Calyx 1–1.5 cm long, tomentose; lobes ± narrowly attenuate-triangular, 2–3 times as long as the tube. Standard subcircular, yellow, often tinged reddish-brown outside, fading orange-red, pubescent at least medially and apically outside; wings exceeding the keel; keel semicircular, shortly and bluntly beaked, 1.1–1.5(1.8) cm long, pubescent in a thin band towards the lower margin. Pod subsessile, 2.5–3.2 × 1.2–1.5 cm, shortly and broadly cylindrical, with a dense shaggy tomentum, (12)20–26-seeded. Seeds 3.5–4 mm long, obliquely oblong-cordiform, minutely granulate, usually shiny, pale yellow or ochre.

Zambia. N: Mbala Distr., Kasulo Dam, fl. 11.i.1955, *Richards* 4027 (K). E: Isoka Distr., Nyika Plateau, Kangampande Mt., fr. 6.v.1952, *White* 2739 (FHO; K). **Zimbabwe**. E: Chimanimani Distr., Muwengezi (Wengezi) R., fl. & fr. 18.iii.1951, *Chase* 3620 (BM; COI; LISC; SRGH). S: Masvingo Distr., 6 km east of Great Zimbabwe, fl. 1.vii.1930, *Hutchinson & Gillett* 3353 (K). **Malawi**. N: Mzimba Distr., Mzuzu, Marymount, fl. & fr. 6.viii.1970, *Pawek* 3671 (K). S: Mt. Zomba, fl. & fr. 25.i.1959, *Robson & Jackson* 1317 (BM; K; LISC; SRGH). **Mozambique**. N: Lichinga (Vila Cabral), fl. iii.1934, fr. v.iii.1934, *Torre* 19 (BM; COI; K; LISC). MS: Serra de Chôa para Catandica (Vila Gouveia), fr. 17.ix.1942, *Mendonça* 275 (BM; K; LISC).

Also northwards in the highlands of eastern Africa to Ethiopia and slightly westwards to eastern Dem. Rep. Congo. Montane grassland, *Brachystegia*, *Uapaca* woodland, riverbanks, dambos and disturbed places; 780–1850 m.

30. **Crotalaria polytricha** Polhill, Crotalaria Africa & Madagascar: 120, fig. 21 (1982). —Lock, Leg. Afr. Check-list: 198 (1989). Type: Zambia, near Ndola, Lake Chirengwa (Chilengwa), 17.vi.1972, *Fanshawe* 11467 (BR; K, holotype; LISC; NDO; SRGH).

Bushy herb or small shrub, up to 1–1.6 m tall, densely covered with rather long spreading hairs. Leaves shortly petiolate, 3-foliolate; leaflets 2.5–6 × 1–3 cm, elliptic to obovate, pilose on both surfaces; stipules 4–6 mm long, filiform. Racemes laxly 2–4-flowered; bracts 4–8 mm long, filiform; bracteoles inserted near middle of pedicel, similar. Calyx 1.2–1.4 cm long, pilose with irregularly arranged hairs not completely obscuring the surface; lobes attenuate, 2.5–3 times as long as the tube. Standard obovate-circular, yellow, thinly pilose towards the apex outside; wings ± as long as the keel; keel 1–1.2 cm long, semicircular, crested on the upper margin, with a rather short narrow acute beak, sparsely pilose in a band near the lower margin. Pod very shortly stipitate, mostly 2.5 × 1 cm, subcylindrical, densely villous outside, sparsely pilose inside, mostly (15)20–28-seeded. Seeds 3–3.5 mm long, obliquely oblong-cordiform, smooth, shiny, yellow.

Zambia. W: Ndola Distr., Lake Chirengwa (Chilengwa), sunken lake, 12.vii.1953, *Fanshawe* 142 (BR; K; NDO).

Known only from the type locality, at fringes of evergreen thicket near Lake Chirengwa.

31. **Crotalaria densicephala** Welw. ex Baker in F.T.A. **2**: 21 (1871). —E.G. Baker in J. Linn. Soc., Bot. **42**: 381 (1914). —Torre in C.F.A. **3**: 62 (1962). —Polhill, Crotalaria Africa & Madagascar: 121 (1982). —Lock, Leg. Afr. Check-list: 174 (1989). Type from Angola.

Crotalaria dumetorum Baker in F.T.A. **2**: 22 (1871). —E.G. Baker in J. Linn. Soc., Bot. **42**: 382 (1914). Type from Angola.

Crotalaria lachnoclada Harms in Warburg, Kunene-Samb.-Exped. Baum: 254 (1903). —E.G. Baker in J. Linn. Soc., Bot. **42**: 382 (1914). Type from Angola.

Crotalaria vasculosa sensu Hutchinson, Botanist South. Africa: 497 (1946), non Wall. ex Benth.

Bushy herb, up to 1.2 m tall, but sometimes flowering when no more than 20 cm tall, densely and coarsely hairy, sometimes tomentose. Leaves shortly petiolate, all 3-

foliolate or sometimes the uppermost 1-foliolate; leaflets 3.8 × 1.2–4.5 cm, elliptic-oblong or elliptic to obovate, densely hairy on both surfaces or only beneath, sometimes tomentose and felty; stipules mostly 1.5–5 mm long, subulate, sometimes hidden by indumentum. Racemes dense, few–many-flowered; bracts 2–4 mm long, linear to linear-lanceolate; bracteoles on the short pedicels, rather smaller. Calyx 7–9 mm long, pilose on the tube but the lobes often subglabrous except for the densely ciliate margins; lobes attenuate, 2.5–4 times as long as the tube. Standard elliptic or elliptic-obovate, yellow, glabrous outside; wings mostly a little shorter than the keel; keel semicircular, with a rather small narrow beak, not strongly crested, 7.5–10 mm long, lanate-ciliate but not hairy on lateral surfaces. Pod shortly stipitate, 1.3–2 cm long, oblong-ellipsoid to oblong-obovoid, rather densely and irregularly covered with long spreading hairs, mostly 4–6-seeded. Seeds c. 6 mm long, obliquely oblong-cordiform, smooth or granulate, ochre.

Zambia. B: Kabwe Distr., Chibombo R., fr. 15.vi.1953, *Fanshawe* 89 (K; NDO). W: Mwinilunga, fl. & fr. immat. 22.v.1969, *Mutimushi* 3424 (K; NDO; SRGH); Kitwe, fr. 20.v.1955, *Fanshawe* 2295 (K; NDO; SRGH). C: Mkushi Distr., 8 km east of Chiwefwe, fr. vii.1930, *Hutchinson & Gillett* 3684 (K).

Also in Angola and Dem. Rep. Congo (Katanga). Miombo woodland, sometimes in thickets or damp places; c. 1100–1400 m.

32. **Crotalaria chrysotricha** Polhill in Kew Bull. **22**: 233, fig. 9 (1968); in F.T.E.A., Leguminosae, Pap.: 880 (1971); Crotalaria Africa & Madagascar: 121 (1982). —Lock, Leg. Afr. Check-list: 171 (1989). Type from Tanzania (Ufipa).

Crotalaria amoena sensu Hutchinson, Botanist South. Africa: 503 (1946), non Welw. ex Baker.

Low bushy slightly woody herb, up to 70 cm tall, densely usually rather spreading pubescent. Leaves shortly petiolate, mostly 3-foliolate with the uppermost often 1-foliolate; leaflets mostly 4–7.5 × 2–3.6 cm, elliptic or rarely obovate-elliptic, glossy and glabrous above apart from midrib, closely golden appressed pubescent beneath; stipules 3–11 mm long, subulate to linear-lanceolate. Racemes with 4–16 flowers fairly closely arranged above; bracts 3.5–6 mm long, subulate to filiform; pedicels 4–10 mm long, apically recurved, with filiform bracteoles on the upper part. Calyx c. 7 mm long, appressed pubescent; lobes attenuate, ± twice as long as the tube. Standard elliptic, dull yellow, glabrous outside; wings shorter than the keel; keel 9–11 mm long, semicircular, crested behind the short beak. Pod including the 4–5 mm long stipe up to 1.8–2.2 cm long, oblong-obovoid, densely subappressed pubescent outside, glabrous to sparsely pilose inside, (4)6–8-seeded. Seeds 3.5–4 mm long, obliquely oblong-cordiform, minutely granulate, pale yellow.

Zambia. N: Mbala–Lunzua road, fl. & fr. 19.ii.1967, *Richards* 22101 (K; P; SRGH); Chinsali, fr. 14.viii.1965, *Fanshawe* 9243 (K; NDO).

Extending over the Zambia border just into Tanzania and Dem. Rep. Congo (Katanga). Miombo woodland; 1200–1650 m.

33. **Crotalaria tristis** Polhill, Crotalaria Africa & Madagascar: 122, fig. 22 (1982). —Lock, Leg. Afr. Check-list: 207 (1989). Type: Zambia, Mbala–Kambole, *Richards* 8239 (K, holotype).

Bushy shrub, up to 1.3 m tall, dullish tomentose on the branches. Leaves 3-foliolate, uppermost sometimes very reduced in size; leaflets mostly 2–5 × 0.8–1.5 cm, oblanceolate to narrowly elliptic, dull-coloured with a short rather dense appressed pubescence on both surfaces; petioles 1–3 mm long; stipules inconspicuous. Racemes laxly few-flowered, sometimes aggregated into pseudopanicles by reduction of upper lateral flowering branches; bracts up to 2 mm long, linear; bracteoles very small. Calyx 6–8 mm long, densely appressed puberulous; upper lobes on either side oblong, obliquely truncate, coherent with the short bluntly triangular lateral lobe, shorter than the tube. Standard oblate, yellow, shaded brown and tomentellous outside; wings shorter than the keel; keel 1–1.2 cm long, semicircular, crested behind the short beak. Pod including the 5–7 mm long stipe c. 2.5 cm long, narrowly oblong-clavate, rather densely appressed pubescent, 10–12-seeded. Seeds 3.5–4 mm long, oblique-cordiform, smooth, pale yellow.

Zambia. N: Mbala Distr., Mbala (Abercorn) to Kambole road, fl. & fr. immat. 19.ii.1957, *Richards* 8239 (K); Kambole, fr. 5.vi.1957, *Richards* 10006 (K).

Known only from the cited specimens. "Bush" – presumably secondary miombo associations; 1500 m.

34. **Crotalaria amoena** Welw. ex Baker in F.T.A. **2**: 26 (1871). —E.G. Baker in J. Linn. Soc., Bot. **42**: 355 (1914). —Verdoorn in Bothalia **2**: 396 (1928). —Wilczek in F.C.B. **4**: 161 (1953). —Torre in C.F.A. **3**: 58 (1962). —Polhill in F.T.E.A., Leguminosae, Pap.: 881 (1971); Crotalaria Africa & Madagascar: 122, fig. 23 (1982). —Lock, Leg. Afr. Check-list: 165 (1989). TAB. 3,7: **25**. Type from Angola.

Bushy woody herb or small shrub, up to 2 m tall; branches tomentose. Leaves shortly petiolate, mostly 3-foliolate with the uppermost often 1-foliolate; leaflets mostly 3–10 × 1.2–3.5 cm, elliptic-lanceolate or elliptic to oblong-ovate, chartaceous, with slightly prominent venation, glossy and glabrous above except on midvein, sericeous beneath; stipules 2.5–6 mm long, subulate or linear. Racemes terminal, on reduced uppermost lateral branches and sometimes axillary, with 6–16 flowers rather closely arranged above; bracts 4–7 mm long, attenuate-subulate or linear; bracteoles near top of pedicel, filiform. Calyx 1–1.2 cm long, densely puberulous to tomentellous; lobes narrowly attenuate-triangular, ± twice as long as the tube. Standard subcircular or oblate, yellow tinged brownish or purplish, tomentellous outside; wings shorter than the keel; keel semicircular, crested behind the short beak, 1.2–1.6 cm long, sometimes hairy towards the lower margin. Pod 2.5–3.2 cm long, orange-brown velvety tomentellous outside, densely pubescent inside, maturing 3–8(12) seeds. Seeds 3–4 mm long, oblique-cordiform, granulate or smooth, brown.

Zambia. B: Mongu Distr., Lui R. Pan, fl. & fr. 15.vii.1964, *Verboom* 1073 (K; LISC; SRGH). N: Mbala (Abercorn), fr. 12.vii.1964, *Mutimushi* 895 (K; NDO; SRGH). W: Kitwe, fl. & fr. 10.v.1955, *Fanshawe* 2267 (BR; K; NDO; SRGH).

Also in Angola, Dem. Rep. Congo (Katanga) and Tanzania. Miombo woodland; 1050–1750 m.

There is a marked tendency for plants from Angola and Barotseland to have a band of hairs towards the lower margin of the keel, a feature not observed elsewhere.

35. **Crotalaria criniramea** Baker f. ex Polhill, Crotalaria Africa & Madagascar: 124, fig. 18/1 (1982). —Lock, Leg. Afr. Check-list: 173 (1989). Type: Zambia, Mwinilunga, *Marks* 16 (BM, fragment; K, holotype).

Prostrate herb with well-branched tomentose stems. Leaves very shortly petiolate, mostly 3-foliolate but uppermost generally 1-foliolate; leaflets 3–5 × 1.5–2.6 cm, elliptic, silky tomentose beneath; stipules 3–4 mm long, subulate to linear-lanceolate. Racemes curved-ascending, terminal and on short lateral branches bearing reduced foliage, laxly 4–8-flowered; bracts 4–7 mm long, linear or linear-lanceolate; bracteoles inserted on pedicel, filiform. Calyx 7–8 mm long, coarsely tomentose; lobes narrowly triangular, longer than the tube. Standard subcircular, bright yellow, glabrous outside; wings a little shorter than the keel; keel 12–14 mm long, semicircular with a short slightly incurved beak. Ovary long-stipitate, hairy especially along the upper and lower sides, c. 12-ovulate. Pod and seeds unknown.

Zambia. W: Mwinilunga, fl. ii.1929, *Marks* 16 (BM, fragment; K).

Known only from the type, collected in a plain in the Mwinilunga area; 1350–1500 m.

Vegetatively this species simulates *C. variegata*, but the flower arrangement, colour and shape of the keel are quite different.

36. **Crotalaria umbellifera** R.E. Fr., Wiss. Ergebn. Schwed. Rhod.-Kongo-Exped., part 1: 73 (1914). —E.G. Baker, Legum. Trop. Africa: 35 (1926). —Polhill, Crotalaria Africa & Madagascar: 125 (1982). —Lock, Leg. Afr. Check-list: 208 (1989). Type: Zambia, Katwe, 6.xi.1911, *R.E. Fries* 1207 (UPS, holotype).

Prostrate perennial, with slender densely strigose-pubescent stems up to c. 40 cm long. Leaves 3-foliolate; leaflets (1)1.6–3.8 × (0.5)0.8–1.8 cm, obovate-rhombic or oblanceolate, apiculate, subdensely appressed puberulous and prominently veined

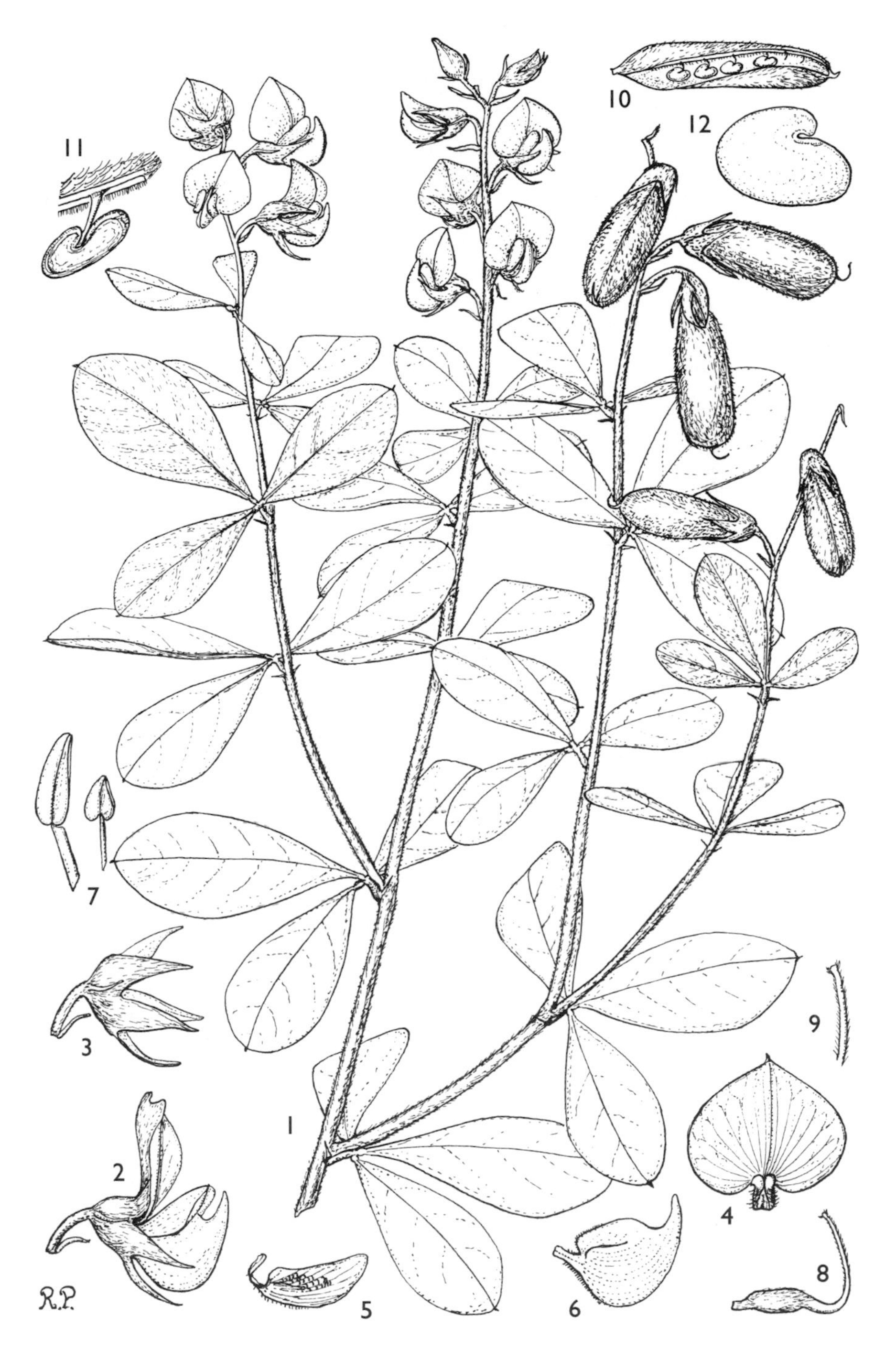

Tab. 3,7: **25**. CROTALARIA AMOENA. 1, flowering and fruiting branch (× $^2/_3$), from *Greenway & Polhill* 11696 and *McCallum-Webster* 735; 2, flower (× $1\,^1/_3$); 3, calyx (× $1\,^1/_2$); 4, standard (× $1\,^1/_3$); 5, wing (× $1^1/_3$); 6, keel (× $1^1/_3$); 7, anthers (× $3^1/_3$); 8, gynoecium (× $1\,^1/_2$); 9, tip of style (× 3), 2–9 from *McCallum-Webster* 735; 10, valve of dehisced pod (× 1); 11, detail of same showing hairy placenta (× 3), 10 & 11 from *Greenway & Polhill* 11696; 12, seed (× $3\,^1/_3$), from *Verboom* 1073. Drawn by Roger Polhill. From Crotalaria Africa & Madagascar.

beneath; petioles 5–12 mm long; stipules 3–8 mm long, subulate to linear-lanceolate. Racemes subumbelliform, pedunculate, 4–8-flowered; bracts 4–7 mm long, linear; pedicels 6–10 mm long, slender, ascending, with filiform bracteoles near the middle or below. Calyx 5–7 mm long, densely and shortly appressed pubescent; lobes attenuately or acuminately triangular, ± twice as long as the tube. Standard oblate, royal blue to mauve or ?whitish, with scattered hairs only at the apex and along the edge outside; wings a little shorter than the keel, yellow; keel 7.5–10 mm long, rounded, with a crest behind the short slightly incurved beak. Ovary 2-ovulate. Pod and seeds unknown.

Zambia. N: Mbala Distr., Kambole Escarpment, fl. 1.ii.1959, *Richards* 10840 (K); same locality, fl. 12.xi.1964, *Richards* 19260 (K). W: Mwinilunga, fl. xi.1929, *Marks* 11 (BM, fragment & sketch; K).

Recorded only from the Mbala and Mwinilunga Districts of Zambia. Miombo woodland; 1350–1650 m.

Not to be confused with *C. variegata*, which is similar in habit, flower arrangement and petal coloration, but with quite different keel and indumentum. The flowers of the type were said to be yellowish, but may have been polychromatic with the standard whitish rather than bluish.

37. **Crotalaria goetzei** Harms in Bot. Jahrb. Syst. **28**: 399 (1900). —E.G. Baker in J. Linn. Soc., Bot. **42**: 411 (1914). —Verdoorn in Bothalia **2**: 380 (1928). —Brenan, Check-list For. Trees Shrubs Tang. Terr.: 413 (1949); in Mem. New York Bot. Gard. **8**: 248 (1953). —Binns, First Check List Herb. Fl. Malawi: 79 (1968). —Polhill in F.T.E.A., Leguminosae, Pap.: 882 (1971); Crotalaria Africa & Madagascar: 126 (1982). —Lock, Leg. Afr. Check-list: 180 (1989). —F. White, Dowsett-Lemaire & Chapman, Evergreen For. Fl. Malawi: 324 (2001). Type from S Tanzania.

Crotalaria rotundicarinata Baker f. in J. Linn. Soc., Bot. **42**: 396 (1914). —Verdoorn in Bothalia **2**: 396 (1928). Type: Malawi, without precise locality, *Buchanan* 821 (BM; K, holotype).

Bushy shrub, 1–3 m tall; branches ± densely covered with mostly slightly spreading hairs. Leaves 3-foliolate; leaflets (2.5)3.5–5.5(7.5) × 0.8–2.5 cm, elliptic, elliptic-oblanceolate or obovate, pubescent beneath, rather conspicuously veined; petioles (0.5)1–2.5 cm long; stipules 4–15 × 0.5–12 mm, linear to foliaceous (with an elliptic-lanceolate to subcircular blade). Racemes usually numerous, both terminal and leaf-opposed, laxly few-flowered; bracts 3.5–7 mm long, filiform to linear-lanceolate; bracteoles inserted on the pedicel, smaller. Calyx 9–13 mm long, thinly pubescent; lobes attenuate-triangular, ± twice as long as the tube. Standard subcircular, yellow, sometimes flushed dull purple, pubescent at least medially and apically outside; wings as long as the keel; keel 1.1–1.4 cm long, semicircular, with a crest behind the short beak. Pod 2.5–3.8 cm long, shortly stipitate, subcylindrical, densely covered with rather long fine ± spreading hairs outside, sometimes with a few hairs along the suture inside, 10–14-seeded. Seeds c. 4 mm long, oblique-cordiform, smooth or obscurely papillose, brown or dark grey-green.

Zambia. N: Mbala Distr., Lake Chila, fl. & fr. 5.xi.1958, *Robson & Fanshawe* 498 (BM; K; LISC; SRGH). E: Nyika, fl. 24.xii.1962, *Fanshawe* 7219 (K; NDO). **Malawi**. N: Rumphi Distr., Livingstonia Escarpment, Manchewe Falls, fr. 5.vii.1970, *Pawek* 3593 (K). S: Mulanje Distr., Mt. Mulanje, Lichenya (Luchenya) Plateau, fl. & fr. 14.vii.1946, *Brass* 16837 (K; SRGH). **Mozambique**. N: Malema Distr., monte Cucuteia, junto as quedas do rio, c. 750 m, fl. & fr. 16.iii.1964, *Torre & Paiva* 11222 (LISC). Z: Gurué, R. Licungo, fl. 7.iv.1943, *Torre* 5108 (BM; K; LISC).

Also in southern Tanzania. Montane forest margins and associated bushland or grassland, extending down valleys into miombo woodland but probably never far from rivers and lakes; 750–2200 m.

38. **Crotalaria gazensis** Baker f. in J. Linn. Soc., Bot. **40**: 51 (1911); in J. Linn. Soc., Bot. **42**: 396 (1914). —Eyles in Trans. Roy. Soc. South Africa **5**: 370 (1916). —Verdoorn in Bothalia **2**: 388 (1928). —Drummond in Kirkia **8**: 218 (1972). —Polhill, Crotalaria Africa & Madagascar: 126, fig. 18/3 (1982). —Lock, Leg. Afr. Check-list: 179 (1989). Type: Zimbabwe, Nyahode R., *Swynnerton* 1493 (BM, holotype; K).

Shrubby, up to 1–2 m tall, or a slender-stemmed straggling or trailing perennial herb; branches variously pubescent. Leaves 3-foliolate; leaflets 1–4 × 0.3–1.8 cm, narrowly elliptic-oblong to elliptic, oblanceolate or obovate, sometimes slightly inrolled at the margins, subglabrous to pubescent beneath; petiole 0.5–16 mm long; stipules 1–4 mm long, subulate to triangular-lanceolate, recurved. Racemes on main and usually numerous relatively short lateral branches (which often bear smaller leaves), few-flowered or the flowers solitary; bracts up to 2.5 mm long, linear to linear-lanceolate; bracteoles inserted on the pedicel, small and filiform. Calyx 7–9(11) mm long, glabrous to pubescent; lobes narrowly attenuate-triangular, ± twice as long as the tube. Standard subcircular, yellow, sometimes with darker lines or tinged reddish, glabrous outside except sometimes along the midvein; wings longer than the keel; keel 7–10 mm long, rather strongly rounded and crested behind the short beak. Pod 2–3 cm long, shortly stipitate, subcylindrical, pubescent, 8–12-seeded. Seeds c. 3.5 mm long, oblique-cordiform, smooth, ochre.

Subsp. **gazensis** —Polhill, Crotalaria Africa & Madagascar: 126, fig. 18/3 (1982). —Lock, Leg. Afr. Check-list: 179 (1989).

Shrubby; leaflets mostly rounded at the apex, often drying blackish; petioles mostly 3–16 mm long, much exceeding the stipules.

Zimbabwe. N: Gokwe, fl. & fr. 12.iii.1964, *Bingham* 1339 (K; SRGH). E: Mutare Distr., Himalayas, Banti North, fl. 4.iii.1954, *Wild* 4501 (K; LISC; SRGH). S: Bikita Distr., Mt. Hozvi (Horzi), fl. & fr. 9.v.1969, *Moyo* in *GHS* 194457 (K; SRGH). **Malawi**. S: Mt. Mulanje, Linji Falls, 21.ii.1986, *Chapman* 7236 (K; MO). **Mozambique**. MS: Sussundenga Distr., Mavita, fl. 8.iv.1948, *Barbosa* 1382 (BM; LISC); Serra Mocuta (Macuta), south side, fl. 3.vi.1971, *Pope* 447 (K; SRGH).

Principally along the mountainous Zimbabwe–Mozambique border but with records as far west as Gokwe and east to Gorongosa. Edges of montane forest and associated bushland or grassland, extending into miombo woodland probably mainly along the rivers; 800–2200 m.

Subsp. **herbacea** Polhill, Crotalaria Africa & Madagascar: 126 (1982). —Lock, Leg. Afr. Check-list: 179 (1989). Type: Zimbabwe, Shurugwi (Selukwe) Peak, 19.iii.1964, *Wild* 6437 (K, holotype; LISC; SRGH).

Crotalaria gazensis sensu Verdoorn in Bothalia **2**: 388 (1928) pro parte. —sensu Burtt Davy, Fl. Pl. Ferns Transvaal, pt. 2: 398 (1932).

Herbs with slender straggling or trailing stems; leaflets mostly pointed, drying greenish; petioles mostly not exceeding 3 mm in length, often scarcely longer than the stipules.

Zimbabwe. C: Marondera (Marandellas), Grasslands Research Station, fl. & fr. 5.ii.1967, *Corby* 1735 (K; SRGH). E: Chipinge (Chipinga), fl. & fr. 29.ii.1956, *Drummond* 5121 (BR; K; LISC; SRGH). S: Mberengwa Distr., Mt. Buhwa, fl. 27.iv.1973, *Pope* 951 (K; SRGH). **Mozambique**. MS: 24 km de Catandica (Vila Gouveia) para a fronteira do Zimbabwe, fl. & fr. 26.iii.1966, *Torre & Correia* 15421 (LISC).

Also in South Africa (Northern Province). Grassland and disturbed places; 1100–1800 (2100) m.

It seems unlikely at first sight to include prostrate slender-stemmed herbs in the same species as shrubs up to 2 m tall and extreme forms are vegetatively very dissimilar even as herbarium specimens. Nevertheless subsp. *gazensis* is decidedly variable in the Zimbabwe/Mozambique border region and the size and shape of the leaf parts sometimes approximate quite closely. For example *F.R. Williams* 81A, from Chimanimani (Melsetter) and without clear habit notes, is not certainly referable to either subspecies, though probably subsp. *gazensis*; conversely *Goldsmith* 38/68, from the same area, is clearly subsp. *herbacea* but with more rounded leaflets than usual.

39. **Crotalaria phylicoides** Wild in Kirkia **4**: 140 (1964). —Drummond in Kirkia **8**: 218 (1972). —Polhill, Crotalaria Africa & Madagascar: 126 (1982). —Lock, Leg. Afr. Check-list: 196 (1989). Type: Zimbabwe, Chimanimani Mts., 16.iii.1957, *Phipps* 683 (K; SRGH, holotype).

Shrubby or suffrutescent, up to 90 cm tall; branches densely spreading hairy. Leaves closely arranged, subsessile, 3-foliolate; leaflets mostly 7–15 × 1–3 mm, linear to narrowly oblong-elliptic, with conspicuous recurved apiculum and inrolled margins, thinly hairy; stipules 1–1.5 mm long, subulate-triangular. Flowers solitary or in pairs; bracts 2–3 mm long, linear; bracteoles inserted on the pedicel, small. Calyx 7–10 mm long, rather densely and coarsely pubescent; lobes attenuate-triangular, up to twice as long as the tube. Standard subcircular or oblate, yellow, sometimes tinged reddish, pubescent at least medially and apically outside; wings exceeding the keel; keel strongly rounded, crested behind the short beak, 8–9 mm long. Pod 2–2.4 cm long, subsessile, oblong-fusiform, densely pubescent, 8–12-seeded. Mature seeds not seen.

Zimbabwe. E: Chimanimani Mts., fl. 6.vi.1949, *Wild* 2866 (K; LISC; SRGH); same locality, fl. & fr. 15.iii.1965, *Corby* 1277 (K; SRGH). **Mozambique**. MS: Sussundenga Distr., Chimanimani Mts., fl. 8.vi.1949, *Wild* 2914 (K; LISC; SRGH).

Known only from the Chimanimani Mts. Rocky places in montane grassland; 1700–1950 m.

The affinity of this species is with *C. gazensis*, not with *C. caudata* and allies as suggested in comments quoted by Wild with the original description.

40. **Crotalaria dura** Wood & Evans in J. Bot. **35**: 487 (1897). —E.G. Baker in J. Linn. Soc., Bot. **42**: 295 (1914) excl. syn. —Verdoorn in Bothalia **2**: 396 (1928). —Dyer, Fl. Pl. South Africa **22**: t. 878 (1942). —Polhill, Crotalaria Africa & Madagascar: 127 (1982). —Lock, Leg. Afr. Check-list: 176 (1989). Syntypes from South Africa (KwaZulu-Natal).

Virgate perennial herb or subshrub with a number of ascending stems to 60 cm tall, branched above, appressed pubescent to tomentellous. Leaves shortly petiolate, 3-foliolate; leaflets mostly 8–25 × 3–10 mm, narrowly oblong-elliptic to elliptic or oblanceolate to obovate, appressed pubescent to sericeous at least beneath; stipules narrow, variously developed. Racemes mostly well developed, with a number of flowers laxly to rather closely arranged towards the top; bracts 2–6 mm long, linear to linear-lanceolate; bracteoles on upper part of the pedicel, small, filiform. Calyx 6–7 mm long, appressed pubescent; lobes narrowly attenuate-triangular, longer than the tube. Standard elliptic or elliptic-obovate, yellow, glabrous or with small scattered hairs near the apex outside; wings a little shorter than the keel; keel 7.5–8.5 mm long, strongly rounded and crested behind the small beak. Pod 8–11 mm long, shortly stipitate, obliquely obovoid to oblong-ellipsoid, sericeous, 2–4-seeded. Seeds c. 3.5 mm long, oblique-cordiform, smooth, brown.

Subsp. **mozambica** Polhill, Crotalaria Africa & Madagascar: 128 (1982). —Lock, Leg. Afr. Check-list: 176 (1989). Type: Mozambique, de Moamba a Pessene, *Torre* 7364 (K, holotype; LISC).

Stipules 1–4 mm long, uncinate, persistent; leaflets sometimes hairy above. Ovary 2-ovulate. Pod c. 8 mm long, rather densely pubescent or silky-tomentose.

Mozambique. GI: Chokwè Distr., entre Inchobane (Bilene) e Muianga, fr. 8.vii.1948, *Torre* 8056 (LISC); Inharrime Distr., de Inharrime a Chacane, fl. & fr. 28.i.1941, *Torre* 2588 (K; LISC). M: Marracuene, fl. 25.i.1946, *Gomes e Sousa* 3360 (BM; COI; K; LISC; PRE; SRGH).

Also in South Africa (north KwaZulu-Natal). Coastal grassland, dunes and disturbed places on sandy soils.

Subsp. *dura* occurs in the KwaZulu-Natal midlands and has 4–6 ovules, slightly larger appressed pubescent pods, very small stipules and the leaflets glabrous above.

41. **Crotalaria huillensis** Taub. in Engler, Pflanzenw. Ost-Afrikas **C**: 205 (1895) in obs. —E.G. Baker in J. Linn. Soc., Bot. **42**: 295 (1914). —Torre in C.F.A. **3**: 46 (1962). —Polhill, Crotalaria Africa & Madagascar: 128 (1982). —Lock, Leg. Afr. Check-list: 181 (1989). Type from Angola.

Crotalaria ramosissima Baker in F.T.A. **2**: 26 (1871) non Roxb. (1832). Type as for species.

Erect virgate or well-branched perennial herb or small shrub up to 1 m tall; branches densely appressed to spreading pubescent. Leaves 3-foliolate; leaflets 1–3.5 × 0.4–1.3 cm, oblanceolate to oblong-elliptic or obovate, rounded to slightly emarginate, appressed pubescent beneath to pilose on both surfaces; petioles 1.5–10 mm long; stipules up to 2 mm long, subulate to triangular, recurved. Racemes usually with a number of laxly to rather closely arranged flowers; bracts usually less than 2 mm long (longer in var. *cacondensis*), linear to linear-lanceolate; bracteoles on the pedicel, minute. Calyx 4–6 mm long, subglabrous to pubescent; lobes longer than the tube, narrowly triangular. Standard broadly elliptic, yellow, glabrous or with a few hairs near the apex outside; wings a little shorter than the keel; keel 6–8 mm long, strongly rounded, crested behind the short beak. Pod 7–9 mm long, shortly stipitate, subglobose to subcylindrical, densely pilose, 2–6-seeded. Mature seeds not seen.

Subsp. **zambesiaca** Polhill, Crotalaria Africa & Madagascar: 128 (1982). —Lock, Leg. Afr. Check-list: 182 (1989). Type: Caprivi Strip, Katimo Mulilo–Ngoma, *Killick & Leistner* 3017 (K, holotype; PRE).

Stems virgate, with few ascending branches above. Keel 8 mm long. Ovules 2.

Caprivi Strip. 5 km south of Katimo Mulilo on main road to Ngoma, fl. 22.xii.1958, *Killick & Leistner* 3017 (K; PRE). **Zambia**. B: Senanga Distr., Ngonye (Sioma) Falls, fl. 1.ii.1975, *Brummitt, Chisumpa & Polhill* 14205 (K; LISC; NDO; PRE); Sesheke Distr., Masese, fr. 20.vi.1962, *Fanshawe* 6890 (K; LISC; NDO).

Known only from the cited specimens. Miombo and *Baikiaea* woodland, by rivers and dambos, on sandy soils; 900–1050 m.

The status of this taxon should be reappraised when more material is available. *C. huillensis* in south-western Angola is quite variable with two distinct varieties (see Torre, loc. cit.), both of which differ from subsp. *zambesiaca* in being lower bushy plants, with rather smaller flowers and 4–6 ovules.

42. **Crotalaria modesta** Polhill, Crotalaria Africa & Madagascar: 129, fig. 26 (1982). —Lock, Leg. Afr. Check-list: 192 (1989). Type: Zambia, Mweru Wantipa, *Richards* 9146 (K, holotype).

Procumbent or ascending rather bushy annual or short-lived perennial; branches slender, pubescent. Leaves 3-foliolate; leaflets 1–2.8 × 0.4–1 cm, oblanceolate to elliptic-obovate, thinly appressed pubescent above, more densely so beneath; petioles 3–10 mm long; stipules 1–2 mm long, linear to linear-lanceolate. Racemes laxly 2–6-flowered, with slender rhachis; bracts 1–2 mm long, filiform to linear-lanceolate; bracteoles on the pedicel, minute. Calyx 3.5–5 mm long, appressed pubescent; lobes narrow, more than twice as long as the tube. Standard broadly elliptic, pale yellow, with a few hairs apically; wings distinctly shorter than the keel; keel 4–5 mm long, strongly rounded, crested behind the small circumflexed beak. Pod 1.2–1.3 cm long, shortly stipitate, subcylindrical, slightly broadened upwards, appressed pubescent, 8–12-seeded. Seeds 2.5–3 mm long, oblique-cordiform, smooth, brownish or ochre.

Zambia. N: 96 km east of Kasama, fl. & fr. 6.v.1962, *E.A. Robinson* 5142 (K; M; SRGH). C: Serenje Distr., Kundalila Falls, fl. & fr. 12.iii.1975, *Hooper & Townsend* 691 (K; LUS; NDO; SRGH).

Also in Dem. Rep. Congo (Katanga). Seasonally wet grassland or damp disturbed places on sandy soils; 900–1400 m.

43. **Crotalaria glaucifolia** Baker in F.T.A. **2**: 38 (1871). —E.G. Baker in J. Linn. Soc., Bot. **42**: 353 (1914). —Torre in C.F.A. **3**: 56 (1962). —Polhill in F.T.E.A., Leguminosae, Pap.: 883, fig. 122 (1971); Crotalaria Africa & Madagascar: 132, fig. 27 (1982). —Lock, Leg. Afr. Check-list: 180 (1989). TAB. 3,7: **26**. Type from Angola.

Crotalaria longifoliolata De Wild. in Ann. Mus. Congo, Sér. IV, Bot. [Études Fl. Katanga] **1**: 187, t. 45 fig. 1–10 (1903). —E.G. Baker in J. Linn. Soc., Bot. **42**: 354 (1914). —Wilczek in F.C.B. **4**: 127 (1953). —Hepper in F.W.T.A., ed. 2, **1**: 550 (1958). —Torre in C.F.A. **3**: 57 (1962). Type from Dem. Rep. Congo (Katanga).

Erect laxly branched annual to 85 cm, with somewhat zig-zag appressed pubescent stem. Leaves 3-foliolate; leaflets mostly 9–16 × 1.5–5 cm (lower ones shorter and relatively broad), linear-lanceolate to ovate or elliptic-obovate, appressed puberulous beneath; petiole 2.5–6 cm long, rather broad; stipules 1–3.5 cm long, filiform to linear-lanceolate, caudate-spathulate or leaf-like. Racemes laxly few–many-flowered; bracts 2–4 mm long, linear-subulate to linear-lanceolate; bracteoles on the pedicel, small. Calyx (1.1) 1.2–1.5 cm long, glabrous to appressed puberulous; lobes narrowly lanceolate-triangular, up to 2–3 times as long as the tube. Standard broadly elliptic or subcircular, violet-blue with a yellow basal patch, brown puberulous along the midvein outside; wings exceeding the keel; keel 1.1–1.4 cm long, strongly rounded, crested behind the short narrow beak and with an obliquely lateral spur from the leading edge. Pod 4–8 cm long, subcylindrical, rather markedly inflated and somewhat broadened distally, abruptly contracted into a 5–15 mm long stipe, sparsely appressed puberulous, 24–32-seeded. Seeds 3.5–5 mm long, oblong-cordiform, reticulate-rugose, dark brown or black.

Zambia. N: Kasama Distr., Mwamba's Village, fl. 25.ii.1962, *E.A. Robinson* 4951 (K; M; SRGH).

Tab. 3,7: **26**. CROTALARIA GLAUCIFOLIA. 1, flowering branch (× 1), from *Richards* 928; 2 & 3, flower, front and side views (× 1$^1/_3$), from *Milne-Redhead & Taylor* 9329; 4, standard (× 1$^1/_3$); 5, wing (× 1$^1/_3$); 6–8, keel, different views (× 1$^1/_3$); 9, anthers (× 3$^1/_3$); 10, gynoecium (× 1$^1/_3$); 11, upper part of style (× 3$^1/_3$), 4–11 from *Richards* 928; 12, pod (× 1); 13, seed (× 3$^1/_3$), 12 & 13 from *Fanshawe* 1030. Drawn by Roger Polhill. From F.T.E.A.

W: 8 km east of Solwezi, fl. & fr. 17.iii.1961, *Drummond & Rutherford-Smith* 6981 (BR; K; LISC; SRGH). C: 5 km Serenje to Kapiri Mposhi, fl. & fr. 3.ii.1973, *Strid* 2836 (K). E: Chipata Distr., Ngoni area, fl. & fr. ii.1962, *Verboom* 473 (K; SRGH). **Malawi**. N: Viphya Plateau, fl. 24.v.1973, *Pawek* 6734 (K). **Mozambique**. N: 42 km de Cuamba (Nova Freixo) para Mandimba, fl. & fr. immat. 24.ii.1964, *Correia* 138 (LISC).

Also in Cameroon highlands, Uganda (doubtful record), Angola, Dem. Rep. Congo (Katanga) and southern Tanzania. Miombo woodland, often in grassy places around rock outcrops, by roads and dambos on sandy soils; 700–1800 m.

44. **Crotalaria anisophylla** (Hiern) Welw. ex Baker f. in J. Linn. Soc., Bot. **42**: 260 (1914). —Torre in C.F.A. **3**: 28 (1962). —Drummond in Kirkia **8**: 217 (1972). —Polhill, Crotalaria Africa & Madagascar: 134 (1982). —Lock, Leg. Afr. Check-list: 165 (1989). Type from Angola.

Crotalaria glauca var. ?*anisophylla* Hiern, Cat. Afr. Pl. Welw. **1**: 195 (1896).

Crotalaria hislopii Corbishley in Bull. Misc. Inform., Kew **1922**: 27 (1922). —E.G. Baker, Legum. Trop. Africa: 25 (1926). —Verdoorn in Bothalia **2**: 416 (1928). —Merxmüller in Proc. & Trans. Rhodesia Sci. Assoc. **43**: 19 (1951). Type: Zimbabwe, ?Rusape, *Hislop* 155 (K, holotype).

Erect annual, 15–50 cm tall, with long procumbent lower branches, pubescent with appressed slightly biramous hairs. Leaves shortly petiolate, 1-foliolate; leaflets mostly 9–14.5 × 0.5–2 cm, linear to linear-lanceolate, attenuate, acute, but lower ones up to 7 × 3.5 cm, often lanceolate to elliptic, shorter and rounded at the apex, thinly appressed pubescent beneath; stipules 3–10 mm long, linear or subulate. Racemes very lax, rather few-flowered; bracts 2–4.5 mm long, linear to linear-lanceolate, bracteoles on the pedicel, filiform. Calyx (9)11–12 mm long, appressed puberulous; lobes narrowly triangular-lanceolate, 2–3 times as long as the tube. Standard subcircular, blue to violet, with a few hairs apically outside; wings oblong-obovate, exceeding the keel; keel, (8)9–11 mm long, strongly rounded, with a crest behind the small beak and an oblique spur from the leading edge. Pod including the 0.5–1 cm long stipe 3–4.2 cm long, subcylindrical, slightly broadened distally, shortly appressed pubescent, with numerous seeds. Seeds 4–5 mm long, obliquely oblong-cordiform, reticulately rugulose, brownish.

Zambia. C: Lusaka Distr., Chakwenga Headwaters, fl. & fr. 14.ii.1965, *E.A. Robinson* 6373 (BR; K; M; SRGH). S: Choma Distr., Siamambo Forest Reserve, fl. 24.i.1960, *White* 6375 (FHO; K). **Zimbabwe**. N: Gokwe, fl. & fr. 12.iii.1962, *Bingham* 153 (K; LISC; SRGH). W: Matobo Distr., Matopos, fl. & fr. 14.iv.1931, *Stent* in *GHS* 3964 (SRGH). C: Marondera Distr., Wilton Farm, fl. 18.ii.1967, *Corby* 1768 (K; SRGH). E: Mutare Distr., Himalayas, fl. & fr. 1.iii.1967, *Drewe* 1 (K).

Also in Angola. Grassland, disturbed and damp places on sandy soils, in miombo and *Baikiaea* woodland areas.

45. **Crotalaria adamsonii** Baker f. in J. Linn. Soc., Bot. **42**: 261 (1914). —Verdoorn in Bothalia **2**: 416 (1928). —Wilczek in F.C.B. **4**: 87, t. 5 (1953). —Binns, First Check List Herb. Fl. Malawi: 78 (1968). —Polhill in F.T.E.A., Leguminosae, Pap.: 886 (1971); Crotalaria Africa & Madagascar: 135 (1982). —Lock, Leg. Afr. Check-list: 163 (1989). Type: Malawi, Shire Highlands, Mt. Soche, *G. Adamson* 321 (BM, holotype; K; P).

Crotalaria noldeae Rossberg in Repert. Spec. Nov. Regni Veg. **39**: 159 (1936). —Torre in C.F.A. **3**: 33 (1962). Type from Angola (Malange).

Suffrutex with small woody rootstock and several laxly branched and ascending 4-angled sparsely appressed puberulous stems, up to 50 cm tall. Leaves subsessile, simple; blade 3.5–8 × 0.7–2(2.6) cm, lanceolate, oblong-lanceolate or elliptic, acute, apiculate, rounded to subcordate at the base, sparsely appressed puberulous beneath; stipules 1.5–4 mm long, filiform to subulate. Racemes mostly terminal, laxly few-flowered; bracts 2–4 mm long, linear-subulate; bracteoles at top of the pedicel, 2–4 mm long, subulate to linear-lanceolate. Calyx 8–11(14) mm long, shortly appressed pubescent; lobes narrowly triangular-lanceolate, 2–3 times as long as the tube. Standard subcircular, pale yellow, veined reddish-purple and apically pubescent outside; wings ± as long as the keel; keel 8–12(14) mm long, rounded, with a short slightly incurved beak. Pod 3.4–4 cm long, subsessile, fusiform-cylindrical, appressed pubescent, many-seeded. Seeds 2–2.5 mm long, obliquely cordiform, rugulose.

Zambia. N: Mbala Distr., Kambole–Mbala, fl. 18.ii.1957, *Richards* 8222 (K). W: 12 km west of

Mwinilunga, Kanjima Dambo, fl. & fr. 24.i.1975, *Brummitt, Chisumpa & Polhill* 14062 (BR; K; LISC; LUS; MO; NDO; P; SRGH). **Malawi.** N: Chitipa Distr., Misuku Hills, Kanjera, fl. 19.iv.1976, *Pawek* 11133 (K; MAL; MO). S: Blantyre Distr., Upper Hynde Dam, 2 km north of Limbe, fl. & fr. 14.ii.1970, *Brummitt* 8556 (K; LISC; MAL; SRGH). **Mozambique**. N: Maúa Distr., andados 36 km de Maúa para Marrupa, c. 600 m, fl. & fr. 19.ii.1964, *Torre & Paiva* 10660 (LISC). Z: Gurué, fl. 5.iv.1943, *Torre* 5080 (BM; K; LISC).

Also in Angola, Dem. Rep. Congo (Katanga) and SE Tanzania, but known from rather few collections. Marshes, ditches and other seasonally wet sandy places in miombo woodland; c. 1000–1650 m.

46. **Crotalaria cornetii** Taub. & Dewèvre in Bull. Soc. Roy. Bot. Belgique **34**: 94 (1895). —E.G. Baker in J. Linn. Soc., Bot. **42**: 268 (1914). —Wilczek in F.C.B. **4**: 99, fig. 6A (1953). —Duvigneaud & Timperman in Bull. Soc. Roy. Bot. Belgique **91**: 139 (1959). —Polhill, Crotalaria Africa & Madagascar: 136, fig. 28 (1982). —Lock, Leg. Afr. Check-list: 172 (1989). Type from Dem. Rep. Congo (Katanga).

Suffrutex with woody rootstock and numerous spreading or ascending subangular stems up to 75 cm tall, glabrous overall except for minute puberulence on the youngest parts. Leaves estipulate, subsessile, simple; blade 2.5–5.5 × 1.2–3.5 cm, ovate-elliptic to elliptic or elliptic-oblong, bluntly pointed to rounded at the apex, shortly cordate at the base, chartaceous, ± glaucous, prominently venose on both surfaces. Racemes terminal and sometimes also axillary with the few flowers rather crowded towards the top; bracts 3–4(8) mm long, linear to linear-lanceolate; bracteoles on the pedicel, filiform. Calyx 1.2–1.6(1.7) cm long, drying blackish; upper and lateral lobes on either side coherent, narrowly attenuate-triangular, more than twice as long as the tube. Standard subcircular, pink to mauve or purplish with darker lines; wings oblong-obovate, exceeding the keel; keel 1.4–1.6 cm long, strongly rounded about the middle, with a fairly short somewhat projecting pointed beak. Pod subsessile, 3–3.5 cm long, oblong-clavate to subcylindrical, with c. 10 seeds. Mature seeds not seen.

Zambia. N: Chinsali Distr., Great North Road, 64 km south of Shiwa Ngandu turnoff, fl. & fr. 29.iii.1961, *Angus* 2568 (FHO; K; SRGH). W: Ndola, fl. 25.ii.1969, *Mutimushi* 3012 (NDO; K). C: Serenje Distr., Kundalila Falls, fl. & fr. immat. 6.iv.1961, *Richards* 14965 (K; SRGH).

Also in Dem. Rep. Congo (Katanga). Miombo woodland and grassland, persisting in disturbed places; 1200–1700 m.

In the Katanga it is commonly on metalliferous soils, see Duvigneaud & Timperman, loc. cit. (1959). One of the more remarkably unusual looking miombo woodland species of the genus.

47. **Crotalaria glauca** Willd., Sp. Pl. **3**: 974 (1802). —J.G. Baker in F.T.A. **2**: 12 (1871). —Taubert in Engler, Pflanzenw. Ost-Afrikas **C**: 204 (1895). —E.G. Baker in J. Linn. Soc., Bot. **42**: 259 (1914). —Verdoorn in Bothalia **2**: 415 (1928). —Hutchinson, Botanist South. Africa: 520 (1946). —Wilczek in F.C.B. **4**: 102, t. 6 (1953). —Brenan in Mem. New York Bot. Gard. **8**: 245 (1953). —Hepper in F.W.T.A., ed. 2, **1**: 548 (1958). —Binns, First Check List Herb. Fl. Malawi: 79 (1968). —Polhill in F.T.E.A., Leguminosae, Pap.: 885 (1971). —Drummond in Kirkia **8**: 218 (1972). —Tredgold & Biegel, Rhod. Wild Fl.: 26, pl. 17, fig. 7 (1979). —Polhill, Crotalaria Africa & Madagascar: 136, fig. 29 (1982). —Lock, Leg. Afr. Check-list: 179 (1989). Type from Ghana.

Crotalaria glauca var. *mildbraedii* Baker f. in J. Linn. Soc., Bot. **42**: 260 (1914). Type from Dem. Rep. Congo (Kivu).

Crotalaria glauca var. *welwitschii* Baker f. in J. Linn. Soc., Bot. **42**: 260 (1914). —Verdoorn in Bothalia **2**: 415 (1928). —Wilczek in F.C.B. **4**: 105 (1953). —Torre in C.F.A. **3**: 27 (1962). Type from Angola.

Crotalaria glauca var. *elliotii* Baker f. in J. Linn. Soc., Bot. **42**: 260 (1914). —Verdoorn in Bothalia **2**: 415 (1928). —Wilczek in F.C.B. **4**: 104 (1953). Type from Kenya.

Erect annual or short-lived perennial, laxly branched above, up to 0.5–1.2(1.8) m tall, with a rather slender glabrous or subglabrous stem. Leaves almost always estipulate, subsessile, simple; blade (1.2)2–8.5 × 0.2–1(1.5) cm, linear to oblong-lanceolate, oblong or rarely elliptic-obovate, glabrous or practically so, ± glaucous, sometimes drying blackish. Racemes laxly 3–12-flowered; bracts 1–9 mm long, linear to narrowly elliptic; bracteoles on the pedicel, filiform, small. Calyx 4–7 mm long, glabrous or sparsely puberulous; lobes narrow, longer than the tube. Standard subcircular, pale yellow, veined brownish, glabrous outside; wings exceeding the keel, bright yellow; keel 6–8 mm long, semicircular, crested behind the small beak. Pod with the (3)4–7 mm long stipe (1.8)2.4–3.6 cm long, cylindrical, glabrous, 24–32-seeded. Seeds 2–3 mm long, obliquely oblong-cordiform, smooth, shiny, bottle-green to brown.

Zambia. B: Zambezi (Balovale), fl. & fr. 8.vii.1963, *E.A. Robinson* 5574 (K; M; SRGH). N: 40 km Mbala–Tunduma, fl. & fr. 12.iv.1962, *Richards* 16317 (K; SRGH). W: Kitwe, fl. & fr. 25.iii.1966, *Mutimushi* 1327 (K; NDO; SRGH). C: Lusaka, 28.iv.1957, *Noak* 223 (K; SRGH). S: Mumbwa, fl. & fr. 25.iii.1964, *van Rensburg* 2874 (K; SRGH). **Zimbabwe**. C: Harare, fl. & fr. 3.iii.1929, *Eyles* 6179 (K; SRGH). E: Mutare (Umtali), fl. & fr. 8.iii.1960, *Chase* 7282 (BM; K; SRGH). **Malawi**. N: Rumphi Distr., Nchenachena, fl. & fr. 24.iv.1969, *Pawek* 2350 (K). C: Lilongwe Distr., Malingunde, Kamuzu Dam, fl. & fr. immat. 6.iii.1968, *Salubeni* 1008 (K). S: Mangochi Distr., Jalasi, fl. 15.iii.1955, *Exell, Mendonça & Wild* 904 (BM; LISC; SRGH). **Mozambique**. N: Mecubúri Distr., Namina, fl. & fr. 8.v.1948, *Pedro & Pedrógão* 3236 (LMA). Z: Lugela Distr., Namagoa Estate, fl. & fr. 27.vi.1949, *Faulkner* Kew No. 452 (COI; K; SRGH). MS: Barué Distr., Catandica (Vila Gouveia), fl. & fr. 28.iii.1966, *Torre & Correia* 15516 (LISC).

Widespread in tropical Africa. Catholic opportunist, but generally in locally damp sandy ± disturbed places in otherwise moderately dry areas; 400–1750 m.

A polymorphic species, any natural subdivision of which seems quite impracticable with the range of material now available. The most obvious variation is in the foliage. Normally the short-lived stem-leaves are relatively broad, elliptic-oblong to ± elliptic, and longer linear to lanceolate leaves predominate as the branches develop above. Sometimes practically all the leaves are narrow (var. *elliotii*), the pods then often rather small, and sometimes with evidence that the plants are growing under rather unfavourable conditions. Sometimes practically all the leaves are broad (var. *welwitschii*), in this case some plants are juvenile (flowering can commence at an early stage), some with obviously arrested or damaged leading shoots (which causes numerous broad-leaved ± spreading branches to grow out), some shade forms, and others, particularly in W Zambia, retaining broad leaves for an exceptionally long time for no obvious reason. The conditions and degrees to which plants dry black (another character sometimes used to define var. *welwitschii*) are also variable.

48. **Crotalaria orthoclada** Welw. ex Baker in F.T.A. **2**: 29 (1871). —E.G. Baker in J. Linn. Soc., Bot. **42**: 369 (1914). —Wilczek in F.C.B. **4**: 265, fig. 16 (1953). —Torre in C.F.A. **3**: 60 (1962). —Polhill in F.T.E.A., Leguminosae, Pap.: 887 (1971); Crotalaria Africa & Madagascar: 138, fig. 18/4 (1982). —Lock, Leg. Afr. Check-list: 194 (1989). Type from Angola.

Shrubby, usually with a single slightly woody stem and numerous ascending branches above, up to 1–3 m tall; branches ± densely covered with rather long fine subappressed to somewhat spreading hairs. Leaves estipulate, shortly petiolate, 3-foliolate; leaflets 1.5–2.5(4) × 0.4–0.7(1.3) cm, oblanceolate, ± rounded and apiculate, ± densely silky pubescent beneath. Racemes mostly terminal, shortly pedunculate, with 5–12 laxly or (in West Africa) rather closely arranged flowers; bracts 4–8 mm long, linear-lanceolate to lanceolate; bracteoles inserted on the pedicel, linear or filiform. Calyx 7–9 mm long, densely silky pubescent; upper lobes attenuate-triangular, 2–3 times as long as the tube. Standard subcircular, dull yellow, often tinged reddish, rather densely pubescent outside; wings brighter yellow, ± as long as the keel; keel 7–9 mm long, strongly rounded, with a broad slightly incurved beak very shortly twisted at the tip. Pod 2.5–3 cm long, subsessile, cylindrical, densely covered with fine mostly spreading hairs, c. 20-seeded. Seeds c. 3 mm long, oblique-cordiform, smooth, dark brown.

Zambia. N: Mbala (Abercorn), fl. & fr. 10.vii.1964, *Mutimushi* 836 (K; NDO; SRGH). W: 95 km Mwinilunga–Solwezi, fl. & fr. 18.ix.1952, *White* 3291 (FHO; K). **Malawi**. N: near Chitipa (Fort Hill), fr. 11.vi.1938, *Pole Evans & Erens* 707 (BR; K; PRE).

Principally in the highlands from N Nigeria and Cameroon to Kenya (Elgon), down the Western Rift Valley and into Angola, but also extending into the Zambia–Angolan plateau regions along the rivers. Grassland and evergreen forest by lakes, rivers and marshes; (?1200)1350–1800 m.

49. **Crotalaria caudata** Welw. ex Baker in F.T.A. **2**: 18 (1871). —E.G. Baker in J. Linn. Soc., Bot. **42**: 370 (1914). —Wilczek in F.C.B. **4**: 119, fig. 9 (1953). —Torre in C.F.A. **3**: 60 (1962). —Binns, First Check List Herb. Fl. Malawi: 79 (1968). —Polhill in F.T.E.A., Leguminosae, Pap.: 887 (1971). —Drummond in Kirkia **8**: 217 (1972). —Polhill, Crotalaria Africa & Madagascar: 139, fig. 30 (1982). —Lock, Leg. Afr. Check-list: 170 (1989). TAB. 3,7: **27**, fig. A. Type from Angola.

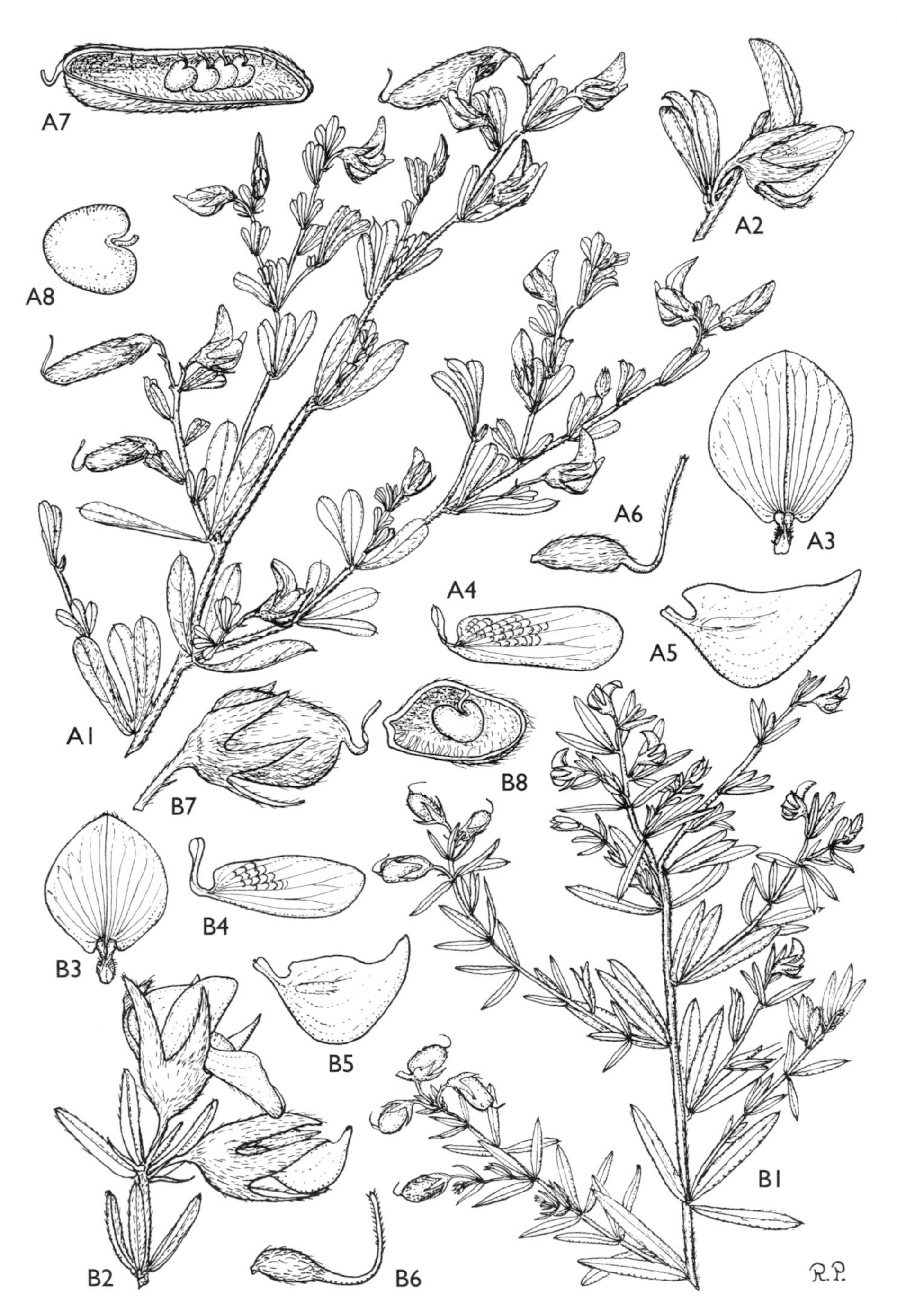

Tab. 3,7: **27**. A. —CROTALARIA CAUDATA. A1, flowering and fruiting branch (× 1); A2, axillary flower (× 2); A3, standard (× 3); A4, wing (× 3); A5, keel (× 3); A6, gynoecium (× 3); A7, valve of dehisced pod (× 2), A1–A7 from *Richards* 5496; A8, seed (× 6), from *Fanshawe* 2327. B. —CROTALARIA CISTOIDES subsp. ORIENTALIS. B1, flowering and fruiting branch (× 1); B2, tip of flowering branch (× $3^1/_3$); B3, standard (× 4); B4, wing (× 4); B5, keel (× 4); B6, gynoecium (× 4); B7, pod (× $3^1/_3$); B8, valve of dehisced pod (× 3), B1–B8 from *Richards* 11245. Drawn by Roger Polhill. From Crotalaria Africa & Madagascar.

Crotalaria harmsiana Taub. in Engler, Pflanzenw. Ost-Afrikas **C**: 205 (1895). —E.G. Baker in J. Linn. Soc., Bot. **42**: 397 (1914). —Verdoorn in Bothalia **2**: 392 (1928). —Brenan, Check-list For. Trees Shrubs Tang. Terr.: 413 (1949). Syntypes from Tanzania.

Crotalaria harmsiana var. *congoensis* Baker f. in J. Bot. **58**: 76 (1920). Type from Dem. Rep. Congo (Katanga).

Erect or decumbent slightly woody short-lived perennial, sometimes shrubby, up to 0.7–1.7 m tall, much branched, variously hairy. Leaves rather closely arranged, estipulate, sessile, 3-foliolate; leaflets mostly held erect, 0.7–1.7(3) × 0.2–0.7(1.1) cm, linear-oblanceolate to oblanceolate, rounded and apiculate, with slightly inrolled margins, thinly hairy on both surfaces. Racemes numerous on main branches and reduced lateral branches, sometimes axillary, short, subsessile, 2–3-flowered or the flowers solitary; bracts 1.5–5 mm long, filiform to linear-lanceolate; bracteoles at top of the pedicel, 2–3 mm long, filiform. Calyx 5–8 mm long, densely silky-pubescent; lobes narrowly attenuate-triangular, 2–3 times as long as the tube. Standard obovate-elliptic to subcircular, bright yellow, often marked or tinged reddish, pubescent outside at least towards the apex; wings shorter than to as long as the keel; keel 7–9 mm long, rather narrow, shortly rounded, with a rather projecting sharp beak. Pod subsessile, 1.6–2.5 cm long, cylindrical, densely spreading hairy outside, pubescent inside at least along the lower suture, 14–20-seeded. Seeds 2–3 mm long, oblique-cordiform, smooth, straw-coloured.

Zambia. N: Mbala Distr., 72 km Mbala–Kasama, fl. & fr. 30.iii.1955, *Exell, Mendonça & Wild* 1349 (BM; LISC; SRGH). W: Ndola, fl. & fr. 20.iii.1954, *Fanshawe* 984 (BR; K; LISC; NDO; SRGH). C: Serenje Distr., L. Lusiwasi, fl. & fr. 5.iv.1961, *Richards* 14958 (K; SRGH). **Zimbabwe**. E: Chimanimani (Melsetter), Kasipiti, fl. & fr. 20.ix.1964, *Loveridge* 1157 (K; LISC; SRGH). **Malawi**. N: Nkhata Bay, fl. & fr. 7.ix.1955, *Jackson* 1749 (K; SRGH). C: Mzimba Distr., Lusangadzi, fl. & fr. 24.iv.1967, *Salubeni* 675 (K; SRGH). **Mozambique**. N: Marrupa Distr., 68 km Maúa–Marrupa, Messalo R., fl. & fr. 14.viii.1981, *Jansen, de Koning & de Wilde* 280 (K; WAG). Z: Gurué, próximo do R. Malema, fl. & fr. 6.xi.1967, *Torre & Correia* 15961 (LISC). MS: Sussundenga Distr., Chimanimani Mts., Gossamer Falls, fl. & fr. 26.iv.1974, *Pope & Müller* 1313 (K; SRGH).

Also in Cameroon, Dem. Rep. Congo, Rwanda, Uganda, Tanzania and Angola. Miombo woodland, also evergreen forest and grassland associations near swamps, rivers and dambos, disturbed places and shrub savanna at higher altitudes; 450–1800 m.

C. caudata is variable in habit and foliage, but is generally well known and easily recognized. In NW Zambia, however, the two related and much less common species which follow could be easily confused; the differences are noted under each.

50. **Crotalaria kwengeensis** R. Wilczek in Bull. Jard. Bot. État **23**: 136 (1953); in F.C.B. **4**: 116 (1953). —Polhill, Crotalaria Africa & Madagascar: 140 (1982). —Lock, Leg. Afr. Check-list: 185 (1989). Type from Dem. Rep. Congo.

Suffrutex 15–50 cm tall; stems ultimately numerous, ascending from a small woody rootstock, rather slender, appressed puberulous to subappressed pubescent. Leaves estipulate, sessile, 3-foliolate; leaflets 10–20 × (1)2–7 mm, linear-oblanceolate to oblong-oblanceolate or narrowly elliptic-oblong (lower ones ± oblong-obovate), obtuse, glossy, slightly coriaceous, flat or slightly incurved at the margins, appressed puberulous to pubescent on both surfaces; venation rather inconspicuous, but basal lateral nerves running close to the margin nearly to the apex. Racemes terminal on main and rather short lateral branches, short, 2–6-flowered; bracts 2.5–5 mm long, linear to lanceolate; bracteoles on the upper part of the pedicel, linear, 1–2 mm long. Calyx 3.5–7 mm long, puberulous; lobes narrowly attenuate-triangular, 2–3 times as long as the tube. Standard elliptic-ovate, apiculate, bright yellow, puberulous medially or over the upper half; wings ± as long as the keel; keel 5–9 mm long, shortly rounded in the lower half, with a straight rather projecting sharp beak. Pod subsessile, 7–12 × 3–4 mm, cylindrical, densely pubescent outside, pilose inside, 4–12-seeded. Seeds up to 2.5 mm long, oblique-cordiform, smooth, pale yellow.

Closely related to *C. caudata* but fairly easily recognized in Zambia by the short fine indumentum, the slightly coriaceous leaflets which dry brown and have strongly ascending submarginal basal nerves (best seen on older leaves) and by the smaller pods. The flowers are similar, but when dried those of *C. kwengeensis* turn a characteristic red or orange overall, whereas those of *C. caudata* remain yellow except for the usual red flush on the standard or turn dull brown.

Var. **kwengeensis** —Polhill, Crotalaria Africa & Madagascar: 141 (1982).

Keel 8–9 mm long. Pod 9–12 mm long, 8–12-seeded.

Zambia. W: Mwinilunga Distr., Ikelenge, fl. 16.iv.1965, *E.A. Robinson* 6606 (K; LISC; SRGH); Mwinilunga, fr. 16.v.1969, *Mutimushi* 3446 (K; NDO; SRGH).
Also in southern Dem. Rep. Congo. Seasonally damp grassland; c. 1350–1500 m.

Var. **parviflora** Polhill, Crotalaria Africa & Madagascar: 141 (1982). Type from Dem. Rep. Congo.

Keel 5–6 mm long. Pod 7–9 mm long, 4–8-seeded.

Zambia. W: Mwinilunga road, between Kanyama turnoff and Samuteba, fl. & fr. 25.iii.1975, *Hooper & Townsend* 383 (K; SRGH).
Also in Dem. Rep. Congo. Sandy pans; c. 1400 m.

51. **Crotalaria kerkvoordei** R. Wilczek in Bull. Jard. Bot. État **23**: 134 (1953); in F.C.B. **4**: 114 (1953). —Polhill, Crotalaria Africa & Madagascar: 140 (1982). —Lock, Leg. Afr. Check-list: 184 (1989). Type from Dem. Rep. Congo (Katanga).

Suffrutex or bushy subshrub 40–70 cm tall, several-stemmed from a small woody rootstock, with numerous floriferous ascending branches above, covered in a fine silky tomentum giving a silvery (drying golden) sheen on the younger parts. Leaves estipulate, sessile, 3-foliolate; leaflets 10–30 × 1–6 mm, linear to narrowly oblong-elliptic, pointed and apiculate, slightly incurved at the margins, finely silky tomentose on both surfaces; venation inconspicuous, but basal lateral nerves ascending close to the margin nearly to the apex. Racemes terminal, subsessile, short, 2–7-flowered or the flowers solitary; bracts 4–5 mm long, linear-lanceolate; bracteoles on the pedicel near or below the middle, 2–2.5 mm long, linear. Calyx 7–9 mm long, silky tomentose; lobes narrowly attenuate-triangular, 2–3 times as long as the tube. Standard broadly elliptic-ovate, apiculate, orange-yellow, ± densely pubescent on the upper half outside; wings ± as long as the keel; keel 9–12 mm long, falcate to strongly rounded in the lower half, with a narrow slightly incurved pointed beak. Pods subsessile, 1.2–1.6 cm long, oblong-ellipsoid to shortly cylindrical, tomentose outside, pilose inside, 6–8-seeded. Seeds 2.5 mm long, oblique-cordiform, smooth, yellow to orange.

Zambia. N: Mbereshi–Kawambwa, fl. & fr. 18.i.1960, *Richards* 12409 (K; SRGH); Kawambwa Distr., west of Kasonkomona Fly Picket, fl. 19.xi.1958, *Johnsen* 34 (BM).
Also in the Katanga Province of Dem. Rep. Congo on the Biano Plateau. Damp grassy places in marshes, dambos and roadside ditches; 1200–1300 m.
Similar to *C. caudata*, but young parts with a short silky tomentum, the leaflets more pointed with strongly ascending basal lateral nerves (best seen on older leaves), the flowers orange-yellow and rather larger with a more incurved beak and the pods rather smaller and few-seeded.

52. **Crotalaria sylvicola** Baker f. in J. Linn. Soc., Bot. **42**: 397 (1914). —Torre in C.F.A. **3**: 66 (1962). —Polhill, Crotalaria Africa & Madagascar: 144 (1982). —Lock, Leg. Afr. Check-list: 206 (1989). Type from Angola.

Perennial herb or suffrutex, with numerous slender branched stems ascending to 15–40 cm, densely subappressed to spreading hirsute. Leaves sessile, estipulate, 3-foliolate; leaflets 7–15 × 2–4 mm, lanceolate to elliptic-lanceolate, coriaceous, usually thinly covered with short spreading hairs; nerves beneath pale, prominent, with a marginal nerve on either side running from base to apex. Flowers 1–2, leaf-opposed near the ends of the branches, with or without a short peduncle; bracts 1.5–2 mm long, linear. Calyx 5–8 mm long, hirsute; upper lobes narrowly oblong, pointed, 3–4 times as long as the tube. Standard subcircular, pale greeny-yellow, pubescent medially or over the upper part outside; wings subequalling the keel; keel c. 6 mm long, half-round, scarcely beaked. Pod subsessile, 18–20 × 6–8 mm, oblong-ellipsoid, pilose outside and on the suture inside, c. 6–8-seeded. Seeds 3 mm long, oblique-cordiform.

Zambia. W: 10 km east of Mwinilunga, fl. & fr. 23.i.1975, *Brummitt, Chisumpa & Polhill* 14048 (K).

Also in Angola and SW Dem. Rep. Congo. Grassy places on Kalahari Sands, termite mounds in dambos; 1360 m.

The raised marginal nerve on the leaflets is a distinctive feature.

53. **Crotalaria leptoclada** Harms in Warburg, Kunene-Samb.-Exped. Baum: 255 (1903). —R.E. Fries, Wiss. Ergebn. Schwed. Rhod.-Kongo-Exped. **1**: 75 (1914). —E.G. Baker in J. Linn. Soc., Bot. **42**: 397 (1914). —Verdoorn in Bothalia **2**: 385 (1928). —Wilczek in F.C.B. **4**: 112 (1953). —Torre in C.F.A. **3**: 66 (1962). —Polhill, Crotalaria Africa & Madagascar: 141 (1982). —Lock, Leg. Afr. Check-list: 188 (1989). Type from Angola.

Low herb of varied habit, producing from an ultimately woody rootstock numerous slender pubescent stems, which may be trailing in open places, ascending in other vegetation or rather strictly erect after burning, up to 15–30 cm tall, ultimately copiously branched. Leaves estipulate, sessile, 3-foliolate; leaflets 6–14 × 1.5–3(5) mm, oblanceolate to narrowly elliptic-oblong, ± pointed, sparsely to quite densely pilose on both surfaces. Flowers 1–2 on filiform leaf-opposed peduncles scattered along the branches; peduncle usually c. 1 cm long, rarely up to 3 cm; bracts 1–3 mm long, linear or linear-lanceolate; bracteoles inserted at or near the top of the pedicel, linear. Calyx 3–7 mm long, shortly appressed to rather coarsely pubescent; lobes narrowly attenuate-triangular, usually twice as long as the tube. Standard elliptic-obovate to subcircular, white becoming flushed red outside or rarely yellow, apically pubescent outside; wings exceeding the keel; keel 3–6 mm long, abruptly rounded with a practically straight tapered beak. Pod subsessile, 1–1.6 cm long, cylindrical, spreading pubescent outside and also inside (at least along lower suture), 6–8-seeded. Seeds 2–2.5 mm long, oblique-cordiform, smooth.

Var. **leptoclada**

Leaflets up to 3 mm wide. Calyx 3–5 mm long. Keel 3–4 mm long.

Zambia. B: Mongu, fl. & fr. 30.ix.1962, *Fanshawe* 7078 (K; NDO; SRGH). N: 80 km south of Kasama, fl. & fr. 16.iv.1961, *E.A. Robinson* 4606 (K; M; PRE; SRGH). W: Mwinilunga, fl. & fr. 16.v.1969, *Mutimushi* 3418 (K; NDO).

Also in Congo (Brazzaville), southern Dem. Rep. Congo and Angola. Damp grassland by dambos, lakes and rivers, usually on sandy soils, sometimes on termite mounds; 800–1650 m.

As indicated in the description, the habit varies considerably in relation to the site and incidence of burning. Variants with yellow flowers have been found with the normal white forms at Ishiba Ngandu (Shiwa Ngandu) and Mbala in northern Zambia. Var. *bianoensis* R. Wilczek, with larger flowers and rather larger leaflets, is well known on the Biano Plateau of Dem. Rep. Congo (Katanga), but not recorded elsewhere.

54. **Crotalaria abscondita** Welw. ex Baker in F.T.A. **2**: 19 (1871). —E.G. Baker in J. Linn. Soc., Bot. **42**: 281 (1914). —Torre in C.F.A. **3**: 37 (1962). —Polhill, Crotalaria Africa & Madagascar: 142 (1982). —Lock, Leg. Afr. Check-list: 163 (1989). Type from Angola.

Perennial herb, with a number of ultimately well-branched stems ascending to 30–70 cm from a well developed taproot, densely villous (hairs often up to c. 2 mm long, very fine, becoming twisted and appearing torulose). Leaves closely arranged, estipulate, sessile, 3-foliolate; leaflets mostly held erect, mostly 8–20 × 2–5(8) mm, lanceolate or narrowly elliptic, pointed, villous. Racemes terminal, sessile, dense, shortly cylindrical, 8–16-flowered; bracts 6–10 mm long, linear-lanceolate to narrowly elliptic-oblong, attenuate at either end; pedicels 0.5–1.5 mm long; bracteoles inconspicuous. Calyx 6–10 mm long, villous; lobes narrowly attenuate-triangular, 2–3 times as long as the tube. Standard broadly elliptic to obovate, yellow, apically pilose outside; wings exceeding the keel; keel 5–6 mm long, rounded, with a small very slightly incurved pointed beak. Pod practically sessile, 6–8 mm long, subglobose-ellipsoid, densely villous outside, sometimes pilose on the suture inside, 2–4-seeded. Seeds 2–3 mm long, broadly oblique-cordiform, smooth to minutely papillose, yellow to dark brown.

Zambia. N: 32 km south of Mpika, fr. 30.iii.1984, *Brummitt, Chisumpa & Nshingo* 17005 (K;

NDO; SRGH). C: Serenje Distr., Kundalila Falls, fl. & fr. 4.ii.1973, *Strid* 2830 (K).
Also in Dem. Rep. Congo (Katanga) and Angola. Seasonally wet places in miombo woodland; 1200–1500 m.

55. **Crotalaria cistoides** Welw. ex Baker in F.T.A. **2**: 18 (1871) excl. specim. *Welwitsch* 1928. —E.G. Baker in J. Linn. Soc., Bot. **42**: 282 (1914) excl. specim. *Welwitsch* 1928. —Torre in C.F.A. **3**: 37 (1962). —Polhill in F.T.E.A., Leguminosae, Pap.: 888 (1971); Crotalaria Africa & Madagascar: 143 (1982). —Lock, Leg. Afr. Check-list: 171 (1989). Lectotype, selected by Torre loc. cit., from Angola.

Low bushy herb or small shrub, up to 20–80 cm tall, with a well developed system of branches covered in a tomentum of short ± appressed to long fine mostly spreading hairs. Leaves rather closely arranged, estipulate, sessile, 3-foliolate; leaflets 6–20 × 1.5–6(8) mm, oblanceolate to narrowly elliptic-oblong, with the apiculum slightly recurved and the margins flat to slightly inrolled, appressed to spreading pilose on both surfaces, sometimes sericeous. Flowers numerous, solitary or in small very shortly stalked clusters terminal on main and very reduced lateral branches or axillary; bracts 2–4 mm long, linear to linear-lanceolate; bracteoles at the top of the short pedicel, small. Calyx 4.5–6 mm long, densely hairy; lobes narrowly attenuate-triangular to oblong-triangular, c. 3 times as long as the tube. Standard broadly elliptic or obovate-elliptic, yellow, often tinged reddish or purplish veined, sericeous at least towards the apex outside; wings exceeding the keel; keel 3.5–5 mm long, abruptly rounded, with a short straight or slightly incurved pointed beak. Pod practically sessile, 5–7 mm long, shortly ovoid-ellipsoid to oblong-ellipsoid, pilose outside and also inside at least along the lower suture, 2–4-seeded. Seeds c. 3 mm long, obliquely oblong-cordiform, smooth.

Subsp. **cistoides**

Leaflets oblanceolate, shortly rounded to the apiculate tip, with generally flat margins, often appressed pilose. Pedicels rarely more than 1 mm long.

Zambia. B: Kabompo–Chizela (Chizera), 1.5 km SW of Nkulwashi (Kabompo R. Pontoon), fl. & fr. 23.iii.1961, *Drummond & Rutherford-Smith* 7232 (K; P; SRGH). W: Solwezi, fl. & fr. 15.v.1969, *Mutimushi* 3321 (K; NDO).
Also in southern Angola. Miombo woodland on sandy soils; c. 1000–1300 m.
The Zambian specimens have larger leaflets and longer calyces than any Angolan material yet seen, but this is probably explicable in terms of rather more favourable growth conditions.

Subsp. **orientalis** Polhill in Kew Bull. **25**: 175 (1971); in F.T.E.A., Leguminosae, Pap.: 889 (1971); Crotalaria Africa & Madagascar: 143, fig. 30/9–16 (1982). —Lock, Leg. Afr. Check-list: 171 (1989). TAB. 3,7: **27**, fig. B. Type: Zambia, Mpulungu–Mbala, *Richards* 8338 (BR; K, holotype; SRGH).

Leaflets narrowly elliptic-oblong, pointed, with the margins slightly inrolled, spreading pilose. Pedicels mostly 2–3 mm long.

Zambia. N: Mbala Distr., Uningi Pans, fl. & fr. 16.v.1968, *Richards* 23262 (K; SRGH); Chinsali, fl. & fr. 14.viii.1965, *Fanshawe* 9244 (K; NDO).
Also in Tanzania (Ufipa). Miombo woodland, often on sandy soils, secondary bushland near rivers, roads and other disturbed places; 1000–1500 m.

56. **Crotalaria variegata** Welw. ex Baker in F.T.A. **2**: 19 (1871). —E.G. Baker in J. Linn. Soc., Bot. **42**: 400 (1914). —Wilczek in F.C.B. **4**: 150 (1953). —Torre in C.F.A. **3**: 68 (1962). —Polhill in Kew Bull. **22**: 236 (1968); in F.T.E.A., Leguminosae, Pap.: 889 (1971). —Drummond in Kirkia **8**: 219 (1972). —Gonçalves in Garcia de Orta, Sér. Bot. **5**: 72 (1982). —Polhill, Crotalaria Africa & Madagascar: 145, fig. 18/5 (1982). —Lock, Leg. Afr. Check-list: 209 (1989). Type from Angola.
Crotalaria sericifolia var. *gweloensis* Baker f. in J. Linn. Soc., Bot. **42**: 398 (1914). —Eyles in Trans. Roy. Soc. South Africa **5**: 372 (1916). —Verdoorn in Bothalia **2**: 385 (1928). —Binns, First Check List Herb. Fl. Malawi: 79 (1968). Type: Zimbabwe, Gweru (Gwelo), *Rand* 50 (BM, holotype).
Crotalaria gweloensis (Baker f.) Milne-Redh. in Kew Bull. **15**: 166 (1961). —Torre in C.F.A. **3**: 68 (1962).

Crotalaria variegata var. *humpatenis* Torre in Mem. Junta Invest. Ultramar, sér. 2, **19**: 45, t. 24 (1960); in C.F.A. **3**: 68 (1962). Type from Angola.

Trailing perennial herb, with a number of radiating flexuous stems up to 1 m long; younger parts with a ± dense subappressed or generally shortly spreading indumentum. Leaves 3-foliolate; leaflets mostly 1.5–6.5 × 0.8–3.2 cm, oblanceolate, elliptic or obovate, acute to rounded or retuse with a short recurved apiculum, chartaceous, glabrous and glossy to densely hairy above, ± silky and often silvery tomentose beneath; petiole 3–10(20) mm long; stipules 3–8 mm long, subulate to acuminately triangular, spreading or recurved. Racemes terminal and leaf-opposed, sometimes aggregated into pseudopanicles by reduction of the foliage on upper branches, up to 2.5–5 cm long, subumbelliform, 3–6(10)-flowered; bracts mostly 2–4 mm long, linear or linear-lanceolate; bracteoles inserted on the pedicel, similar to the bract. Calyx 5–9 mm long, with a dense covering of short mostly appressed hairs; lobes narrow, 2–4 times as long as the tube. Standard subcircular or oblate, blue or white inside, reddish and usually pubescent outside; wings yellow, practically as long as the keel; keel 7–9 mm long, abruptly rounded to subangular in the lower half, with a short only slightly incurved beak. Pod subsessile, 1.2–1.6 cm long, oblong-ellipsoid, with a short dense appressed or spreading indumentum outside, usually with hairs inside at least along the suture, 4–6-seeded. Seeds c. 3 mm long, oblique-cordiform, smooth.

Zambia. B: 59 km Kabompo–Zambezi (Balovale), fl. & fr. 24.iii.1961, *Drummond & Rutherford-Smith* 7301 (K; LISC; SRGH). N: Mbala Distr., Kawimbe, fl. & fr. 6.iv.1959, *McCallum-Webster* 728 (BR; K; LISC; SRGH). W: Solwezi Distr., Chifubwa National Monument, fl. 8.i.1969, *Mutimushi* 2923 (K; NDO). C: 32 km east of Kapiri Mposhi, fl. 12.vi.1960, *Leach & Brunton* 10018 (K; LISC; SRGH). E: Lundazi, fl. i.1962, *Verboom* 448 (K; SRGH). S: 19 km east of Choma, fl. & fr. 28.v.1955, *E.A. Robinson* 132 (K; SRGH). **Zimbabwe**. N: Gokwe, fl. 9.ii.1962, *Bingham* 132 (K; SRGH). C: Harare Distr., Enterprise, fl. & fr. iii 1924, *Eyles* 6919 (K; SRGH). E: Chimanimani Distr., Nyamzure Hill (Pork Pie), fl. 25.xii.1948, *Chase* 1419 (BM; COI; K; LISC; SRGH). **Malawi**. N: near Mzimba, Mbawa, fl. 13.iii.1953, *Jackson* 1146 (BR; K). C: Dzalanyama Forest Reserve, near Chionjeza (Chiungiza), fl. 9.ii.1959, *Robson* 1517 (BM; BR; K; LISC; SRGH). **Mozambique**. N: Ngauma Distr., Massangulo, fl. 15.v.1948, *Pedro & Pedrógão* 3505 (LMA). T: Macanga Distr., a 39 km de Furancungo para Vila Coutinho, fl. 15.vii.1949, *Barbosa & Carvalho* 3620 (K; LISC; LMA). MS: Barué Distr., Serra de Chôa, a 26 km de Catandica (Vila Gouveia), fl. & fr. 26.iii.1966, *Torre & Correia* 15402 (LISC).

Also in Angola, Dem. Rep. Congo (Katanga) and southern Tanzania. Miombo woodland, less often in grassy ± damp places on sand, persisting in disturbed places; c. 800–1700 m.

57. **Crotalaria lasiocarpa** Polhill in Kew Bull. **22**: 236, fig. 11 (1968); in F.T.E.A., Leguminosae, Pap.: 890 (1971); Crotalaria Africa & Madagascar: 147 (1982). —Lock, Leg. Afr. Check-list: 187 (1989). Type from S Tanzania.

Annual, with short erect stem to 25 cm or so tall and long slender pilose radiating branches prostrate or somewhat ascending in other vegetation. Leaves 3-foliolate; leaflets mostly 1.5–5 × 0.7–2.5 cm, elliptic to obovate, but those of the leading shoot sometimes longer and relatively narrow, appressed pilose beneath; petiole 8–24 mm long; stipules 1.5–5 mm long, filiform to subulate-caudate. Racemes mostly leaf-opposed, long-pedunculate, up to 3–5.5 cm long, with 2–5 flowers towards the top of the very slender axis; bracts 1–3 mm long, linear; bracteoles at the base of the calyx, small. Calyx 3–4.5 mm long, pilose; lobes narrow, 2–3 times as long as the tube. Standard subcircular, pale yellow, veined maroon or mauve, glabrous outside; wings ± as long as the keel; keel 5–6 mm long, angled in the lower half, with a relatively long narrow slightly recurved beak. Pod subsessile, (1)1.2–1.3 cm long, ellipsoid-oblong, rarely smaller and subglobose, densely covered with fine spreading hairs, (10)14–16-seeded. Seeds 2–2.5 mm long, oblique-cordiform, smooth, dull orange.

Zambia. N: Mbala Distr., Kawimbe, fl. & fr. 18.iv.1959, *McCallum-Webster* 782 (K); Mwambe Village, fl. & fr. 15.iii.1960, *Richards* 12742 (K).

Also in southern Tanzania. Grassland in sandy sometimes damp places, disturbed ground; 1500–1750 m.

58. **Crotalaria reptans** Taub. in Engler, Pflanzenw. Ost-Afrikas **C**: 204 (1895). —E.G. Baker in J. Linn. Soc., Bot. **42**: 304 (1914). —Verdoorn in Bothalia **2**: 404 (1928) excl. *Flanagan* 3091 —Binns, First Check List Herb. Fl. Malawi: 79 (1968). —Polhill in Kew Bull. **22**: 242, fig. 12/1–3 (1968); in F.T.E.A., Leguminosae, Pap.: 890 (1971). —Drummond in Kirkia **8**: 218 (1972). —Polhill, Crotalaria Africa & Madagascar: 148, fig. 31/1–9 (1982). —Lock, Leg. Afr. Check-list: 201 (1989). TAB. 3,7: **28**, fig. A. Type from Tanzania.

Annual or short-lived perennial, with a short erect stem and many long slender trailing much-divided pilose branches up to 30–90 cm long. Leaves 3-foliolate; leaflets mostly 2–4.5 × 1.2–3 cm, elliptic or obovate, with those of the leading shoot sometimes up to 7 cm long and relatively narrow, appressed pilose beneath; petiole 0.8–4 cm long; stipules 1–4(8) mm long, linear-subulate. Racemes leaf-opposed, subumbelliform, usually longer than the opposed leaf, (2.5)4–14 cm long, 2–8-flowered; bracts 1–2 mm long, linear-subulate; bracteoles inserted at the top of the pedicel, small. Calyx (4.5)6–8 mm long, pubescent or shortly pilose; lobes triangular-subulate, 1.7–3 times as long as the tube. Standard obovate or subcircular, yellow, veined brown and glabrous outside; wings shorter than the keel; keel 7–12 mm long, angled in the lower third, with a long straight tapered beak. Pod subsessile, 7–12 × 3.5–6 × 6–9 mm, discoidal to shortly ellipsoid or even subglobose, but usually distinctly compressed laterally when young, thick-walled and tardily dehiscent, usually rugose with the vein-network impressed to slightly raised, shortly but conspicuously pubescent, 1–3-seeded. Seeds c. 4 mm across, subcircular, slightly rugulose, brown.

Zambia. C: 8 km SE of Lusaka, fl. & fr. 3.iii.1957, *Noak* 139 (K; SRGH). E: 3 km west of Luangwa R. Bridge, fl. & fr. 30.v.1961, *Leach & Rutherford-Smith* 11092 (K; LISC; SRGH). S: Livingstone, fl. & fr. 22.iii.1961, *Fanshawe* 6453 (BR; K; LISC; NDO; SRGH). **Zimbabwe**. N: Hurungwe Distr., Zambezi Valley, Mhenza (Menswa) Pan, fl. & fr. 26.ii.1953, *Wild* 4068 (K; LISC; SRGH). **Malawi**. N: Nkhata Bay Distr., Likoma Island, fl., *W.P. Johnson* (K). S: Machinga Distr., Mlomba, fl. 2.ii.1955, *Jackson* 1437 (K; SRGH). **Mozambique**. N: 10 km Nampula–Corrane, near R. Colave, fl. & fr. 14.iv.1961, *Balsinhas & Marrime* 396 (BM; COI; K; LMA).

Also in Uganda and Tanzania. Miombo and sometimes mopane woodland areas, generally in disturbed, sandy places; c. 300–800(1200) m.

Normally easily distinguished from *C. pisicarpa* by the larger flowers and by the different shape, ornamentation and indumentum of the pod, but there are two difficulties discussed more fully by Polhill, loc. cit. (1968). First, unusually small flowers may be produced at the end of the growing season, but at that stage the more reliable fruit characters are available. Secondly, intermediates occur in the vicinity of the R. Zambezi and its tributaries, e.g. *Fanshawe* 6453, *Wild* 4068, cited above, and Kariba, *Jarman* B+C 27 (K; SRGH), apparently of secondary introgressive origin, and the allocation of plants has to be somewhat arbitrary with more reliance on fruit characters.

59. **Crotalaria pisicarpa** Welw. ex Baker in F.T.A. **2**: 16 (1871). —E.G. Baker in J. Linn. Soc., Bot. **42**: 303 (1914). —Eyles in Trans. Roy. Soc. South Africa **5**: 371 (1916). —Verdoorn in Bothalia **2**: 404 (1928). —Schreiber in Mitt. Bot. Staatssamml. München **2**: 290 (1957); in Merxmüller, Prodr. Fl. SW. Afrika, fam. 60: 25 (1970). —Torre in C.F.A. **3**: 48 (1962). —Polhill in Kew Bull. **22**: 239, fig. 12/4, 5 (1968); in F.T.E.A., Leguminosae, Pap.: 891 (1971). —Drummond in Kirkia **8**: 218 (1972). —Polhill, Crotalaria Africa & Madagascar: 148, fig. 31/10–17 (1982). —Lock, Leg. Afr. Check-list: 197 (1989). TAB. 3,7: **28**, fig. B. Type from Angola.

Crotalaria pilulicarpa Taub. in Engler, Pflanzenw. Ost-Afrikas **C**: 204 (1895). —E.G. Baker in J. Linn. Soc., Bot. **42**: 303 (1914). Syntypes from Tanzania.

Crotalaria pilulicarpa var. *schinzii* Baker f. in J. Linn. Soc., Bot. **42**: 303 (1914). —Verdoorn in Bothalia **2**: 404 (1928). —Burtt Davy, Fl. Pl. Ferns Transvaal, pt. 2: 400 (1932). Syntypes from South Africa and Namibia.

Crotalaria reptans sensu Verdoorn in Bothalia **2**: 404 (1928) quoad *Flanagan* 3091 tantum.

Annual or short-lived perennial, with a short erect stem and long trailing slender much-divided spreading-pubescent or pilose branches radiating from the base, up to 60 cm or more long. Leaves 3-foliolate; leaflets mostly 1.2–4.2 × 0.5–2.5 cm, narrowly to broadly elliptic or oblanceolate to obovate, with those of the leading shoot up to 7 cm long and relatively narrow, appressed pilose at least beneath; petiole 6–35 mm long; stipules 1–6 mm long, linear-subulate. Racemes leaf-opposed, subumbelliform, shorter or longer than the opposed leaf, 1–9 cm long, 3–8-flowered; bracts up to 1.5 mm long, linear; bracteoles at the top of the pedicel, filiform, small. Calyx

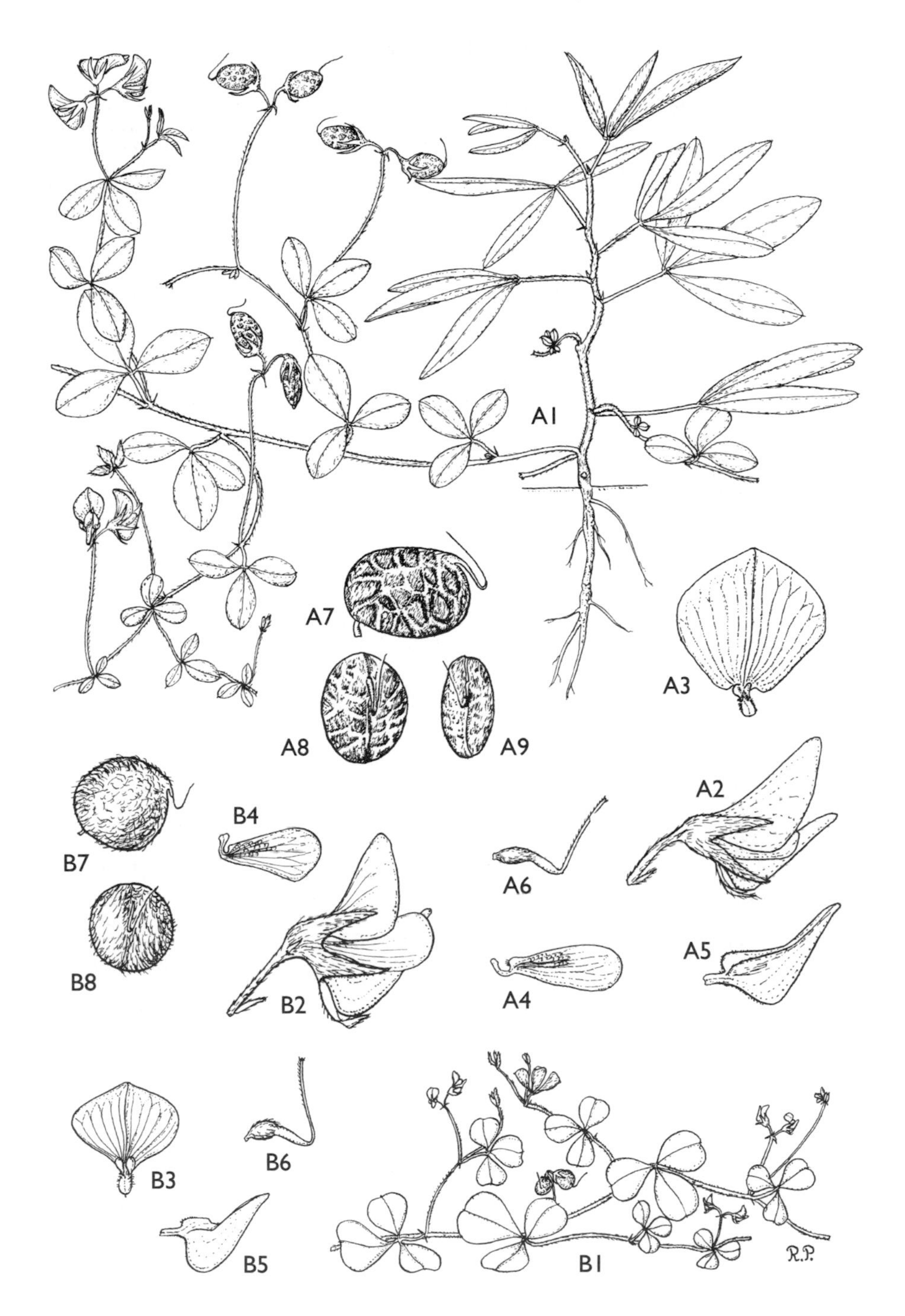

Tab. 3,7: **28**. A. —CROTALARIA REPTANS. A1, habit (× $^{2}/_{3}$) from *Polhill & Paulo* 2045 and 2147; A2, flower (× 2); A3, standard (× 2); A4, wing (× 2); A5, keel (× 2); A6, gynoecium (× 2); A7 & A8, pod (× 2), A2–A8 from *Polhill & Paulo* 2147; A9, pod before maturity (× 2), from *Polhill & Paulo* 2045. B. —CROTALARIA PISICARPA. B1, habit (× $^{2}/_{3}$), from *E.A. Robinson* 1376; B2, flower (× 4); B3, standard (× 4); B4, wing (× 4); B5, keel (× 4); B6, gynoecium (× 4), B2–B6 from *E.A. Robinson* 1446; B7 & B8, pod (× 2), from *Plowes* 1689. Drawn by Roger Polhill. From Crotalaria Africa & Madagascar.

3.5–4.5(5) mm long, pubescent; lobes subulate or narrowly triangular-acuminate, 1.25–2 times as long as the tube. Standard obovate or subcircular, yellow, finely veined reddish-brown and glabrous outside; wings shorter than the keel; keel angled in the lower third, with a long straight tapered beak, 4.5–6(7.5) mm long. Pod practically sessile, 5–6(7) × 4–6 × 4.5–6 mm, subglobose, thick-walled and tardily dehiscent, densely covered with hairs obscuring the surface, ultimately somewhat glabrescent and sometimes rugulose, 1–3-seeded. Seeds 3–3.5 mm across, subcircular in outline, slightly rugulose, terra-cotta or pale brown.

Botswana. N: Xaru, fl. & fr. 23.ii.1984, *P.A. Smith* 4381 (K; SRGH). SE: Kgatleng Distr., Mochudi, fl. & fr. i–iv.1914, *Harbor* in *Rogers* 6904 (K). **Zambia**. S: 8 km north of Mapanza, Masonsa, fl. & fr. 18.iii.1956, *E.A. Robinson* 1376 (K; SRGH). **Zimbabwe**. N: Zvimba Distr., Trelawney, fl. & fr. 27.iii.1943, *Jack* 138 (K; LISC; SRGH). W: Umguza Distr., Nyamandhlovu Research Station, fl. & fr. 25.iii.1954, *Plowes* 1689 (K; SRGH). C: Kwekwe (Que Que), fl. & fr. 8.ii.1967, *Corby* 1743 (K; SRGH). S: 32 km north of Masvingo (Fort Victoria), fl. & fr. 4.v.1962, *Drummond* 7957 (K; SRGH). **Malawi**. N: Karonga Distr., Chaminade School, fl. & fr. 15.iv.1976, *Pawek* 11041 (K; MO; SRGH). C: Lilongwe Distr., Chitedze, fl. & fr. 22.iii.1955, *Exell, Mendonça & Wild* 1109 (BM; BR; LISC; SRGH). **Mozambique**. MS: Sussundenga Distr., Mavita, fl. & fr. 28.iv.1948, *Barbosa* 1659 (BM; K; LISC).

Also in Tanzania, Angola, Namibia and South Africa (Northern Province, Gauteng and Mpumalanga). Woodland, shrub savanna and grassland, generally in disturbed sandy places; 550–1300 m.

See note above under *C. reptans*.

60. **Crotalaria ononoides** Benth. in Hooker, London J. Bot. **2**: 572 (1843). —J.G. Baker in F.T.A. **2**: 22 (1871). —E.G. Baker in J. Linn. Soc., Bot. **42**: 307 (1914). —Verdoorn in Bothalia **2**: 384 (1928). —Wilczek in F.C.B. **4**: 136 (1953) excl. var. *grandiflora*. —Hepper in F.W.T.A., ed. 2, **1**: 549 (1958). —Torre in C.F.A. **3**: 49 (1962). —Polhill in F.T.E.A., Leguminosae, Pap.: 892, fig. 123 (1971); Crotalaria Africa & Madagascar: 154, fig. 33 (1982). —Lock, Leg. Afr. Check-list: 194 (1989). TAB. 3,7: **29**. Type from Sierra Leone.

Much branched spreading annual, procumbent or ascending, sometimes up to 0.5–1 m tall; branches pilose with subappressed or spreading hairs. Leaves shortly petiolate, 3-foliolate; leaflets mostly 2.5–5(6) × 0.8–1.8(2.4) cm, lanceolate to elliptic or obovate, the upper ones often longer and proportionally narrower, glabrous to sparsely pilose above, subappressed pilose beneath; stipules simple or 2–3(5)-fid, up to 5–13 mm long, with both the simple stipules and segments linear-subulate. Flowers numerous in sessile heads; bracts basally fused to the pedicel, spreading from just below the calyx, 3-fid with linear segments up to 10–13 mm long; bracteoles basally fused to the calyx tube, the free part directed forwards, 8–12 mm long, linear. Calyx 7–10(12) mm long, deeply divided, slightly 2-lipped, pilose; lobes subulate, caudate, 4–5 times as long as the tube. Standard elliptic-obovate to subcircular, yellow, finely veined or reddish flushed with age, glabrous except along the midvein outside; wings ± as long as the keel; keel 7–9 mm long, subangular, narrowly beaked. Pod sessile, 8–12 mm long, ellipsoid, glabrous except for a few hairs at the apex, 8–14-seeded. Seeds 1.5–2 mm long, oblique-cordiform, smooth, pale orange-brown.

Zambia. N: Mbala Distr., Kalambo R., Sansia Falls, fl. & fr. 8.v.1961, *Richards* 15120 (K; SRGH). W: Solwezi Distr., 5 km east of Solwezi, Kifubwa R., fl. 17.iii.1961, *Drummond & Rutherford-Smith* 6965 (K; LISC; SRGH). C: 9.5 km SE of Lusaka, fl. 22.ii.1957, *Noak* 129 (K; SRGH). S: Mumbwa, fl. & fr. 22.iii.1963, *van Rensburg* 1783 (K; SRGH).

Widespread in areas of moderate to high rainfall over most of tropical Africa, except the NE and SE, also in Madagascar. Grassland in dambos, flood plains, by rivers and in disturbed places; 1000–1700 m.

Var. *pubescens* R. Wilczek, with the leaflets pilose instead of glabrous above, occurs in Dem. Rep. Congo and East Africa, and might be found in the Flora Zambesiaca area.

61. **Crotalaria nigricans** Baker in F.T.A. **2**: 21 (1871). —Oliver in Trans. Linn. Soc., London **29**: 51, t. 26A (1872). —E.G. Baker in J. Linn. Soc., Bot. **42**: 308 (1914). —Verdoorn in Bothalia **2**: 384 (1928). —Hutchinson, Botanist South. Africa: 506 (1946). —Brenan, Check-list For. Trees Shrubs Tang. Terr.: 413 (1949). —Wilczek in F.C.B. **4**: 110, fig. 8 (1953). —Binns, First Check List Herb. Fl. Malawi: 79 (1968). —Polhill in F.T.E.A., Leguminosae, Pap.: 894 (1971); Crotalaria Africa & Madagascar: 155 (1982). —Lock, Leg. Afr. Check-list: 193 (1989). Type from Tanzania.

Tab. 3,7: **29**. CROTALARIA ONONOIDES. 1, flowering branch (× 1); 2, flower (× 3); 3, standard (× 3); 4, wing (× 3); 5, keel (× 3); 6, anthers (× 6); 7, gynoecium (× 3), 1–7 from *Richards* 1524; 8, pod (× 3), from *Fanshawe* 3059; 9, seed (× 10), from *Milne-Redhead* 3117. Drawn by Roger Polhill. From F.T.E.A.

Perennial herb; stems erect or prostrate, much branched, 0.3–1.5 m long, with fine spreading hairs; plants often drying blackish. Leaves 3-foliolate; leaflets (1)1.5–3 × 0.5–1.2 cm, oblanceolate to oblong-obovate, usually sparsely pubescent above, thinly hairy beneath; petiole 2–6 mm long; stipules 3–6 mm long, lanceolate. Flowers 4–12 in sessile or shortly pedunculate terminal heads; bracts 4–7(9) mm long, ovate, acuminate; pedicels short; bracteoles inserted just below the calyx, similar to the bract but a little smaller. Calyx 4-lobed, 8–13 mm long, almost enclosing the corolla, slightly accrescent, pubescent; lobes ovate, 2–3 times as long as the tube. Standard circular-ovate to oblate, usually pale yellow, reddish-tinged and puberulous along the midvein outside towards the apex; wings longer than the keel; keel 8–10 mm long, rounded about the middle, with a short incurved beak. Pod enveloped by the calyx, sessile, 7–10 mm long, ellipsoid, spreading pubescent outside, sparsely hairy inside, 4–8-seeded. Seeds c. 2.5 mm long, oblique-cordiform, smooth, mauvish-brown.

Var. **nigricans**

Stems prostrate, radiating from a rather woody taproot; inflorescences never more than a few centimetres from the ground.

Zambia. N: Chambeshi R., fl. & fr. 17.vii.1930, *Hutchinson & Gillett* 3794 (BM; COI; K; LISC; P; SRGH). W: Luanshya, fl. 4.iv.1954, *Fanshawe* 1068 (K; NDO; SRGH). C: Kabwe Distr., 3 km south of Kapiri Mposhi, fl. & fr. 12.vi.1960, *Leach & Brunton* 10007 (K; LISC; SRGH). **Malawi**. N: Mzimba Distr., Mzuzu, Marymount, fl. 6.vii.1975, *Pawek* 9816 (K; MAL; MO; SRGH). C: near Kasungu, Lisasadzi, fl. 9.iv.1955, *Jackson* 1622 (BM; K; LISC; SRGH). **Mozambique**. N: 60 km east of Mandimba, fl. 15.v.1961, *Leach & Rutherford-Smith* 10862 (K; LISC; SRGH).

Also in Dem. Rep. Congo (Katanga) and Tanzania. Miombo woodland, particularly sandy places where a little additional water collects by roads, etc., also grassland by streams and on flood plains; 650–1500 m.

Var. **erecta** Milne-Redh. in Kew Bull. **15**: 157 (1961). —Polhill in F.T.E.A., Leguminosae, Pap.: 895 (1971); Crotalaria Africa & Madagascar: 156, fig. 34 (1982). Type from Tanzania.

Stems ± erect, bushy, forming a clump, up to 0.7–1.5 m tall; inflorescences always well above the ground.

Zambia. W: Mwinilunga Distr., west of R. Kasanjiku (Kasingiko), fl. 5.viii.1930, *Milne-Redhead* 832 (K).

Also in Dem. Rep. Congo (Katanga) and Tanzania, in the northern and western part of the species range. Apparently growing in sites similar to those occupied by var. *nigricans*.

62. **Crotalaria torrei** Polhill, Crotalaria Africa & Madagascar: 164, fig. 38 (1982). —Lock, Leg. Afr. Check-list: 207 (1989). TAB. 3,7: **30**. Type: Mozambique, próximo do Pico Namuli, 9.iv.1943, *Torre* 5145 (LISC, holotype).

Small rhizomatous shrub up to 1.5 m tall; branches densely spreading pilose. Leaves 3-foliolate; leaflets 2.5–5 × 1.3–2.5 cm, elliptic or elliptic-obovate, pilose on both surfaces with irregularly arranged hairs, more densely so beneath particularly on the nerves; petiole 3–10 mm long; stipules (2)3–8 mm long, subulate-caudate. Racemes rather short, densely 12–24-flowered; bracts 4–5 mm long, subulate-caudate; bracteoles inserted just below the calyx, 3–4 mm long, linear-caudate. Calyx 7–8 mm long, densely pilose; lobes narrowly attenuate-triangular, ± twice as long as the tube. Standard oblate, yellow, finely lined with brown, glabrous outside; wings broad, longer than the keel; keel 1–1.2 cm long, strongly rounded about the middle, with a narrow acute upwardly directed beak. Pod sessile, 1.5–2 cm long (including the beak), ovoid-fusiform, produced into a well-formed beak (including style-base), densely spreading pilose, probably 6–8-seeded. Seeds not seen.

Mozambique. Z: Gurué Distr., próximo do Pico Namuli, fl. 9.iv.1943, *Torre* 5145 (LISC); Serra do Gurué, próximo da nascente do R. Malema, fr. 4.i.1968, *Torre & Correia* 16875 (LISC).

Known only from the Gurué mountain range of Mozambique. Damp, sometimes rocky places in montane savanna; c. 1700 m.

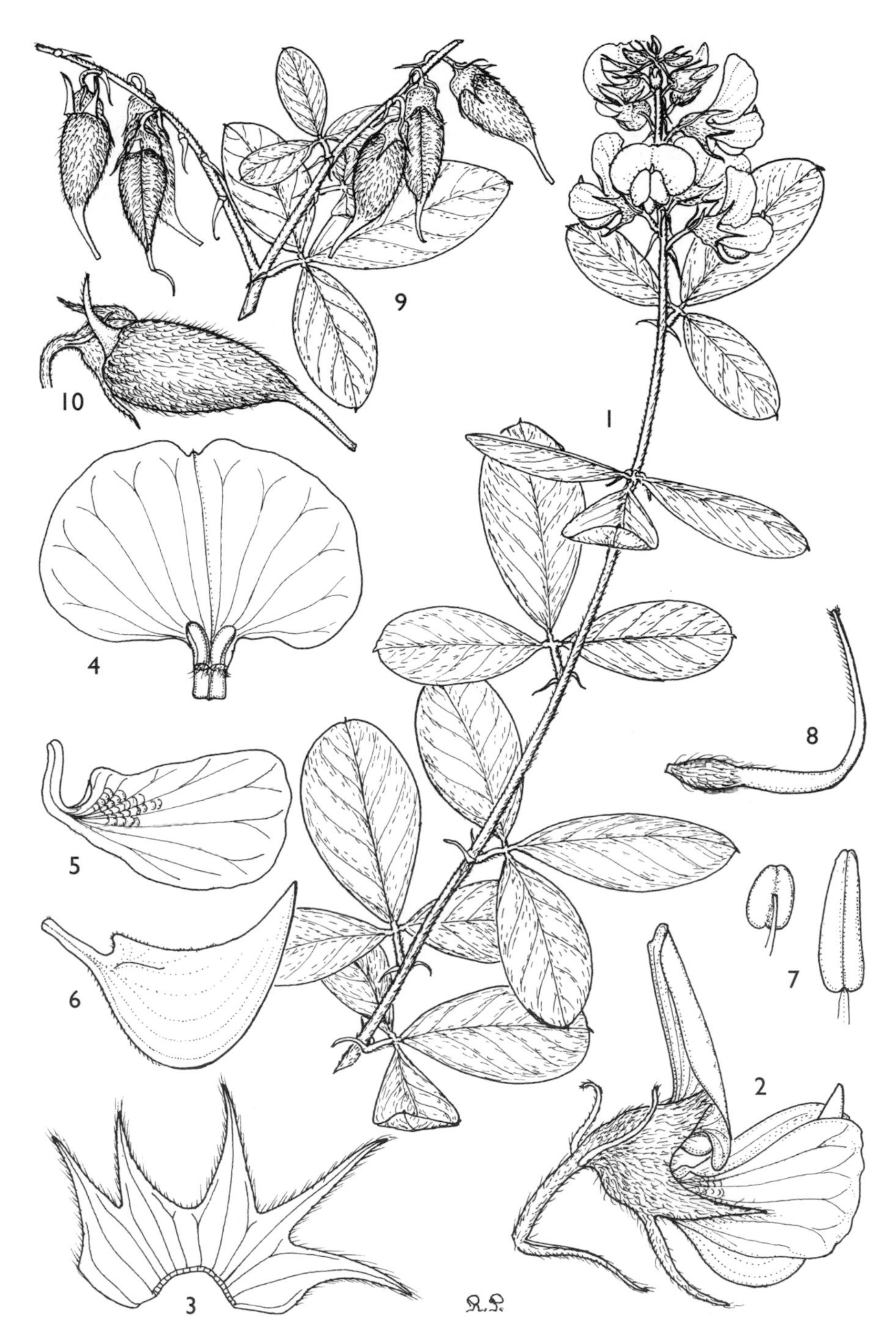

Tab. 3,7: **30**. CROTALARIA TORREI. 1, flowering branch (× 1); 2, flower (× 3); 3, calyx, opened out (× 3); 4, standard (× 3); 5, wing (× 3); 6, keel (× 3); 7, anthers (× 10); 8, gynoecium (× 3), 1–8 from *Torre* 5145; 9, fruiting branchlet (× 1); 10, pod (× 2), 9 & 10 from *Torre & Correia* 16875. Drawn by Roger Polhill. From Crotalaria Africa & Madagascar.

63. **Crotalaria collina** Polhill, Crotalaria Africa & Madagascar: 165, fig. 39 (1982). —Lock, Leg. Afr. Check-list: 171 (1989). Type: Zimbabwe, Nyanga (Inyanga)–Rusape, Liverpool, *Challenger* in *GHS* 140747 (K, holotype; LISC; SRGH).

Bushy herb 0.6–1 m tall; young branches densely covered with longish subappressed to retrorsely spreading hairs. Leaves 3-foliolate; leaflets 2.5–10.5 × 1.5–5 cm, elliptic or elliptic-obovate, appressed pilose beneath; petiole shorter than the leaflets; stipules 2–5(10) mm long, linear to subulate-caudate, often recurved. Racemes shortly pedunculate, ± dense, many-flowered; bracts 1.5–3.5 mm long, subulate, deflexed; bracteoles at the base of the calyx, sometimes curved upwards, 1.5–2.5 mm long, triangular to linear-lanceolate. Calyx reflexed against the ascending pedicel, 3.5–4.5 mm long, coarsely hairy; lobes narrowly triangular, shorter to longer than the tube. Standard broadly elliptic to subcircular, bright yellow, glabrous outside; wings shorter than the keel; keel 8–11 mm long, rounded about the middle, with a narrow sharply pointed incurved beak. Pod including the 2–3 mm long stipe 12–18 × 6–8 mm, ellipsoid, densely covered with longish irregularly arranged slightly spreading hairs, 2–4-seeded. Seeds 4–5 mm long, very obliquely cordiform, with the radicular lobe strongly incurved, smooth, greyish or brown.

Zimbabwe. C: Goromonzi Distr., Chinamora Reserve, summit of Ngomakurira (Ngoma Kurira), fl. 27.i.1956, *Drummond* 5103 (K; LISC; SRGH). E: Nyanga (Inyanga), Selbourne, fl. & fr. 7.iii.1966, *Plowes* 2761 (K; SRGH). S: Masvingo Distr., Mutirikwi (Kyle) Dam, fl. & fr. 27.ii.1971, *Corby* 2176 (K; LISC; SRGH). **Mozambique**. MS: Barué Distr., Catandica (Vila Gouveia), Serra de Chôa, fr. 4.vii.1941, *Torre* 3003 (BM; K; LISC).

Known only from Zimbabwe and adjacent parts of Mozambique. Forest edges and upland grassland, usually on rocky hillsides or outcrops, less often in woodland, on termite mounds and roadsides; 1200–1950 m.

64. **Crotalaria uncinella** Lam., Encycl. Méth. Bot. **2**: 200 (1786). —Peltier in J. Agric. Trop. Bot. Appl. **6**: 185 (1959). —Polhill in Kew Bull. **25**: 281, fig. 3 (1971); in F.T.E.A., Leguminosae, Pap.: 895 (1971); Crotalaria Africa & Madagascar: 166, fig. 40 (1982). —Lock, Leg. Afr. Check-list: 208 (1989). Type from Réunion.

Crotalaria fischeri Taub. in Engler, Pflanzenw. Ost-Afrikas **C**: 205 (1895); E.G. Baker in J. Linn. Soc., Bot. **42**: 288 (1914). —Polhill in Kew Bull. **22**: 248 (1968). Syntypes from Tanzania.

Shrub 1–3(5) m tall, with tomentose branches in the Flora Zambesiaca area. Leaves 3-foliolate; leaflets mostly 2.5–5(8) × 1–2(2.7) cm, oblanceolate-elliptic, elliptic or obovate-oblong, subglabrous to pilose above, rather densely silky pilose beneath; petioles mostly 0.8–3 cm long; stipules up to 1 mm long, subulate, or lacking. Racemes with many fairly closely arranged flowers; bracts 1–3 mm long, shortly caudate-lanceolate; bracteoles on the upper part of the pedicel, small. Calyx 3.5–5 mm long, densely appressed pilose; lobes narrowly triangular, ± as long as the tube. Standard broadly elliptic to subcircular, yellow, glabrous; wings ± as long as the keel or a little shorter; keel rounded about the middle (except in Asia), with an incurved narrow beak, 9–11 mm long in Africa, smaller elsewhere. Pod with a 2–3 mm long stipe 8–11 × 5–7 mm in Africa (sometimes smaller elsewhere), subglobose-ellipsoid, tomentose in Africa, 2-seeded; style very persistent. Seeds 3–4 mm long, oblique-cordiform, smooth, light brown.

Mozambique. N: Lalaua Distr., Lalaua, fl. & fr. 24.iv.1937, *Torre* 1465 (COI; LISC). Z: Montes de Ile, fr. 26.vi.1943, *Torre* 5581 (BM; LISC).

Also in eastern Tanzania, Madagascar and the Mascarene Is., with a subspecies in Asia. Forest margins, rock outcrops and streamsides, or sometimes in miombo woodland; 850–1100 m.

65. **Crotalaria schlechteri** Baker f. in J. Linn. Soc., Bot. **42**: 365 (1914). —Verdoorn in Bothalia **2**: 395 (1928). —Polhill, Crotalaria Africa & Madagascar: 171, fig. 35/12 (1982). —Lock, Leg. Afr. Check-list: 202 (1989). Type: Mozambique, Incanine, *Schlechter* 12037 (BM, holotype; BR; COI; K; L; MO; P).

Perennial herb, with an erect laxly branched stem, 40–80 cm tall; branches silvery tomentellous. Leaves rather shortly petiolate, 3-foliolate; leaflets 3–6 × 1.5–3 cm, elliptic or obovate, densely silky pubescent beneath; stipules 4–6 mm long, linear or subulate-caudate. Racemes with generally numerous flowers closely arranged at least on the upper part; bracts 2–3 mm long, shortly lanceolate-caudate; bracteoles at the base of the calyx, 1.5–3 mm long, obliquely elliptic-oblong. Calyx 4–5 mm long, appressed pubescent; lobes narrowly triangular-acuminate, a little shorter than the tube. Standard broadly elliptic or subcircular, yellow, glabrous or with a few hairs along the midvein outside; wings as long as the keel or a little shorter; keel 1–1.3 cm long, rather shallowly rounded, with a tapered acute projecting beak. Pod shortly stipitate, 2.5–3 cm long, asymmetrically oblong-clavate, rather densely appressed pubescent, 16–20-seeded. Seeds c. 3 mm long, broadly cordiform, smooth, brown.

Mozambique. GI: Chibuto, fl. 9.ix.1952, *Pedro* 3987 (LMA). M: Magude, fl. & fr. 22.i.1948, *Torre* 7189 (BM; K; LISC).

Also in South Africa (Mpumalanga). Coastal plain savanna and grassland.

66. **Crotalaria pallida** Aiton, Hort. Kew. **3**: 20 (1789). —Klotzsch in Peters, Naturw. Reise Mossambique **6**, part 1: 88 (1861)*. —E.G. Baker in J. Linn. Soc., Bot. **42**: 348 (1914). —Polhill in Kew Bull. **22**: 262 (1968); in F.T.E.A., Leguminosae, Pap.: 905 (1971); Crotalaria Africa & Madagascar: 184, fig. 45 (1982). —Lock, Leg. Afr. Check-list: 195 (1989). Type from Ethiopia.

Erect well-branched sometimes robust herb 1–2(3) m tall; branches densely appressed pubescent. Leaves 3(4)-foliolate; leaflets mostly 3–13 × 2.5–7 cm, elliptic to obovate, appressed puberulous beneath; petiole ± as long as leaflets or shorter; stipules up to 3 mm long, filiform, deciduous, or absent. Racemes shortly pedunculate, mostly 15–30 cm long, many-flowered; bracts up to 5 mm long, linear, caducous before the flowers open; bracteoles at the base of the calyx, similar but smaller. Calyx 6–7.5 mm long, becoming basally truncate and deflexed against the pedicel, ± densely appressed pubescent; lobes narrowly triangular-acuminate, subequal to or longer than the tube. Standard elliptic, yellow, usually veined reddish-brown, glabrous outside; wings much shorter than the keel; keel 1.2–1.4(1.5) cm long, shallowly rounded, with a narrow somewhat projecting pointed beak (slightly incurved only at the tip), yellowish-green, sometimes conspicuously lined brown or purple. Pod 3.8–5 × 0.6–0.8 cm, shortly stipitate, subcylindrical, sometimes slightly curved, puberulous, usually glabrescent, 30–40-seeded. Seeds c. 3.5 mm long, oblique-cordiform, smooth to minutely papillose, brown or mottled ochre and dark grey-green.

Var. **pallida** —Polhill in Kew Bull. **22**: 262 (1968); in F.T.E.A., Leguminosae, Pap.: 905 (1971). —Drummond in Kirkia **8**: 218 (1972). —Polhill, Crotalaria Africa & Madagascar: 186, fig. 45 (1982).

Crotalaria mucronata Desv. in J. Bot. Agric. **3**: 76 (1814). —Senn in Rhodora **41**: 355 (1939). —Wilczek in F.C.B. **4**: 270, fig. 17 (1953). —Mogg in Macnae & Kalk, Nat. Hist. Inhaca Isl., Moçamb.: 146 (1958). —Torre in C.F.A. **3**: 54 (1962). —Binns, First Check List Herb. Fl. Malawi: 79 (1968). Type from Jamaica.

Crotalaria striata DC., Prodr. **2**: 131 (1825). —J.G. Baker in F.T.A. **2**: 38 (1871). —E.G. Baker in J. Linn. Soc., Bot. **42**: 345 (1914). —Eyles in Trans. Roy. Soc. South Africa **5**: 372 (1916). —Verdoorn in Bothalia **2**: 399 (1928). —Burtt Davy, Fl. Pl. Ferns Transvaal, pt. 2: 399 (1932). —Hutchinson, Botanist South. Africa: 462 (1946), but not. p. 526. —Merxmüller in Proc. & Trans. Rhodesia Sci. Assoc. **43**: 20 (1951). —Martineau, Rhod. Wild Fl.: 34 (1953). —Binns, First Check List Herb. Fl. Malawi: 80 (1968) as "*stricta*". Type from India.

Leaflets mostly 6–13 cm long, but commonly rather small in the Flora Zambesiaca area, elliptic, widest at a point 0.45–0.6 of the length from base to apex, acute or rounded at the apex.

Zambia. N: Mbala Distr., Mpulungu, fl. & fr. 20.xi.1950, *Bullock* 3491 (BR; K), not typical. W: Kitwe, fl. & fr. immat. 29.viii.1959, *Fanshawe* 5186 (K; NDO). C: South Luangwa National Park (Luangwa South Game Reserve), fr. 12.xi.1965, *Astle* 4092 (SRGH). **Zimbabwe**. N: Mutoko (Mtoko) Reserve, Chitora R. road crossing, fl. & fr. immat. iv.1956, *R.M. Davies* 1914 (K; LISC;

* Treated as a new name, but this seems more likely to have been an error in presentation than the coincidence of choosing a homonym for the same species.

SRGH). C: Marondera (Marandellas), fl. & fr. 6.iv.1950, *Wild* 3320 (K; M; SRGH). E: Mutare Distr., Marange Communal Land (Maranke T.T.L.), c. 24 km south of Odzi, fl. & fr. 10.iv.1969, *Plowes* 3184 (K; SRGH). S: 15 km NW of Buhera, Chikawakawa Hill (Chigavagava), fl. & fr. 20.iv.1969, *Biegel* 2928 (K; SRGH). **Malawi**. C: Nkhotakota Distr., 6 km north of Benga turnoff by Lipyodzi Bridge, fl. & immat. fr. 15.vi.1970, *Brummitt* 11419 (K; LISC; MAL; SRGH). S: Zomba Distr., Nchisi (Chilwa) Island, fl. 11.iii.1955, *Exell, Mendonça & Wild* 804 (BM; LISC; SRGH). **Mozambique**. N: Nampula, fl. & fr. 11.v.1937, *Torre* 1421 (COI; LISC). Z: Lugela Distr., Namagoa, fl. & fr. iv–v.1943 (also ix.1946), *Faulkner* Pretoria No. 293 (BR; K; LM; PRE), not typical. MS: Gorongosa Distr., Parque Nacional da Gorongosa (Parque Nacional de Caça), margem direita do rio Urema, a 26 km do Acampamento de Chitengo, c. 35 m, fl. & fr. 2.v.1964, *Torre & Paiva* 12219 (LISC). GI: Mandlakazi (Manjacaze) to Chongoéne, Missão de São Benedito dos Muchopes, fl. & fr. 2.iv.1959, *Barbosa & Lemos* 8462 (COI; K; LISC; LMA). M: Matutuíne Distr., Catuane, R. Maputo, fl. & fr. 15.v.1964, *Moura* 68 (COI).

Pantropical, in part adventive; commonly cultivated as a green manure and readily naturalized. Disturbed places in forest, woodland and savanna, along rivers, roads and on cultivated ground; 0–1500 m.

The typical form of var. *pallida*, with large elliptic pointed leaflets, occurs in the Flora Zambesiaca area largely or perhaps wholly as an introduced plant which has become naturalized in places. The usual, seemingly native, form in south-central and southern Africa approaches var. *obovata* and has elliptic to elliptic-obovate leaflets, mostly 5–8 cm long, bluntly pointed to rounded at the apex. Typical var. *obovata* is virtually restricted to the vicinity of the Great Lakes, apart from its introduction to experimental stations (mainly in Zimbabwe and southern Mozambique), and readily recognized by the mostly small obovate frequently retuse leaflets. Nevertheless the differences between the varieties are less clear in the Flora Zambesiaca area than in many parts of the world and a relatively high proportion of intermediates can be expected.

Within the Flora Zambesiaca area and excluding introduced plants, var. *pallida* generally has rather faint lines on the standard and often none on the keel, whereas var. *obovata* generally (not always) has conspicuous reddish-brown or purple venation on these petals. The pods of var. *obovata* are more often slightly curved. See also note under 67. *C. rogersii*.

Var. **obovata** (G. Don) Polhill in Kew Bull. **22**: 265 (1968); in F.T.E.A., Leguminosae, Pap.: 906 (1971); Crotalaria Africa & Madagascar: 186 (1982). Type from Ghana.

Crotalaria falcata DC., Prodr. **2**: 132 (1825). —J.G. Baker in F.T.A. **2**: 40 (1871). —E.G. Baker in J. Linn. Soc., Bot. **42**: 344 (1914). —Wilczek in F.C.B. **4**: 169, fig. 11 (1953). —Hepper in F.W.T.A., ed. 2, **1**: 551 (1958). Type from Senegal.

Crotalaria obovata G. Don, Gen. Syst. **2**: 138 (1832). —E.G. Baker in J. Linn. Soc., Bot. **42**: 345 (1914). Type as for var. *obovata*.

Crotalaria mucronata sensu Brenan in Mem. New York Bot. Gard. **8**: 247 (1953) et auct. al. pro parte, non Desv.

Leaflets 3–7 cm long, obovate-elliptic to obovate, widest at a point 0.6–0.8 of the length from base to apex, rounded or retuse at the apex.

Zambia. N: Mbala Distr., L. Tanganyika, Kasaba (Cassava) Sands, fl. 17.ii.1959, *Richards* 10940 (BR; K; LISC; SRGH). W: Kitwe Distr., Mindolo, fl. & fr. 6.iii.1964, *Mutimushi* 658 (K; NDO; SRGH). **Malawi**. N: 27 km north of Chilumba, Ngara (Ngala) Beach, fl. & fr. 4.vii.1970, *Pawek* 3566 (K). S: lower Mwanza R., fl. & fr. 6.x.1946, *Brass* 18021 (K; NY). **Mozambique**. N: Lago Distr., L. Niassa, Metangula, fl. & fr. 10.x.1942, *Mendonça* 749 (BM; LISC). T: Zumbo, fl. & immat. fr. 6.i.1974, *Macêdo* 5502 (LMU).

Pantropical; in Africa principally in the W African coastal region, across the Congo basin and then common all around the Great Lakes and rivers in their vicinity, but also cultivated and locally naturalized elsewhere. Lake shores, river margins, often on sand, sometimes in swamp forest and marshes; 200–1750 m.

Cultivated in Zimbabwe, e.g. Henderson Research Station, *Wild* 3819 (K; LISC; SRGH) & 3999 (K; LISC; SRGH), and southern Mozambique, e.g. Umbelúzi Experimental Station, *Myre* 759 (LISC; LMA).

Both varieties are recorded for Mozambique, Tete, by Gonçalves in Garcia de Orta, Sér. Bot. **5**: 69 (1982).

67. **Crotalaria rogersii** Baker f. in J. Linn. Soc., Bot. **42**: 347 (1914). —Eyles in Trans. Roy. Soc. South Africa **5**: 371 (1916). —Verdoorn in Bothalia **2**: 399 (1928). —Wilczek in F.C.B. **4**: 167 (1953). —Binns, First Check List Herb. Fl. Malawi: 79 (1968). —Polhill in F.T.E.A., Leguminosae, Pap.: 906 (1971). —Drummond in Kirkia **8**: 218 (1972). —Jacobsen in Kirkia **9**: 160 (1973). —Polhill, Crotalaria Africa & Madagascar: 186 (1982). —Lock, Leg. Afr. Check-list: 201 (1989). Lectotype, selected by Polhill (1971): Zambia, Mazabuka, *F.A. Rogers* 8330 (K, lectotype; SRGH).

Crotalaria rogersii forma *kilwaensis* R. Wilczek in Bull. Jard. Bot. État **23**: 151 (1953); in F.C.B. **4**: 168 (1953). Type from Dem. Rep. Congo (Katanga).
Crtotalaria pallida sensu Moriarty, Wild Fl. Malawi: 128, t. 64, fig. 3 (1975).

Erect or rarely decumbent perennial herb, with 1–several simple to much branched stems from a woody rootstock (but also flowering in first year of growth), up to 2.7 m tall; branches densely covered with short appressed or crisped hairs. Leaves 3-foliolate; leaflets mostly 4–8.5 × 1.5–5.5 cm, oblanceolate-elliptic to broadly elliptic or obovate-elliptic, appressed puberulous beneath, often rather glaucous; petiole shorter to longer than the leaflets; stipules 1–4 mm long, filiform to linear-caudate, caducous, or lacking. Flowers similar to *C. pallida*, but the standard and more particularly the keel often flushed purple; beak of the keel slightly more incurved. Pod shortly stipitate, 2.8–3.5 × 0.6–0.9 cm, subcylindrical, appressed puberulous, often glabrescent, 18–28-seeded. Seeds 3.5–4 mm long, oblique-cordiform, minutely papillose, greenish-brown or mottled.

Zambia. B: Kalabo, fl. & fr. 16.xi.1959, *Drummond & Cookson* 6511 (K; LISC; SRGH). N: Mbala Distr., L. Chila, fl. & fr. 13.v.1955, *Richards* 5707 (K; SRGH). W: Kitwe, fl. & fr. 18.vi.1955, *Fanshawe* 2338 (K; NDO; SRGH). C: Lusaka Distr., Chilanga, fl. & fr. immat. 13.ix.1963, *Farrell* 388 (K; SRGH). E: Lundazi, fl. & fr. 17.x.1967, *Mutimushi* 2276 (K; NDO; SRGH). S: Choma Distr., Mapanza, fl. & fr. 13.viii.1958, *E.A. Robinson* 2876 (K; SRGH). **Zimbabwe**. N: Mazowe (Mazoe) R., fl. viii.1905, *Eyles* 181 (BM; SRGH). C: 14 km Harare–Domboshawa, Hatcliffe South, fl. 26.ix.1955, *Drummond* 4878 (BR; K; LISC; SRGH). **Malawi**. N: Karonga Distr., Nyungwe R., fl. 15.x.1930, *Migeod* 981 (BM). C: Lilongwe Distr., Chankhandwe Dambo, fl. 26.viii.1950, *Jackson* 83 (BM; BR; K). S: Zomba, fl. *Sharpe* 36 (K).

Also in Tanzania and Dem. Rep. Congo (Katanga). Disturbed places in woodland and savanna, also damp places by lakes, rivers and in dambos or flood plains, persisting on cultivated ground; 600–1700 m.

Plants in their first year of growth are not always easy to distinguish from *C. pallida* until the characteristic pods are formed. The petals are often characteristically flushed purple, but pure yellow or faintly lined flowers may be found within single populations, see for example *Brenan & Greenway* in *Brenan* 7716 (K) or *Richards* 5463 (K). The species is fairly resistant to burning and grazing and shows accordingly considerable variation in habit and vegetative features.

68. **Crotalaria ochroleuca** G. Don, Gen. Syst. **2**: 138 (1832). —J.G. Baker in F.T.A. **2**: 37 (1871). —E.G. Baker in J. Linn. Soc., Bot. **42**: 327 (1914). —Wilczek in F.C.B. **4**: 221 (1953). —Hepper in F.W.T.A., ed. 2, **1**: 551 (1958). —Torre in C.F.A. **3**: 51 (1962). —Binns, First Check List Herb. Fl. Malawi: 79 (1968). —Polhill in F.T.E.A., Leguminosae, Pap.: 908 (1971). —Drummond in Kirkia **8**: 218 (1972). —Jacobsen in Kirkia **9**: 160 (1973). —Gonçalves in Garcia de Orta, Sér. Bot. **5**: 69 (1982). —Polhill, Crotalaria Africa & Madagascar: 188, fig. 46 (1982). —Lock, Leg. Afr. Check-list: 193 (1989). TAB. 3,7: **31**. Type from São Tomé.
Crotalaria intermedia sensu J.G. Baker in F.T.A. **2**: 37 (1871) pro parte. —sensu E.G. Baker in J. Linn. Soc., Bot. **42**: 327 (1914) pro parte. —sensu Eyles in Trans. Roy. Soc. South Africa **5**: 370 (1916). —sensu Verdoorn in Bothalia **2**: 394 (1928). —sensu Merxmüller in Proc. & Trans. Rhodesia Sci. Assoc. **43**: 19 (1951) non Kotschy.

Erect annual or short-lived perennial, up to 0.5–2.7 m tall, with lax ascending ribbed branches covered in short fine appressed hairs. Leaves 3-foliolate, estipulate; leaflets mostly 5–13(18) × 0.4–3 cm, mostly linear or linear-lanceolate, but also lanceolate or elliptic-lanceolate, with the upper ones longer and relatively narrow, appressed puberulous beneath; petiole shorter than the leaflets. Racemes up to 10–50 cm long, rather laxly many-flowered; bracts 2–4 mm long, subulate or linear-caudate, expanded at the base; bracteoles at the base of the calyx or immediately below, 1–2 mm long, linear. Calyx 6–8 mm long, becoming basally truncate and deflexed against the pedicel, glabrous; lobes acuminate, the upper ones at least shorter than the tube. Standard ovate, cream or pale yellow, veined reddish or maroon; wings usually shorter than the keel; keel (1.7)1.8–2.1(2.3) cm long, shortly rounded, with a rather long projecting beak. Pod subsessile, (4)5–7 × (1.2)1.5–2 cm, broadly cylindrical, appressed puberulous, up to 100-seeded. Seeds rather variable in size, but mostly c. 3.5 mm long, oblique-cordiform, smooth, pale yellow to orange.

Botswana. N: Thaoge (Toakhe) R. Swamp, fl. & fr. 21.iii.1961, *Richards* 14819 (K). **Zambia**. B: Mongu, fl. & fr. 26.iii.1964, *Verboom* 1016 (K; SRGH). N: 25 km west of Kasama, fr. 7.v.1961, *E.A. Robinson* 4638 (BR; K; M; SRGH). W: Kitwe, fl. & fr. 28.ii.1954, *Fanshawe* 881 (BR; K; NDO; SRGH). C: Lusaka–Kapiri Mposhi, fl. & fr. 4.iv.1961, *Richards* 14917 (K; SRGH). E: Lundazi, fl.

Tab. 3,7: **31**. CROTALARIA OCHROLEUCA. 1, flowering and fruiting branch (× $^{2}/_{3}$), from *Richards* 5278 and s.n.; 2, flower (× 2); 3, standard (× 2); 4, wing (× 2); 5, keel (× 2); 6, gynoecium (× 2), 2–6 from *Richards* 5278; 7, seed (× 6$^{1}/_{2}$), from *Richards* s.n. Drawn by Roger Polhill. From Crotalaria Africa & Madagascar.

& fr. 1.vi.1954, *E.A. Robinson* 809 (K). S: Mazabuka, edge of Kafue Flats, fl. & fr. 7.iv.1955, *Exell, Mendonça & Wild* 1431 (BM; LISC; SRGH). **Zimbabwe**. N: Gokwe South Distr., Tari (Tara) R. at Gokwe–Binga road crossing, fl. & fr. 30.iii.1962, *Bingham* 205 (K; LISC; SRGH). W: Hwange Distr., Victoria Falls, south bank, fl. v.1915, *Rogers* 13126 (K). C: Harare, fl. & fr. iii.1920, *Eyles* 2118 (COI; K; SRGH). E: Mutare Distr., near Odzi, fl. & fr. 11.iii.1969, *Corby* 2119 (K; SRGH). **Malawi**. N: near Chitimba, fl. & fr. 24.iv.1969, *Pawek* 2319 (K). C: Lilongwe, fl. 1.iv.1955, *Jackson* 1566 (BR; K; SRGH). S: Blantyre Distr., Limbe, fl. & fr. 25.iii.1969, *Patel* 21 (K; SRGH). **Mozambique**. N: Lichinga (Vila Cabral), fl. & fr. v–vi.1934, *Torre* 192 (BM; COI; K; LISC). T: 92 km Vila Mouzinho–Zóbuè, fl. & fr. 19.vii.1949, *Barbosa & Carvalho* 3704 (K; LISC; LMA). MS: Gondola Distr., Chimoio (Vila Pery, Mandigo), fl. 19.ix.1911, *Rogers* 4504 (BM). GI: Chibuto Distr., Maniquenique, fr. 15.iv.1955, *L. Cabral* 10 (LMA). M: Namaacha Distr., Umbelúzi, Estação Experimental, fl. & fr. immat. 26.v.1949, *Myre & Balsinhas* 720 (LMA).*

Widespread in tropical Africa except the NE and southernmost parts. Damp grassy places along rivers, in flood plains and dambos, at roadsides and sometimes a weed of cultivated ground; up to 1750 m.

Under trial in various experimental stations as a green manure and ± naturalized.

69. **Crotalaria trichotoma** Bojer in Ann. Sci. Nat., Bot., Sér. 2, **4**: 265 (1835). —Polhill in Fl. Mascar., 80 Légumineuses: 200, t. 71 (1990). Type grown in Mauritius from seed collected in NW Madagascar (? introduced from E. Africa).

Crotalaria zanzibarica Benth. in Hooker, London J. Bot. **2**: 548 (1843). —J.G. Baker in F.T.A. **2**: 35 (1871). —E.G. Baker in J. Linn. Soc., Bot. **42**: 347 (1914). —Torre in C.F.A. **3**: 55 (1962). —Polhill in F.T.E.A., Leguminosae, Pap.: 911, fig. 126 (1971); Crotalaria Africa & Madagascar: 191, fig. 47 (1982). —Lock, Leg. Afr. Check-list: 210 (1989). Type from Zanzibar.

Crotalaria usaramoensis Baker f. in J. Linn. Soc., Bot. **42**: 346 (1914). Types from Tanzania.

Erect annual or short-lived perennial, up to 0.7–2.7 m tall, well branched at least above; branches ribbed, appressed puberulous. Leaves 3-foliolate, estipulate; leaflets mostly (4.5)6–14 × 1–4 cm, lanceolate to elliptic-oblong, lower ones sometimes elliptic, appressed puberulous beneath, rarely also above; petioles shorter than the leaflets. Racemes up to 30–40 cm long, many-flowered; bracts (1)2–4 mm long, linear-caudate; bracteoles inserted at the base of the calyx or immediately below, slightly smaller than the bract. Calyx 4–6 mm long, becoming basally truncate and deflexed against the pedicel, glabrous or rather thinly appressed puberulous; lobes generally reduced to small widely spaced subulate teeth. Standard elliptic to obovate-elliptic, yellow, veined reddish-purple outside; wings ± as long as the keel, with a dark mark at the base; keel 1.2–1.4 cm long, shortly rounded about the middle, with a sharp slightly incurved beak. Pod (3.2)3.5–4.5 × 0.7–1.2 cm, shortly stipitate, subcylindrical, appressed puberulous, 50–70-seeded. Seeds 2.2–2.8 mm long, oblique-cordiform, smooth, orange-buff or terra-cotta.

Mozambique. N: Mogincual, fl. & fr. 15.xi.1936, *Torre* 1031 (COI; LISC); Mogovolas Distr., Nametil (Namatil), fl. & fr. 12.vii.1948, *Pedro & Pedrógão* 4428 (LMA).

Also native to eastern and southern Tanzania, possibly also NW Madagascar, but widely cultivated as a green manure and locally naturalized elsewhere. Open woodland, grassland and disturbed places; 0–1200 m.

Without fruits it is sometimes difficult to differentiate between *C. trichotoma* and *C. kirkii*, though *C. kirkii* can usually be recognized by the weaker development of lateral branches, the predominantly narrow leaflets and the small bracts (the bracts of *C. trichotoma* are more variable in length). The problem is complicated because both species extend inland to Lake Malawi and in this region both show modifications of the calyx, appearing more similar to related species. In the coastal region the calyx is usually glabrous, occasionally puberulous, and the lobes are reduced to small widely spaced subulate teeth. Inland the calyx is consistently puberulous and the lobes are better developed, narrowly triangular-acuminate. The pods of *C. trichotoma* also appear to be relatively small (little fruiting material is available). It is possible that the species originated from this inland area or else that some introgression has occurred with related species such as *C. lanceolata* and *C. subcapitata*.

C. cleomoides Klotzsch in Peters, Naturw. Reise Mossambique **6**, part 1: 59 (1861), is based on a plant without fruits from rather further south, collected by Peters at Sena. A single flower is preserved at BM and from this and the description *C. cleomoides* appears to be synonymous with either *C. trichotoma* or *C. kirkii*, but in view of the difficulties mentioned above, it is not possible to determine which. Verdoorn in Bothalia **2**: 394 (1928) places it as a synonym of *C.*

* Perhaps introduced, see *Myre* 756 (LISC; LMA).

lanceolata, which is common around Sena, and this may be correct but the flower seems too large for that species.

The form of *C. trichotoma* most commonly cultivated has relatively broad pods, e.g. Zimbabwe, Harare, *Arnold* in *GHS* 6346 (K; SRGH); Mozambique, Quelimane, Medal Plantations, *Wild & Pedro* 5887 (K; LISC; SRGH) and Umbelúzi Experimental Station, *Myre* 754 (LISC; LMA). This form was found as an escape near the C.I.C.A. Experimental Camp at Nhacongo, 12.x.1945, *Pedro* 347 (LMA), and may become more widely naturalized.

70. **Crotalaria kirkii** Baker in F.T.A. **2**: 36 (1871). —E.G. Baker in J. Linn. Soc., Bot. **42**: 346 (1914). —Brenan, Check-list For. Trees Shrubs Tang. Terr.: 416 (1949). —Polhill in F.T.E.A., Leguminosae, Pap.: 913 (1971); Crotalaria Africa & Madagascar: 192 (1982). —Lock, Leg. Afr. Check-list: 185 (1989). Type from Tanzania.

Erect annual or short-lived perennial, up to 60 cm tall, with few ascending ribbed appressed puberulous branches. Leaves 3-foliolate (rarely a few of the uppermost 1-foliolate), estipulate; leaflets mostly 6–13 × 0.4–1 cm, linear-lanceolate, but the lower ones up to 2 cm wide, elliptic to ovate, appressed puberulous beneath; petiole shorter than the leaflets. Racemes up to 17–30 cm long, many-flowered; bracts filiform, 0.5–2 mm long; bracteoles at the base of the calyx, less than 1 mm long. Flowers similar to *C. trichotoma*. Pods subsessile, 2.4–3 × 0.5–0.6 cm, cylindrical, appressed puberulous, c. 50-seeded. Seeds c. 2.5 mm long, oblique-cordiform, smooth, terra-cotta.

Malawi. N: Nkhata Bay Distr., Likoma Island, 1900, *Kenyon* 52 (K). **Mozambique**. N: aredores de Mocímboa da Praia, c. 10 m, fl. 14.iv.1964, *Torre & Paiva* 11917 (LISC); Ngauma Distr., Massangulo, fl. & fr. vi.1933, *Gomes e Sousa* 1445 (BM; COI).

Also coastal regions of Kenya and Tanzania, extending inland to Lake Malawi. Open woodland, grassland and disturbed places; 0–800 m.

See note under 69. C. *trichotoma*.

71. **Crotalaria lanceolata** E. Mey., Comment. Pl. Afr. Austr.: 24 (1836). —Harvey in F.C. **2**: 43 (1862). —J.G. Baker in F.T.A. **2**: 36 (1871) pro parte. —E.G. Baker in J. Linn. Soc., Bot. **42**: 343 (1914) excl. var. —Verdoorn in Bothalia **2**: 394 (1928) pro majore parte. —Burtt Davy, Fl. Pl. Ferns Transvaal, pt. 2: 399 (1932). —Merxmüller in Proc. & Trans. Rhodesia Sci. Assoc. **43**: 19 (1951). —Mogg in Macnae & Kalk, Nat. Hist. Inhaca Isl., Moçamb.: 146 (1958). —Binns, First Check List Herb. Fl. Malawi: 79 (1968). —Polhill in F.T.E.A., Leguminosae, Pap.: 913 (1971). —Drummond in Kirkia **8**: 218 (1972). —Moriarty, Wild Fl. Malawi: 128, t. 64, fig. 1 (1975). —Gonçalves in Garcia de Orta, Sér. Bot. **5**: 68 (1982). —Polhill, Crotalaria Africa & Madagascar: 193 (1982). —Lock, Leg. Afr. Check-list: 186 (1989). Type from South Africa (KwaZulu-Natal).

Erect laxly branched annual or short-lived perennial, 0.2–1.7 m tall; branches mostly ascending, lower ones sometimes spreading, ribbed, covered at least when young with short appressed or rarely slightly spreading hairs. Leaves 3-foliolate (rarely a few of the uppermost 1-foliolate), estipulate; leaflets mostly 4–11(17) × 0.3–2(3.5) cm, usually predominantly linear-lanceolate to lanceolate, the lower ones sometimes elliptic, but sometimes (particularly subsp. *exigua*) all elliptic-oblong to elliptic, thinly appressed puberulous beneath, rarely also above; petiole shorter than the leaflets. Racemes up to 8–35 cm long, many-flowered; bracts 0.5–3 mm long, subulate or filiform, expanded at the base; bracteoles at the base of the calyx, 0.3–1 mm long. Calyx 2–4 mm long, becoming basally truncate and deflexed against the pedicel, puberulous; lobes subulate to triangular-acuminate, usually shorter than the tube. Standard broadly elliptic to subcircular, yellow, veined reddish-purple outside; keel 5–11 mm long, rounded about the middle, with an incurved or projecting beak. Pod (1.6)1.8–3.8 × 0.4–0.6 cm, subsessile, cylindrical, sometimes curved up at the tip, appressed puberulous, 20–50-seeded. Seeds 2–3 mm long, oblique-cordiform, smooth, pale yellow to orange-brown.

1. Keel with a projecting beak, longer than the wings, 9–11 mm long; pod (1.8)2.2–3.1(3.3) cm long · iii) subsp. *prognatha*
– Keel with an incurved beak, shorter than the wings, 5–9 mm long · · · · · · · · · · · · · · · · 2
2. Keel 6.5–9 mm long; pod usually (2.5)2.8–3.5(3.8) cm long, but sometimes smaller in vicinity of other subspecies · i) subsp. *lanceolata*
– Keel 5–6 mm long; pod (1.6)1.8–2.2 cm long · ii) subsp. *exigua*

i) Subsp. **lanceolata**

Crotalaria mossambicensis Klotzsch in Peters, Naturw. Reise Mossambique **6**, part 1: 60, t. 10 (1861). Syntypes: Mozambique, Cabaceira, Sena; and Inhambane, *Peters* (B†, syntypes; K; P).*

Keel 6.5–9 mm long, shallowly rounded, with a rather short incurved beak, shorter than the wings. Pods usually (2.5)2.8–3.5(3.8) cm long, sometimes smaller in intermediate forms.

Zimbabwe. C: Marondera (Marandellas), 20.iii.1942, *Dehn* 656 (M; SRGH). E: Vumba road, roadside to Maitland Smith's homestead, fl. 13.iii.1960, *Chase* 7469 (K; LISC; SRGH). S: locality and date not recorded, *Cleghorn* 1869 (SRGH). **Malawi**. S: Zomba Distr., Mtwiche (Mtwice), fl. & fr. 10.iii.1955, *Exell, Mendonça & Wild* 772 (BM; LISC; SRGH). **Mozambique**. N: Nampula, fl. & fr. 13.ii.1937, *Torre* 1377 (COI; LISC). Z: Mocuba, Posto Agrícola, fl. 6.vi.1949, *Barbosa & Carvalho* 2984 (K; LISC; LMA). MS: 5 km Chemba–Nhacolo (Tambara), fl. & fr. 23.iv.1960, *Lemos & Macuácua* 140 (BM; COI; K; LISC; LMA; SRGH). GI: Inhambane, *Peters* (B†). M: Namaacha, fl. & fr. 3.iii.1948, *Torre* 7469 (BM; K; LISC).

Also in Tanzania, South Africa (southwards to the Eastern Cape) and Madagascar. Open woodland, grassland and disturbed ground, sometimes in damp places; 0–1450(1600) m.

ii) Subsp. **exigua** Polhill, Crotalaria Africa & Madagascar: 193 (1982). —Lock, Leg. Afr. Check-list: 186 (1989). Type: Mozambique, entre Mocuba e Milange, a 46.7 km de Mocuba, *Barbosa & Carvalho* 2989 (K; LISC; LMA, holotype).

Plants generally small with slender branches, the lower ones of which are often spreading. Leaflets often all similar, narrow to exceptionally broad. Keel 5–6 mm long, shallowly rounded, with a rather short narrow slightly incurved beak, shorter than the wings. Pods (1.6)1.8–2.2 cm long.

Malawi. S: Mwanza Distr., Thambani Forest Reserve, fl. & fr. 29.vii.1984, *Patel & Nachamba* 1529 (K; MAL). **Mozambique**. N: Malema Distr., Mutuáli, base do monte Cucuteira, junto ao rio Neuce, c. 650 m, fl. & fr. 16.iii.1964, *Torre & Paiva* 11198 (LISC); 38 km east of Ribáuè, fr. 17.v.1961, *Leach & Rutherford-Smith* 10908 (K; LISC; SRGH). Z: 47 km Mocuba–Milange, fl. & fr. 19.vii.1949, *Barbosa & Carvalho* 3669 (K; LISC; LMA).

Known only in the area from the southern end of Lake Malawi east to Ribáuè and south to Mocuba. Open woodland, sometimes on termite mounds; c. 100–650 m.

iii) Subsp. **prognatha** Polhill in Kew Bull. **22**: 270, fig. 20/5,6 (1968); Crotalaria Africa & Madagascar: 194 (1982). —Lock, Leg. Afr. Check-list: 186 (1989). Type: Zambia, 27 km Lusaka to Kabwe (Broken Hill), *Drummond & Rutherford-Smith* 6907 (K, holotype; SRGH).

Keel 9–11 mm long, shallowly rounded, with the beak projecting forward beyond the wings. Pods (1.8)2.2–3.1(3.3) cm long.

Zambia. C: Lusaka Distr., Mt. Makulu, fl. & fr. 24.iii.1969, *van Rensburg* 3128 (K; SRGH). E: Katete, fl. & fr. 24.iii.1955, *Exell, Mendonça & Wild* 1155 (BM; LISC; SRGH). S: Choma Distr., Siamambo Forest Reserve, fl. 13.iii.1960, *White* 7765 (FHO; K). **Zimbabwe**. N: Glendale, Paridya Farm, fl. & fr. 23.iii.1953, *Barrett* in *GHS* 50835 (K; SRGH). C: Chegutu (Hartley), Poole Farm, fl. & fr. immat. 25.iii.1948, *R.M. Hornby* 2858 (K; SRGH). E: 19 km Nyanga (Inyanga)–Mutare (Umtali), fl. & fr. 6.v.1965, *Corby* 1317 (K; SRGH). **Malawi**. C: Lilongwe Distr., Dzalanyama Forest Reserve, fl. & fr. 22.iii.1970, *Brummitt* 9308 (K; LISC; MAL; PRE; SRGH). S: Blantyre Distr., Ndirande Mt., NW side of summit, fl. & fr. 3.iii.1970, *Brummitt* 8874 (K). **Mozambique**. MS: Manica Distr., Serra de Vumba, Manica (Macequece), fl. 8.iii.1948, *Garcia* 663 (LISC).

Not known elsewhere. Open woodland, grassland and disturbed ground, sometimes in damp places; 1000–1700 m.

The ranges of subsp. *lanceolata* and subsp. *prognatha* are divided approximately by a line along the Zimbabwe/Mozambique border north to Lake Malawi, but interdigitate to some extent and intermediates are fairly common where the subspecies meet. Subsp. *exigua* occurs in a small area within the range of subsp. *lanceolata* and comprises a rather heterogeneous series of specimens, all with small flowers and pods, but some similar in habit and foliage to subsp. *lanceolata*, but others small and slender with uniform and sometimes unusually broad elliptic leaflets. Subsp. *lanceolata* normally has longer pods than both the

---

*Localities not recorded on K and P sheets.

other subspecies, but in the vicinity of the other subspecies smaller fruits occur quite commonly, e.g. Malawi, near Balaka, *Exell, Mendonça & Wild* 922 (BM; LISC; SRGH); Mozambique, Muecate–Nacaroa, *Torre & Paiva* 9905 (LISC); Quixaxe, R. Metapa, *Pedro & Pedrógão* 4715 (LMA). All subspecies meet in southern Malawi and a rather high proportion of plants are atypical in one respect or another. There appears to be no very obvious geographical, altitudinal or ecological separation of the subspecies in this area, but further information is needed. To avoid confusion of subsp. *exigua* with *C. bernieri* and *C. onobrychis* it is important to observe the shape of the keel.

72. **Crotalaria subcapitata** De Wild. in Ann. Mus. Congo, Sér. IV, Bot. [Études Fl. Katanga] **1**: 186 (1903). —E.G. Baker in J. Linn. Soc., Bot. **42**: 378 (1914). —Wilczek in F.C.B. **4**: 183 (1953). —Binns, First Check List Herb. Fl. Malawi: 80 (1968). —Polhill in Kew Bull. **22**: 274 (1968); in F.T.E.A., Leguminosae, Pap.: 915 (1971). —Gonçalves in Garcia de Orta, Sér. Bot. **5**: 72 (1982). —Polhill, Crotalaria Africa & Madagascar: 196 (1982). —Lock, Leg. Afr. Check-list: 206 (1989). Type from Dem. Rep. Congo (Katanga).

Procumbent or straggling erect annual or perennial, sometimes up to 0.5–1.3 m tall; branches somewhat ribbed, shortly appressed or spreading pubescent. Leaves 3-foliolate; leaflets mostly 1.5–8(11) × 0.3–2.5(3) cm, very variable, linear-lanceolate to elliptic or obovate, often varying up the stem, narrowest towards the top, appressed pubescent beneath, sometimes also above; petiole shorter than the leaflets; stipules (0.5)1–3 mm long, linear or subulate, often recurved. Racemes 5–40(60) cm long, usually with a long peduncle and numerous flowers crowded above (but sometimes peduncle shorter than the rhachis); bracts 1–3(5) mm long, subulate or linear-caudate, expanded at the base; bracteoles inserted at the base of the calyx or just below, 0.5–2 mm long, linear. Calyx 3–4(5) mm long, becoming basally truncate and deflexed against the pedicel, variously appressed or less commonly spreading pubescent; lobes narrowly attenuate-triangular, a little longer than the tube. Standard obovate-circular, yellow, veined reddish-brown or purplish; wings ± as long as the keel; keel (5)6–9.5(10.5) mm long, strongly rounded about the middle, with an incurved attenuate sharp beak, often lined and blotched purplish. Pod practically sessile, 12–24 × 3.5–5 mm, subcylindrical, usually rather densely appressed or spreading pubescent, 20–30-seeded. Seeds 1.5–2.5 mm long, oblique-cordiform, smooth, orange or brown.

Var. **subcapitata** —Polhill in Kew Bull. **22**: 274 (1968); in F.T.E.A., Leguminosae, Pap.: 916 (1971); Crotalaria Africa & Madagascar: 197 (1982).

Crotalaria lanceolata var. *malangensis* Baker f. in J. Linn. Soc., Bot. **42**: 344 (1914). —Torre in C.F.A. **3**: 54 (1962). Type from Angola.

Crotalaria nicholsonii Baker f. in J. Linn. Soc., Bot. **42**: 346 (1914). —Verdoorn in Bothalia **2**: 400 (1928). —Wilczek in F.C.B. **4**: 180 (1953). —Torre in C.F.A. **3**: 54 (1962) pro majore parte. —Binns, First Check List Herb. Fl. Malawi: 79 (1968). Lectotype, selected by Polhill (1971): Zambia, Luangwa Valley, *J.G. Nicholson* (K, lectotype).

Crotalaria acervata Baker f. in J. Bot. **58**: 74 (1920). —Wilczek in F.C.B. **4**: 176 (1953). Type from Dem. Rep. Congo (Katanga).

Crotalaria lanceolata sensu Hutchinson, Botanist South. Africa: 503 & 531 (1946) non E. Mey.

Crotalaria striata sensu Hutchinson, Botanist South. Africa: 526 (1946) non DC.

Crotalaria longipedunculata De Wild. ex R. Wilczek in Bull. Jard. Bot. État **23**: 154 (1953); in F.C.B. **4**: 181 (1953). Type from Dem. Rep. Congo (Kivu).

Crotalaria longipedunculata forma *glabra* R. Wilczek in Bull. Jard. Bot. État **23**: 156 (1953); in F.C.B. **4**: 182 (1953). Type from Dem. Rep. Congo (Bas-Congo).

Crotalaria sp. sensu Brenan in Mem. New York Bot. Gard. **8**: 248 (1953) quoad specim. *Brass* 16989, & 249 (1953) quoad specim. *Brass* 17625.

Crotalaria fwamboensis sensu Binns, First Check List Herb. Fl. Malawi: 79 (1968) as "*furamboensis*".

Stems straggling-ascending, up to 0.5–1.3 m tall, generally with the leaflets varied in shape up the stem, longer and relatively narrow above; petioles mostly 1.5–4 cm long. Pod (1.6)1.8–2.4 cm long.

Zambia. N: Mpika Distr., Kaloswe, fl. & fr. 24.vii.1930, *Hutchinson & Gillett* 4066 (BM; COI; K; LISC; SRGH). W: Chingola, fl. & fr. 5.viii.1955, *Fanshawe* 2414 (BR; K; LISC; NDO; SRGH). C: Serenje Distr., SE of Kanona, Kundalila Falls, fl. 15.x.1967, *Simon & Williamson* 1027 (K; SRGH).

Malawi. N: Mzimba Distr., Mzuzu, Katoto, fl. & fr. 1.viii.1970, *Pawek* 3640 (K). C: Ntchisi Mt., fl. 20.ii.1959, *Robson & Steele* 1678 (BM; K; LISC; SRGH). S: Ntcheu Distr., Lower Kirk Range, Chipusiri, fl. & fr. immat. 17.iii.1955, *Exell, Mendonça & Wild* 943 (BM; LISC; SRGH). **Mozambique**. N: Ngauma Distr., Massangulo, fl. iv.1933, *Gomes e Sousa* 1392 (COI), not typical. T: Chifunde Distr., 5 km de Mualadze (Vila Gamito) para Furancungo, fl. & fr. 11.vii.1949, *Barbosa & Carvalho* 3577 (K; LISC; LMA).

Widespread in grassy places from Ghana to Uganda and south through Angola and Dem. Rep. Congo to the Zambezi. Upland grassland, commonly in damp places, descending into miombo woodland mainly along rivers and dambos; 900–2300 m.

Readily distinguished from related species in the Flora Zambesiaca area by the presence of stipules and by the shape of the keel, but very variable in habit, foliage and in the size of the flowers and pods, so that many names have been applied. The only real difficulties seem to occur along the SE coast of Lake Malawi, where a significant proportion of specimens seen of *C. kirkii*, *C. lanceolata* and *C. subcapitata* are atypical in one respect or another and introgression seems the most probable explanation. *Gomes e Sousa* 1392, cited above, lacks stipules (as far as can be seen), the calyx is only sparsely puberulous, with unusually narrow widely spaced lobes, but the keel is unmistakably that of *C. subcapitata*.

Var. **fwamboensis** (Baker f.) Polhill in Kew Bull. **22**: 275 (1968); in F.T.E.A., Leguminosae, Pap.: 916 (1971); Crotalaria Africa & Madagascar: 197 (1982). Type: Zambia, Lake Tanganyika, Fwambo, *Carson* 120 (BM, fragment; K, holotype).

Crotalaria fwamboensis Baker f. in J. Linn. Soc., Bot. **42**: 377 (1914).

Stems numerous, prostrate or shortly ascending from a well developed rootstock, with the leaflets mostly of similar shape along the stems; petioles mostly less than 1 cm long. Pod usually 1.2–1.6 cm long.

Zambia. N: Mbala Distr., Saisi, Kellet's Farm, fl. & fr. 13.iv.1959, *McCallum-Webster* 760 (K). **Malawi**. N: 28 km south of Chikangawa, fr. 3.ix.1978, *E. Phillips* 3858 (K; MO).

Also in Dem. Rep. Congo (Katanga) and southern Tanzania. Dambos and marshes; c. 1500–1750 m.

It is not intended to include here plants of var. *subcapitata* growing out again after burning, e.g. Malawi, Dedza, *Brass* 17625 (BR; K), but on the definitions given they could be technically included. The record for Botswana given by Polhill, loc. cit. (1968), was based on a wrongly labelled sheet at K. At SRGH the Richards specimen with a tag numbered 14786 has been corrected by the collector to 13786 and the data in her writing refers to a plant from Mbala.

73. **Crotalaria onobrychis** A. Rich., Tent. Fl. Abyss. **1**: 154 (1847). —F.W. Andrews, Fl. Pl. Anglo-Egypt. Sudan **2**: 188 (1952). —Polhill, Crotalaria Africa & Madagascar: 200, fig. 49/1–6 (1982). —Lock, Leg. Afr. Check-list: 194 (1989). Type from Ethiopia.

Crotalaria astragalina subsp. *onobrychis* (A. Rich.) Baker f. in J. Linn. Soc., Bot. **42**: 375 (1914).

Crotalaria astragalinoides Baker f. in Rev. Zool. Bot. Africaines **23**: 180 (1933). Type from Dem. Rep. Congo.

Crotalaria astragalina sensu Wilczek in F.C.B. **4**: 179 (1953) non Hochst. ex A. Rich.

Crotalaria impressa subsp. *onobrychis* (A. Rich.) Cufod. in Bull. Jard. Bot. État **39**, Suppl. [Enum. Pl. Aethiop. Sperm.]: xxiii (1969).

Erect annual, 14–40 cm tall; branches slender, appressed puberulous. Leaves 3-foliolate, estipulate; leaflets mostly 2–6 × 0.4–2.5 cm, elliptic to narrowly elliptic-oblong, appressed puberulous beneath, sometimes also above; petiole shorter than the leaflets. Racemes 2.5–14 cm long, shortly pedunculate, dense at first, elongating to sublax or lax; bracts 1–2.5 mm long, subulate-caudate; bracteoles at the base of the calyx, minute. Calyx 2–3 mm long, becoming basally truncate and deflexed against the pedicel, sparsely to conspicuously puberulous; lobes narrowly triangular, as long as the tube or shorter. Standard subcircular to oblate, yellow veined purple; wings longer than the keel; keel 4–6 mm long, rounded, with a very short blunt beak. Pod practically sessile, 1–1.4 × 0.4–0.5 cm, subcylindrical, subglabrous to appressed puberulous, 10–20-seeded. Seeds 2–2.5 mm in diameter, subcircular-cordiform, minutely punctate, pale brown.

Malawi. S: Mangochi Distr., Lake Malombe (Pamalombe), Mvela (Mvera), fl. & fr. 1.vii.1955, *Jackson* 1667 (K).

Also from NE Dem. Rep. Congo to Ethiopia, possibly brought into Malawi from the upper Nile region by migrating birds. Lakeside alkaline swamp.

74. **Crotalaria bernieri** Baill. in Bull. Mens. Soc. Linn. Paris **1**: 444 (1885). —Peltier in J. Agric. Trop. Bot. Appl. **6**: 284 (1959). —Polhill in F.T.E.A., Leguminosae, Pap.: 918 (1971); Crotalaria Africa & Madagascar: 201 (1982). —Lock, Leg. Afr. Check-list: 168 (1989). Syntypes from Madagascar.

Erect laxly branched annual 20–70(100) cm tall; branches thinly appressed puberulous, glabrescent. Leaves 3-foliolate, estipulate; leaflets mostly 3–6.5 × 0.4–1.2 cm, linear-lanceolate to oblong-lanceolate, thinly appressed puberulous beneath; petiole much shorter than the leaflets. Racemes 5–15 cm long, often long-pedunculate, with numerous flowers crowded above; bracts 0.5–2 mm long, subulate or triangular-caudate; bracteoles at the base of the calyx, smaller. Calyx 1.5–2.5 mm long, becoming basally truncate and deflexed against the pedicel, appressed puberulous; lobes triangular, slightly acuminate, shorter than the tube. Standard oblate, pale yellow, veined reddish-brown or purple; wings longer than the keel; keel 4–5 mm long, rounded about the middle, with a very short blunt slightly incurved beak. Pod subsessile, 2–2.5 × 0.6–0.8 cm, ellipsoid-cylindrical, puberulous, 20–40-seeded. Seeds 2–2.5 mm long, oblong-reniform, smooth, mauve, pale brown or orange-red.

Zimbabwe. E: Chipinge Distr., Lower Save (Sabi), east bank near Hippo Mine, fl. 12.iii.1957, *Phipps* 585 (K; SRGH). **Mozambique**. N: andados 5 km de Mocímboa da Praia para Diaca, c. 20 m, fl. 14.iv.1964, *Torre & Paiva* 11931 (LISC). M: Namaacha, fl. & fr. 25.vi.1948, *Torre* 7900 (LISC).

Also in W Uganda, coastal regions of Tanzania and Kenya, the Comoro Is. and Madagascar. Wooded grassland and disturbed places; 0–350 m.

75. **Crotalaria chrysochlora** Baker f. ex Harms, Wiss. Ergebn. Deutsch. Zentr.-Afrika Exped., Bot.: 244 (1911). —E.G. Baker in J. Linn. Soc., Bot. **42**: 375 (1914). —Wilczek in F.C.B. **4**: 267 (1953). —Polhill in F.T.E.A., Leguminosae, Pap.: 920, fig. 127 (1971); Crotalaria Africa & Madagascar: 202, fig. 50 (1982). —Lock, Leg. Afr. Check-list: 171 (1989). Type from Rwanda.

Perennial with a thick woody rootstock and a number of curved-ascending or decumbent stems, sometimes up to 30 cm tall, often less; stems pubescent with short appressed or spreading hairs. Leaves 3-foliolate, estipulate; leaflets mostly 1.5–4 × 0.6–1.8 cm, elliptic to obovate, appressed puberulous beneath; petiole shorter than the leaflets. Racemes up to 4–10 cm long, pedunculate, with 8–20 flowers rather closely arranged above; bracts 1.5–3 mm long, subulate; bracteoles at the base of the calyx, less than 1 mm long. Calyx 3.5–4.5 mm long, basally truncate and deflexed against the pedicel, puberulous; lobes triangular, slightly acuminate, ± as long as the tube. Standard obovate-circular to oblate, yellow, veined reddish-brown or purple; wings exceeding the keel; keel 5–6(7) mm long, rounded, with a short blunt beak. Pod practically sessile, 15–18 × 4–6 mm, subcylindrical, appressed puberulous, 10–16-seeded. Seeds 2.5–3 mm long, oblique-cordiform, smooth, dark brown.

Zambia. W: Mwinilunga Distr., Matonchi Farm, fl. & fr. 1.ix.1930, *Milne-Redhead* 1012 (K); slope east of Matonchi Farm, fl. & fr. 6.x.1937, *Milne-Redhead* 2601 (K).

Also in Cameroon, Dem. Rep. Congo, Sudan, Uganda, Kenya, Tanzania and Angola. Miombo woodland, in open places subject to annual fires; c. 1400 m.

76. **Crotalaria cleomifolia** Welw. ex Baker in F.T.A. **2**: 43 (1871). —E.G. Baker in J. Linn. Soc., Bot. **42**: 350 (1914). —Verdoorn in Bothalia **2**: 398 (1928). —Hutchinson, Botanist South. Africa: 508 (1946). —Brenan in Mem. New York Bot. Gard. **8**: 247 (1953). —Wilczek in F.C.B. **4**: 158 (1953). —Topham, Check List For. Trees Shrubs Nyasaland Prot.: 74 (1958). —Hepper in F.W.T.A., ed. 2, **1**: 552 (1958). —Torre in C.F.A. **3**: 55 (1962). —White, F.F.N.R.: 147 (1962). —Binns, First Check List Herb. Fl. Malawi: 79 (1968). —Polhill in F.T.E.A., Leguminosae, Pap.: 922 (1971). —Drummond in Kirkia **8**: 217 (1972). —Brummitt in Wye Coll. Malawi Proj. Rep.: 65 (1973). —Gonçalves in Garcia de Orta, Sér. Bot. **5**: 66 (1982). —Polhill, Crotalaria Africa & Madagascar: 203, fig. 51 (1982). —Lock, Leg. Afr. Check-list: 171 (1989). TAB. 3,7: **32**. Type from Angola.

Shrubby much branched perennial 1–4.5 m tall; branches variously appressed puberulous to spreading yellowish tomentose. Leaves 3–5-foliolate; leaflets 4–12 × 1.5–5.5 cm, elliptic, sometimes narrowly so, often acuminate, hairy at least beneath, discolorous, dark green above; petiole shorter than the leaflets; stipules 1–5 mm

Tab. 3,7: **32**. CROTALARIA CLEOMIFOLIA. 1, flowering and fruiting branch (× $^2/_3$), from *Milne-Redhead & Taylor* 11018; 2, flower (× 2); 3, standard (× 2); 4, wing (× 2); 5, keel (× 2); 6, gynoecium (× 2), 2–6 from *Polhill & Paulo* 1751; 7, seed (× 6$^1/_2$), from *Bogdan* 5478. Drawn by Roger Polhill. From Crotalaria Africa & Madagascar.

long, subulate to linear-caudate. Racemes mostly 10–30(40) cm long, rather shortly pedunculate, subdensely many flowered; bracts 2–8(10) mm long, subulate or linear-caudate; bracteoles at the base of the calyx, 2–8(10) mm long, linear-caudate. Calyx 6–7 mm long, becoming basally truncate and deflexed against the pedicel, appressed puberulous to spreading tomentose; lobes narrowly triangular-acuminate, as long as the tube or longer. Standard elliptic, bright yellow, usually veined reddish-brown; wings markedly shorter than the keel; keel 1.2–1.7 cm long, rather strongly rounded about the middle, with a narrow beak incurved towards the tip. Pod shortly stipitate, (3.5)3.8–5(6) cm long, narrowly cylindrical, sometimes shortly curved upwards at the tip, sparsely puberulous to densely spreading pubescent, 30–40-seeded. Seeds 3–4 mm long, oblique-cordiform, smooth, brown.

Zambia. N: 70 km south of Mbala (Abercorn), fl. & fr. 18.vii.1930, *Hutchinson & Gillett* 3836 (BM; K; LISC; SRGH). W: Copperbelt Distr., Mufulira, fl. & fr. 5.vi.1934, *Eyles* 8172 (BM; K; SRGH). E: 10 km Katete–Chadiza, fr. 8.x.1958, *Robson* 10 (BR; K; LISC; SRGH). **Zimbabwe**. E: Mutasa Distr., Honde Valley, fl. 17.iv.1958, *Phipps* 1097 (K; SRGH). **Malawi**. N: Mzimba Distr., Mzuzu, Marymount, fl. & fr. 9.vii.1970, *Pawek* 3597 (K). C: Dedza Forest, fl. & fr. immat. 13.vi.1968, *Salubeni* 1102 (K; SRGH). S: Zomba Mt., fl. & fr. 20.vi.1956, *Banda* 270 (BM; LISC; SRGH). **Mozambique**. N: Ribáuè Mt., south face, fl. & fr. immat. 19.viii.1962, *Leach & Schelpe* 11410 (K; LISC; SRGH). Z: Maganja da Costa, fl. & fr. immat. 24.vi.1946, *Pedro* 1458 (LMA). T: entre Zóbuè e Vila Coutinho, fl. 2.vi.1962, *Gomes e Sousa* 4765 (COI; K). MS: Barué Distr., Catandica (Vila Gouveia), fl. & fr. 3.vii.1941, *Torre* 2982 (LISC).

Widespread in tropical Africa. Forest margins and streamsides, extending into miombo woodland mainly along the rivers; (100)750–1650 m.

Occasionally cultivated, e.g. Harare, *Arnold* in *GHS* 6474 (K; SRGH).

77. **Crotalaria chirindae** Baker f. in J. Linn. Soc., Bot. **42**: 377 (1914). —Eyles in Trans. Roy. Soc. South Africa **5**: 370 (1916). —Verdoorn in Bothalia **2**: 399 (1928). —Brenan in Mem. New York Bot. Gard. **8**: 248 (1953). —Binns, First Check List Herb. Fl. Malawi: 79 (1968). —Polhill in F.T.E.A., Leguminosae, Pap.: 923 (1971). —Drummond in Kirkia **8**: 217 (1972). —Polhill, Crotalaria Africa & Madagascar: 205 (1982). —Lock, Leg. Afr. Check-list: 171 (1989). Syntypes: Zimbabwe, Chirinda, *Swynnerton* 397; Mozambique, Manica e Sofala, Inhamadzi Valley, *Swynnerton* 1498 (BM, syntypes; K).

Erect, straggling-ascending or decumbent well-branched herb, up to 1–1.5 m tall; branches appressed puberulous. Leaves all 3-foliolate or sometimes uppermost 1-foliolate; leaflets mostly 5–15 × 0.8–4 cm, usually narrowly oblong-lanceolate above to elliptic lower down, sometimes all elliptic, often acuminate, puberulous beneath; petiole shorter than the leaflets; stipules mostly 2–4 mm long, subulate or filiform, sometimes caducous. Racemes 10–23(40) cm long, rather densely many-flowered; bracts mostly 4–8 mm long, subulate-caudate; bracteoles at the base of the calyx, 4–9 mm long, linear-caudate. Calyx 3.5–4.5 mm long, becoming basally truncate and deflexed against the pedicel, appressed pubescent; lobes narrowly acuminate-triangular, at least as long as the tube. Standard broadly elliptic, yellow, veined purple or reddish-brown; wings markedly shorter than the keel; keel 8–10 mm long, rounded about the middle, with a narrow incurved sharp beak. Pod shortly stipitate, 2.4–3 × 0.6–0.7 cm, narrowly cylindrical, densely pubescent, 20–26-seeded. Seeds 2.5–3 mm long, oblique-cordiform, smooth, yellow to orange.

Zambia. E: Chipata (Fort Jameson), fl. 4.v.1962, *Verboom* 872 (SRGH). **Zimbabwe**. C: Hwedza (Wedza) Mt., fl. & fr. 27.iv.1961, *Corby* 1000 (SRGH). E: Chipinge (Chipinga), fl. 29.ii.1956, *Drummond* 5122 (BR; COI; K; SRGH). **Malawi**. N: Mzimba Distr., 3 km SW of Mzuzu, fl. & fr. 21.vi.1972, *Pawek* 5475 (K). C: Dedza Mt., fl. 24.v.1968, *Salubeni* 1099 (K; SRGH). S: Zomba Plateau, fl. & fr. 5.vi.1946, *Brass* 16242 (K; NY; SRGH). **Mozambique**. N: Lichinga (Vila Cabral), fl. & fr. immat. 17.v.1948, *Pedro & Pedrógão* 3651 (LMA). Z: Gurué Mts., fl. & fr. 4.viii.1979, *de Koning* 7569 (K; LMU; WAG). MS: Sussundenga Distr., Mavita, fl. 8.iv.1948, *Barbosa* 1381 (LISC).

Also in southern Tanzania. Forest margins, streamsides, open miombo woodland and tree savanna; 750–1900 m.

78. **Crotalaria comosa** Baker in F.T.A. **2**: 34 (1871). —E.G. Baker in J. Linn. Soc., Bot. **42**: 380 (1914). —Wilczek in F.C.B. **4**: 268 (1953). —Hepper in F.W.T.A., ed. 2, **1**: 550 (1958). —Torre in C.F.A. **3**: 102 (1962). —Binns, First Check List Herb. Fl. Malawi: 79 (1968). —Polhill, Crotalaria Africa Madagascar: 208 (1982). —Lock, Leg. Afr. Check-list: 172 (1989). Type from Angola.

Erect generally well-branched annual or short-lived perennial, up to 0.5–2 m tall; branches appressed pubescent to tomentellous. Leaves 3-foliolate, estipulate; leaflets variable, mostly 3–17 × 0.8–4.5 cm, linear-lanceolate to elliptic-lanceolate or elliptic, appressed puberulous or pubescent beneath; petiole shorter than the leaflets. Racemes 6–35 cm long, usually shortly pedunculate, dense, many-flowered; bracts 3–12 mm long, caudate; bracteoles inserted at base of the calyx, 3–5 mm long, filiform. Calyx 3.5–5 mm long, becoming basally truncate and deflexed against the pedicel, appressed pubescent; lobes narrowly triangular-acuminate, ± as long as the tube or longer. Standard obovate-circular or obovate-elliptic, yellow, veined brown or maroon; wings ± as long as the keel; keel 6–8 mm long, strongly rounded about the middle, with a narrow sharp beak incurved at the tip. Pod very shortly stipitate, 1.2–1.6 × 0.5–0.7 cm, ellipsoid-oblong, shortly pubescent to tomentellous, 10–16-seeded. Seeds c. 3 mm long, oblique-cordiform, smooth, brownish.

Zambia. N: Mbala Distr., Inona (Inono) Valley, fl. & fr. 13.iv.1955, *Richards* 5434 (K; SRGH). W: 11 km west of Solwezi, fl. 18.iii.1961, *Drummond & Rutherford-Smith* 7015 (K; LISC; SRGH). E: Chipata (Fort Jameson), fr. ii.1962, *Verboom* 422 (SRGH). S: near Mumbwa, fr. 1911, *Macaulay* 323 (K). **Malawi**. C: Lilongwe, fl. 31.iii.1955, *Jackson* 1531 (K; SRGH). **Mozambique**. T: Macanga Distr., 38 km de Mualadze (Vila Gamito) para Furancungo, fl. & fr. 11.vii.1949, *Barbosa & Carvalho* 3571 (LMA).

Widespread in tropical Africa except the easternmost and southernmost parts. Miombo woodland, grassland and disturbed ground, sometimes in damp places; 900–1400 m.

79. **Crotalaria vasculosa** Wall. ex Benth. in Hooker, London J. Bot. **2**: 584 (1843). —E.G. Baker in J. Linn. Soc., Bot. **42**: 279 (1914). —Verdoorn in Bothalia **2**: 395 (1928). —Mogg in Macnae & Kalk, Nat. Hist. Inhaca Isl., Moçamb.: 146 (1958). —Binns, First Check List Herb. Fl. Malawi: 80 (1968). —Polhill in F.T.E.A., Leguminosae, Pap.: 925 (1971). —Drummond in Kirkia **8**: 219 (1972). —Tredgold & Biegel, Rhod. Wild Fl.: 27, pl. 16, fig. 4 (1979). —Polhill, Crotalaria Africa & Madagascar: 209 (1982). —Lock, Leg. Afr. Check-list: 209 (1989). Type allegedly from Mauritius (not otherwise recorded there), grown in Calcutta.

Shortly erect annual, with long decumbent lower branches, hirsute with tawny spreading hairs. Leaves shortly petiolate, 3-foliolate; leaflets very variable, 2.5–7.5(11.5) × 1–2.5 cm, linear-lanceolate to elliptic or obovate, with those of the leading shoot often longer and narrower, conspicuously pilose on both surfaces; stipules up to 3–4 mm long, filiform. Racemes mostly terminal, shortly pedunculate, up to 3–17 cm long, densely many-flowered; bracts up to 4 mm long, filiform; bracteoles inserted at the base of the calyx, similar. Calyx 3–4.5 mm long, becoming basally truncate and deflexed against the pedicel, subglabrous except for scattered bristly hairs along the lobe margins; lobes acuminately triangular, a little longer than the tube. Standard subcircular, pale yellow, purplish veined; wings usually shorter than the keel; keel 6.5–8.5 mm long, rounded about the middle, with a sharp slightly projecting beak. Pod sessile, 1.6–1.8(2) cm long, subcylindrical, somewhat pointed, densely covered with long brown spreading hairs over a furry tomentum, c. 20-seeded. Seeds c. 2.5 mm long, oblique-cordiform, smooth, pale greenish-buff, mottled greyish-mauve.

Zimbabwe. E: Mutare (Umtali), aerodrome road, fl. 20.iii.1963, *Chase* 7965 (K; SRGH). S: Chivi Distr., 6.5 km north of Runde (Lundi) R. Bridge, near Madzivire Dip, fr. 3.v.1962, *Drummond* 7887 (K; LISC; SRGH). **Malawi**. N: Nkhata Bay Distr., Likoma Island, fl., s.d. *W.P. Johnson* (K). C: Salima–Balaka, 12 km south of Chipoka, fr. 4.v.1980, *Blackmore, Brummitt & Banda* 1453 (K; MAL). S: Mulanje Mt., fl. s.d., *Forbes* 98 (EA). **Mozambique**. N: Nampula, Campo Experimental do C.I.C.A., fl. 11.iv.1961, *Balsinhas & Marrime* 378 (BM; COI; K; LISC; LMA). Z: Lugela Distr., Namagoa Estate, fl. & fr. vii.1946, *Faulkner* Kew No. 25 (COI; K). MS: Manica Distr., entre o rio Revue e Manica (Macequece), fl. 9.iii.1948, *Garcia* 548 (LISC). GI: Mandlakazi (Manjacaze), fl. & fr. 25.iii.1948, *Torre* 7563 (BM; K; LISC). M: Maputo Distr., Inhaca I., Estação de Biologia Marítima, fl. & fr. 11.vii.1957, *Barbosa* 7666 (K; LMA; PRE; SRGH).

Also in East Africa and South Africa (KwaZulu-Natal). Open woodland, grassland, streamsides and disturbed places; 0–1100 m.

80. **Crotalaria abbreviata** Baker f. in J. Linn. Soc., Bot. **42**: 278 (1914). —Wilczek in F.C.B. **4**: 131 (1953). —Polhill in F.T.E.A., Leguminosae, Pap.: 926 (1971). —Drummond in Kirkia **8**: 217 (1972). —Polhill, Crotalaria Africa & Madagascar: 209 (1982). Type from Dem. Rep. Congo (Lake Tanganyika).

Crotalaria fulvella Merxm. in Proc. & Proc. & Trans. Rhodesia Sci. Assoc. **43**: 92 (1951). —Lock, Leg. Afr. Check-list: 163 (1989). Type: Zimbabwe, Marondera (Marandellas), *Dehn* 177 (M, holotype; SRGH).

Shortly erect annual with long prostrate lower branches, pilose or hirsute with rather tawny spreading hairs. Leaves shortly petiolate, 3-foliolate; leaflets 10–50 × 4–12 mm, oblanceolate to obovate-elliptic, subappressed pilose on both surfaces; stipules 2–4.5 mm long, filiform to linear-caudate. Racemes shortly pedunculate, 2–3(7) cm long, with 5–16(28) crowded flowers; bracts 2–4 mm long, subulate-caudate or filiform; bracteoles inserted at the base of the calyx, 1.5–2 mm long, similar to the bract. Calyx 2.5–4 mm long, becoming basally truncate and deflexed against the pedicel, pilose; lobes triangular, slightly acuminate, ± as long as the tube. Standard subcircular, pale yellow, veined purplish-brown; wings exceeding the keel; keel 4–5 mm long, rounded about the middle, with a short straight beak. Pod sessile, 6–8 mm long, subglobose, slightly compressed laterally, densely covered with long brown spreading hairs over a felty tomentum, rather thick-walled, (1)2–4-seeded. Seeds 3–3.8 mm long, oblique-cordiform, smooth, pale ochre-brown.

Zambia. N: Mbala Distr., Lunzua R., fl. & fr. 6.iv.1959, *Richards* 11076 (K; SRGH). **Zimbabwe**. C: Marondera (Marandellas), Delta Farm, fl. & fr. 5.iv.1950, *Wild* 3291 (K; SRGH). **Malawi**. N: Chitipa Distr., Nyika Plateau, Nganda, fl. & fr. 7.iv.1997, *Patel, Ludlow, Scott & Jarman* 5102 (K).

Also in Dem. Rep. Congo (Katanga) and Tanzania. Grassland and disturbed places usually on sand, sometimes in damp sites; 750–2250 m.

81. **Crotalaria anthyllopsis** Welw. ex Baker in F.T.A. **2**: 15 (1871). —E.G. Baker in J. Linn. Soc., Bot. **42**: 263 (1914). —Eyles in Trans. Roy. Soc. South Africa **5**: 370 (1916). —Verdoorn in Bothalia **2**: 414 (1928). —Merxmüller in Proc. & Trans. Rhodesia Sci. Assoc. **43**: 18 (1951). —Brenan in Mem. New York Bot. Gard. **8**: 245 (1953). —Wilczek in F.C.B. **4**: 81 (1953). —Hepper in F.W.T.A., ed. 2, **1**: 548 (1958). —Torre in C.F.A. **3**: 29 (1962). —Binns, First Check List Herb. Fl. Malawi: 78 (1968). —Polhill in F.T.E.A., Leguminosae, Pap.: 927 (1971). —Drummond in Kirkia **8**: 217 (1972). —Polhill, Crotalaria Africa & Madagascar: 210 (1982). —Lock, Leg. Afr. Check-list: 165 (1989). Lectotype, selected by Torre, loc. cit., from Angola.

Crotalaria anthyllopsis var. *albopilosa* R.E. Fr., Wiss. Ergebn. Schwed. Rhod.-Kongo-Exped. **1**: 73 (1914). Type: Zambia, Luapula R., *R.E. Fries* 563 (UPS, holotype).

Shortly erect annual with long decumbent lower branches, thinly covered with long bristly spreading hairs. Leaves shortly petiolate, 1-foliolate; leaflets mostly 3–10 × 0.8–3 cm, ovate to oblong-lanceolate, the upper ones longer and narrower, pilose, usually thinly so, on both surfaces; stipules 6–10(12) mm long, linear or filiform. Racemes up to 2–5 cm long, subsessile, dense, many-flowered; bracts up to 7–14(17) mm long, linear or filiform; bracteoles inserted at base of the calyx, similar. Calyx 5–6 mm long, becoming basally truncate and deflexed against the pedicel, sparsely hirsute; lobes attenuate-triangular, much longer than the tube. Standard subcircular, pale yellow, purplish veined; wings exceeding the keel; keel 5–6(8) mm long, strongly rounded about the middle, with a straight or slightly incurved beak. Pod 6–7 mm long, sessile, subglobose, slightly compressed laterally, densely hirsute, (1)3–6-seeded. Seeds 2–3 mm long, oblique-cordiform, wrinkled, reddish-brown.

Zambia. N: Mbala Distr., Ndundu, fl. & fr. 4.iii.1959, *McCallum-Webster* 723 (BR; K; SRGH). W: Kitwe, fl. & fr. 24.iv.1963, *Mutimushi* 283 (K; NDO; SRGH). C: Lusaka, fl. 18.iv.1955, *Best* 80 (K; SRGH). E: Chipata (Fort Jameson), fl. 19.iv.1960, *Wright* 252 (K). S: Choma Distr., Mapanza Mission, fl. & fr. 5.iv.1953, *E.A. Robinson* 159 (K). **Zimbabwe**. N: Hurungwe Distr., 10 km NW of Makuti, fl. 14.iii.1961, *Drummond & Rutherford-Smith* 6893 (K; SRGH). W: Matobo Distr., Farm Besna Kobila, fl. & fr. iii.1957, *O.B. Miller* 4198 (BR; K; LISC; SRGH). C: Chegutu Distr., Poole Farm, fl. 3.iv.1948, *R.M. Hornby* 2859 (BR; K; LISC; SRGH). E: Mutare (Umtali), fl. 15.iii.1960, *Chase* 7304 (BM; K; LISC; SRGH). S: Masvingo Distr., Makaholi Experimental Station, fl. 13.iii.1978, *Senderayi* 223 (K; SRGH). **Malawi**. N: Mzimba Distr., Mbawa Experimental Station, fl. 5.iv.1955, *Jackson* 1589 (K; SRGH). C: Lilongwe Agricultural Research Station, fl. 30.iii.1953, *Jackson* 1190 (BM; LISC). S: Kirk Range, fl. & fr. 17.iii.1955, *Exell, Mendonça & Wild* 1003 (BM; LISC; SRGH). **Mozambique**. N: 38 km east of Ribáuè, fl. 17.v.1961, *Leach & Rutherford-Smith* 10899 (K; LISC; SRGH). MS: Sussundenga Distr., Mavita, fl. 8.iv.1948, *Barbosa* 1387 (BM; LISC).

Also in Mali, Nigeria, Central African Republic, Dem. Rep. Congo, Rwanda, Burundi, Ethiopia, East Africa and Angola. Open woodland, grassland and disturbed places; 700–1800 m.

82. **Crotalaria cylindrostachys** Welw. ex Baker in F.T.A. **2**: 15 (1871). —E.G. Baker in J. Linn. Soc., Bot. **42**: 257 (1914) pro majore parte. —Eyles in Trans. Roy. Soc. South Africa **5**: 370 (1916). —Verdoorn in Bothalia **2**: 414 (1928). —Torre in C.F.A. **3**: 27 (1962). —Polhill in F.T.E.A., Leguminosae, Pap.: 927 (1971). —Drummond in Kirkia **8**: 217 (1972). —Polhill, Crotalaria Africa & Madagascar: 211 (1982). —Lock, Leg. Afr. Check-list: 173 (1989). Type from Angola.

Erect annual, 0.3–1.2 m tall, with usually numerous hirsute ascending branches. Leaves shortly petiolate, 1-foliolate; leaflets mostly 3–8 × 1.2–3 cm, ovate-lanceolate to elliptic or sometimes ovate, pilose, often sparsely so, on both surfaces; stipules 4–8 mm long, filiform. Racemes up to 17(23) cm long, usually pedunculate, dense, many-flowered; bracts 5–7 mm long, filiform; bracteoles inserted at base of the calyx, similar. Calyx 2–4 mm long, becoming basally truncate and deflexed against the pedicel, sparsely hirsute; lobes triangular, ± as long as the tube. Standard subcircular, dull yellow, purplish-brown veined; wings exceeding the keel; keel 4–4.5 mm long, rounded about the middle, with a short rather blunt slightly incurved beak, 4–4.5 mm long. Pod sessile, 6–8 mm long, subglobose, somewhat compressed laterally, densely brown villous, 1–2-seeded. Seeds c. 3 mm long, oblique-cordiform, wrinkled, brown.

Zambia. C: Mpika Distr., South Luangwa National Park, 5 km south of Lubi R., fr. 10.iv.1967, *Prince* 460 (K; SRGH); Kafue R. north of the Kafue road bridge, fl. 1.iii.1963, *van Rensburg* 1556 (K; SRGH). S: Kalomo Distr., Bombwe Forest, fl. & fr. immat. 1933, *Martin* 604/33 (K). **Zimbabwe**. N: Hurungwe Distr., Chirundu, fl. & fr. 15.iii.1966, *Simon* 722 (K; LISC; SRGH). W: Hwange Distr., Gwayi (Gwai)–Lutope R. junction, fl. & fr. 26.ii.1963, *Wild* 6005 (K; LISC; SRGH). S: Masvingo (Victoria), fr. 1909, *Monro* 1045 (BM). **Malawi**. N: Karonga Distr., Chaminade School, fl. & fr. 23.iv.1975, *Pawek* 9525 (K; MAL; MO).

Also in Tanzania and Angola. Miombo and mopane savanna woodland or disturbed places, often on sand, sometimes in damp spots; 350–1200 m.

83. **Crotalaria schinzii** Baker f. in J. Linn. Soc., Bot. **42**: 370 (1914). —Verdoorn in Bothalia **2**: 386 (1928). —Burtt Davy, Fl. Pl. Ferns Transvaal, pt. 2: 398 (1932). —Drummond in Kirkia **8**: 218 (1972). —Polhill, Crotalaria Africa & Madagascar: 217 (1982). —Lock, Leg. Afr. Check-list: 202 (1989). Type from South Africa (Northern Province).

Small trailing perennial; stems rather slender, laxly branched, 20–60 cm long, quite densely spreading hirsute. Leaves 3-foliolate; leaflets 1.5–4 × 0.7–1.5 cm, mostly oblong-elliptic to elliptic-obovate, sometimes some at least oblanceolate to narrowly oblong-lanceolate, pilose at least beneath; petiole shorter than the leaflets; stipules 3–6 mm long, linear-lanceolate, ± caudate. Racemes leaf-opposed, 6–12(20) cm long, pedunculate, lax (except at the tip), 8–24-flowered; bracts 1.5–3 mm long, lanceolate to ovate-caudate; bracteoles inserted just below the calyx, similar to the bract but a little smaller. Calyx 4–6 mm long, appressed pilose; lobes narrowly triangular, ± acuminate, 1–2 times as long as the tube. Standard subcircular, yellow, veined purplish-brown outside; wings markedly shorter than the keel; keel 7.5–9 mm long, rounded, with a fairly short but narrow slightly incurved beak. Pod shortly stipitate, 1.1–2.4 × 0.45–0.9 cm, oblong, fusiform to subcylindrical, slightly broadened upwards, shortly pubescent, (10)16–24-seeded. Seeds 2–3 mm long, oblique-cordiform, rugulose, brown.

Botswana. SE: 20 km NW of Molepolole, fl. & fr. 2.xii.1954, *Codd* 8930 (K; PRE; SRGH). **Zimbabwe**. S: Beitbridge, fl. & fr. 17.iii.1967, *Corby* 1839 (K; SRGH). **Mozambique**. GI: Magude, 28 km de Mapulanguene, fl. & fr. 17.ii.1953, *Myre & Balsinhas* 1551 (K; LMA).

Also in South Africa (Northern Province in the Limpopo Valley basin). Woodland and savanna; 100–1000 m.

84. **Crotalaria sp. A**

Trailing or weakly ascending perennial; stems slender, laxly branched, up to 40 cm long, shortly appressed pubescent. Leaves 3-foliolate; leaflets mostly 15–35 × 3–6 mm, mostly narrowly oblong-elliptic, upper ones longer and relatively narrow, lowermost elliptic-oblanceolate, shortly appressed pubescent beneath; petiole shorter than the leaflets; stipules 0.5–1 mm long, subulate. Racemes leaf-opposed, up to 30 cm long, laxly 16–24-flowered; bracts 1–2 mm long, subulate; bracteoles at the base of the calyx, similar. Calyx 4–5 mm long, appressed pubescent; lobes triangular, slightly acuminate, ± as long as the tube. Standard elliptic-obovate to subcircular, yellow, lined brownish; wings a little shorter than the keel; keel c. 8 mm long, rounded about the middle, with an incurved beak, conspicuously lined brown. Pods very immature, probably maturing oblong-ellipsoid to subcylindrical, shortly stipitate and with numerous seeds, pubescent.

Zambia. N: Chinsali Distr., Shiwa Ngandu, Machipara Hill, fl. 16.i.1959, *Richards* 10685 (K; SRGH).

Known only from the above gathering. Woodland, among grass and rocks; 1500 m.

More material, particularly in fruit, is desirable before this species is formally described. It is most similar to *C. schinzii*, differing in the indumentum, the narrow leaflets and much smaller stipules.

85. **Crotalaria steudneri** Schweinf. in Verh. K. K. Zool.-Bot. Ges. Wien **18**: 651 (1868). —J.G. Baker in F.T.A. **2**: 30 (1871). —E.G. Baker in J. Linn. Soc., Bot. **42**: 368 (1914). —Schreiber in Merxmüller, Prodr. Fl. SW. Afrika, fam. 60: 27 (1970). —Polhill in F.T.E.A., Leguminosae, Pap.: 936 (1971). —Drummond in Kirkia **8**: 219 (1972). —Polhill, Crotalaria Africa & Madagascar: 218 (1982). —Lock, Leg. Afr. Check-list: 205 (1989). Type from Eritrea.

Crotalaria hispida Schinz in Bull. Herb. Boissier, sér. 2, **3**: 820 (1903). —E.G. Baker in J. Linn. Soc., Bot. **42**: 367 (1914). —Verdoorn in Bothalia **2**: 386 (1928). —Burtt Davy, Fl. Pl. Ferns Transvaal, pt. 2: 398 (1932). —Binns, First Check List Herb. Fl. Malawi: 79 (1968). Type from Namibia.

Ascending and spreading annual, much branched from the base, with the lower branches mostly long and decumbent, sometimes to 65(100) cm tall, usually much smaller; branches hispid, often thinly, with rather long spreading hairs. Leaves 3-foliolate; leaflets mostly 1.5–7 × 0.4–1.7 cm, linear-lanceolate to lanceolate or oblanceolate, with the lower ones sometimes elliptic or obovate, thinly appressed pilose beneath; petiole shorter than the leaflets; stipules 3–9 mm long, subulate to linear-lanceolate, caudate. Racemes up to 8–16(25) cm long, long and slenderly pedunculate, laxly few–many-flowered; bracts 1–2 mm long, subulate; bracteoles usually at the base of the calyx, less than 1 mm long. Calyx 3–4(4.5) mm long, subglabrous to appressed pubescent; lobes acuminately triangular, longer than the tube. Standard subcircular, yellow, reddish-brown veined outside; wings shorter than the keel to ± as long; keel 5–7(9) mm long, rounded below the middle, with a short straight or less commonly incurved beak. Pod sessile, 10–13(17) × (6)7–8 mm, subglobose-ellipsoid to broadly ellipsoid or rarely oblong-ellipsoid, often mottled brown, pubescent, 20–24-seeded. Seeds 2.5–3 mm long, oblique-cordiform, often slightly rugulose, reddish-brown.

Botswana. N: Ngamiland Distr., 22 km south of Nokaneng on road to Tsao (Tsau), fl. & fr. 11.iii.1965, *Wild & Drummond* 6860 (K; LISC; SRGH). **Zambia**. B: Sesheke Distr., Machili, fl. & fr. immat. 1.i.1961, *Fanshawe* 6062 (K; NDO; SRGH). C: Lusaka Distr., Iolanda, north bank of Kafue R., near Kafue town, fl. & fr. 14.iii.1965, *E.A. Robinson* 6432 (BR; K; LISC; M; SRGH). E: Chipata Distr., Masumba Village, fl. & fr. 24.iii.1963, *Verboom* 806 (K; LISC; SRGH). S: Mazabuka, fl. & fr. 6.ii.1963, *van Rensburg* 1334 (K; SRGH). **Zimbabwe**. N: Hurungwe Distr., Mhenza (Mensa) Pan, 18 km ESE of Chirundu Bridge, fl. 29.i.1958, *Drummond* 5335 (BR; K; LISC; SRGH). W: Umguza Distr., Redbank, Whinburn, fl. & fr. 20.i.1950, *Plowes* 1483 (K; LISC; SRGH). C: Chegutu (Hartley), fl. & fr. 15.ii.1962, *R.M. Hornby* 3270 (K; SRGH). E: Nyanga Distr., Cheshire, fl. 15.i.1931, *Norlindh & Weimarck* 4434 (K; LD; M). S: Masvingo (Victoria), fl. & fr. *Monro* 1047 (K). **Malawi**. N: Karonga Distr., Ngara (Ngala), 24 km north of Chilumba, fl. & fr. 22.iv.1969, *Pawek* 2280 (K). C: Salima Distr., Lifidzi Goat Breeding Centre, fl. & fr. 13.iv.1985, *Banda, Balaka, Tawakali & Kwatha* 2483 (K; MAL). S: Machinga Distr., Shire R. near Liwonde ferry, fl. & fr. 13.iii.1955, *Exell, Mendonça & Wild* 841 (BM; LISC; SRGH). **Mozambique**. N: Namapa, base da Serra Malala, fl. 30.iii.1961, *Balsinhas & Marrime* 340 (K; LISC).

Well distributed in eastern Africa, excluding the most easterly and southerly parts, from Sudan and N Ethiopia south to Namibia and the Northern Province of South Africa. Mopane woodland (rarely in miombo), savanna and disturbed places, often on seasonally damp sandy ground near streams, roads and dambos; 250–1300 m.

The species is reasonably uniform over most of its range, but seems to be genetically unstable in Tanzania and northern Mozambique, where forms with an incurved beak to the keel, unusually elongate pods or a glabrous calyx occur sporadically, without any consistent correlation of the character differences. See also note under *C. preladoi*, below.

86. **Crotalaria preladoi** Baker f. in J. Linn. Soc., Bot. **42**: 367 (1914). —Verdoorn in Bothalia **2**: 386 (1928). —Polhill, Crotalaria Africa & Madagascar: 218 (1982). —Lock, Leg. Afr. Check-list: 198 (1989). Type: Mozambique, Cabaceira Grande, *Prelado* 38 (B†, holotype; BM, fragment).

Annual, with weak slender thinly appressed pilose branches. Leaves 3-foliolate; leaflets 10–40 × 2–7 mm, variable in shape, narrowly oblong to linear-lanceolate or elliptic, thinly appressed pilose beneath; petioles 1–2 cm long; stipules 2–5 mm long, linear-lanceolate. Racemes leaf-opposed, laxly 5–8(?–many)-flowered; bracts up to 2 mm long, linear; bracteoles inserted at the base of the calyx, c. 1 mm long. Calyx 4–5 mm long, thinly appressed pilose; lobes acuminately triangular, somewhat longer than the tube. Standard subcircular-obovate, white(?), lined reddish-purple; wings as long as the keel; keel c. 5 mm long, rounded, with a short slightly incurved beak. Pod sessile, 9–10 × 6 mm, ovoid-ellipsoid, pubescent.

Mozambique. N: Cabaceira Grande, fl. & fr. 30.iii.1894, *Prelado* 38 (B†; BM, fragment).
Imperfectly known from the type gathering alone.
Crotalaria steudneri is rather variable in this part of Mozambique, as indicated above, and *C. preladoi* is probably a synonym. Nevertheless no other specimen exactly matches the description, particularly in terms of the indumentum, and decision on its status seems better left until the type locality has been explored more thoroughly.

87. **Crotalaria nudiflora** Polhill, Crotalaria Africa & Madagascar: 219, fig. 55 (1982). —Lock, Leg. Afr. Check-list: 193 (1989). Type: Zambia, Kasama Distr., Luombi R., *Exell, Mendonça & Wild* 1381 (BM, holotype; LISC; SRGH).

Erect widely branching annual, 10–60 cm tall; stem slender, subglabrous to sparsely puberulous above. Leaves 3-foliolate; leaflets mostly 15–45 × 3–15 mm, variable in shape, linear-lanceolate to lanceolate or oblanceolate to elliptic or obovate-elliptic, sparsely appressed puberulous beneath; petiole shorter (sometimes only slightly) than the leaflets; stipules 1–1.5 mm long, subulate, deciduous. Racemes numerous, the main ones 4.5–11 cm long, but also smaller ones on reduced branchlets appearing at first glance axillary from the upper leaves, all long pedunculate with 6–12 flowers widely spaced on the slender rhachis; bracts 1–1.5 mm long, subulate; bracteoles inserted on the pedicel, small. Calyx c. 4 mm long, glabrous; lobes narrowly acuminate-triangular, shorter to a little longer than the tube. Standard obovate-circular, yellow, veined brown or purplish; wings exceeding the keel; keel 6–7 mm long, abruptly rounded near the middle, with a narrow attenuate beak incurved at the tip. Pod shortly stipitate, (0.8)1.2–1.5 × 0.35–0.4 cm, narrowly subcylindrical, pointed at either end, puberulous, (6)10–16-seeded. Seeds 1.5–2 mm long, oblique-cordiform, smooth to minutely rugulose, brownish.

Zambia. N: Kawambwa, fl. & fr. 23.vi.1957, *E.A. Robinson* 2404 (K; SRGH); 67 km Mporokoso–Kasama, fl. & fr. 2.iv.1984, *Brummitt, Chisumpa & Nshingo* 17095 (K; NDO; SRGH).
Known only from northern Zambia. Miombo woodland, often in disturbed grassy places or near rivers; c. 1200–1350 m.

88. **Crotalaria spartea** Baker in F.T.A. **2**: 12 (1871) pro parte. —E.G. Baker in J. Linn. Soc., Bot. **42**: 262 (1914). —Verdoorn in Bothalia **2**: 416 (1928). —Burtt Davy, Fl. Pl. Ferns Transvaal, pt. 2: 401 (1932). —Wilczek in F.C.B. **4**: 91 (1953). —Hepper in Kew Bull. **11**: 115 (1956). —Torre in C.F.A. **3**: 29 (1962). —Binns, First Check List Herb. Fl. Malawi: 80 (1968). — Polhill in F.T.E.A., Leguminosae, Pap.: 937, fig. 129 (1971). —Drummond in Kirkia **8**: 218 (1972). —Gonçalves in Garcia de Orta, Sér. Bot. **5**: 71 (1982). —Polhill, Crotalaria Africa & Madagascar: 220, fig. 56 (1982). —Lock, Leg. Afr. Check-list: 204 (1989). TAB. 3,7: **33**. Lectotype, selected by Hepper, loc. cit., from Angola.

Erect or straggling annual 0.2–1.3 m tall; branches well spaced, curved-ascending, sparsely appressed puberulous, glabrescent. Leaves very shortly stalked, simple; blades (40)60–130 × (3)5–8(15) mm, mostly linear-lanceolate to lanceolate, the lower sometimes shorter and ± elliptic, sparsely appressed puberulous beneath; stipules 1–2.5 mm long, filiform to subulate-caudate. Racemes laxly 10–16(24)-flowered; bracts 1–2 mm long, ± filiform; bracteoles inserted at the base of the calyx, similar. Calyx 3–4 mm long, sparsely appressed puberulous; lobes acuminately

Tab. 3,7: **33**. CROTALARIA SPARTEA. 1, habit (× ²⁄₃); 2, flower (× 4); 3, standard (× 4); 4, wing (× 4); 5, keel (× 4); 6, anthers (× 15); 7, gynoecium (× 4), 1–7 from *Milne-Redhead & Taylor* 9394; 8, pod (× 1); 9, seed (× 6), 8 & 9 from *Milne-Redhead & Taylor* 9394B. Drawn by Roger Polhill. From Crotalaria Africa & Madagascar.

triangular, ± as long as the tube. Standard elliptic to obovate-elliptic, bright yellow, reddish-brown veined outside; wings much shorter than the keel; keel 7–9 mm long, shallowly curved just below the middle, with a narrow sharply pointed only slightly incurved beak. Pod practically sessile, 16–25 × 4–5 mm, subcylindrical, slightly broadened upwards, glabrous except for a few hairs along the upper suture, green mottled dull purple before maturity, 20–32-seeded. Seeds 1.5–2 mm long, oblique-cordiform, blue-green, smooth.

Zambia. N: Mbala Distr., Kawimbe Leper Camp, fl. 3.iv.1959, *McCallum-Webster* 744 (BR; K; LISC; SRGH). W: Solwezi, fl. & fr. 9.iv.1960, *E.A. Robinson* 3469 (K; M; SRGH). **Zimbabwe**. C: Goromonzi Distr., Ruwa, fl. 22.iii.1953, *Aylen* in *GHS* 42101 (K; SRGH). E: 8 km north of Nyanga (Inyanga), fl. & fr. 28.iv.1967, *Rushworth* 895 (K; LISC; SRGH). **Malawi**. C: Dedza Distr., Chongoni Forest Reserve, fl. & fr. 8.xii.1967, *Salubeni* 921 (K; SRGH). S: Zomba Plateau, Mlunguzi (Mulungusi) Stream, fl. 27.iii.1937, *Lawrence* 305 (K). **Mozambique**. T: Angónia Distr., road towards Furancungo near Lifidzi (Lipidzi) Mission, fl. 8.iii.1964, *Correia* 184 (LISC).

Extends northwards along the Western Rift Valley in Tanzania, Burundi, Rwanda and Dem. Rep. Congo, also southwards to South Africa (Mpumalanga) and west to Nigeria and Angola. Damp grassland by lakes and rivers, in marshes and dambos, also disturbed places; 1000–1750 m.

89. **Crotalaria paraspartea** Polhill, Crotalaria Africa & Madagascar: 221 (1982). —Lock, Leg. Afr. Check-list: 195 (1989). Type: Mozambique, Namaita, entre Nampula e Murrupula, na area da Gafaria, 26.iii.1964, *Torre & Paiva* 11391 (COI; K; LISC, holotype; LMU; SRGH).

Erect annual up to c. 70 cm tall; branches ascending, the lower ones curved, crisped pubescent. Leaves 1-foliolate; leaflets mostly 3–10.5 × 0.3–2.4 cm, elliptic at the base, varying to lanceolate and linear-lanceolate towards the branch tips, thinly appressed puberulous beneath; petioles mostly 3–6 mm long; stipules c. 0.5 mm long, triangular. Racemes mostly 8–18 cm long, laxly 6–12-flowered; bracts 1–2.5 mm long, linear-caudate; bracteoles at the base of the calyx, similar. Calyx 3.5–4 mm long, appressed pubescent; lobes triangular-acuminate, a little longer than the tube. Standard elliptic-obovate, yellow lined reddish-purple; wings much shorter than the keel; keel 7–8 mm long, strongly rounded, with a narrow slightly incurved beak. Pod 2–2.2 × 0.7 cm, subsessile, subcylindrical, appressed pubescent, glabrescent on the underside, up to c. 18-seeded. Mature seeds not seen.

Mozambique. N: andados 27 km na estrada antiga de Nampula para Muecate, c. 380 m, fl. & fr. 2.iv.1964, *Torre & Paiva* 11576 (LISC; MO).

Known only from the collections cited. Miombo woodland, on sandy soils; 350–400 m.

90. **Crotalaria incompta** N.E. Br. in Bull. Misc. Inform., Kew **1909**: 101 (1909). —E.G. Baker in J. Linn. Soc., Bot. **42**: 262 (1914). —Verdoorn in Bothalia **2**: 416 (1928). —Schreiber in Merxmüller, Prodr. Fl. SW. Afrika, fam. 60: 24 (1970). —Polhill, Crotalaria Africa & Madagascar: 222 (1982). —Lock, Leg. Afr. Check-list: 182 (1989). Type: Botswana, Botletle Valley, ii.1897, *E.J. Lugard* 205 (K, holotype).

Erect annual with long spreading and ascending branches from the base upwards, up to 50 cm tall; branches well spaced above, sparsely appressed puberulous, glabrescent. Leaves 1-foliolate; leaflets 35–80 × 2–8 mm, linear-lanceolate, appressed puberulous beneath; petioles mostly 6–12 mm long; stipules 1–2 mm long, subulate. Racemes laxly few-flowered; bracts 1–2 mm long, linear; bracteoles on the pedicel, similar. Calyx c. 5 mm long, appressed puberulous; lobes narrowly attenuate-triangular, nearly twice as long as the tube. Standard elliptic, yellow (?veining); wings little more than half as long as the keel; keel c. 8 mm long, shallow, shortly rounded about the middle, with a narrow almost straight sharply pointed beak. Pod 1.5–2 cm long, shortly subcylindrical, slightly broadened upwards, abruptly rounded to the narrow 2–3 mm long stipe at the base, appressed puberulous, mottled before maturity, c. 16-seeded. Seeds c. 3–3.5 mm long, rather narrowly oblique-cordiform, rugose particularly peripherally.

Botswana. N: Botletle Valley, fl. ii.1897, *E.J. Lugard* 205 (K).

Also in Namibia. Imperfectly known, probably in seasonally wet places.

91. **Crotalaria germainii** R. Wilczek in Bull. Jard. Bot. État **23**: 161, fig. 24 (1953); in F.C.B. **4**: 97, fig. 5 (1953). —Polhill, Crotalaria Africa & Madagascar: 225 (1982). —Lock, Leg. Afr. Check-list: 179 (1989). Type from Dem. Rep. Congo (Kivu).
Crotalaria robinsoniana Torre in Mem. Junta Invest. Ultramar, sér. 2, **19**: 23 (1960); in C.F.A. **3**: 28 (1962). Type from Angola (Moxico).

Small erect annual, up to 10–25 cm tall, soon developing relatively long spreading and weakly ascending branches particularly from near the base; branches very slender, puberulous at the tips, soon glabrescent. Leaves 1-foliolate, estipulate; leaflets 15–60 × 2–6 mm, mostly linear-lanceolate, the lowermost often shorter and ± elliptic-oblong, appressed puberulous beneath, often only sparsely so; petioles mostly 2–4 mm long. Racemes laxly few-flowered; rhachis very slender; bracts 0.5–1 mm long, linear-subulate; bracteoles inserted at the base of the calyx, very small. Calyx 1.5–2 mm long, sparsely appressed puberulous; lobes triangular, slightly acuminate, slightly shorter than the tube. Standard obovate or obovate-circular, yellow, usually veined purple; wings exceeding the keel; keel 3–3.5 mm long, rounded below the middle, with a narrow slightly incurved beak. Pod 10–14 × 4–5 mm, sessile, oblong, slightly tapered, thinly puberulous to glabrous, 12–14-seeded. Seeds 1.5–2 mm long, oblique-cordiform, slightly rugulose, olive-green to brown.

Zambia. B: Senanga Distr., Mutomema, fl. & fr. 10.iv.1994, *Bingham* 10072 (K). W: banks of Kafue R., 11 km north of Chingola, fl. & fr. 4.v.1960, *E.A. Robinson* 3697 (K; M; SRGH). S: Namwala, fl. & fr. immat. 9.i.1957, *E.A. Robinson* 2094 (BR; K; SRGH).
Also in Dem. Rep. Congo and Angola. Dambos, usually the drier parts in sandy places; 950–1350 m.

92. **Crotalaria involutifolia** Polhill in Kew Bull. **22**: 293, fig. 27/1–11 (1968); in F.T.E.A., Leguminosae, Pap.: 942 (1971); Crotalaria Africa & Madagascar: 225, fig. 57/1–11 (1982). —Lock, Leg. Afr. Check-list: 183 (1989). Type from Tanzania (Ufipa).

Delicate ascending annual, up to 10–50 cm tall; stem very slender, laxly branched, sparsely appressed puberulous, glabrescent. Leaves sessile, simple; blade 10–40 × 0.5–2 mm, linear to linear-oblanceolate, with the margins usually at least partially inrolled, sparsely puberulous, often glabrescent; stipules minute or apparently lacking. Racemes with 2–8 widely spaced flowers on the ± filiform rhachis; bracts 0.5–1 mm long, setaceous; bracteoles inserted at base of the calyx, minute. Calyx 2.5–3.5 mm long, puberulous; lobes acuminately triangular, usually a little longer than the tube. Standard obovate-circular, yellow, veined reddish-brown outside; wings a little shorter to a little longer than the keel; keel (3.5)4–6.5 mm long, rather abruptly rounded below the middle, with a short straight or slightly incurved beak. Pod subsessile, 9–14 × 2–2.5 mm, narrowly cylindrical, pubescent, ultimately glabrescent, 18–22-seeded. Seeds 1–1.2 mm long, oblique-cordiform, smooth, brown.

Zambia. N: Mbala Distr., Old Katwe Road, fl. & fr. 9.v.1955, *Richards* 5603 (BR; K; LISC; LMA; P; SRGH). W: Mwinilunga Distr., Kalenda Plain, fl. & fr. 16.iv.1960, *E.A. Robinson* 3618 (K; SRGH).
Also in Cameroon, Dem. Rep. Congo (Katanga) and southern Tanzania. Damp places in marshes, dambos and rock hollows; 1200–1650 m.

93. **Crotalaria flavicarinata** Baker f. in J. Linn. Soc., Bot. **37**: 437 (1906); in J. Linn. Soc., Bot. **42**: 333 (1914). —Eyles in Trans. Roy. Soc. South Africa **5**: 370 (1916). —Verdoorn in Bothalia **2**: 400 (1928). —Wild in Clark, Victoria Falls Handb.: 149 (1952). —White, F.F.N.R.: 147 (1962). —Schreiber in Merxmüller, Prodr. Fl. SW. Afrika, fam. 60: 24 (1970). —Drummond in Kirkia **8**: 217 (1972). —Polhill, Crotalaria Africa & Madagascar: 229, fig. 58 (1982). —Lock, Leg. Afr. Check-list: 179 (1989). Syntypes: Zimbabwe, Victoria Falls, *Gibbs* 163 (BM, syntype) and *C.E. Allen* 31 (K, syntype).
Crotalaria cataractarum Baker f. in J. Bot. **58**: 74 (1920). Type: Zimbabwe, Victoria Falls, *F.A. Rogers* 13290 (BM, holotype; K).

Erect herb or short-lived shrub, 0.6–2 m tall; branches somewhat angular, ribbed, shortly hairy, glabrescent. Leaves 3-foliolate; leaflets 1.5–6 × 0.5–3 cm, elliptic to oblong-obovate, shortly and often sparsely hairy at least beneath; petiole shorter than the leaflets; stipules 1–3(4) mm long, linear-subulate. Racemes lax, 8–30-

flowered; bracts 1–2 mm long, linear or subulate; bracteoles on the pedicel, smaller. Calyx 4.5–6 mm long, puberulous; lobes subulate or narrowly triangular-acuminate, shorter than the tube, reflexed. Standard elliptic, greenish-white, lined reddish-purple, subglabrous to extensively pubescent outside; wings brighter yellow, shorter than the keel; keel 1.1–1.3(1.5) cm long, abruptly angled in the lower part, with a long straight beak darker coloured at the tip. Pod subsessile, (12)16–22 × 7–8 mm, puberulous, ultimately glabrescent, up to 10–16-seeded. Seeds 3 mm long, broadly oblique-cordiform, smooth, ochre or brown.

Caprivi Strip. Singalamwe–Lizauli, fl. & fr. 2.i.1959, *Killick & Leistner* 3251 (K; PRE). **Botswana**. N: 77 km north of Aha Hills, fl. & fr. 13.iii.1965, *Wild & Drummond* 6987 (BR; K; LISC; SRGH). **Zambia**. B: Senanga Distr., Shangombo, fl. & fr. 8.viii.1952, *Codd* 7448 (BM; COI; K; PRE; SRGH). W: Kitwe Distr., Ichimpe (Ichimpi), fl. 7.xii.1968, *Mutimushi* 2862 (K; NDO). S: Namwala, Kasha Dambo, fl. 17.ii.1963, *C.D. Simpson* 16/63 (BR; K; LISC; SRGH). **Zimbabwe**. W: Hwange (Wankie) Game Reserve, Dett Road, fl. 20.ii.1956, *Wild* 4779 (BR; K; LISC; SRGH).

Also in Namibia. Practically restricted to the drainage basins of the upper Zambezi and Kafue Rivers, generally on Kalahari Sands, in a variety of deciduous and evergreen woodland or savanna formations, also on sandy riverbanks and at edges of riverine forest; 900–1250 m.

94. **Crotalaria stenoptera** Welw. ex Baker in F.T.A. **2**: 12 (1871). —E.G. Baker in J. Linn. Soc., Bot. **42**: 268 (1914). —Harms in Engler, Pflanzenw. Afrikas [Veg. Erde 9] **3**: 549 (1915). —Torre in C.F.A. **3**: 31 (1962). —Polhill, Crotalaria Africa & Madagascar: 231 (1982). —Lock, Leg. Afr. Check-list: 205 (1989). Type from Angola.

Crotalaria stenoptera var. *latifolia* Baker f. in J. Linn. Soc., Bot. **42**: 269 (1914). —Torre in C.F.A. **3**: 31 (1962). Lectotype, selected by Polhill, loc. cit. (1982), from Angola.

Perennial herb, ultimately with a number of erect variably branched stems from a woody rootstock, but flowering from first year of growth, up to 0.5–1 m tall, glabrous overall; stems 3–4-angled, narrowly winged; wings 0.5–2 mm wide at leaf insertion, with stipules on outer points, decurrent into angle of the stem lower down. Leaves simple or 1-foliolate on a petiole not more than 1 mm long, the uppermost sometimes much reduced; blades up to 4–14 × 0.7–5 cm, linear-lanceolate to lanceolate, often long-attenuate, or elliptic, the upper ones longer and narrower, rather leathery; lateral nerves strongly ascending, successively running up close and parallel to the margin for some distance; stipules 3–10 mm long, subulate. Racemes laxly few–many-flowered; bracts 1–2.5 mm long, linear-subulate to linear-lanceolate, 1–2.5 mm long; bracteoles inserted near the middle of the pedicel or below, smaller. Calyx 6–8 mm long; lobes subulate or narrowly attenuate-triangular, longer than the tube, becoming reflexed. Standard elliptic, yellow, lined and later flushed reddish-brown or purplish outside; wings shorter than the keel; keel 0.9–1.3 cm long, angular, sharply bent in the lower part with a long straight beak. Pod shortly stipitate, 2–2.8 × 0.6–1.2 cm, cylindrical, venose, glabrous, 14–18-seeded. Seeds not seen.

Zambia. B: Kabompo–Chizela (Chizera) road, 1.5 km SW of Nkulwashi (Kabompo R. Pontoon), fl. & fr. 23.iii.1961, *Drummond & Rutherford-Smith* 7229 (K; LISC; SRGH).

Also in Dem Rep. Congo (Katanga) and Angola. Woodland on Kalahari Sands; c. 900–1000 m.

95. **Crotalaria trinervia** Polhill, Crotalaria Africa & Madagascar: 231, fig. 59 (1982). —Lock, Leg. Afr. Check-list: 207 (1989). Type: Zambia, Mwinilunga, Kalenda Plain, *Milne-Redhead* 3914 (BR; K, holotype; LISC; P; SRGH).

Perennial herb, with 1–many stems from a horizontal woody rootstock; stems erect, up to 1.5 m tall, simple or branched above, 4-angular, strongly ribbed, glabrous. Leaves mostly 1-foliolate, the uppermost reduced to undifferentiated linear structures; leaflets up to 5–15 × 0.3–2 cm, mostly linear to attenuate-lanceolate, the upper ones the narrower, the lowermost oblanceolate to oblong-obovate and smaller again, mostly long-apiculate, rather leathery, glabrous or with a few scattered hairs beneath; lateral nerves steeply ascending, the second pair running near the margin far into the upper half; petiole 1–3 mm long; stipules 3–10 mm long, linear-subulate. Racemes laxly 6–16-flowered; bracts linear-subulate, 1.5–3 mm long; bracteoles variously inserted on the pedicel, smaller. Calyx 7–8 mm long, glabrous; lobes narrowly attenuate-triangular, nearly twice as long as the tube, becoming reflexed. Standard elliptic, pale violet inside and marginally outside, yellowish-brown with deep brown veins medially outside, glabrous; wings deep violet, shorter than the keel; keel 1.8–2.1 cm long, angular, sharply rounded near the base with a long practically straight beak. Young legumes shortly stipitate, oblong-ellipsoid, sparsely pubescent. Seeds not seen.

Zambia. W: Mwinilunga Distr., west end of Kalenda Plain, fl. & fr. immat. 1.i.1938, *Milne-Redhead* 3914 (BR; K; LISC; P; SRGH).

Known only from the type gathering. Miombo woodland; c. 1300 m.

96. **Crotalaria angulicaulis** Harms in Warburg, Kunene-Samb.-Exped. Baum: 253 (1903). —E.G. Baker in J. Linn. Soc., Bot. **42**: 269 (1914). —Torre in C.F.A. **3**: 32 (1962). —Polhill, Crotalaria Africa & Madagascar: 232 (1982). —Lock, Leg. Afr. Check-list: 165 (1989). Type from Angola.

Erect annual herb, often with several stems from the base, up to 0.5–1.5 m tall; stems 3–4-angled, strongly ribbed to slightly winged, pubescent, sometimes soon glabrescent. Leaves mostly 1-foliolate, with a 1–6 mm long petiole, but the lowermost sometimes simple and subsessile, the uppermost often reduced to small subulate structures; leaflets up to 7–15(19) × 0.5–2 cm, linear-lanceolate to attenuately lanceolate or oblong-oblanceolate to oblong-obovate, pungent, rather leathery and somewhat glaucous, finely pubescent beneath, sometimes glabrescent; lateral nerves steeply ascending, the second pair running near the margin far into the upper half; stipules 2–5 mm long, subulate. Racemes terminal on the main and relatively short lateral branches, the latter subtended by reduced leaves, thus forming a pseudopanicle, lax, with few to numerous flowers; bracts 1.5–2 mm long, linear-subulate or linear-lanceolate; bracteoles inserted on the lower half of the pedicel, smaller. Calyx 4.5–6 mm long, glabrous to finely appressed pubescent; lobes subulate to narrowly triangular-acuminate, as long as or longer than the tube. Standard elliptic or elliptic-oblong, violet-blue inside, lined bluish-brown and at least sparsely pubescent outside; wings somewhat shorter than the keel; keel 0.8–1.1 cm long, angular, sharply bent in the lower part with a long beak, often slightly incurved towards the tip, often thinly pubescent at least towards the base. Pod shortly stipitate, 18–22 × 6–8 mm, oblong-ellipsoid or cylindrical, glabrous to finely appressed pubescent, 12–16-seeded. Seeds up to 4 mm long, oblong-cordiform, smooth, glossy.

Zambia. B: Kabompo to Chizela (Chizera) road, 1.5 km SW of Nkulwashi (Kabompo R. Pontoon), fl. & fr. 23.iii.1961, *Drummond & Rutherford-Smith* 7249 (K; SRGH).

Also in southern Dem. Rep. Congo and southern Angola. Woodland and grassland on Kalahari Sands; c. 900–1000 m.

97. **Crotalaria orientalis** Burtt Davy ex Verdoorn in Bothalia **2**: 417 (1928). —Burtt Davy, Fl. Pl. Ferns Transvaal, pt. 2: 401 (1932). —Schreiber in Merxmüller, Prodr. Fl. SW. Afrika, fam. 60: 25 (1970). —Polhill & Schreiber in Mitt. Bot. Staatssamml. München **12**: 173 (1975). —Polhill, Crotalaria Africa & Madagascar: 233 (1982). —Lock, Leg. Afr. Check-list: 194 (1989). Lectotype, selected by Polhill & Schreiber, loc. cit. (1975), from South Africa (Free State).

Erect virgate herb or subshrub, 0.4–1 m tall; branches glaucous, ribbed, subglabrous to sparsely strigulose. Leaves all 1-foliolate or a few middle ones 3-foliolate, estipulate; leaflets 7–50 × 0.8–9 mm, linear to linear-lanceolate or -oblanceolate, with basal lateral nerves ascending well into the upper half; petioles mostly 6–18 mm long. Racemes laxly 6–12-flowered; bracts 1–4 mm long, linear to linear-lanceolate; bracteoles variously inserted on the 6–10 mm long pedicel, small. Calyx 5–7 mm long, glabrous to sparsely puberulous; lobes subulate to narrowly triangular-acuminate, ± as long as the tube. Standard whitish inside, white or yellow and lined reddish-brown outside, glabrous to pubescent; wings yellow, shorter than the keel; keel (1.3)1.4–1.7 cm long, angled in the lower part with a long straight beak, darker at the tip. Pod subsessile, 10–16 × 6–7 mm, oblong-ellipsoid to shortly cylindrical, 4–8-seeded. Seeds c. 3 mm long, oblique-cordiform, papillate, brown.

Subsp. **orientalis**

Crotalaria spartioides sensu Harvey in F.C. **2**: 41 (1862) pro parte. —sensu E.G. Baker in J. Linn. Soc., Bot. **42**: 270 (1914) pro parte.

Leaflets usually mostly 0.8–2.5 mm wide. Standard glabrous to sparsely pubescent outside. Pods glabrous or early glabrescent (ovary glabrous to densely hairy).

Botswana. SE: 19 km Molepolole–Letlhakeng (Letlaking), fl. 15.ii.1960, *Wild* 4953 (K; M; SRGH); 32 km west of Kanye, fl. & fr. 18.i.1960, *Leach & Noel* 207 (K; LISC; SRGH).

Also in Namibia and South Africa (North-West Province, Free State, Northern Cape Province). Shrub savanna on Kalahari Sands; c. 1200 m.

Subsp. **allenii** (Verdoorn) Polhill & Schreiber in Mitt. Bot. Staatssamml. München **12**: 174 (1975). —Polhill, Crotalaria Africa & Madagascar: 234 (1982). —Lock, Leg. Afr. Check-list: 194 (1989). Lectotype, selected by Polhill & Schreiber, loc. cit. (1975): Zimbabwe, Victoria Falls, *C.E.F. Allen* 226 (K, lectotype; SRGH).

Crotalaria spartioides sensu E.G. Baker in J. Linn. Soc., Bot. **42**: 270 (1914) pro parte. —sensu Eyles in Trans. Roy. Soc. South Africa **5**: 372 (1916).

Crotalaria allenii Verdoorn in Bothalia **2**: 417 (1928). —Wild in Clark, Victoria Falls Handb.: 149 (1952). —Drummond in Kirkia **8**: 217 (1972).

Leaflets usually mostly 2–9 mm wide. Standard extensively pubescent outside. Pods persistently pubescent.

Zimbabwe. N: Gokwe South Distr., Charama Plateau, near turnoff on road from Gokwe to Charama, fl. 18.iii.1962, *Bingham* 175 (K; SRGH). W: Hwange National Park (Wankie Game Reserve), Gwayi Corridor (Gwaai), fl. & fr. 14.ii.1956, *Wild* 4723 (COI; K; L; LISC; SRGH); Nkayi Distr., Gwampa Forest Land (Reserve), fl. xii.1954, *Goldsmith* 119/55 (K; LISC; SRGH).

Also in Namibia (Okavango area) and South Africa (North-West and Northern Provinces). Grassland, often at edges of wet pans or "vleis", or *Baikiaea* woodland, on Kalahari Sand; 1200–1600 m.

98. **Crotalaria spartioides** DC., Prodr. **2**: 128 (1825). —Harvey in F.C. **2**: 40 (1862) pro parte. —E.G. Baker in J. Linn. Soc., Bot. **42**: 270 (1914) pro parte. —Verdoorn in Bothalia **2**: 417 (1928). —Polhill & Schreiber in Mitt. Bot. Staatssamml. München **12**: 175 (1975). —Polhill, Crotalaria Africa & Madagascar: 234 (1982). —Lock, Leg. Afr. Check-list: 204 (1989). Type from South Africa (Northern Cape Province).

Crotalaria virgultalis sensu E.G. Baker in J. Linn. Soc., Bot. **42**: 269 (1914) pro parte. —sensu Schreiber in Merxmüller, Prodr. Fl. SW. Afrika, fam. 60: 28 (1970) pro parte.

Erect virgate shrubby plant, 0.4–2 m tall; branches glaucous, ribbed, mostly rather slender, sparsely strigulose to finely appressed pubescent. Leaves 1-foliolate, estipulate; leaflets usually mostly 10–30 × 0.8–2.5 mm, needle-like, linear or less often linear-oblanceolate, subglabrous to shortly hairy; petiole 0.5–4(6) mm long. Racemes laxly 6–24-flowered; bracts 1–3 mm long, linear or linear-lanceolate; bracteoles usually on the upper part of the 3–6 mm long pedicel, minute. Calyx 5–7 mm long, subglabrous to appressed puberulous; lobes narrowly triangular or subulate, ± as long as the tube. Standard elliptic, bright yellow, usually at least sparsely hairy outside; wings shorter than the keel; keel angled in the lower part with a long straight beak, usually not darker coloured at the tip, 11–14 mm long. Pod subsessile, 10–12 × 5–6 mm, oblong-ellipsoid to shortly cylindrical, at least sparsely appressed puberulous, 6–8-seeded. Seeds 2.5–3 mm long, broadly oblique-cordiform, papillose, brownish.

Botswana. SW: Kgalagadi Distr., 5 km NW of Hukuntsi, fl. 23.ii.1969, *Blair Rains & Yalala* 2 (K; SRGH). SE: Kalahari Desert near Malichwë, fl. ii.1897, *E.J. Lugard* 232 (K).

Also in Namibia and South Africa (Northern Cape and North-West Provinces). Tree and shrub savanna on Kalahari Sand; 1000–1200 m.

Rather uniform in Botswana, but further west and south appears to have introgressed extensively with *Crotalaria virgultalis* Burch. ex DC., see Polhill & Schreiber, loc. cit. (1975).

99. **Crotalaria heidmannii** Schinz in Mém. Herb. Boissier No. 1: 128 (1900). —E.G. Baker in J. Linn. Soc., Bot. **42**: 334 (1914). —Verdoorn in Bothalia **2**: 411 (1928). —Torre in C.F.A. **3**: 53 (1962). —Schreiber in Merxmüller, Prodr. Fl. SW. Afrika, fam. 60: 24 (1970). —Drummond in Kirkia **8**: 218 (1972). —Polhill, Crotalaria Africa & Madagascar: 235 (1982). —Lock, Leg. Afr. Check-list: 181 (1989). Syntypes from Namibia.

Erect annual, 30–80 cm tall, much branched from the base upwards; branches ascending, terete, pubescent with longish rather stiff appressed hairs. Leaves 3-foliolate; leaflets 20–70 × 1.5–4 mm, linear-lanceolate, appressed pubescent at least beneath; petiole much shorter than leaflets; stipules 1–4 mm long, linear or subulate. Racemes laxly 4–6(12)-flowered; rhachis slender; bracts 1–2 mm long, linear; bracteoles lacking. Calyx 5–7 mm long, thinly appressed pubescent; lobes subulate, usually at least twice as long as the tube. Standard elliptic to oblong-obovate, yellow, lined brownish and thinly pubescent outside along the veins especially medially; wings shorter than the keel; keel 9–13 mm long, bent at right angles in the lower part, with a long practically straight beak. Pod shortly stipitate, (8)10–12 × 4.5–6 mm, oblong-ellipsoid to shortly cylindrical, appressed puberulous, 8–12-seeded. Seeds c. 2 mm long, bluntly triangular, rugulose, brown.

Botswana. N: 16 km west of Dukwe on Francistown–Maun road, fl. & fr. 22.ii.1966, *Blair Rains* 7 (K). **Zimbabwe**. W: Bulawayo, fl. i.1898, *Rand* 44 (BM). E: Chipinge, near Hot Springs, fl. & fr. immat. 10.iii.1969, *Corby* 2113 (K; SRGH).

Also in southern Angola and Namibia. Mopane woodland and savanna, on sandy soils; 900–1050 m.

Easily confused with large flowered forms of *C. sphaerocarpa*, but the racemes are laxer and fewer-flowered, the lateral surfaces of the keel glabrous, and the pods are larger with more numerous seeds.

100. **Crotalaria sphaerocarpa** Perr. ex DC., Prodr. **2**: 133 (1825). —J.G. Baker in F.T.A. **2**: 23 (1871). —E.G. Baker in J. Linn. Soc., Bot. **42**: 289 (1914) pro majore parte. —Eyles in Trans. Roy. Soc. South Africa **5**: 372 (1916). —Hutchinson, Botanist South. Africa: 501 (1946), but not p. 509. —Schreiber in Mitt. Bot. Staatssamml. München **2**: 290 (1957). —Hepper in F.W.T.A., ed. 2, **1**: 549 (1958). —Torre in C.F.A. **3**: 42 (1962). —Schreiber in Merxmüller, Prodr. Fl. SW. Afrika, fam. 60: 27 (1970). —Polhill in F.T.E.A., Leguminosae, Pap.: 942 (1971). —Drummond in Kirkia **8**: 218 (1972). —Polhill, Crotalaria Africa & Madagascar: 235, fig. 60 (1982). —Lock, Leg. Afr. Check-list: 204 (1989). Type from Senegal.

Subsp. **sphaerocarpa** —Polhill, Crotalaria Africa & Madagascar: 236, fig. 60 (1982). —Lock, Leg. Afr. Check-list: 204 (1989). TAB. 3,7: **34**.

Crotalaria nubica Benth. in Hooker, London J. Bot. **2**: 581 (1843). —J.G. Baker in F.T.A. **2**: 23 (1871). —Verdoorn in Bothalia **2**: 384 (1928). —Burtt Davy, Fl. Pl. Ferns Transvaal, pt. 2: 398 (1932). Lectotype, selected by J.G. Baker, loc. cit. (1971), from Sudan.

Crotalaria nutans Welw. ex Baker in F.T.A. **2**: 24 (1871). —E.G. Baker in J. Linn. Soc., Bot. **42**: 284 (1914). Type from Angola.

Crotalaria sphaerocarpa var. *lanceolata* Schinz in Verh. Bot. Vereins Prov. Brandenburg **30**: 160 (1888). —E.G. Baker in J. Linn. Soc., Bot. **42**: 290 (1914). —Schreiber in Mitt. Bot. Staatssamml. München **2**: 291 (1957). Type from Namibia.

Crotalaria cernua Schinz in Mém. Herb. Boissier No. 1: 127 (1900). —E.G. Baker in J. Linn. Soc., Bot. **42**: 297 (1914). —Torre in C.F.A. **3**: 47 (1962). Syntypes from Namibia.

Crotalaria sphaerocarpa var. *angustifolia* Hochst. ex Baker f. in J. Linn. Soc., Bot. **42**: 289 (1914). —Eyles in Trans. Roy. Soc. South Africa **5**: 372 (1916). —Schreiber in Mitt. Bot. Staatssamml. München **2**: 291 (1957). Type as for *Crotalaria nubica*.

Erect annual (10)30–100(170) cm tall, much branched from the base upwards, covered in longish appressed to somewhat spreading hairs. Leaves 3-foliolate; leaflets very variable, mostly 2–6 × 0.2–2 cm, linear to lanceolate, oblanceolate or elliptic, with longish ± appressed hairs at least beneath; petiole shorter than the leaflets; stipules 1.5–5 mm long, subulate to linear-lanceolate. Racemes laxly 12–many-flowered; bracts 1–1.5 mm long, linear-lanceolate; bracteoles on pedicel, inconspicuous. Calyx 2.5–3.5(4) mm long, pubescent; lobes subulate, 2–3 times as long as the tube. Standard elliptic to oblong-obovate, usually pale yellow, pubescent outside; wings shorter than the keel; keel (4)5–7(9) mm long, angled in the lower part with a long narrow beak, pubescent on the sides. Pod shortly stipitate, 5–7 × 3.5–4.5 mm, shortly oblong-ellipsoid or oblong-obovoid, 1–4(6)-seeded, pubescent. Seeds 2.5–3 mm long, bluntly triangular to very obliquely cordiform, rugulose to nearly smooth, brown.

Botswana. N: Nata R. near Madsiara Drift, fl. & fr. 21.iv.1957, *Drummond & S.C. Seagrief* 5180 (K; SRGH). SW: 27 km north of Kang, fl. & fr. 18.ii.1960, *Wild* 5033 (COI; K; SRGH). SE: Central Distr., Palapye, Molete (Malete), fl. & fr. 7.ii.1958, *de Beer* 698 (K; LISC; SRGH).

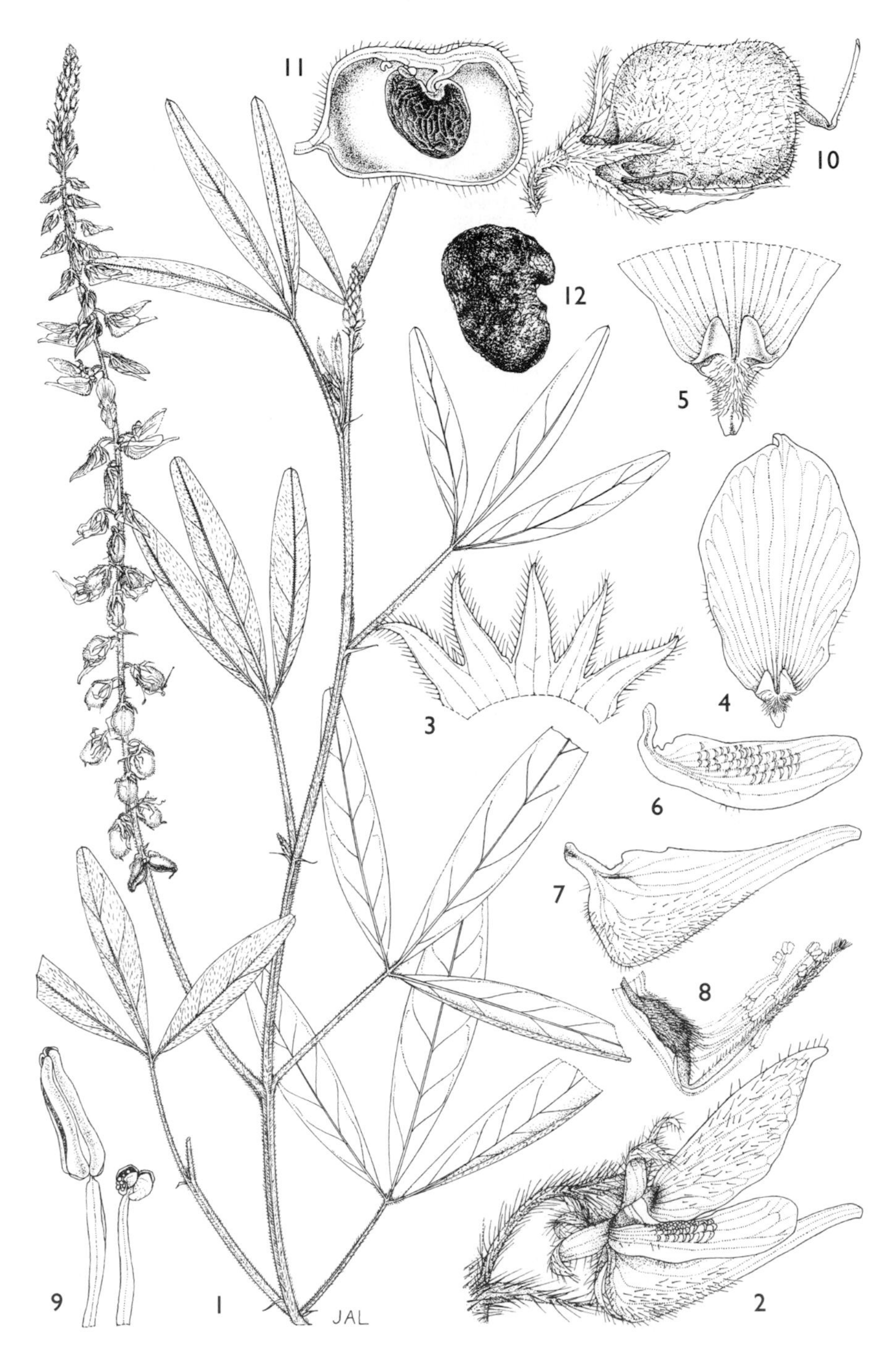

Tab. 3,7: **34**. CROTALARIA SPHAEROCARPA subsp. SPHAEROCARPA. 1, flowering and fruiting branch (× 1); 2, flower (× 6); 3, calyx, opened out (× 6); 4, standard (× 6); 5, base of standard (× 10); 6, wing (× 6); 7, keel (× 6); 8, androecium and gynoecium (× 6), 9, anthers (× 18); 10, pod (× 6); 11, valve of pod (× 6), 1–11 from *Polhill & Paulo* 1222; 12, seed (× 6), from *Wild* 6058. Drawn by Joanna Langhorne. From Crotalaria Africa &

Zambia. B: Sesheke Distr., Masese, fl. & fr. 10.v.1961, *Fanshawe* 6547 (BR; K; LISC; LMA; NDO; SRGH) – large-flowered form. C: Mkushi, fl. & fr. 2.v.1957, *Fanshawe* 3271 (BR; K; NDO; SRGH); Luangwa Game Reserve, Mfuwe, fl. & fr. 17.v.1965, *Mitchell* 2915 (K; SRGH). E: Luangwa Valley, fl. & fr. 24.iii.1963, *Verboom* 831 (SRGH). S: Namwala, fl. & fr. 9.i.1957, *E.A. Robinson* 2104 (K; SRGH). **Zimbabwe**. N: Hurungwe Distr., Urungwe Reserve, Musukwi (Msuku) R., fl. & fr. iv.1956, *R.M. Davies* 1868 (K; LISC; SRGH). W: Hwange (Wankie) Reserve, near Matetsi R., fl. & fr. 28.ii.1963, *Wild* 6058 (K; LISC; SRGH). C: Harare (Salisbury) Experimental Station, fl. & fr. 22.v.1943, *Arnold* in *GHS* 10093 (K; SRGH). E: Mutare Distr., near Odzi, fl. & fr. 11.iii.1969, *Corby* 2120 (K; SRGH). S: Beitbridge, fl. & fr. 16.ii.1955, *Exell, Mendonça & Wild* 435 (BM; LISC; SRGH). **Malawi**. S: Mangochi, near airfield, fl. & fr. 24.ii.1979, *Brummitt & Patel* 15455 (K; LISC; MAL; SRGH). **Mozambique**. MS: 5 km Chemba–Nhacolo (Tambara), fl. & fr. 23.iv.1960, *Lemos & Macuácua* 141 (BM; K; LISC; LMA; SRGH). GI: R. Chicomo, near Chicomo, fl. & fr. 8.xii.1944, *Mendonça* 3333 (BM; K; LISC). M: Matutuíne (Bela Vista) to Salamanga, fl. & fr. 23.iv.1948, *Torre* 7700 (BM; K; LISC).

Tropical Africa, in savanna areas of moderately low rainfall, in a narrow northern zone (c. 12°N) from Senegal to the Sudan, then throughout the Zambezian Domain to 30°S, disjunct between. Woodland and savanna, usually on sand (sometimes on loam or clay), along rivers, on dunes and waste ground; 0–1350 m.

The various varieties based on differences in leaflet shape seem worthless. In a zone of varying width either side of the Cunene and upper Zambezi Valleys plants with large flowers (keel 7–9 mm long) and ovules varying freely from 2–6 in number are quite frequent. Some incipient divergence seems to be occurring, but is insufficiently marked for any formal taxonomic recognition to be practicable. The names *C. nutans* and *C. cernua* are based on larger flowered specimens from this region.

Subsp. *polycarpa* (Benth.) Hepper is entirely glabrous and restricted to Senegal.

101. **Crotalaria microcarpa** Hochst. ex Benth. in Hooker, London J. Bot. **2**: 573 (1843). —J.G. Baker in F.T.A. **2**: 16 (1871). —Taubert in Engler, Pflanzenw. Ost-Afrikas **C**: 204 (1895). —E.G. Baker in J. Linn. Soc., Bot. **42**: 402 (1914). —Verdoorn in Bothalia **2**: 401 (1928). —Wilczek in F.C.B. **4**: 145 (1953). —Hepper in F.W.T.A., ed. 2, **1**: 549 (1958). —Torre in C.F.A. **3**: 70 (1962). —Binns, First Check List Herb. Fl. Malawi: 79 (1968). —Polhill in F.T.E.A., Leguminosae, Pap.: 943 (1971). —Drummond in Kirkia **8**: 218 (1972). —Jacobsen in Kirkia **9**: 160 (1973). —Polhill, Crotalaria Africa & Madagascar: 237 (1982). —Lock, Leg. Afr. Check-list: 191 (1989). Type from Sudan.

Much branched spreading annual or short-lived perennial, 10–70 cm tall, covered with longish appressed or spreading hairs. Leaves shortly petiolate, 3-foliolate; leaflets 10–30 × 3–8 mm, narrowly lanceolate to oblanceolate, pilose on both surfaces; stipules 0.5–2 mm long, setaceous. Flowers few to numerous in heads or short dense (rarely lax) racemes on a slender relatively well developed peduncle; bracts 1.5–3.5 mm long, linear or linear-lanceolate; bracteoles inserted at the base of the calyx, smaller. Calyx 2.5–3.5 mm long, pilose; lobes triangular, ± as long as the tube. Standard elliptic-obovate, yellow, usually marked (or flushed) reddish, pubescent outside; wings ± as long as the keel; keel 4.5–7.5 mm long, angled in the lower part, with a straight or slightly recurved narrow beak. Pod sessile, 4–7.5 × 2–3 mm, ellipsoid-oblong or conoid-oblong, pubescent, c. 8–12-seeded. Seeds c. 1–1.2 mm long, oblique-cordiform, smooth, brown.

?**Botswana**. Latitude 23°S, fl. & fr., *Chapman & Baines* (K). **Zambia**. B: Sesheke Distr., Machili, fl. & fr. 14.iii.1961, *Fanshawe* 6437 (K; NDO; SRGH). C: Lusaka Distr., Iolanda, north bank of Kafue R. near Kafue Town, fl. & fr. 14.iii.1965, *E.A. Robinson* 6430 (K; SRGH). E: Katete, fl. & fr. 24.iii.1955, *Exell, Mendonça & Wild* 1156 (BM; LISC; SRGH). S: Mazabuka, fl. & fr. 28.ii.1963, *van Rensburg* 1518 (K; SRGH). **Zimbabwe**. N: 10 km NW of Makuti, fl. & fr. 14.iii.1961, *Drummond & Rutherford-Smith* 6891 (BR; K; LISC; SRGH). C: Goromonzi Distr., Chinamora (Chindamora), fl. iv.1953, *R.M. Davies* 494 (K; SRGH). E: Mutare Distr., Marange C.L. (Maranke Reserve), fl. & fr. 10.ii.1953, *Chase* 4766 (BM; K; SRGH). S: Masvingo Distr., Makaholi Experimental Farm, fl. & fr. 13.iii.1948, *D.A. Robinson* 308 (K; SRGH). **Malawi**. N: Karonga Distr., 24 km north of Chilumba, Ngara (Ngala), fl. & fr. 22.iv.1969, *Pawek* 2281 (K). C: Lilongwe Distr., Chankhandwe Dambo, fl. & fr. 11.iv.1956, *Jackson* 1835 (K). S: Shire R. near Liwonde, fl. 13.iii.1955, *Exell, Mendonça & Wild* 837 (BM; BR; LISC; SRGH). **Mozambique**. N: entre Cuamba e Mutuáli, R. Lúrio, fl. & fr. 24.iv.1961, *Balsinhas & Marrime* 432 (BM; COI; K; LISC; LMA; SRGH).

Widespread in savanna regions of tropical Africa from Mali to Ethiopia and south to Angola, Zimbabwe and ?Botswana. In a variety of habitats, but generally in open places on poor soils, in rocky or sandy sites or on hard pans, often ruderal; 150–1250(2250) m.

Also recorded for Mozambique, Tete, by Gonçalves in Garcia de Orta, Sér. Bot. **5**: 68 (1982).

102. **Crotalaria bongensis** Baker f. in J. Linn. Soc., Bot. **42**: 256 (1914) excl. var. *shirensis*. —Wilczek in F.C.B. **4**: 96 (1953). —Hepper in F.W.T.A., ed. 2, **1**: 548 (1958). —Torre in C.F.A. **3**: 26 (1962). —Polhill in F.T.E.A., Leguminosae, Pap.: 945 (1971); Crotalaria Africa & Madagascar: 240 (1982). —Lock, Leg. Afr. Check-list: 168 (1989). Syntypes from Sudan.

Shortly erect annual, 10–45 cm tall, generally with many spreading branches, particularly from the base, hirsute. Leaves estipulate, 1-foliolate; leaflets 20–80 × 2–6(8) mm, linear to oblong-lanceolate, pilose, at least beneath; petiole 2–4 mm long. Racemes laxly 3–12-flowered; bracts up to 1.5 mm long, setaceous; bracteoles near the top of the pedicel, inconspicuous. Calyx 2–2.5 mm long, pilose, often with purplish nerves; lobes triangular, slightly acuminate, ± as long as the tube. Standard elliptic-obovate, yellow, veined reddish-purple, glabrous outside; wings shorter than the keel; keel 4–5 mm long, angled in the lower part, with a long narrow beak. Pod sessile, 7–10 × 2–3 mm, oblong-ellipsoid to shortly cylindrical, pubescent, c. 12–16-seeded. Seeds c. 1.5–1.8 mm long, oblique-cordiform, smooth, brown.

Zambia. N: Mbala Distr., Iruna Village, fl. & fr. 18.iv.1959, *Richards* 11274 (K); Kasama to Mpika road, Chambeshi Pontoon, fr. 29.iv.1962, *Richards* 16388 (K; SRGH).

Tropical Africa, in higher rainfall savanna zone around the Congo Basin, from Liberia and Nigeria to southern Sudan, south through Uganda, W Kenya, Tanzania to Dem. Rep. Congo (Katanga) and NE Angola. Short grassland and disturbed places, often in sandy seasonally damp spots; 1200–1750 m.

103. **Crotalaria simoma** Polhill, Crotalaria Africa & Madagascar: 240, fig. 61 (1982). —Lock, Leg. Afr. Check-list: 203 (1989). Type: Zambia, Mbala–Mpulungu, close to Inona (Inono) Bridge, *Richards* 5591 (BR; K, holotype; SRGH).

Erect laxly branched graceful annual 25–45 cm tall; stem slender, inconspicuously appressed puberulous, glabrescent. Leaves subsessile, simple; blades 30–60 × 1–2 mm, linear, the upper ones mostly with inrolled margins, sparsely puberulous beneath, glabrescent; stipules 1–2 mm long, setaceous. Racemes numerous, terminal, with 1–3 laxly inserted flowers above a long very slender peduncle; bracts 1–2 mm long, setaceous; bracteoles at top of the pedicel, similar. Calyx 5–6 mm long, appressed puberulous; lobes subulate, twice as long as the tube. Standard elliptic, yellow, lined brown and pubescent outside; wings brighter yellow, much shorter than the keel; keel 11–14 mm long, angled in the lower third, with a long narrow beak incurved slightly towards the darker lined tip. Pod stipitate, 10–14 × 3–4 mm, narrowly subcylindrical, only slightly broadened apically, appressed pubescent, c. 40-seeded; stipe c. 3 mm long. Seeds 1.2–1.5 mm long, cordiform, slightly rugulose, brown.

Zambia. N: Mbala–Mpulungu, near Inona (Inono) Stream, fl. 5.iv.1955, *Richards* 5322 (BR; K; SRGH).

Known only from the Mbala District of Zambia between Mbala and Mpulungu. Roadside, sandy soil; 1050–1300 m.

104. **Crotalaria vanmeelii** R. Wilczek in Bull. Jard. Bot. État **23**: 131 (1953). —Polhill in F.T.E.A., Leguminosae, Pap.: 945 (1971); Crotalaria Africa & Madagascar: 241 (1982). —Lock, Leg. Afr. Check-list: 209 (1989). Type: Zambia, Mbala Distr., Mpulungu, *van Meel* 1309 (BR, holotype).

Erect laxly branched graceful annual, 30–80 cm tall; stem slender, thinly pubescent with short appressed or rather spreading hairs. Leaves subsessile, simple; blade up to 3–6 × 0.2–0.6 cm, linear to lanceolate-oblong, thinly appressed pubescent beneath; stipules up to 1.5 mm long, setaceous. Racemes laxly 3–12-flowered; bracts 1–3 mm long, setaceous; bracteoles at the top of the pedicel, similar. Calyx 4–7 mm long, appressed or rather spreading pubescent; lobes subulate, 2–3 times as long as the tube. Standard elliptic, yellow, veined brown, turning purplish overall, pubescent outside; wings much shorter than the keel; keel 8.5–12 mm long, bent at right angles in the lower third, with a long straight narrow beak. Pod sessile, 10–15 × 3–3.5 mm, narrowly cylindrical, thinly pubescent, up to 22-seeded. Seeds 1.5–2 mm long, cordiform, rugulose, light brown.

Zambia. N: Kaputa Distr., escarpment road to Sumbu, fl. & fr. 5.iv.1957, *Richards* 9039 (BR; K; SRGH); Mbala Distr., Namkolo (Niamkolo), fl. 22.iii.1960, *Richards* 12778 (BR; K; LISC; SRGH).

Also in Tanzania (Ufipa); known only from around the southern end of Lake Tanganyika. Open disturbed places on sandy soils; 750–1500 m.

105. **Crotalaria decora** Polhill in Kew Bull. **22**: 296, t. 28 (1968); in F.T.E.A., Leguminosae, Pap.: 945 (1971); Crotalaria Africa & Madagascar: 241 (1982). —Lock, Leg. Afr. Check-list: 174 (1989). Type: Zambia, Mweru Wantipa, Kangiri, *Richards* 9088 (BR; EA; K, holotype; SRGH).

Erect laxly branched graceful annual up to 55 cm tall; stem slender, sparsely appressed pubescent. Leaves subsessile, simple; blades up to 15–30 × 1.5–5 mm, linear-oblanceolate to oblong-oblanceolate, thinly appressed pubescent beneath; stipules c. 1–1.5 mm long, setaceous. Racemes numerous, laxly 3–12-flowered; rhachis filiform; bracts 0.5–1 mm long, setaceous, bracteoles at top of the pedicel, very small. Calyx 2–3 mm long, thinly pubescent or practically glabrous on the tube; lobes subulate or acuminately triangular, a little longer than the tube. Standard elliptic-oblong, yellow, finely veined brown, glabrous outside; wings distinctly shorter than the keel; keel 6–8 mm long, bent at right angles in the lower third, with a long narrow almost straight to slightly recurved beak. Pod subsessile, 5–8 × 2–2.5 mm, shortly cylindrical, thinly appressed pubescent, c. 8–12-seeded. Seeds c. 1.5 mm long, cordiform, slightly rugulose, brown.

Zambia. N: Kaputa Distr., Mweru Wantipa, near Muzombwe, fl. & fr. 15.iv.1961, *Phipps & Vesey-FitzGerald* 3209 (K; SRGH); Kalungwishi R., Lumangwe Falls, fl. & fr. 14.iv.1989, *Goyder, Pope & Radcliffe-Smith* 3026 (K).

Also in Tanzania (Ufipa). Wet grassland and drier sandy places in miombo or mateshi thicket; 1050–1100 m.

106. **Crotalaria shirensis** (Baker f.) Milne-Redh. in Kew Bull. **15**: 159 (1961). —Torre in C.F.A. **3**: 26 (1962). —Binns, First Check List Herb. Fl. Malawi: 80 (1968). —Polhill in F.T.E.A., Leguminosae, Pap.: 946, fig. 130/1–11 (1971). —Drummond in Kirkia **8**: 218 (1972). —Moriarty, Wild Fl. Malawi: 128, t. 64, fig. 2 (1975). —Polhill, Crotalaria Africa & Madagascar: 242, fig. 62/1–11 (1982). —Lock, Leg. Afr. Check-list: 203 (1989). TAB. 3,7: **35**, fig. A. Type: Malawi, Shire Highlands, *Buchanan* 57 (K, holotype).

Crotalaria bongensis var. *shirensis* Baker f. in J. Linn. Soc., Bot. **42**: 256 (1914). —Verdoorn in Bothalia **2**: 415 (1928).

Small erect annual, 5–25(40) cm tall, sometimes much branched; stem very slender, with small scattered appressed or spreading hairs. Leaves subsessile, simple; blades c. 2–6(8) × 0.2–0.5 cm, mostly linear-oblong to narrowly elliptic, but the basal ones often broader, elliptic to ovate, sparsely puberulous beneath; stipules up to 1–3 mm long, setaceous. Racemes slenderly pedunculate, lax, mostly 1–3-flowered; bracts up to 1 mm long, setaceous; bracteoles at the top of the pedicel, similar. Calyx 2.5–4 mm long, puberulous; lobes acuminately triangular, a little longer than the tube. Standard elliptic, pale yellow, with a darker red-fringed mark at the base inside, reddish veined and glabrous outside; wings brighter yellow, shorter than the keel; keel 4–6 mm long, bent at right angles in the lower third, with long narrow forwardly projecting beak. Pod 8–13 × 2–3 mm, sessile, narrowly cylindrical, rather sparsely pubescent, c. 15–20-seeded. Seeds c. 1 mm long, cordiform, smooth to slightly rugulose, reddish-brown.

Zambia. N: Mbala Distr., Chilongowelo Escarpment, fl. & fr. 6.iv.1962, *Richards* 16268 (K; SRGH). W: Kitwe, fl. & fr. 10.iii.1963, *Fanshawe* 7737 (K; NDO; SRGH). C: Lusaka Distr., 100–129 km east of Lusaka, Chakwenga Headwaters, fl. & fr. 5.iii.1965, *E.A. Robinson* 6391 (K; M; SRGH). E: Chipata Distr., Lunkwakwa Forest Reserve, fl. & fr. 30.iii.1963, *Verboom* 692 (K; SRGH). **Zimbabwe**. N: Hurungwe Distr., Mwami (Miami), fl. & fr. iv.1926, *Rand* 23 (BM). W: locality and date not recorded, *Drummond* 6794 (SRGH). C: Goromonzi Distr., Chinamora Reserve, Domboshawa Hills, fl. & fr. 3.iv.1965, *Corby* 1300 (K; SRGH) — not typical. **Malawi**. N: Rumphi Distr., Livingstonia Escarpment, fl. & fr. 24.iv.1969, *Pawek* 2341 (K); Kasungu Distr., Chimaliro Forest, fl. & fr. 14.iv.1971, *Pawek* 4657 (K). S: Zomba Distr., Malosa, fl. & fr. 10.iii.1955, *Exell, Mendonça & Wild* 784 (BM; LISC; SRGH). **Mozambique**. N: 28 km Ribáuè–Malema, fl. & fr. 21.iii.1964, *Correia* 221 (LISC).

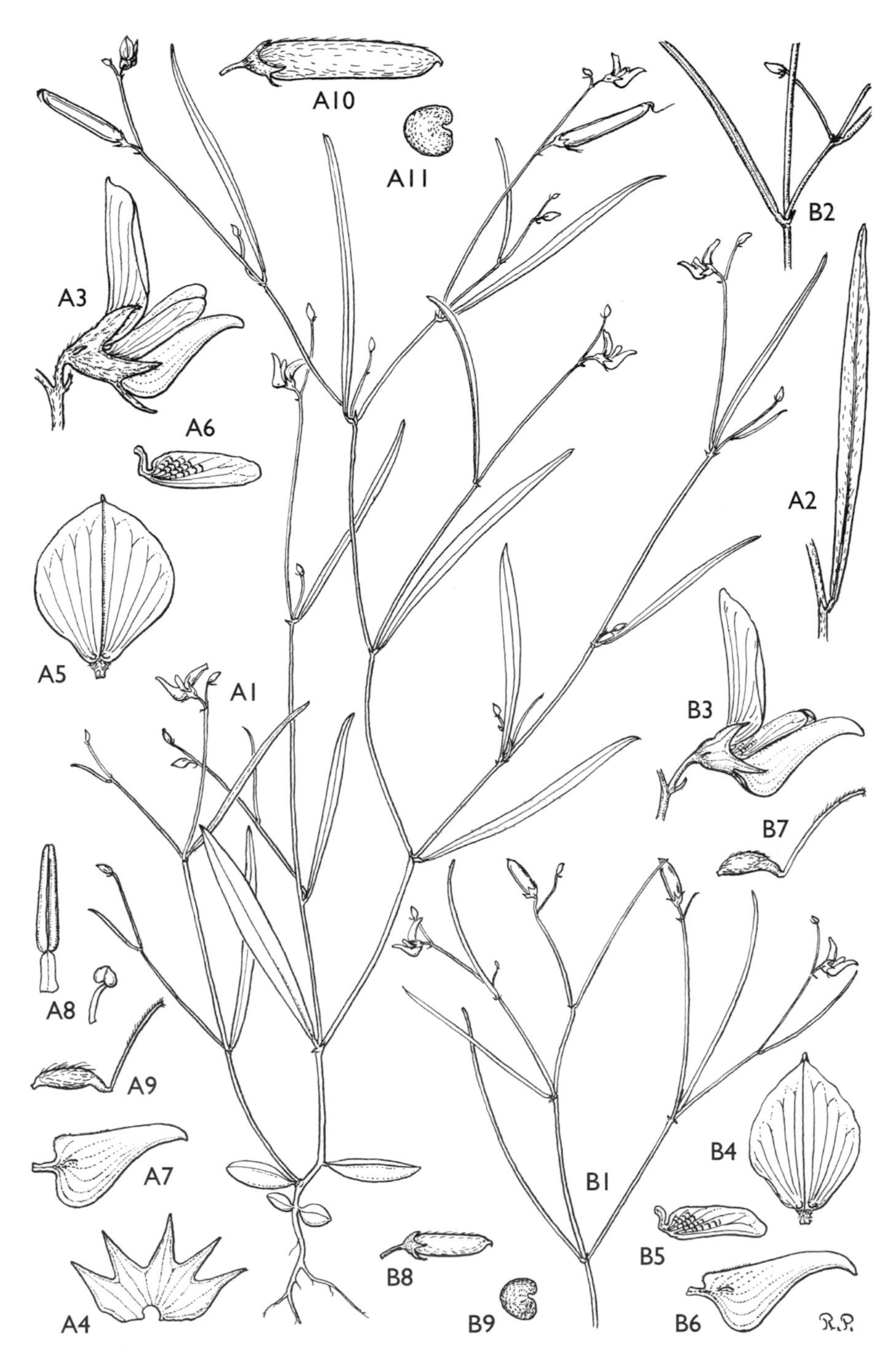

Tab. 3,7: **35**. A. —CROTALARIA SHIRENSIS. A1, habit (× 1); A2, leaf (× 2); A3, flower (× 4); A4, calyx, opened out(× 4); A5, standard (× 4); A6, wing (× 4); A7, keel (× 4); A8, anthers (× 20); A9, gynoecium (× 40), A1–A9 from *Milne-Redhead & Taylor* 9500; A10, pod (× 2), from *Stolz* 1989; A11, seed (× 6), from *E.A. Robinson* 6637. B. —CROTALARIA TERETIFOLIA. B1, flowering and fruiting branch (× 1); B2, detail of same (× 2); B3, flower (× 4); B4, standard (× 4); B5, wing (× 4); B6, keel (× 4); B7, gynoecium (× 4); B8, pod (× 2); B9, seed (× 6), B1–B9 from *Milne-Redhead & Taylor* 9327B. Drawn by Roger Polhill. From F.T.E.A.

Also in Cameroon, Central African Republic, Burundi, W Kenya, Tanzania, Dem. Rep. Congo (Katanga) and Angola. Miombo woodland, short grassland and disturbed places, also pockets of soil on rock outcrops; on sandy and clay soils; 500–2000 m.

The differences between *C. shirensis* and *C. minutissima* are slight, but seem to hold where the ranges overlap. At the southern extreme of the range of *C. shirensis* in central Zimbabwe, outside the area of *C. minutissima,* the size of the flowers and pods reach their lowest dimensions, thus approaching *C. minutissima* morphologically, but presumably as a result of independent divergence.

107. **Crotalaria teretifolia** Milne-Redh. in Kew Bull. **15**: 161 (1961). —Polhill in F.T.E.A., Leguminosae, Pap.: 946, fig. 130/12–20 (1971); Crotalaria Africa & Madagascar: 243, fig. 62/12–20 (1982). —Lock, Leg. Afr. Check-list: 207 (1989). TAB. 3,7: **35**, fig. B. Type from Tanzania.

Inconspicuous erect wiry annual, up to 25 cm tall, sometimes much branched; stem very slender, glabrous. Leaves subsessile, simple; blades c. 20–40 × 0.3–0.8 mm, mostly terete, with a median groove along the upper surface, sometimes linear-oblanceolate, glabrous; stipules up to 1 mm long, setaceous. Flowers and fruits similar to *C. shirensis.* Calyx 1.5–2.5 mm long, with a few appressed hairs. Keel 4.5–5.5 mm long. Pod (6)7–8 × 2 mm, c. 12–15-seeded.

Zambia. N: Mbala Distr., Nkali Dambo, fl. & fr. 4.iv.1955, *Richards* 5300 (K; LISC; SRGH). W: Solwezi Distr., Zambia/Dem. Rep. Congo border, Mulenga Protected Forest Area, fl. & fr. 19.iii.1961, *Drummond & Rutherford-Smith* 7066 (BR; K; LISC; M; SRGH). C: 55 km NE of Serenje, fl. & fr. 2.iii.1962, *E.A. Robinson* 4981 (K). **Malawi**. N: 16 km east of Chitipa, Kaseye Mission, fl. & fr. 5.iv.1969, *Pawek* 1955 (K). **Mozambique**. N: Mandimba–Massangulo, fl. & fr. 5.iv.1961, *Carvalho* 486 (K).

Also in S Tanzania and Dem. Rep. Congo (Katanga). Miombo woodland, short grassland and disturbed places, on sandy soils; c. 900–1800 m.

108. **Crotalaria minutissima** Baker f. in J. Linn. Soc., Bot. **42**: 255, t. 9 (1914). —Wilczek in F.C.B. **4**: 91 (1953). —Polhill, Crotalaria Africa & Madagascar: 243 (1982). —Lock, Leg. Afr. Check-list: 192 (1989). Type from Dem. Rep. Congo (Katanga).

Small erect annual, 10–35 cm tall, sometimes much branched; stem very slender, sparsely appressed puberulous. Leaves sessile, simple; blades mostly 15–35 × (0.8)1–3 mm, linear or linear-oblanceolate, with the margins sometimes partially inrolled, the lower ones apparently always narrow, thinly appressed puberulous beneath; stipules 0.2–0.5 mm long, setaceous. Flowers and fruits similar to *C. shirensis*, but smaller. Calyx 1.5–2 mm long. Keel 3–4 mm long. Pod 6–7.5(8) × 1.8–2 mm, thinly puberulous.

Zambia. N: 95 km east of Kasama, fl. & fr. 3.iv.1961, *E.A. Robinson* 4576 (BM; BR; K; LISC; M; SRGH). W: 7 km east of Chizela (Chizera), fl. & fr. 27.iii.1961, *Drummond & Rutherford-Smith* 7425 (K; P; SRGH).

Also in Dem. Rep. Congo (Katanga) and possibly in Burundi. Short grassland, laterite pans, soil pockets on rock outcrops, disturbed places; c. 1200–1500 m.

See note under *C. shirensis* above.

109. **Crotalaria calycina** Schrank in Pl. Rar. Hort. Monac.: t. 12 (1817). —J.G. Baker in F.T.A. **2**: 15 (1871). —E.G. Baker in J. Linn. Soc., Bot. **42**: 271 (1914). —Verdoorn in Bothalia **2**: 415 (1928). —Wilczek in F.C.B. **4**: 84, t. 4 (1953). —Hepper in F.W.T.A., ed. 2, **1**: 547 (1958). —Torre in C.F.A. **3**: 32 (1962). —Binns, First Check List Herb. Fl. Malawi: 78 (1968). —Polhill in F.T.E.A., Leguminosae, Pap.: 949 (1971); Crotalaria Africa & Madagascar: 250, fig. 64 (1982). —Lock, Leg. Afr. Check-list: 170 (1989). Type grown in Europe, origin uncertain.

Erect annual, 0.15–1 m tall, usually little branched; stem densely covered with long coarse mostly appressed hairs. Leaves simple; blade 4–15 × 0.4–1.5 cm, linear to oblong-lanceolate, acute, glabrous except along midvein above, appressed pilose beneath; petiole 1–4 mm long; stipules very small, obscured by indumentum. Flowers few in terminal racemes and in the upper axils; bracts 6–18 mm long, lanceolate, caudate; bracteoles near the top of the pedicel, similar. Calyx 2–2.3 cm long, accrescent, longer than the corolla, deeply divided into 2 lips, densely covered with long tawny spreading hairs; upper lobes oblong-lanceolate, 3–4 times as long as the tube. Standard elliptic-oblong, pale yellow, with a few hairs apically outside; wings nearly as long as the keel; keel 1.5–1.7 cm long, subangular, with a straight twisted beak. Pod sessile, 2–2.5 cm long, subcylindrical, glabrous, 16–36-seeded. Seeds 2.5–3 mm long, subtriangular, shiny, smooth, pale yellow to light brown.

Zambia. W: Kitwe, fl. & fr. 21.iii.1954, *Fanshawe* 988 (K; LISC; SRGH). S: Mumbwa Distr., near Nangoma, fl. & fr. 20.iii.1963, *van Rensburg* 1733 (K; SRGH). **Malawi**. C: Nkhota Kota Distr., Malunda Village, fl. & fr. 11.v.1986, *Patel & Kwatha* 3120 (K; MAL). S: Machinga Distr., SE of Ntaja, Nilankhandwe Dambo, fl. 16.ii.1979, *Blackmore, Brummitt & Banda* 438 (MAL). **Mozambique**. N: Lago Distr., andados 13 km de Metangula para Nova Coimbra, c. 600 m, fl. & fr. 29.ii.1964, *Torre & Paiva* 10930 (LISC). Z: Lugela Distr., Namagoa, fl. & fr. ii–iii.1943, *Faulkner* PRE 165 (COI; K; PRE; SRGH). MS: Gondola Distr., Matsinho (Chimoio), estrada para Tete, fl. & fr. 23.iii.1948, *Garcia* 707 (BM; K; LISC).

Widespread in tropical Africa, except S and NE, extending to Asia and N Australia. Seasonally wet grassland and disturbed places; 600–1300 m.

110. **Crotalaria occidentalis** Hepper in Kew Bull. **11**: 113, fig. 1 (1956); in F.W.T.A., ed. 2, **1**: 547 (1958). —Polhill, Crotalaria Africa & Madagascar: 253 (1982). —Lock, Leg. Afr. Check-list: 193 (1989). Type from Sierra Leone.

Erect, sometimes diffuse, annual, 15–50 cm tall; branches slender, appressed pubescent. Leaves subsessile, simple; blade (7)10–22 × 3–6 mm, oblong, rounded at both ends, appressed pubescent beneath; stipules mostly 1–2 mm long, subulate, but enlarged on flowering branches (leaflet proportionally reduced), 2–4 mm long, linear-lanceolate to obliquely lanceolate-acuminate. Racemes mostly on rather short lateral branches with modified leaves, 1–3 cm long, 1–2-flowered (second flower sometimes aborting), the axis filiform; bracts 1–1.5 mm long, linear or linear-subulate (upper one paired by a filiform rhachis-extension); bracteoles at top of a rather short pedicel, smaller. Calyx 3.5–4 mm long, deeply divided into 2 lips, appressed pubescent; lobes narrowly triangular-lanceolate, c. 3 times as long as the tube. Standard broadly elliptic, yellow, lined red, puberulous medially near apex outside; wings as long as the keel; keel 5 mm long, subangular, with a rather short twisted beak. Pod shortly stipitate, 1.2–1.6 × 0.35–0.45 cm, oblong-fusiform, glabrous, 10–16-seeded. Seeds 1.2–1.5 mm long, very obliquely cordiform, minutely pitted, greenish to reddish-brown.

Zambia. N: Kawambwa Distr., Mushota, fr. 28.vi.1961, *Astle* 765 (K; SRGH).

Also in Mali, ?Senegal, Guinea-Bissau, Sierra Leone and Cameroon; only the one record for Zambia. Dambo; c. 1200 m.

Possibly spread on the feet of migrating birds.

111. **Crotalaria juncea** L., Sp. Pl.: 714 (1753). —Binns, First Check List Herb. Fl. Malawi: 79 (1968). —Polhill in F.T.E.A., Leguminosae, Pap.: 950 (1971). —Drummond in Kirkia **8**: 218 (1972). —Polhill, Crotalaria Africa & Madagascar: 372, fig. 105 (1982). —Lock, Leg. Afr. Check-list: 184 (1989). Type from India.

Erect laxly branched annual, up to 1.5(3) m tall; stem ribbed, appressed pubescent. Leaves simple; blade 6–15 × 0.5–3 cm, oblong-lanceolate, finely appressed pubescent on both surfaces, more conspicuously so beneath; petiole 3–5 mm long; stipules 1–2 mm long, filiform. Racemes 10–50 cm long, laxly 6–20-flowered; bracts 3–5 mm long, elliptic, acuminate; bracteoles at base of the calyx, 2–5 mm long, linear. Calyx 1.6–2 cm long, slightly 2-lipped, brownish tomentellous, with longer hairs interspersed; upper lobes narrowly attenuate-triangular, up to 3–4 times as long as the tube. Standard elliptic to subcircular, bright yellow, faintly reddish marked or tinged, with scattered hairs outside; wings a little shorter than the keel; keel 1.7–2.2 cm long, subangular, with a long slightly incurved twisted beak. Pod subsessile, 3–5.5 × 1.2–1.7 cm, cylindrical, velvety tomentose, 6–12-seeded. Seeds 6–7 mm long, oblique-cordiform, with the radicular lobe strongly incurved, smooth to papillose around the hilum, dark brown to black.

Zambia. W: Ndola, fl. 20.iii.1954, *Fanshawe* 975 (BR; K). C: 11 km SE of Lusaka, fl. 6.i.1952, *Best* 11 (K). E: Petauke Distr., R. Nyamadzi, fl. 25.iii.1955, *Exell, Mendonça & Wild* 1171 (BM; LISC; SRGH). S: Choma to Masuku Mission, fl. 21.ii.1963, *van Rensburg* 1389 (K; SRGH).

Zimbabwe. N: Nyadekese Dam, fl. & fr. 12.iv.1972, *Loveridge* 1853 (K; SRGH). W: Inyathi (Inyati) Mission, fl. & fr. 23.iv.1947, *Keay* in *FHI* 21224 (K; SRGH). E: Chimanimani Distr., Melsetter Pasture Research Station, fl. 19.ii.1950, *F.R. Williams* 75 (SRGH). **Malawi**. S: Zomba, Nakawere, fl. 16.i.1935, *Clements* 435 (MAL). **Mozambique**. N: Mogovolas Distr., Nametil, fr. 12.vii.1948, *Pedro & Pedrógão* 4434 (LMA). Z: Lugela Distr., Namagoa Estate, fl. vi.1946, *Faulkner* K 72 (K). MS: Sussundenga Distr., Rotanda, fl. 17.xi.1746, *Pedro & Pedrógão* 261 (LMA).

An Indian species, widely grown as a green manure or for fibre (sunnhemp), becoming naturalized in places.

112. **Crotalaria tabularis** Baker f. in J. Linn. Soc., Bot. **42**: 324 (1914). —Verdoorn in Bothalia **2**: 399 (1928). —White, F.F.N.R.: 147 (1962). —Binns, First Check List Herb. Fl. Malawi: 80 (1968). —Polhill in Kew Bull. **22**: 307 (1968) excl. syn. *C. stanerana*; in F.T.E.A., Leguminosae, Pap.: 951 (1971); Crotalaria Africa & Madagascar: 258, fig. 66/2 and 67/2 (1982). —Lock, Leg. Afr. Check-list: 206 (1989). —F. White, Dowsett-Lemaire & Chapman, Evergreen For. Fl. Malawi: 324 (2001). Type: Malawi, Nyika Plateau, *Whyte* s.n. (K, holotype).

Shrub 1–3.5 m tall; young branches puberulous. Leaves 3-foliolate; leaflets mostly 4–9 × 2–3.5 cm, elliptic or obovate-elliptic, appressed puberulous beneath; petiole 3–11 cm long, slender; stipules filiform or lacking. Racemes sublaxly 8–30-flowered; bracts c. 3–9 mm long, linear, caducous; bracteoles near middle of pedicel or above, 1–6 mm long, filiform. Calyx 8–11 mm long, appressed puberulous; lobes narrowly triangular, slightly acuminate, lanate along margins inside. Standard broadly obovate-elliptic, bright yellow; wings usually as long as the keel; keel (1.3)1.5–2 cm long, rounded, with a slightly incurved twisted beak. Pod shortly stipitate, c. 4.5–6 × 1–1.3 cm, oblong-clavate, puberulous, becoming venose, c. 16–30-seeded. Seeds c. 4 mm long, oblique-cordiform, smooth, dull ochre.

Zambia. E: Isoka Distr., Nyika Plateau, Kangampande, fl. 6.v.1952, *White* 2732 (BM; FHO; K); Chama Distr., Chowo Forest, fl. 28.iv.1973, *Pawek* 6687 (K). **Malawi**. N: Chitipa Distr., Mugesse (Mughesse), fr. 13.ix.1977, *E. Phillips* 2841 (K; MO). **Mozambique**. N: Mts. east of Lago Niassa (Lake Nyasa), 1900, *W.P. Johnson* (K).

Also in Tanzania and Kenya. Montane forest edges; 1650–2150 m.

113. **Crotalaria macrocarpa** E. Mey., Comment. Pl. Afr. Austr.: 24 (1836). —Harvey in F.C. **2**: 45 (1862). —E.G. Baker in J. Linn. Soc., Bot. **42**: 324 (1914) pro parte. —Verdoorn in Bothalia **2**: 406 (1928). —Drummond in Kirkia **8**: 218 (1972). —Polhill, Crotalaria Africa & Madagascar: 261, fig. 67/5 (1982). —Lock, Leg. Afr. Check-list: 190 (1989). Type from South Africa (KwaZulu-Natal).

Subsp. **matopoensis** Polhill, Crotalaria Africa & Madagascar: 262 (1982). —Lock, Leg. Afr. Check-list: 190 (1989). Type: Zimbabwe, Matobo Distr., Kobila Reserve, *O.B. Miller* 1574 (K, holotype; MO; SRGH).

Shrub 2–3 m tall; branches finely strigulose (youngest parts with weaker spreading hairs as well). Leaves 3-foliolate; leaflets mostly 2–4 × 0.8–2 cm, elliptic to obovate, sparsely puberulous beneath; petiole 1.5–3.5 cm long, slender; stipules 0.5–2 mm long, linear-subulate, becoming recurved. Racemes fairly closely 6–16-flowered; bracts 1–2.5 mm long, subulate to linear-lanceolate, spreading, subpersistent; bracteoles on the pedicel, setaceous. Calyx 1–1.1 cm long, appressed puberulous; upper lobes narrowly triangular, 1–1.5 times as long as the tube. Standard elliptic or elliptic-obovate, yellow, later tinged brown; wings nearly as long as the keel, ciliolate proximally on the lower edge; keel 1.5–1.8 cm long, abruptly rounded in the lower half, with a slightly incurved twisted beak. Pod 3–4.5 × 1.4–1.8 cm, fatly ellipsoid, abruptly contracted to the 0.3–0.8 cm long stipe, appressed pubescent, c. 8–14-seeded. Seeds 6 mm long, reniform, granular, dark.

Zimbabwe. W: Matobo Distr., Matopos, Silozwi Hill (Mt. Silorzwe), i.1963, *Wild* 5969 (K; LISC; SRGH); Farm Chesterfield, fl. & fr. i.1956, *O.B. Miller* 3289 (K; PRE).

Known only from the Matopo Hills and adjacent parts of the Matobo District, in dense woodland and in rocky places; c. 1400–1500 m.

Subsp. *macrocarpa,* which occurs in lowland bushveld regions of KwaZulu-Natal, is a smaller plant, with rather longer pedicels and calyx lobes, glabrous wings and rather more numerous ovules.

114. **Crotalaria axillaris** Aiton, Hort. Kew. **3**: 20 (1789). —E.G. Baker in J. Linn. Soc., Bot. **42**: 388 (1914). —Verdoorn in Bothalia **2**: 389 (1928). —Wilczek in F.C.B. **4**: 138, t. 8 (1953). —Topham, Check List For. Trees Shrubs Nyasaland Prot.: 74 (1958). —Torre in C.F.A. **3**: 64 (1962). —Binns, First Check List Herb. Fl. Malawi: 78 (1968). —Polhill in F.T.E.A., Leguminosae, Pap.: 954 (1971); Crotalaria Africa & Madagascar: 268, fig. 67/6 (1982). —Lock, Leg. Afr. Check-list: 166 (1989). Type from Ghana, cultivated in England.
Crotalaria hildebrandtii Vatke in Oesterr. Bot. Z. **29**: 220 (1879). —Taubert in Engler, Pflanzenw. Ost-Afrikas **C**: 207 (1895). Type from Kenya.

Bushy herb or shrub, 1–3 m tall; branches rather thinly to densely covered with short appressed or crisped hairs. Leaves 3-foliolate; leaflets mostly 4–10 × 2–5 cm, elliptic, shortly hairy beneath, with appressed or more often slightly spreading hairs most dense along the nerves; petiole 3–10 cm long; stipules 0.5–2 mm long, linear. Flowers 2–6(12) in the axils, clustered or very shortly racemose; bracts up to 1.5 mm long, linear; bracteoles minute. Calyx 1–1.3(1.5) cm long, puberulous to pubescent; lobes narrowly attenuate-triangular, twice as long as the tube. Standard subcircular-obovate, clear yellow, tinged reddish with age, medially puberulous outside; wings as long as the keel; keel 1.5–1.8 cm long, rounded a little below the middle, with a twisted beak. Pod 4.5–6 × 0.8–1.5 cm, oblong-clavate, narrowed into a stipe 0.6–1(1.5) cm long, puberulous or pubescent, becoming venose, c. 16–20-seeded. Seeds 4–5 mm long, subreniform, smooth, orange-brown or buff.

Zambia. N: Mbala Distr., Lunzua Valley, above Kafukulu (Kafakulu) Village, fl. & fr. 5.iii.1955, *Richards* 4782 (K). **Malawi**. S: Zomba, fl. xi.1900, *Purves* 12 (K). **Mozambique**. N: Mueda Distr., andados 30 km de Nairoto (Nantulo) para Mueda, c. 310m, fl. & fr. 10.iv.1964, *Torre & Paiva* 11856 (LISC); Mossuril Distr., Goa I., fl. 26.i.1954, *Gomes e Sousa* 4171 (COI; K; LMA).

Quite widespread in parts of tropical Africa with a higher and more equable distribution of rainfall, from Ghana to southern Ethiopia, south to Angola and the Flora Zambesiaca area, but with a number of disjunctions in the range. Riverine forest or woodland, coastal bushland on coral; 0–1150 m.

Plants from the coastal region of eastern Africa, including specimens from Goa I., have a very short appressed white indumentum on the branches; inland the hairs are usually at least slightly spreading, often crisped, slightly yellowish and rather variable in density.

115. **Crotalaria recta** Steud. ex A. Rich., Tent. Fl. Abyss. **1**: 152 (1847). —J.G. Baker in F.T.A. **2**: 40 (1871). —E.G. Baker in J. Linn. Soc., Bot. **40**: 52 (1911); in J. Linn. Soc., Bot. **42**: 352 (1914). —Eyles in Trans. Roy. Soc. South Africa **5**: 371 (1916). —Verdoorn in Bothalia **2**: 393 (1928). —Brenan, Check-list For. Trees Shrubs Tang. Terr.: 416 (1949). —Wilczek in F.C.B. **4**: 153, t. 9 (1953). —Topham, Check List For. Trees Shrubs Nyasaland Prot.: 75 (1958). —Hepper in F.W.T.A., ed. 2, **1**: 550 (1958). —Torre in C.F.A. **3**: 56 (1962). —White, F.F.N.R.: 147 (1962). —Binns, First Check List Herb. Fl. Malawi: 79 (1968). —Polhill in F.T.E.A., Leguminosae, Pap.: 957 (1971). —Gonçalves in Garcia de Orta, Sér. Bot. **5**: 70 (1982). —Drummond in Kirkia **8**: 218 (1972). —Polhill, Crotalaria Africa & Madagascar: 272, fig. 73 (1982). —Lock, Leg. Afr. Check-list: 200 (1989). Syntypes from Ethiopia.
Crotalaria simplex A. Rich., Tent. Fl. Abyss. **1**: 153 (1847). —Verdoorn in Bothalia **2**: 395 (1928). Type from Ethiopia.
Crotalaria recta subsp. *simplex* (A. Rich.) Baker f. in J. Linn. Soc., Bot. **42**: 353 (1914).
Crotalaria recta var. *katangensis* Robyns ex R. Wilczek in Bull. Jard. Bot. État **23**: 148 (1953); in F.C.B. **4**: 156 (1953). Type from Dem. Rep. Congo (Katanga).

Perennial, usually with several erect sparingly branched stems, usually 1–2(3) m tall; stems hollow, ribbed, appressed pubescent. Leaves 3(rarely 5)-foliolate; leaflets mostly 2–14(20) × 1–6 cm, linear-lanceolate, oblanceolate-elliptic, elliptic or obovate, pubescent beneath; petiole (0.7)2–6 cm long; stipules (3)6–9 mm long, linear-lanceolate. Racemes 15–50 cm long, many-flowered; bracts 6–11 mm long, linear-lanceolate, caudate; bracteoles on the pedicel, 2–5 mm long, filiform. Calyx 9–12 mm long, glabrous or pubescent; lobes triangular, ± as long as the tube or longer. Standard subcircular, yellow, veined and suffused purplish at the base inside, glabrous or with a few hairs at the apex outside; wings oblong-obovate, longer than the keel, dark mark at the base; keel 1.2–1.5 cm long, rounded, with a short incurved twisted beak. Pod shortly stipitate, 4.5–6.5 × 1.5–2 cm, broadly oblong-clavate, glabrous (in the Flora Zambesiaca area), c. 26–34-seeded. Seeds c. 4 mm long, oblique-cordiform, with the narrow end strongly incurved, smooth or papillose around the hilum, iron-grey.

Zambia. N: Kasama Distr., Mungwi, fl. & fr. 14.iii.1962, *E.A. Robinson* 5025 (K; SRGH). W: 5 km east of Solwezi, near Kifubwa R., fl. 17.iii.1961, *Drummond & Rutherford-Smith* 6963 (K; SRGH). C: 14 km south of Lusaka, fl. & fr. 23.iii.1952, *White* 2327 (BM; FHO; K). **Zimbabwe**. C: Bromley, fl. 18.ii.1949, *Corby* 392 (K; SRGH). E: Mutare Distr., Chipondiomwe (Chipondomwe), fl. 13.iii.1955, *Chase* 5515 (BM; COI; K; LISC; SRGH). **Malawi**. N: Rumphi Distr., Nchenachena (Nchena-Nchena), fl. & fr. 24.iv.1969, *Pawek* 2349 (K). C: Lilongwe Distr., Chitedze, fl. 22.iii.1955, *Exell, Mendonça & Wild* 1124 (BM; LISC; SRGH). S: 10 km Limbe–Chiradzulu, fl. 18.iii.1964, *Salubeni* 291 (K; SRGH). **Mozambique**. N: Ngauma Distr., Massangulo, fr. 15.v.1948, *Pedro & Pedrógão* 3556 (LMA). Z: Montes do Gurué, fl. & fr. 8.iv.1943, *Torre* 5121 (BM; LISC). T: Moatize Distr., Montes de Zóbuè, fl. & fr. 3.x.1942, *Torre* 561 (K; LISC). MS: entre Chimoio (Vila Pery) e Garuso (Garuzo), fl. & fr. immat. 29.iii.1948, *Barbosa* 1266 (BM; K; LISC).

Plateaux and highlands of tropical Africa from Nigeria to Ethiopia, south to Angola, South Africa (Northern Province, Mpumalanga) and Swaziland. Forest margins, upland grassland, miombo in higher rainfall areas, extending into drier areas along streams and dambos, also in disturbed places; 1050–2000 m.

The calyx is usually glabrous or nearly so, but plants from near the Katanga–Zambia border have rather densely sericeous-pubescent calyces. These may be referred to var. *katangensis*, but the variety is not formally upheld here because plants with similarly hairy calyces occur occasionally in Nigeria and Ethiopia; in Ethiopia even the pods may be densely hairy on occasion.

The species has been tried occasionally as a green manure, but is subject to fungal attack by *Fusarium* and has not proved very successful.

116. **Crotalaria retusa** L., Sp. Pl.: 715 (1753). —J.G. Baker in F.T.A. **2**: 13 (1871). —E.G. Baker in J. Linn. Soc., Bot. **42**: 270 (1914). —Wilczek in F.C.B. **4**: 90 (1953). —Torre in C.F.A. **3**: 32 (1962). —Polhill in F.T.E.A., Leguminosae, Pap.: 958 (1971); Crotalaria Africa & Madagascar: 272 (1982). —Lock, Leg. Afr. Check-list: 201 (1989). Syntypes from Sri Lanka.

Erect annual or short-lived perennial, 0.5–1.5 m tall, usually well branched; branches slightly ribbed, subdensely covered with short hairs. Leaves simple; blade 3.5–11 × 1.5–4 cm, oblanceolate to oblong-obovate, finely appressed pubescent beneath; petiole c. 2–4 mm long; stipules 1–5 mm long, linear or subulate. Racemes 10–30 cm long, many-flowered; bracts c. 2–6 mm long, subulate-caudate to lanceolate-caudate; bracteoles on the pedicel, 1–2 mm long, filiform. Calyx 1.1–1.4 cm long, glabrous to appressed puberulous; upper lobes broadly lanceolate-triangular, longer than the tube. Standard obovate-circular, pale yellow, veined and outside often suffused reddish-purple, usually puberulous along the midvein outside; wings oblong-obovate, longer than the keel, bright yellow; keel 1.3–1.5 cm long, rounded, with a fairly short slightly incurved twisted beak. Pod shortly stipitate, (3)4–5 × 1–1.8 cm, oblong-clavate, glabrous, c. 12–20-seeded. Seeds c. 5–5.5 mm long, oblique-cordiform, with the narrow end strongly incurved, finely papillose, yellowish or brown.

Var. **retusa**

Branches with short appressed hairs. Racemes long, with the flowers closely arranged only towards the top. Calyx thinly but uniformly puberulous.

Mozambique. N: Meconta Distr., andados 12 km de Nametil para Nampula, margens do rio Metona, c. 210 m, fl. & fr. 1.iv.1964, *Torre & Paiva* 11550 (LISC). Z: Pebane Distr., 64 km Mualama–Nabúri, fl. & fr. 5.x.1949, *Barbosa & Carvalho* 4317 (K; LMA).

Pantropical, probably Asian in origin, widely distributed as a fibre plant and more recently as a green manure, readily naturalized. Also cultivated, e.g. Zimbabwe, Henderson Research Station, fl. & fr. 1.ii.1953, *Wild* 4000 (K; SRGH); Mozambique, Mocuba, Posto Agrícola, fl. 6.vi.1949, *Barbosa & Carvalho* 2979 (K; LMA).

Var. **tunguensis** (Pires de Lima) Polhill in Kew Bull. **22**: 311 (1968); in F.T.E.A., Leguminosae, Pap.: 958 (1971); Crotalaria Africa & Madagascar: 273, fig. 74 (1982). TAB. 3,7: **36**. Type: Mozambique, Niassa, near Palma, *Pires de Lima* 204 (PO, holotype).

Crotalaria tunguensis Pires de Lima in Brotéria, Sér. Bot. **19**: 120 (1921).

Branches with slightly spreading hairs. Racemes short and dense in flower, the rhachis elongating only as the pods begin to develop. Calyx glabrous, except at the base.

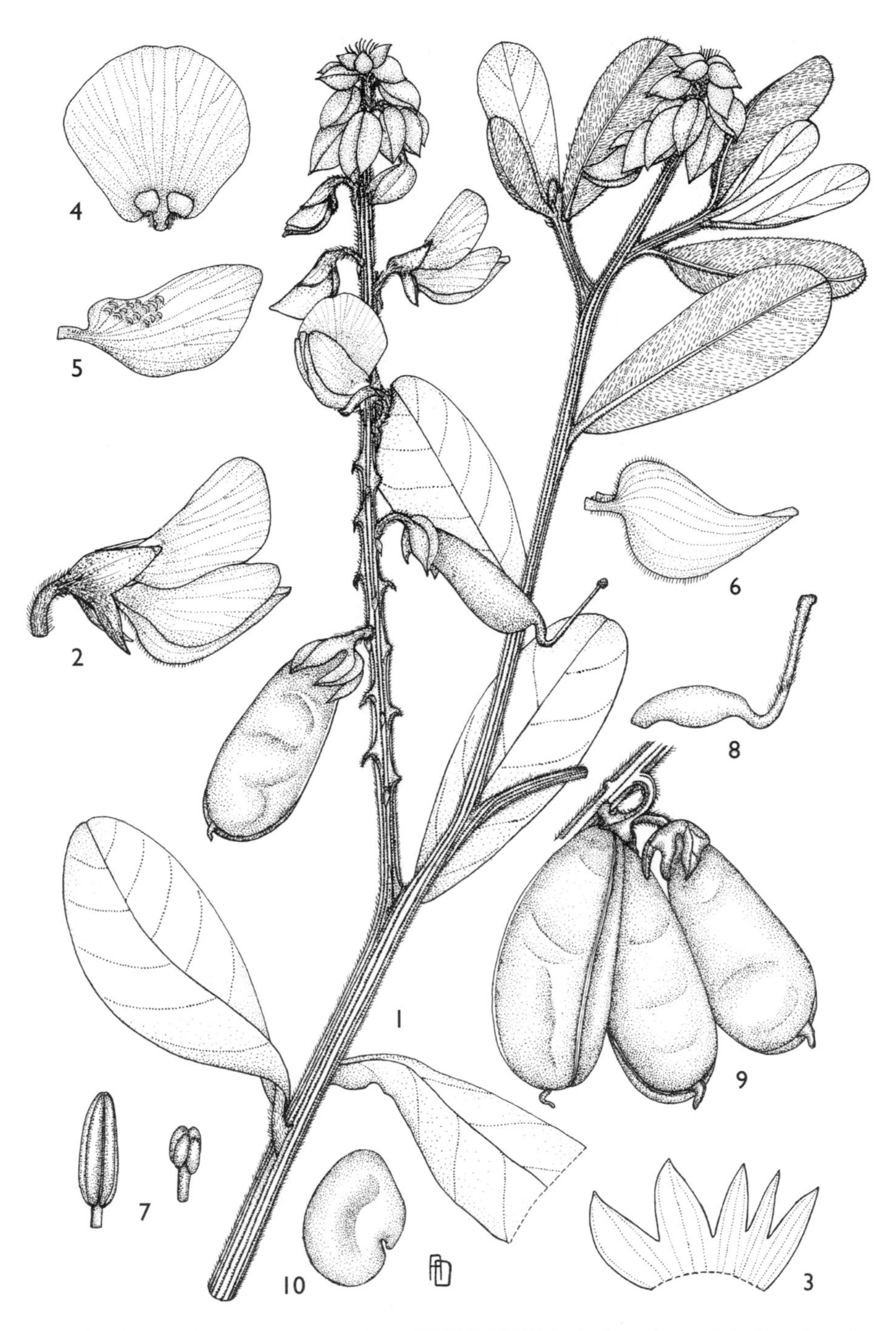

Tab. 3,7: **36**. CROTALARIA RETUSA var. TUNGUENSIS. 1, flowering and fruiting branch (× $^2/_3$); 2, flower (× $1^1/_3$); 3, calyx, opened out (× $1^1/_3$); 4, standard (× $1^1/_3$); 5, wing (× 2); 6, keel (× 2); 7, anthers (× 10); 8, gynoecium (× $1^1/_3$), 1–8 from *Leach & Rutherford-Smith* 10923; 9, pods (× $^2/_3$), from *Drummond & Hemsley* 3619; 10, seed (× 4), from *Leach & Rutherford-Smith* 10923. Drawn by Ann Davies. From Crotalaria Africa & Madagascar.

Mozambique. N: Nampula, fl. & fr. 15.ii.1937, *Torre* 1146 (COI; LISC); Mossuril Distr., Goa I., fl. & fr. 19.v.1961, *Leach & Rutherford-Smith* 10923 (LISC; SRGH).

Also along coasts of Somalia, Kenya and Tanzania, often in less disturbed places than var. *retusa*, apparently native. Coastal bushland, grassland and roadsides; 0–200 m.

117. **Crotalaria deserticola** Taub. ex Baker f. in J. Linn. Soc., Bot. **42**: 383 (1914). —Polhill in F.T.E.A., Leguminosae, Pap.: 960 (1971). —Drummond in Kirkia **8**: 217 (1972). —Polhill, Crotalaria Africa & Madagascar: 277, fig. 76/4 (1982). —Lock, Leg. Afr. Check-list: 174 (1989). Type from Tanzania.

Erect annual, usually 0.5–1.5 m tall; branches thinly appressed puberulous. Leaves 3-foliolate; leaflets variable, mostly 2–7.5 × 0.2–3 cm, linear to oblanceolate or elliptic, thinly appressed puberulous beneath; petiole 1–4(5) cm long; stipules 0.5–4 mm long, subulate-caudate to linear-lanceolate. Racemes laxly 8–24-flowered; bracts 1–3 mm long, linear to linear-lanceolate; bracteoles on the pedicel, setaceous. Calyx 3–7 mm long, glabrous to thinly pubescent; upper lobes narrowly triangular, longer than the tube. Standard broadly elliptic to subcircular, yellow usually veined reddish-brown, glabrous to medially puberulous outside; wings nearly as long as the keel; keel 8–11 mm long in the Flora Zambesiaca area (6–18 mm elsewhere), bent at right angles in the lower half, with a long narrow twisted beak. Pod (1.5)2–3(4) × 0.4–0.7(1.1) cm, oblong-clavate, narrowed into the 3–5(9) mm long stipe, appressed puberulous, c. 16–30-seeded. Seeds c. 2 mm long, smooth to slightly rugulose, yellowish to brown.

Zambia. N: Chinsali Distr., Shiwa Ngandu, fl. & fr. 14.i.1959, *E.A. Robinson* 3212 (K; SRGH). W: Ndola, fl. & immat. fr. 22.iii.1954, *Fanshawe* 1019 (BR; K; NDO). C: Mkushi, fl. & fr. 2.v.1957, *Fanshawe* 3273 (K; NDO). **Zimbabwe**. N: locality not recorded, iv.1965, *Wild* 7346 (SRGH). C: Hwedza (Wedza) Mt., cultivated at Marondera (Marandellas), fl. & immat. fr. 22.iv.1961, *Corby* 1305 (K; LISC; SRGH). **Malawi**. N: Chitipa Distr., Chisenga, fr. 12.vii.1970, *Brummitt* 12067 (K; LISC; MAL; PRE; SRGH). C: 19 km north of Nkhotakota (Nkhota Kota), north bank of Bua R. by road bridge, fl. & fr. 16.vi.1970, *Brummitt* 11454 (K; MAL; SRGH). **Mozambique**. N: 60 km west of Nampula, fl. & fr. 22.v.1961, *Leach & Rutherford-Smith* 10971 (K; SRGH).

Extends northwards to eastern Dem. Rep. Congo, southern Sudan and southern Ethiopia; common and much more variable in East Africa, where a variety and a subspecies are recognized. Grassland, termite mounds, stream banks and disturbed places in deciduous woodland; 250–1750 m.

118. **Crotalaria miranda** Milne-Redh. in Kew Bull. **15**: 165 (1961). —Polhill in F.T.E.A., Leguminosae, Pap.: 965 (1971); Crotalaria Africa & Madagascar: 279 (1982). —Lock, Leg. Afr. Check-list: 192 (1989). Type: Zambia, Mbala, Ndundu to Kawimbe, *McCallum-Webster* 870 (K, holotype).

Annual, erect to 70 cm, laxly branched; branches strigulose with colourless hairs having a brown point of attachment a little above the base. Leaves simple; blade 4–9 × 0.3–1 cm, linear-lanceolate to oblong-elliptic, thinly strigulose with submedifixed hairs beneath and usually also above; petiole 1–3 mm long; stipules 1–2 mm long, linear-subulate. Racemes laxly (4)8–20(26)-flowered; bracts 1–4 mm long, linear to linear-lanceolate, bracteoles usually adnate to the calyx, 1–2.5 mm long, linear. Calyx 5–7 mm long, strigulose; upper lobes narrowly attenuate-triangular, ± twice as long as the tube. Standard obovate, yellow, lined reddish, puberulous apically outside; wings shorter than the keel; keel 1.1–1.3 cm long, angular with a long straight twisted beak. Pod 2.2–2.8 × 0.5–0.6 cm, oblong-clavate, narrowed to the 2–4 mm long stipe, strigulose, c. 20–26-seeded. Seeds 2–2.5 mm long, oblique-cordiform, smooth, greyish or brownish-green.

Zambia. N: Mbala Distr., Ndundu to Nkali Dambo, fl. & fr. 7.iv.1959, *Richards* 11201 (K; LISC); Mbala Distr., Kawimbe–Ndundu, fl. & fr. 2.v.1959, *Richards* 11364 (K).

Also in Tanzania. Miombo woodland, usually in disturbed places; 1500–1750 m.

119. **Crotalaria onusta** Polhill, Crotalaria Africa & Madagascar: 279, fig. 77/7–13 (1982). —Lock, Leg. Afr. Check-list: 194 (1989). Type: Zambia, Mbala Distr., Chilongowelo, *Richards* 1502 (BR; K, holotype).

Erect annual, 30–40 cm tall, with densely subappressed pubescent branches. Leaves 3-foliolate; leaflets 2–5 × 0.7–2 cm, elliptic to elliptic-oblong, appressed pubescent beneath; petiole 1.5–3.5 cm long; stipules 2–3 mm long, linear-subulate. Racemes 4–11 cm long, dense, c. 20–40-flowered, with the peduncle usually longer than the rhachis and often becoming bent over with age; bracts 4–6 mm long, attenuately linear-lanceolate, caducous; bracteoles on the upper part of the pedicel, 2 mm long, setaceous. Calyx 6–7 mm long, appressed pubescent; upper lobes narrowly attenuate-triangular, 2–2.5 times as long as the tube. Standard elliptic, yellow, lined and later flushed reddish-brown, pubescent near the apex outside; wings a little shorter than the keel; keel 1.1–1.3 cm long, angular, with a narrow twisted beak. Pod subsessile, 13–15 × 7–8 mm, oblong-ellipsoid to oblong-obovoid, pubescent, c. 12–16-seeded. Mature seeds not seen.

Zambia. N: Mbala Distr., Chilongowelo, fl. & immat. fr. 14.iv.1952, *Richards* 1458 (K); same locality, fl. & fr. 26.ix.1954, *Richards* 1865 (K).

Known only from Chilongowelo, near Mbala. Disturbed grassy places; 1450 m.

120. **Crotalaria distans** Benth. in Hooker, London J. Bot. **2**: 582 (1843). —Harvey in F.C. **2**: 43 (1862). —Verdoorn in Bothalia **2**: 409 (1928). —Burtt Davy, Fl. Pl. Ferns Transvaal, pt. 2: 400 (1932). —Schreiber in Merxmüller, Prodr. Fl. SW. Afrika, fam. 60: 23 (1970). —Drummond in Kirkia **8**: 217 (1972). —Polhill, Crotalaria Africa & Madagascar: 280, fig. 75/5 and 76/9 (1982). —Lock, Leg. Afr. Check-list: 175 (1989). TAB. 3,7: **37**, fig. A. Syntypes from South Africa (Free State).

Erect or ascending annual, (0.2)0.3–1.3 m tall; branches appressed or subappressed pubescent. Leaves 3-foliolate; leaflets 1.5–6.5 × 0.3–1.5 cm, linear to oblong-oblanceolate, appressed pubescent beneath; petiole 0.5–1.5(2) cm long; stipules 1–4 mm long, linear-subulate. Racemes 6–30(40) cm long, laxly 6–30-flowered; bracts 1–3.5 mm long, linear-subulate to linear-lanceolate, shorter than the pedicel; bracteoles on the calyx or just below, 0.5–3 mm long, linear. Calyx 3–7 mm long, appressed pubescent; lobes triangular, 0.8–1.5 times as long as the tube, occasionally longer in northern parts of Zimbabwe. Standard elliptic, yellow, lined and later often flushed reddish or purplish, pubescent outside; wings a little shorter than the keel; keel 0.8–1.8 cm long, angular, with a narrow twisted beak. Pod with a stipe 0.5–4 mm long, 1.2–2.8 × 0.6–0.8(1) cm, oblong-obovoid, oblong-ellipsoid or subcylindrical, with the upper suture impressed, pubescent, c. 14–40-seeded. Seeds 2–2.5 mm long, oblique-cordiform, smooth, brown.

1. Pod 1.2–1.6 cm long, ± oblong-obovoid, 14–20-seeded; bracteoles at the base of the calyx or just below, 1–2 mm long; keel 1.1–1.3(1.4) cm long · · · · · · · · · · · · · · iii) subsp. *mediocris*
– Pod 1.7–2.8 cm long, subcylindrical to oblong-ellipsoid, only slightly broadened upwards, 18–40-seeded; bracteoles on the calyx tube, 1.5–3 mm long; keel (1.1)1.3–1.8 cm long · · · 2
2. Pod c. 3 times as long as broad, with a stipe 3–4 mm long · · · · · · · · · i) subsp. *macrotropis*
– Pod 1.5–2.5 times as long as broad, with a stipe 1–2.5 mm long · · · ·ii) subsp. *macaulayae*

i) Subsp. **macrotropis** (Baker f.) Polhill, Crotalaria Africa & Madagascar: 280, fig. 76/6 (1982). —Lock, Leg. Afr. Check-list: 175 (1989). Type from Dem. Rep. Congo (Lubumbashi).

Crotalaria macrotropis Baker f. in J. Bot. **58**: 75 (1920); Legum. Trop. Africa: 43 (1926). —Drummond in Kirkia **8**: 218 (1972).

Crotalaria mundyi Baker f., Legum. Trop. Africa: 42 (1926) excl. specim. *Eyles* 694. —Verdoorn in Bothalia **2**: 407 (1928). Type: Zimbabwe, Harare, Salisbury Experimental Station, *Mundy* L.1 (BM, fragment; K, holotype).

Crotalaria macaulayae sensu Wilczek in F.C.B. **4**: 141 (1953).

Stem erect, lower branches almost horizontal, upper ones ascending, usually striate, with short mostly appressed hairs. Bracteoles on the calyx, 1.5–2.5 mm long. Keel (1.2)1.3–1.6 cm long. Pod with a stipe 3–4 mm long, (1.7)2–2.8 × (0.5)0.6–0.8 cm, subcylindrical, 24–40-seeded.

Zambia. W: Solwezi, fl. & fr. 6.i.1969, *Mutimushi* 2875 (K; NDO; SRGH). C: Chisamba, fl. & fr. 15.v.1959, *E.A. Robinson* 3261 (K; SRGH) — not typical. **Zimbabwe**. C: Harare, Industrial Sites, Coventry Road, fl. & fr. 10.iii.1969, *Biegel* 2878 (K; SRGH).

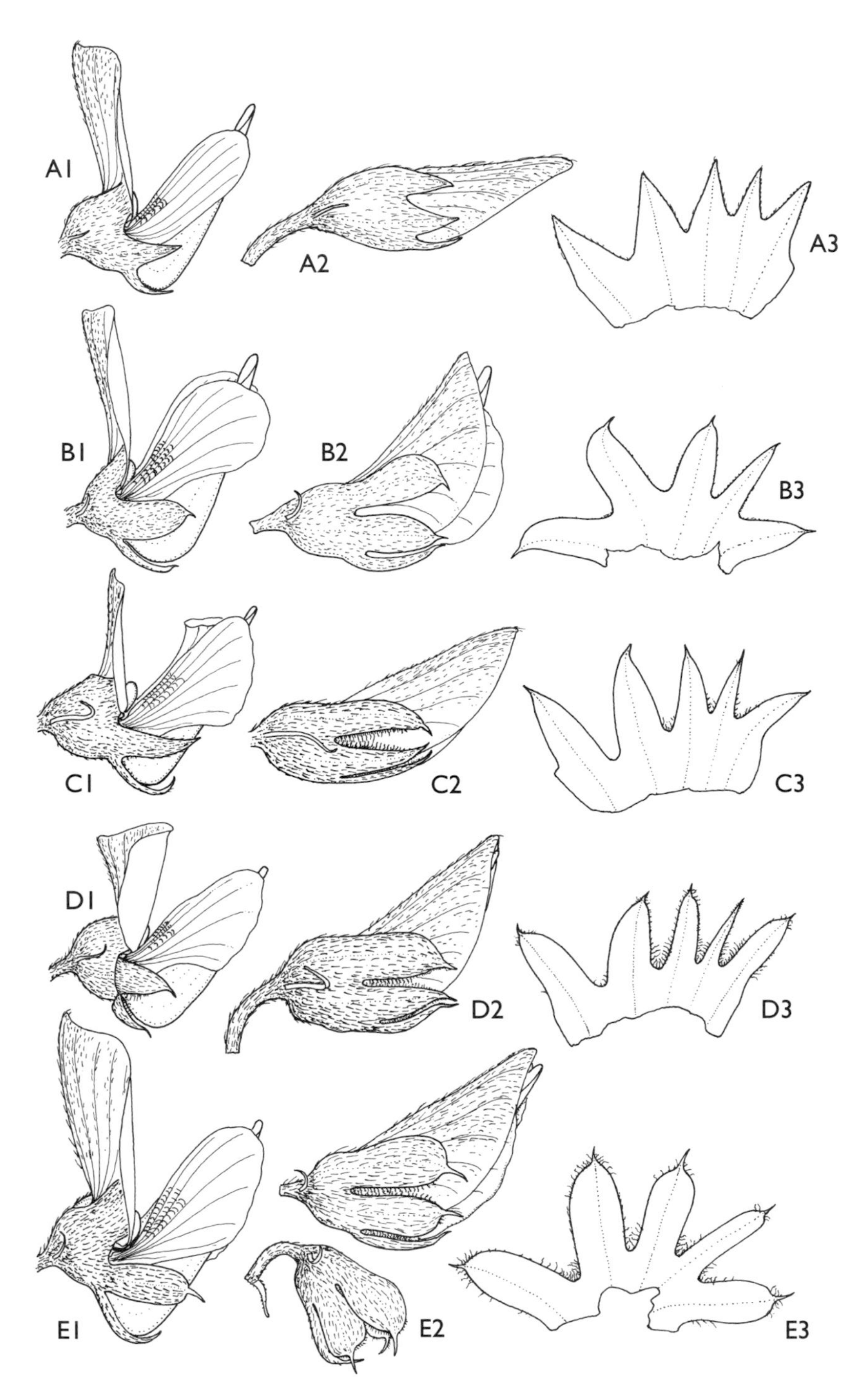

Tab. 3,7: **37**. Flowers of *Crotalaria* subsect. *Longirostres*: 1, flower (× 2 ²/₃); 2, bud (× 4); 3, calyx, opened out (× 2 ²/₃). A. —CROTALARIA DISTANS subsp. DISTANS, 1, 3 from *Williamson* 705; 2 from *Mogg* 8510. B. —CROTALARIA DINTERI, from *W. Giess* jun. 191. C. —CROTALARIA LABURNOIDES, from *Faulkner* 1643. D. —CROTALARIA SENEGALENSIS, from *Burnett* 49/84. E. —CROTALARIA PLATYSEPALA, from *W. Giess* 9856. Drawn by Roger Polhill. From Crotalaria Africa & Madagascar.

Mainly in the Copperbelt of Zambia, just extending into Dem. Rep. Congo (Katanga), also into the Central Province north of Lusaka but intergrading here with subsp. *macaulayae.* Introduced to Harare, Salisbury Experimental Station about 1920 and now naturalized around the city. Wet grassland in dambos and by streams; 1200–1400 m.

ii) Subsp. **macaulayae** (Baker f.) Polhill, Crotalaria Africa & Madagascar: 280, fig. 76/8 (1982). —Lock, Leg. Afr. Check-list: 175 (1989). Type: Zambia, near Mumbwa, *Macaulay* 758 (K, holotype).

Crotalaria macaulayae Baker f. in J. Linn. Soc., Bot. **42**: 326 (1914). —Verdoorn in Bothalia **2**: 411 (1928) pro parte. —Drummond in Kirkia **8**: 218 (1972).

Branches ascending, subterete, with appressed or mostly slightly spreading hairs. Bracteoles on the calyx, 1.5–2.5 mm long. Keel (1.2)1.3–1.8 cm long. Pod with a stipe 1–2.5 mm long, 1.7–2.2 × (0.6)0.7–0.8 cm, oblong-ellipsoid to subcylindrical, slightly broadened upwards, 18–34-seeded.

Zambia. W: 3 km west of Chizela (Chizera), fl. 23.iii.1961, *Drummond & Rutherford-Smith* 7200 (K; LISC; SRGH) — not typical. S: 19 km north of Choma, fl. & fr. 22.iii.1957, *E.A. Robinson* 2159 (K; SRGH). **Zimbabwe**. N: Zvimba Distr., Mutorashangu (Mtoroshanga) Pass, fl. & immat. fr. 26.ii.1961, *Leach* 10740 (K; LISC; SRGH). W: Insiza Distr., Shangani, fl. 1.iii.1943, *Feiertag* in *GHS* 45341 (K; SRGH). C: Makwiro, fl. 2.iv.1931, *E.G. Smith* in *GHS* 3835 (BM; SRGH). **Malawi**. S: Mangochi Distr., L. Malombe (Pamalombe), Mvela (Mvera), fl. & fr. 1.vi.1955, *Jackson* 1666 (BM; K; SRGH).

Drainage area of middle Zambezi and its tributaries. Seasonally wet grassland of flood plains, dambos and riversides; 500–1400 m.

Crotalaria friesii, when flowering in its first season, is likely to be confused with this subspecies (see note under *C. friesii*).

Drummond & Rutherford-Smith 7200, which lacks pods, is cited above, but included with some doubt, the standard being very sparsely hairy and the subspecies is not otherwise known from so far up the Zambezi. Little fruiting material has been seen from northern parts of Zimbabwe, but certain specimens such as *Eyles* 694 (K; SRGH) from a picrite formation and *Wild* 5801 (K; LISC; SRGH) from serpentine or morite soils on the Great Dyke have unusually small fruits, though the number of ovules is normal for the subspecies. In this area the calyx lobes are often larger than normal in the species.

Sometimes cultivated as a green manure, e.g. Zimbabwe, Harare, Salisbury Experimental Station, fl. & fr. 1921, *Mundy* L. 51/20 (K; SRGH); Henderson Research Station, Mazowe (Mazoe), fl. & immat. fr. 1.ii.1953, *Wild* 4004 (K; LISC; SRGH).

iii) Subsp. **mediocris** Polhill, Crotalaria Africa & Madagascar: 281, fig. 76/7 (1982). —Lock, Leg. Afr. Check-list: 175 (1989). Type: Zimbabwe, Beitbridge, between Customs Post and Limpopo R., *Drummond* 6005 (K, holotype; LISC; SRGH).

Crotalaria distans sensu E.G. Baker in J. Linn. Soc., Bot. **42**: 340 (1914) pro minore parte. —sensu Verdoorn in Bothalia **2**: 409 (1928) pro parte. —sensu Burtt Davy, Fl. Pl. Ferns Transvaal, pt. 2: 400 (1932) pro parte.

Crotalaria macaulayae sensu Verdoorn in Bothalia **2**: 411 (1928) pro parte. —sensu Burtt Davy, Fl. Pl. Ferns Transvaal, pt. 2: 400 (1932) pro parte.

Erect with ascending branches. Racemes mostly 16–40-flowered. Bracteoles at the base of the calyx or just below, 1–2 mm long. Keel 1.1–1.3(1.4) cm long. Pod with a stipe 1 mm long, 1.2–1.6 × 0.7–0.8(0.9) cm, ± oblong-obovoid, 14–20-seeded.

Botswana. N: Francistown, fl. & fr. 7.iii.1961, *Richards* 14549 (K). SE: 53 km NW of Serowe, fl. 25.iii.1965, *Wild & Drummond* 7312 (K; SRGH). **Zimbabwe**. W: Matobo Distr., Farm Besna Kobila, fl. & fr. iv.1959, *O.B. Miller* 5901 (K; LISC; SRGH). S: 3 km from Beitbridge on West Nicholson Road, fl. & fr. 15.ii.1955, *Exell, Mendonça & Wild* 388 (BM; LISC; SRGH).

Occurring in the Limpopo Basin, and grading into subsp. *distans* (with generally fewer smaller flowers and oblong-ellipsoid pods, TAB. 3,7: **37**, fig. A) in Northern Province, Mpumalanga and the Northern Cape Province of South Africa; not to be confused with *C. kurtii* Schinz which occurs further west in Namibia. Mopane woodland, *Acacia, Commiphora* bushland and grassland, not obviously associated with wet places; 900–1450 m.

Two collections from the Great Dyke in northern Zimbabwe at the Vanad Pass on the east slopes of the Mvurwi (Umvukwe) Range, *Wild* 7990 (K; SRGH) and *Polhill & Pope* 4755 (K; SRGH) probably belong within a broad concept of *C. distans*. They are small annuals with branches mostly from near the base, with well-spaced leaves, rather narrow leaflets and rather few-flowered inflorescences.

121. **Crotalaria friesii** Verdoorn in Bothalia **2**: 405 (1928). —Polhill, Crotalaria Africa & Madagascar: 281 (1982). —Lock, Leg. Afr. Check-list: 179 (1989). Type: Zambia, Kabwe (Broken Hill), *R.E. Fries* 219 (UPS, holotype).

Stems ultimately numerous from a woody rootstock, but flowering from first year of growth, procumbent to erect, 15–40(180) cm long, subterete, subappressed pubescent; rootstock vertical with rhizome-like branches from the top. Leaves 3-foliolate; leaflets (8)10–20(30) × (2)3–6(8) mm, mostly oblong-oblanceolate, appressed pubescent beneath; petiole 6–12(15) mm long; stipules 1–2 mm long, subulate. Racemes (3)7–15(20) cm long, laxly (4)8–16(24)-flowered; buds ascending; bracts 1–2(2.5) mm long, linear-lanceolate; bracteoles usually near top of the pedicel, 0.5–1.5 mm long. Calyx (3.5)4.5–6 mm long, appressed pubescent; upper lobes narrowly triangular, 1.2–1.5 times as long as the tube. Standard elliptic, yellow, lined or flushed reddish and pubescent outside; wings oblong-obovate, ± as long as the keel; keel 9–11(12) mm long, angular, with a narrow twisted beak. Pod subsessile, 10–14 × 5–7 mm, ellipsoid-oblong, pubescent, 10–12-seeded. Seeds 2–2.5 mm long, oblique-cordiform, smooth, brown.

Zambia. C: Lusaka South Forest Reserve, fr. 6.ix.1963, *Farrell* 366 (K; SRGH). S: Mazabuka, Veterinary Research Station, fl. & fr. 20.vii.1965, *van Rensburg* 3020 (K; SRGH).

Known only from around the Kafue Basin. Miombo and mixed deciduous woodland or grassland, on varied soils, but perhaps generally where there is a fairly high water-table; 1000–1300 m.

When flowering in the dry season as a perennial this species is easily recognized. First year plants, however, tend to flower at the end of the rains and are likely to be confused with the annual *C. distans* subsp. *macaulayae* from the same region, but the lower branches are procumbent, not ascending, the bracteoles are small and inserted lower and the flowers and fruits are slightly smaller.

122. **Crotalaria sp. B**

Crotalaria grantiana sensu Drummond in Kirkia **8**: 218 (1972) quoad *Drummond* 6843.

Short-lived perennial flowering from the first year, with a taproot that may thicken and develop secondary roots; stems numerous from the top of the taproot, slender, diffuse, much branched, procumbent to weakly ascending, up to 20–30 cm long, appressed pubescent. Leaves 3-foliolate; leaflets 7–15 × 2–6 mm, elliptic-oblanceolate, rounded to slightly retuse at the apex, appressed puberulous beneath; petiole 6–8(12) mm long; stipules 1–1.5 mm long, subulate. Racemes 3–6 cm long, laxly 3–6 flowered, with a slender rhachis; bracts 1–2 mm long, linear; bracteoles on the pedicel, 0.5–1 mm long. Calyx 3–4 mm long, appressed pubescent; lobes 2.5–3 mm long, triangular. Standard elliptic-obovate, yellow, probably lined and fading reddish, densely pubescent outside; wings oblong-obovate, a little shorter than the keel; keel 8–8.5 mm long, angular, with a narrow twisted beak. Pod subsessile, 8–9 × 4.5–5.5 mm, obovoid-ellipsoid, pubescent, 6–8-seeded. Mature seeds not seen.

Zimbabwe. N: Guruve Distr., east of Nyarasuswi (Nyaruswiswi), fl. & fr. 16.iii.1960, *Drummond* 6843 (K; LISC; MO; P; PRE; SRGH).

Known only from the above collection from the Great Dyke in northern Zimbabwe. On serpentine at the foot of a grass-covered serpentine hill.

See also 128. *Crotalaria sp. C* from copper bearing soils on the Great Dyke, and also certain anomalous variants of 120. *C. distans* from the same formation.

123. **Crotalaria mocubensis** Polhill, Crotalaria Africa & Madagascar: 282, fig. 77/1–6 (1982). — Lock, Leg. Afr. Check-list: 192 (1989). Type: Mozambique, Mocuba, Posto Agrícola, *Barbosa & Carvalho* 2921 (K; LISC, holotype; LMA).

Annual, erect at first, then spreading, 0.4–1.5 m tall; branches appressed puberulous. Leaves 3-foliolate; leaflets 2–4.5 × 0.6–3 cm, usually elliptic, sometimes oblong-elliptic, appressed puberulous beneath; petiole 1.5–4 cm long; stipules 1–4 mm long, linear-subulate. Racemes (5)10–25(35) cm long, ± laxly 16–40-flowered; buds ascending; bracts 1–3(3.5) mm long, linear-subulate; bracteoles on the calyx, 0.5–1 mm long. Calyx 4–5 mm long, appressed puberulous; lobes narrowly attenuate-triangular, 1.5–2 times as long as the tube, not conspicuously lanate along the lobe margins inside. Standard broadly elliptic, bright yellow, finely reddish lined, pubescent outside; wings only slightly broadened distally, nearly as long as the keel; keel 8–10 mm long, angular with a narrow twisted beak. Pod very shortly stipitate, 9–11(12) × 5–6 mm, ellipsoid to slightly oblong-obovoid, puberulous, 6–8(10)-seeded. Seeds 3 mm long, oblique-cordiform, smooth, brown.

Mozambique. Z: 11.7 km from Régulo Guja to Derre, fl. & fr. 10.vi.1949, *Barbosa & Carvalho* 3018 (K; LMA). T: Tete, 16.8 km from Km 148 and the Junction, fl. & fr. 22.vi.1949, *Barbosa & Carvalho* 3228 (K; LISC; LMA) — not typical. MS: Chupanga (Lacerdónia)–Marromeu, fl. & fr. 8.v.1942, *Torre* 4105 (BM; K; LISC).

Known only from Zambezia Province and adjacent parts of Mozambique. Deciduous woodland and disturbed places; mostly below 200 m.

Barbosa & Carvalho 3228, cited above, from farther inland, is included with some doubt. It looks more like *C. senegalensis* except the upper calyx lobes are scarcely oblong, but this might be explained if some introgression between *C. senegalensis* and *C. vigulata* occurred in this region. The habit is more erect, the leaflets are narrower, the calyx-lobes are more hairy along the margins inside and the pods are longer than in typical *C. mocubensis.*

124. **Crotalaria dinteri** Schinz in Bull. Herb. Boissier, sér. 2, **1**: 875 (1901) non *C. dinteri* Schinz (1903). —E.G. Baker in J. Linn. Soc., Bot. **42**: 302 (1914). —Verdoorn in Bothalia **2**: 404 (1928). —Schreiber in Merxmüller, Prodr. Fl. SW. Afrika, fam. 60: 23 (1970). —Polhill, Crotalaria Africa & Madagascar: 283, fig. 75/4 (1982). —Lock, Leg. Afr. Check-list: 175 (1989). TAB. 3,7: **37**, fig. B. Type from Namibia.

Decumbent or ascending perennial, with generally velutinous stems 0.3–1 m long, occasionally glabrous throughout. Leaves 3-foliolate; leaflets 8–17 × 5–14 mm, elliptic to obovate, rounded to retuse at the apex, cuneate to the base, usually densely puberulous at least beneath; petiole 3–13 mm long; stipules 1–2 mm long, attenuate-triangular. Racemes c. 5–25 cm long, laxly (6)12–20(30)-flowered; buds ascending; bracts 1–1.5 mm long, linear-lanceolate; bracteoles usually on the calyx, 1 mm long. Calyx 4.5–6 mm long, usually appressed puberulous; upper lobes oblong, ± twice as long as the tube, abruptly contracted to an excentric apiculate tip, sometimes with white hairs on margin inside. Standard broadly elliptic, yellow, finely lined reddish and usually pubescent outside; wings oblong-obovate, nearly as long as the keel; keel (7)8–10 mm long, angular, with a narrow twisted beak, lanate along the upper suture. Pod subsessile, 8–9 × 5–6 mm, ellipsoid or ellipsoid-obovoid, pubescent, 6–8-seeded. Mature seeds not seen.

Botswana. N: near Nata R. on Maun Road, fl. & fr. 9.iii.1961, *Richards* 14618 (K). SW: Ghanzi Distr., 4 km north of Dondong Borehole, fl. & fr. 28.xi.1976, *Skarpe* 100 (K; SRGH).

Also in Namibia and South Africa (Northern Cape). Grassland and *Acacia* or *Commiphora, Combretum* savanna on sand or calcrete, persisting in overgrazed places; 900–1150 m.

125. **Crotalaria laburnoides** Klotzsch in Peters, Naturw. Reise Mossambique **6**, part 1: 57 (1861). —E.G. Baker in J. Linn. Soc., Bot. **42**: 338 (1914). —Polhill in F.T.E.A., Leguminosae, Pap.: 966 (1971); Crotalaria Africa & Madagascar: 284, figs 75/2 and 76/15 (1982). —Lock, Leg. Afr. Check-list: 186 (1989). TAB. 3,7: **37**, fig. C. Type from Zanzibar.

Crotalaria bagamoyoensis Baker f. in J. Linn. Soc., Bot. **42**: 332 (1914). Type from Tanzania.

Crotalaria maxillaris sensu E.G. Baker in J. Linn. Soc., Bot. **42**: 336 (1914) pro minore parte, ? sensu Klotzsch (see note below).

Crotalaria junodiana Baker f. in J. Linn. Soc., Bot. **42**: 341 (1914). —Verdoorn in Bothalia **2**: 407 (1928). Type: Mozambique, Delagoa Bay, *Junod* 388 (BM, sketch & fragment; Z, holotype).

Erect annual (0.1)0.3–1.5 m tall; branches appressed pubescent, sometimes glabrescent. Leaves 3-foliolate; leaflets 2–6 × 0.8–2.5 cm, oblong-elliptic to elliptic, usually appressed pubescent beneath; petiole 1–3 cm long; stipules 1.5–4 mm long, linear-subulate. Racemes (4)10–15 cm long, (8)12–24-flowered; buds ascending; bracts 1.5–3 mm long, linear-subulate to linear-lanceolate, bracteoles on the calyx, 0.5–2 mm long. Calyx 5–8 mm long, usually appressed pubescent; upper lobes oblong to triangular-lanceolate, rather abruptly contracted to an excentric apiculate tip, 1.5–2.5 times as long as the tube, usually with white hairs along the margins inside, becoming attenuate and glabrescent inside in mature flowers. Standard broadly elliptic to subcircular, yellow, finely veined and ageing reddish, pubescent outside; wings oblong-obovate, ± as long as the keel; keel 9–11 mm long, angular, with a narrow twisted beak, lanate along the upper suture. Pod subsessile, 1.6–2.4 × (0.8)1–1.4 cm, broadly oblong-ellipsoid to oblong-obovoid, appressed puberulous to glabrous, c. 16–28-seeded. Seeds 2–2.8 mm long, oblique-cordiform, smooth, yellowish to brown.

Malawi. S: Zomba Distr., Cikowi, fl. & fr. 10.v.1956, *Jackson* 1853 (K). **Mozambique**. N: Mossuril Distr., Goa I., fl. & fr. 19.v.1961, *Leach & Rutherford-Smith* 10919 (K; LISC; SRGH). Z: Gurué Distr., andados 22 km do Nintulo para Lioma, junta ao cruzamento para Mutuáli, c. 700 m, fl. & fr. 10.ii.1964, *Torre & Paiva* 10516 (LISC). MS: Beira, fl. & fr. v.1947, *Pedro* 3139 (LMA). GI: Mundiane, fl. & fr. 8.vii.1947, *Pedro & Pedrógão* 1357 (COI; LMA; SRGH). M: Magude, Chobela, fl. & fr. 5.v.1948, *Torre* 7759 (BM; K; LISC).

In coastal regions north to Somalia, also Comoro, Aldabra and Cosmoledo Is., inland to Lake Malawi and disjunctly to Dem. Rep. Congo and Uganda around Lake Albert. Coastal bushland and grassland, dunes and sandy beaches, inland in deciduous woodland on sandy soils, sometimes ruderal; 0–800 m.

The type of *Crotalaria maxillaris* Klotzsch in Peters, Naturw. Reise Mossambique **6**, part 1: 58 (1861), Mozambique, Cabaceira, *Peters* (B†, holotype), has long been equated with *Kirk* specimens from Tete, which are referred here to *C. senegalensis*. As Cabaceira is on the coast it seems more likely that *C. maxillaris* is a synonym of *C. laburnoides*, but the type no longer exists and the original description is ambiguous.

Var. *nudicarpa* Polhill, with glabrous pods and sparse indumentum elsewhere, occurs on beaches of Kenya and Tanzania and may be found subsequently further south.

126. **Crotalaria senegalensis** (Pers.) Bacle ex DC., Prodr. **2**: 133 (1825). —J.G. Baker in F.T.A. **2**: 31 (1871). —E.G. Baker in J. Linn. Soc., Bot. **42**: 337 (1914). —Wilczek in F.C.B. **4**: 144 (1953). —Hepper in F.W.T.A., ed. 2, **1**: 550 (1958). —Torre in C.F.A. **3**: 53 (1962). —Polhill in F.T.E.A., Leguminosae, Pap.: 967 (1971). —Drummond in Kirkia **8**: 218 (1972). —Gonçalves in Garcia de Orta, Sér. Bot. **5**: 71 (1982). —Polhill, Crotalaria Africa & Madagascar: 285, figs 75/1 and 76/16 (1982). —Lock, Leg. Afr. Check-list: 203 (1989). TAB. 3,7: **37**, fig. D. Type from Senegal.

Crotalaria uncinella var. *senegalensis* Pers., Syn. Pl. **2**: 285 (1807).

Crotalaria maxillaris sensu J.G. Baker in F.T.A. **2**: 30 (1871) typo excluso. —sensu E.G. Baker in J. Linn. Soc., Bot. **42**: 336 (1914) typo excluso. —sensu Verdoorn in Bothalia **2**: 407 (1928) typo excluso, ? sensu Klotzsch (see note after *C. laburnoides*).

Crotalaria karongensis Baker in Bull. Misc. Inform., Kew **1897**: 252 (1897). —Binns, First Check List Herb. Fl. Malawi: 79 (1968). Type: Malawi, Kondowe to Karonga, *Whyte* s.n. (K, holotype).

Annual, erect and spreading, (0.2)0.3–1 m tall; branches appressed or subappressed pubescent. Leaves 3-foliolate; leaflets 2–6.5 × 0.6–2.5 cm, mostly elliptic-oblong, varying from lanceolate to elliptic, appressed pubescent beneath; petiole 1.5–5 cm long; stipules 1–3 mm long, linear-subulate. Racemes mostly 12–45 cm long, laxly (10)16–40-flowered; buds ascending; bracts 1–3.5 mm long, linear-subulate to linear-lanceolate; bracteoles on the calyx, 1–3 mm long. Calyx (5)6–7(8) mm long, appressed pubescent; upper lobes oblong, abruptly contracted to an excentric apiculate tip, valvate, lanate along the margin inside. Standard bright yellow, often finely lined and ageing reddish, broadly elliptic to oblate, pubescent outside; wings obovate-oblong, nearly as long as the keel; keel (7)9–13(15) mm long, angular, with a narrow twisted beak, lanate along the upper suture. Pod shortly stipitate, 1–1.7 × 0.5–0.7(0.8) cm, oblong-ellipsoid, pubescent, 6–16-seeded. Seeds 2.5–3 mm long, oblique-cordiform, yellowish to brown, sometimes mottled darker.

Botswana. N: Central Distr., near Rakops, 8 km NE of Toromoja School, fl. & immat. fr. 28.iv.1971, *Pope* 411 (K; SRGH). **Zambia**. N: Mbala Distr., road between the Escarpment and Kambole, above Lunzua Bridge, fl. & fr. 7.iv.1955, *Richards* 5343 (K; SRGH). C: Mkushi, fl. & fr. 2.v.1957, *Fanshawe* 3269 (K; NDO). E: Chipata Distr., east of Machinje Hills, Lutembwe R. Gorge, fl. 13.x.1958, *Robson & Angus* 101 (K; LISC; SRGH). **Zimbabwe**. N: Mudzi Distr., Nyagoko R., fl. 14.ii.1962, *Wild* 5651 (COI; K; SRGH). E: Chimanimani (Melsetter) to Cashel, fl. & fr. 20.iii.1964, *Corby* 1118 (K; LISC; SRGH). **Malawi**. N: Rumphi Distr., St. Patrick's Seminary, fl. & fr. 4.v.1974, *Pawek* 8569 (K; MO). C: Lilongwe, fl. & fr. 23.iii.1970, *Brummitt & Little* 9321 (K; LISC; MAL; PRE; SRGH). S: Shire R., Mpatamanga Gorge, fl. & fr. 28.ii.1961, *Richards* 14501B (K). **Mozambique**. N: Lugenda R. just north of bridge on road from Marrupa to Mecula, fl. & fr. 11.viii.1981, *Jansen, de Koning & J. de Wilde* 210 (K; WAG). T: Angónia Distr., andados 12 km de Vila Coutinho Velha para Ulónguè (Vila Coutinho Nova), c. 1300 m, fl. & fr.

10.iii.1964, *Torre & Paiva* 11121 (LISC). MS: Sussundenga Distr., Mussambuzi (Messambuzi) Valley, fl. & fr. 5.v.1948, *Pedro* 3647 (LMA).

Savanna regions of northern tropical Africa, c. 10–16°N, from Cape Verde Is. and Senegal to Ethiopia and the Red Sea Hills, then in the eastern Rift Valley disjunctly from Ethiopia to southern Tanzania, also Dem. Rep. Congo and Angola. Woodland and grassland, often in disturbed places, usually on sand; 250–1650 m.

127. **Crotalaria platysepala** Harv., Thes. Cap. **1**: 19, t. 29 (1859). —J.G. Baker in F.T.A. **2**: 34 (1871). —E.G. Baker in J. Linn. Soc., Bot. **42**: 339 (1914). —Verdoorn in Bothalia **2**: 406 (1928) excl. syn. *C. ulbrichiana.* —Schreiber in Merxmüller, Prodr. Fl. SW. Afrika, fam. 60: 25 (1970). —Drummond in Kirkia **8**: 218 (1972). —Gonçalves in Garcia de Orta, Sér. Bot. **5**: 70 (1982). —Polhill, Crotalaria Africa & Madagascar: 286, fig. 75/3 (1982). —Lock, Leg. Afr. Check-list: 197 (1989). TAB. 3,7: **37**, fig. E. Type: Botswana, L. Ngami, between Rivers Thamalakane (Tamulakau) and Boteti (Botletle), Zougha, *McCabe* 8 (K, holotype).

Crotalaria apiculata Schinz in Mém. Herb. Boissier No. 1: 126 (1900). Types from Namibia.

Crotalaria polysepala sensu O.B. Miller in J. S. African Bot. **18**: 30 (1952).

Annual, erect and spreading, (0.1)0.3–1.2 m tall; branches subappressed pubescent. Leaves 3-foliolate; leaflets 1.5–7 × 0.4–3 cm, elliptic-oblong to ovate-oblong, appressed pubescent beneath; petiole 1–4.5 cm long; stipules 1.5–4 mm long, linear-caudate. Racemes mostly 10–30(40) cm long, laxly or sublaxly 16–30(50)-flowered, lateral ones sometimes shorter; buds becoming reflexed, then spreading; bracts 1.5–2.5 mm long, linear-subulate; bracteoles on the calyx, 0.5–1.5 mm long. Calyx 4.5–5.5 mm long, thinly pubescent; upper lobes oblong to spathulate, 2–3 times as long as the tube, broadly obtuse, rounded or retuse to the slightly excentric apiculate tip, overlapping lateral lobes in bud, often without woolly hairs along the margins inside. Standard broadly elliptic to oblate, yellow, finely lined reddish-brown, pubescent outside; wings nearly as long as the keel; keel 7.5–11 mm long, angular, with a narrow twisted beak, lanate along upper suture. Pod shortly stipitate, 8–12 × 5–6 mm, oblong-obovoid, pubescent, 6–8-seeded. Seeds c. 4 × 3.5 mm, oblique-cordiform, smooth, brown.

Botswana. N: 110 km west of Nokaneng, fl. & fr. 12.iii.1965, *Wild & Drummond* 6918 (K; LISC; SRGH). SE: Kgatleng Distr., Mochudi R., fl. & immat. fr. 24.iii.1967, *Michison* A.6 (K). **Zambia**. B: Sesheke Distr., Masese, fl. & fr. 24.v.1962, *Fanshawe* 6840 (K; LISC; NDO). S: Gwembe Valley, fl. & fr. iv.1934, *Trapnell* 1477 (K). **Zimbabwe**. N: Hurungwe Distr., Rifa R., fl. & immat. fr. 24.iii.1953, *Wild* 4070 (K; LISC; SRGH). W: 8 km west of Hwange (Wankie), fl. & fr. 31.iii.1963, *Leach* 11622 (K; LISC; SRGH). E: Save (Sabi) Valley, fl. 12.iii.1963, *Pole Evans* 6473 (K; PRE). S: Save (Sabi) R., Birchenough Bridge, i.1938, *Obermeyer* 2425 (BM). **Mozambique**. T: Zumbo Distr., Sisitso Station, fl. & fr. 9.vii.1950, *Chase* 2604 (BM; K; LISC; SRGH).

Also in Namibia. *Baikiaea*, mopane, *Acacia* and *Terminalia* woodlands, or savanna on Kalahari Sands, sometimes in disturbed places; 350–1050 m.

Phipps 1385 (K; LISC; SRGH) from the Chikwatata Hot Springs, Binga, Zimbabwe, is anomalous, dwarfed and rooting at the nodes, with short erect shoots, small leaves and short racemes. This may be simply a physiological response to extreme conditions around the hot spring, or possibly a distinct taxon. *R.M. Davies* 2015 (K; LISC; SRGH) from Kavira Hot Springs in the same district is a more normal specimen of *C. platysepala.*

128. **Crotalaria sp. C**

Short-lived perennial flowering from the first year of growth, with a main taproot and secondary roots spreading from the upper part; stem shortly erect, developing numerous branches from near the woody base; plants 20–30 cm tall; branches up to 60 cm long, puberulous. Leaves 3-foliolate; leaflets 10–20 × 2–4 mm, linear-elliptic to obovate-oblanceolate, puberulous beneath; petiole 4–12 mm long; stipules 1.5 mm long, linear-subulate. Racemes 3–10 cm long, lax, with a slender rhachis bearing up to 16 flowers but only a few open at a time; bracts c. 1 mm long, linear-elliptic; pedicels 1–2 mm long, recurving; bracteoles at the base of the calyx, 1 mm long, linear. Calyx 4–4.5 mm long, appressed puberulous; upper lobes 2.5 mm long, oblong-triangular, with a blunt excentric tip, with wool along the edge adjacent to the triangular lateral lobe. Standard obovate, clear yellow without red lines, thinly puberulous outside; wings ± three quarters as long a the keel; keel bent at right angles in the lower part, with a narrow twisted beak, 8.5 mm long. Pods not seen.

Zimbabwe. N: Gokwe Distr., Sanyati Mining Company, Copper King North Mine, 17°35'S, 29°16'E, fl. 22.iv.1999, *Wengler* 286 (BR); same locality and date, *Wengler* 312 (BR); same locality, *Baudesson, Leteinturier & Malaisse* 761 (BR; K).

Known only from this locality on the Great Dyke. Slope of copper outcrop and at edge of a copper poisoned dambo on the slope; 970 m.

This species will be described by Miss Leteinturier in a forthcoming issue of Systematics and Geography of Plants. *Crotalaria sp. B*, also from the Great Dyke, but on serpentine, is superficially similar, but there are several stems from the top of the taproot, the upper calyx lobes are triangular and the standard is more densely hairy outside.

129. **Crotalaria subcaespitosa** Polhill in Kew Bull. **22**: 321 (1968); Crotalaria Africa & Madagascar: 286 (1982). —Lock, Leg. Afr. Check-list: 205 (1989). Syntypes: Malawi, Mt. Zomba and Mt. Malosa, *Whyte* s.n. (K, syntypes).

Crotalaria caespitosa Baker in Bull. Misc. Inform., Kew **1897**: 252 (1897) non Roxb. (1832). —E.G. Baker in J. Linn. Soc., Bot. **42**: 335 (1914). —Verdoorn in Bothalia **2**: 412 (1928). —Brenan in Mem. New York Bot. Gard. **8**: 247 (1953). —Binns, First Check List Herb. Fl. Malawi: 78 (1968). Types as for species.

Stems numerous, ascending from a vertical woody rootstock, 10–25 cm long, slightly angular and ribbed, puberulous with short appressed and slightly longer spreading hairs. Leaves 3-foliolate; leaflets 10–25 × 4–8 mm, oblanceolate to elliptic-oblong, appressed puberulous beneath; petiole 1–6 mm long; stipules 1–3 mm long, linear-subulate. Racemes lax, 4–12 cm long, c. 8–16-flowered; bracts 1.5–4 mm long, linear-lanceolate; bracteoles on the pedicel, setaceous. Calyx 5–6 mm long, appressed puberulous; lobes triangular, slightly acuminate, 1.3–2 times as long as the tube. Standard elliptic, bright yellow, lined or flushed reddish outside and puberulous near the apex; wings nearly as long as the keel; keel 8–9 mm long, angular, with a narrow twisted beak. Pod sessile, 9–15 × 7–10 mm, obliquely oblong-obovoid, slightly compressed laterally, puberulous, 8–16-seeded. Seeds 2–2.5 mm long, oblique-cordiform, smooth, brown.

Zambia. E: Nyika Plateau, 3 km SW of Rest House, fl. 21.x.1958, *Robson & Angus* 218 (BM; K; LISC; SRGH); same locality, fr. 21.x.1958, *Robson & Angus* 219 (BM; K; LISC; SRGH). **Malawi**. N: Viphya Plateau, fl. & fr. 24.ix.1972, *Pawek* 5802 (K; MO). C: Ntchisi Mt., fl. 5.viii.1946, *Brass* 17138 (K; SRGH). S: Zomba Plateau, opposite Malosa Saddle, fl. 2.viii.1970, *Brummitt & Banda* 12393 (K; MAL; SRGH).

Also southern Tanzania around L. Malawi. Miombo woodland and upland grassland subject to burning, persistent in disturbed places on poor soils; 1400–2250 m.

130. **Crotalaria inyangensis** Polhill, Crotalaria Africa & Madagascar: 287, fig. 78/6–12 (1982). —Lock, Leg. Afr. Check-list: 183 (1989). TAB. 3,7: **38**, fig. B. Type: Zimbabwe, Nyanga Distr., Juliasdale, Farm Cotswold, *Leach* 10771 (K, holotype; SRGH).

Decumbent perennial; stems 15–60 cm long, densely covered with short stiffish irregularly shaped hairs, usually mostly spreading, some (sometimes most) appressed. Leaves 3-foliolate; leaflets 1–3 × 0.5–1.8 cm, elliptic-oblong, elliptic or elliptic-obovate, appressed puberulous beneath; petiole 3–15 mm long; stipules 2–4 mm long, linear-subulate. Racemes 3–15 cm long, pedunculate with 8–20(30) usually rather closely arranged flowers above; buds ascending; bracts 2–4 mm long, linear-subulate, sometimes exceeding the pedicel; bracteoles at the base of the calyx, rarely below, 1–3 mm long. Calyx 4–5 mm long, appressed puberulous; upper lobes triangular, slightly acuminate, not much longer than the tube. Standard elliptic or subcircular, yellow, lined and ageing reddish-brown, puberulous outside; wings only slightly broadened distally, as long as the keel; keel 7–9 mm long, angular, with a narrow twisted beak. Pod subsessile, 7–9 × 4–5 mm, ellipsoid-obovoid, pubescent, 6–8-seeded. Seeds not seen.

Zimbabwe. E: Nyanga Distr., Bonda–Sanyatwe, fl. & immat. fr. 3.iii.1969, *Corby* 2093 (K; SRGH); Mutare Distr., Himalayas, Engwa, fl. 3.iii.1954, *Wild* 4483 (K; SRGH).

Known only from the highlands of eastern Zimbabwe from Nyanga (Inyanga) to Himalayas and west to Makoni. Upland grassland, sometimes in rocky places or by streams; 1800–2200 m.

131. **Crotalaria dedzana** Polhill, Crotalaria Africa & Madagascar: 289, fig. 78/1–5 (1982). —Lock, Leg. Afr. Check-list: 174 (1989). TAB. 3,7: **38**, fig. A. Type: Malawi, Dedza Mt., *Brummitt* 10095 (K, holotype; MAL; SRGH).

Tab. 3,7: **38**. A. —CROTALARIA DEDZANA. A1, flowering and fruiting branch (× 1); A2, flower (× 4); A3, standard (× 4); A4, wing (× 4); A5, keel (× 4), A1–A5 from *Brummitt* 10095. B. —CROTALARIA INYANGENSIS. B1, flowering and fruiting branch (× 1), from *Corby* 2093 and *O.B. Miller* 5611; B2, flower (× 4); B3, standard (× 4); B4, wing (× 4); B5, keel (× 4); B6, anthers (× 8); B7, gynoecium (× 4), B2–B7 from *Corby* 2093. Drawn by Roger Polhill. From Crotalaria Africa & Madagascar.

Erect perennial, 0.6–1 m tall, with short ascending branches above; branches slightly ribbed, covered with fine irregularly arranged hairs. Leaves 3-foliolate; leaflets 5–15 × 2.5–5 mm, oblanceolate to obovate, sparsely hairy above, finely ± appressed pubescent beneath; petioles 3–10 mm long; stipules 1–2 mm long, subulate. Racemes 3–6(8) cm long, subdensely 8–16-flowered; bracts ± as long as the pedicel, 2.5–3 mm long, linear-lanceolate; bracteoles on the pedicel or calyx, 0.5–2 mm long. Calyx 4–5 mm long, appressed pubescent; lobes acuminately triangular, 0.8–1.5 times as long as the tube. Standard elliptic to subcircular, yellow, later lined and flushed orange-red, pubescent outside; wings ± as long as the keel; keel 7–8 mm long, angular, with a narrow twisted beak. Pod shortly stipitate, 7–10 × 4–6 mm, obliquely oblong-obovoid, downy, 6–8-seeded. Seeds not seen.

Malawi. N: Nyika Plateau, 11 km east of Chelinda on road to Kasaramba, fl. 4.iii.1977, *Pawek* 12441 (K; MAL; MO; SRGH). C: Dedza Mt., fl. 20.iii.1955, *Exell, Mendonça & Wild* 1093 (BM; LISC; SRGH).

Known only from Dedza Mt. and the Nyika Plateau. Miombo woodland and montane grassland; 1650–2300 m.

132. **Crotalaria pilosiflora** Baker in Bull. Misc. Inform., Kew **1897**: 251 (1897). —Milne-Redhead in Kew Bull. **15**: 163 (1961). —Binns, First Check List Herb. Fl. Malawi: 79 (1968). —Brummitt in Wye Coll. Malawi Proj. Rep.: 64 (1973). —Polhill, Crotalaria Africa & Madagascar: 290 (1982). —Lock, Leg. Afr. Check-list: 197 (1989). Type: Malawi, Nyika Plateau, *Whyte* s.n. (K, holotype).

Crotalaria oocarpa sensu E.G. Baker in J. Linn. Soc., Bot. **42**: 339 (1914) pro parte. —sensu Verdoorn in Bothalia **2**: 408 (1928) pro parte.

Shrubby herb to 1.5 m; branches spreading pilose. Leaves 3-foliolate, with tufts of smaller leaves in main axils; leaflets (1)1.5–3.5 × 0.6–1.5 cm, elliptic, slightly to distinctly acuminate to the apiculate tip, pilose beneath; petiole of main leaves 1–3 cm long; stipules 2–4 mm long, linear-subulate. Inflorescences of 2 sorts; main racemes 6–12 cm long, ± laxly 6–12-flowered; supplementary inflorescences 1-flowered from leaf-tufts on a peduncle 1–1.5 cm long; bracts longer than the pedicel, 5–6 mm long, linear-lanceolate; bracteoles just below the calyx, setaceous. Calyx 7–9 mm long, densely pilose; lobes attenuately triangular, ± twice as long as the tube. Standard elliptic, yellow, ?finely reddish lined with age, pilose outside; wings ± as long as the keel; keel 1.1–1.3 cm long, angular, with a narrow twisted beak. Pod with a stipe c. 2 mm long, 1.2–1.5 × 0.7–0.8 cm, obliquely obovoid, furry, 12–14-seeded. Mature seeds not seen.

Malawi. N: Chitipa Distr., Nyika Plateau, eastern foot of Nganda, fl. & fr. 31.vii.1972, *Brummitt & Synge* in *Wye College* WC 104 (K; MAL; SRGH); Rumphi Distr., Livingstonia Escarpment, fl. & fr. 3.vii.1970, *Pawek* 3552 (K).

Known only from the Nyika Plateau in northern Malawi. Miombo woodland, often growing in grassy places and along streams; 950–2100 m.

133. **Crotalaria oocarpa** Baker in Bull. Misc. Inform., Kew **1897**: 252 (1897). —E.G. Baker in J. Linn. Soc., Bot. **42**: 339 (1914) pro parte. —Verdoorn in Bothalia **2**: 408 (1928) pro parte. —Wilczek in F.C.B. **4**: 142 (1953). —Milne-Redhead in Kew Bull. **15**: 162 (1961). —Binns, First Check List Herb. Fl. Malawi: 79 (1968) excl. syn. —Polhill in F.T.E.A., Leguminosae, Pap.: 968 (1971); Crotalaria Africa & Madagascar: 290, fig. 76/17 (1982). —Lock, Leg. Afr. Check-list: 194 (1989). Type: Malawi, between Mpata and the commencement of the Nyasa-Tanganyika Plateau, *Whyte* s.n. (BM, sketch; K, holotype, not found).

Similar to *C. pilosiflora* but less robust, leaflets mostly 1–2.5 × 0.4–1 cm, oblanceolate to obovate, apically rounded, pilose on both surfaces, the flowers and pods often rather smaller.

Malawi. N: Mzimba Distr., 6.5 km SW of Chikangawa, fl. & fr. 19.viii.1978, *E. Phillips* 3788 (K; MO).

Also in Burundi and the highlands of Tanzania, with a subspecies in Yemen, Ethiopia, Kenya and N Tanzania. Upland grassland; c. 1700 m.

134. **Crotalaria virgulata** Klotzsch in Peters, Naturw. Reise Mossambique **6**, part 1: 56 (1861). —J.G. Baker in F.T.A. **2**: 30 (1871) as "*virgatula*". —E.G. Baker in J. Linn. Soc., Bot. **42**: 337 (1914). —Verdoorn in Bothalia **2**: 410 (1928). —Burtt Davy, Fl. Pl. Ferns Transvaal, pt. 2: 400 (1932). —Brenan in Mem. New York Bot. Gard. **8**: 247 (1953). —Binns, First Check List Herb. Fl. Malawi: 80 (1968). —Polhill in F.T.E.A., Leguminosae, Pap.: 968 (1971). —Drummond in Kirkia **8**: 219 (1972). —Jacobsen in Kirkia **9**: 160 (1973). —Gonçalves in Garcia de Orta, Sér. Bot. **5**: 73 (1982). —Polhill, Crotalaria Africa & Madagascar: 290, fig. 79 (1982). —Lock, Leg. Afr. Check-list: 210 (1989). Syntypes: Mozambique, Sena, *Peters* (B†, syntypes; K; P).

Decumbent to erect annual or short-lived perennial, up to 1.8 m tall; branches subappressed pubescent to spreading hirsute. Leaves 3-foliolate, sometimes mostly on short shoots; leaflets 4–50 × 2–15 mm, linear or oblanceolate to elliptic-oblong or obovate, hairy on both surfaces or only beneath; petiole 2–30 mm long; stipules 1–3 mm long, subulate to attenuate-triangular. Racemes usually of two sorts; primary racemes terminal or leaf-opposed on long shoots, 2–40 cm long, laxly 3–40-flowered, indeterminate (ending in a bud) or determinate (ending in a bristle-like extension of the rhachis); supplementary inflorescences usually developed (at least later in the season) from short shoots in axils of the main leaves, with 1–2 flowers on a filiform axis 0.2–2.5 cm long; bracts 0.5–4 mm long, linear-subulate to linear-lanceolate, usually shorter than the pedicel; bracteoles on the pedicel, setaceous. Calyx 3.5–9 mm long, pubescent; lobes narrowly attenuate-triangular (rarely somewhat oblong and apiculate in subsp. *pauciflora*), 1.5–2.5 times as long as the tube, lanate along the inner margins (particularly of the lower lobes) when young. Standard broadly elliptic, yellow, often lined red, pubescent or rarely almost glabrous outside; wings broadened upwards, nearly as long as the keel; keel 6–12(14) mm long, angular, with a narrow twisted beak. Pod shortly stipitate, 8–15 × 5–9 mm, ± oblong-obovoid (sometimes only slightly broadened apically), pubescent outside and sometimes also inside, 6–10(12)-seeded. Seeds 2.2–3 mm long, oblique-cordiform, smooth, brown.

1. Primary racemes indeterminate, (8)10–27 cm long, (8)16–40-flowered (TAB. 3,7: **39**, fig. A); supplementary inflorescences often sparse or lacking at least until late in the season, short; bracts 2–4 mm long; leaflets linear to oblong-elliptic, broadest near middle, main ones mostly 2–5 cm long · · · · · · · · · · · · i) subsp. *virgulata*
– Primary racemes determinate, 2–8(12) cm long, 3–8(12)-flowered (TAB. 3,7: **39**, fig. B); supplementary 1–2-flowered inflorescences sometimes predominant; bracts 0.5–2(3) mm long; leaflets (except sometimes in subsp. *longistyla*) usually distinctly broadest above the middle, often smaller · · · · · · · · · · · · 2
2. Leaves expanding on short shoots to form distinct tufts of 5 or more at the nodes; supplementary inflorescences usually predominant and sharply differentiated · · · · · · · 3
– Leaves mostly not expanding until short shoot grows out, the branch system thus more divaricate and the leaves less obviously in tufts · · · · · · · · · · · · 4
3. Axis of supplementary inflorescences shorter than surrounding leaves, 0.2–1 cm long excluding the pedicel; many leaflets often 8–20 mm long · · · · · · · · · ii) subsp. *longistyla*
– Axis of supplementary inflorescences exceeding the leaf-tuft, usually 1–2.5 cm long (if shorter then leaves small and crowded); leaflets 4–12(20) mm long · · · iii) subsp. *forbesii*
4. Supplementary inflorescences often not sharply differentiated, produced on short shoots that have grown out to some extent; hairs often golden-yellow; lateral nerves often obscure on underside of the leaflets · · · · · · · · · · · · iv) subsp. *pauciflora*
– Supplementary inflorescence with a filiform axis from near base of the short shoot, appearing axillary; hairs whitish; lateral nerves usually visible on underside of the leaflets · · · · · · · · · · · · v) subsp. *grantiana*

i) Subsp. **virgulata**. TAB. 3,7: **39**, fig. A.

Crotalaria spinosa sensu J.G. Baker in F.T.A. **2**: 17 (1871) quoad specim. *Stewart*.

Crotalaria shamvaensis Verdoorn in Bothalia **2**: 405 (1928). Type: Zimbabwe, Shamva, *Mainwaring* in *Eyles* 2244 (K, holotype; SRGH).

Erect, 0.6–1.5 m; branches ascending, leafy with many short shoots, subappressed pubescent to spreading hirsute. Leaflets (main ones) mostly 2–5 × 0.2–1.5 cm, linear to oblong-elliptic, usually hairy above when young, glabrescent; petioles of main leaves often 1–3 cm long. Primary racemes (8)10–27 cm long, (8)16–40-

Tab. 3,7: **39**. —Growth habit of CROTALARIA VIRGULATA (× 2/3). A. —subsp. VIRGULATA, from *Lemos & Macuácua* 66; B. —subsp. LONGISTYLA, from *Corby* 1237; C. —subsp. FORBESII, from *Brummitt* 11425; D. —subsp. PAUCIFLORA, from *McCallum-Webster* 752 and *Salubeni* 758; E. —subsp. GRANTIANA, from *McClean* 561 and *J.H. Ross* 849. Drawn by Roger Polhill. From Crotalaria Africa & Madagascar.

flowered, indeterminate; supplementary inflorescences often sparse or lacking until late in the season, short, 1–2-flowered from leaf-tufts; bracts 2–4 mm long. Keel 10–12 mm long.

Zambia. C: Lusaka Distr., Kafue R., Iolanda, fl. & fr. 14.iii.1965, *E.A. Robinson* 6433 (K; SRGH). E: Chipata Distr., Machinje Hills, Jumbe, fl. 4.iii.1966, *Astle* 4632 (K; SRGH). S: Mazabuka Distr., Munali (Nega Nega) Hills, Munali Pass, fl. & fr. 4.iv.1957, *Angus* 1552 (K; SRGH). **Zimbabwe**. N: Mount Darwin Distr., upper reaches of Nyarandi (Nyatandi) R., fl. 27.i.1960, *Phipps* 2424 (K; SRGH). W: Hwange Distr., Kazungula, fl. & fr. iv.1955, *R.M. Davies* 1106 (K; SRGH). E: Mutare Distr. Sungwizi (Tsungwesi) R., fl. & fr. 31.v.1936, *Eyles* 8600 (K; SRGH). S: Chiredzi Distr., Runde (Lundi) R., Fishan, fl. & fr. 26.iv.1962, *Drummond* 7736 (K; LISC; SRGH). **Malawi**. S: Nsanje Distr., hills near Namyale R., fl. 25.iii.1960, *Phipps* 2736 (K; SRGH). **Mozambique**. T: 3 km Luenha (Changara)–Cuchumane, fl. & fr. 21.v.1971, *Torre & Correia* 18543 (K; LISC). MS: Chemba Distr., Chiou, Campo Experimental do C.I.C.A., fl. & fr. 12.iv.1960, *Lemos & Macuácua* 89 (COI; K; LISC); Revuè–Manica (Macequece), fl. 10.iii.1948, *Barbosa* 1163 (K; LISC). GI: Massangena, fl. vii.1932, *J.C. Smuts* P.349 (K; PRE) — poor specimen.

Occurs either side of the Zimbabwe–Mozambique divide and along the Zambezi from Sena to a little above the Victoria Falls, extending a little way up the northern tributaries of the Shire, Luangwa and Kafue. Woodland, savanna and grassland; 150–1200 m.

This is the most distinctive subspecies, but intergrades completely with subsp. *forbesii* in the northeast and to a lesser extent westwards with subsp. *longistyla* and southwards with subsp. *grantiana*. Subspecies *virgulata* and *forbesii* both occur in the fork of the Zambezi and Shire Rivers up to Tete and Blantyre and adjacent regions south of the Zambezi, probably with subsp. *virgulata* usually on the higher ground, but intermediate forms of all sorts seem to predominate. Unfortunately the syntypes come from this region and are nearly intermediate. An association with the form represented by *C. shamvaensis* rather than the form represented by *C. forbesii* seems correct with the many collections now available from the same general area, but long-standing doubts about the circumscription of *C. virgulata* are understandable.

ii) Subsp. **longistyla** (Baker f.) Polhill, Crotalaria Africa & Madagascar: 292, fig. 79/2 (1982). —Lock, Leg. Afr. Check-list: 210 (1989). TAB. 3,7: **39**, fig. B. Type: Zimbabwe, Harare–Mutare, *F.A. Rogers* 4064 (BM, holotype).

Crotalaria longistyla Baker f. in J. Bot. **58**: 75 (1920). —Verdoorn in Bothalia **2**: 409 (1928) pro parte.

Bushy, 1–1.8 m tall; branches ascending, usually very leafy, the internodes mostly short and with tufts of leaves (on short shoots) at the nodes, white subappressed pubescent to spreading pilose. Leaflets often mostly 8–20 × 2–5 mm, oblanceolate to narrowly elliptic-obovate, glabrous to pilose above; petioles of main leaves (2)5–10(15) mm long. Primary racemes few, sometimes terminal, mostly leaf-opposed, 2–6 cm long, 3–6(8)-flowered, determinate; supplementary inflorescences from the tuft-like short shoots, 1–2-flowered on a short axis 0.2–1 cm long; bracts 0.5–2(3) mm. Keel (8)10–12(14) mm long.

Zimbabwe. N: 1.5 km west of Tsatse (Tsatsi) R. bridge on Concession to Mutorashanga (Mtoroshanga) road, fl. 15.iv.1960, *Leach & Brunton* 9843 (K; SRGH). C: Harare (Salisbury) to Marondera (Marandellas), Cleveland Dam Catchment, *Corby* 1237 (K; SRGH). S: Buhera Distr., 24 km Buhera–Birchenough Bridge, fl. 21.iv.1969, *Biegel* 2936 (K; SRGH).

Central plateau of Zimbabwe, extending along spurs towards the Zambezi, contiguous with subsp. *virgulata* and interdigitating in the northeast. Woodland and wooded grassland; 1000–1600 m.

Wild 7336 (K; SRGH) from an outcrop of copper oxide ore at Copper Queen and *Leach* 11160 (K; SRGH) from 43 km NW of Copper Queen at Ganderowe Falls may belong here, but are markedly depauperate and simulate subsp. *forbesii* to some extent.

iii) Subsp. **forbesii** (Baker) Polhill, Crotalaria Africa & Madagascar: 292, fig. 79/3 (1982). — Lock, Leg. Afr. Check-list: 210 (1989). TAB. 3,7: **39**, fig. C. Lectotype, selected by Polhill, loc. cit.: Mozambique, without locality, but probably Zambezi between Sena and coast, *Forbes* (K, lectotype).

Crotalaria laevigata var. *villosior* Benth. in Hooker, London J. Bot. **2**: 577 (1843). Type as for subsp. *forbesii*.

Crotalaria forbesii Baker in F.T.A. **2**: 18 (1871). —E.G. Baker in J. Linn. Soc., Bot. **42**: 340 (1914).

Crotalaria aculeata sensu Brenan in Mem. New York Bot. Gard. **8**: 246 (1953) quoad specim. *Brass* 18020. —sensu Binns, First Check List Herb. Fl. Malawi: 78 (1968).

Erect or rarely decumbent; branches rather long and straight with obvious tufts of small leaves (on short shoots) at the nodes, subappressed white pubescent. Leaflets 4–12(20) × 2–7(10) mm, linear-oblanceolate to obovate, usually hairy above when young, glabrescent; petiole 2–8 mm long. Primary racemes few, sometimes terminal, mostly leaf-opposed, slender, 2–5 cm long, 4–6-flowered, determinate; supplementary inflorescences numerous from tuft-like short shoots, 1–2-flowered on a filiform axis usually 1–2.5 cm (sometimes shorter, but then leaves also small and crowded); bracts 0.5–2 mm. Keel 8–12 mm long.

Malawi. N: Karonga Distr., Lake Malawi, Ngara (Ngala), fr. 8.vii.1973, *Pawek* 7180 (K). C: Nkhotakota (Nkhota Kota), fl. & fr. 15.vi.1970, *Brummitt* 11425 (K; MAL; SRGH). S: Lower Mwanza R., fl. & fr. 6.x.1946, *Brass* 18020 (K; SRGH). **Mozambique**. N: Mecufi–R. Lúrio, fl. 29.x.1942, *Mendonça* 1114 (BM; K; LISC). Z: 45.5 km Mopeia–Régulo Changalaze, fl. & fr. 2.viii.1949, *Barbosa & Carvalho* 3804 (K; LISC; LMA). T: Changara Distr., Boruma (Boroma), fl. viii.1931, *Sofia Pomba* 37 (COI).* MS: Marromeu Distr., Chupanga, fl. & fr. 13.vii.1941, *Torre* 3093 (BM; K; LISC).

Also in eastern Tanzania, extending southwards in Mozambique to Gorongosa, up the Zambezi and Shire Rivers, intergrading with subsp. *virgulata* (see above) in Tete and southern Malawi, then common all around Lake Malawi. Flowers almost all year round. Woodland and wooded grassland; 40–500(650) m.

iv) Subsp. **pauciflora** (Baker) Polhill, Crotalaria Africa & Madagascar: 293, fig. 79/4 (1982). — Lock, Leg. Afr. Check-list: 210 (1989). TAB. 3,7: **39**, fig. D. Type: Malawi, Nyika Plateau, *Whyte* s.n. (K, holotype).

Crotalaria pauciflora Baker in Bull. Misc. Inform., Kew **1897**: 251 (1897). —E.G. Baker in J. Linn. Soc., Bot. **42**: 399 (1914). —Verdoorn in Bothalia **2**: 408 (1928). —Binns, First Check List Herb. Fl. Malawi: 79 (1968).

Crotalaria forbesii subsp. *vanmeelii* R. Wilczek in Bull. Jard. Bot. État **23**: 148 (1953). Type from Tanzania (Ufipa).

Decumbent to erect, divaricately branched; hairs subappressed to spreading, often yellow. Leaflets mostly 8–20 × 3–8 mm, oblanceolate to elliptic-obovate, usually glabrous above; lateral nerves ± obscure beneath; petioles 3–10(12) mm long. Primary racemes terminal and leaf-opposed, 2–8(12) cm long, 3–8(12)-flowered, determinate; 1–2-flowered supplementary inflorescences often not sharply differentiated, produced at or near ends of short shoots that have mostly grown out to some extent; bracts 1–2 mm long. Keel 7–13 mm long.

Zambia. N: Mbala Distr., Kawimbe, fl. 6.iv.1959, *McCallum-Webster* 752 (K). E: Isoka Distr., Nyika Plateau, fr. 7.vi.1962, *Verboom* 641 (K). **Malawi**. N: Nyika Plateau, vi.1896, *Whyte* s.n. (K). C: 11 km south of Ntchisi, fr. 18.vi.1970, *Brummitt* 11545 (K; LISC; MAL; SRGH); Dedza Distr., Chongoni Forest Reserve, 2.vi.1967, *Salubeni* 758 (K; SRGH).

Also in Tanzania (vicinity of Lake Tanganyika) and Dem. Rep. Congo (vicinity of Lake Mweru). Woodland and grassland; (700)1500–2100 m.

Typically from high ground west of Lake Malawi across to the Ufipa Plateau. Plants from lower down round Lakes Tanganyika and Mweru tend to be more robust, with more contracted short shoots, whitish hairs and large flowers, so that they are morphologically less dissimilar from the more southerly subspecies. *Richards* 9426A (K) from Lake Mweru, Kafulwe, is decidedly anomalous, with tufts of small leaves (like subsp. *forbesii*) and short axes bearing the supplementary inflorescences. Compared with *Richards* 9472 (K) from the same locality it appears to be just a depauperate form of this subspecies, but morphologically it seems more like an odd form of subsp. *forbesii.*

v) Subsp. **grantiana** (Harv.) Polhill, Crotalaria Africa & Madagascar: 293, fig. 79/5 (1982). — Lock, Leg. Afr. Check-list: 210 (1989). TAB. 3,7: **39**, fig. E. Type from South Africa (KwaZulu Natal).

Crotalaria grantiana Harv. in F.C. **2**: 43 (1862). —E.G. Baker in J. Linn. Soc., Bot. **42**: 341 (1914) excl. specim. *Menyharth.* —Verdoorn in Bothalia **2**: 409 (1928). —Burtt Davy, Fl. Pl. Ferns Transvaal, pt. 2: 400 (1932).

Decumbent to erect and bushy; hairs whitish. Leaflets 5–20 × 2–8 mm, oblanceolate to elliptic-obovate, glabrous or hairy above; lateral nerves usually visible

---

*Almost all specimens from Tete are intermediates with subsp. *virgulata.*

beneath; petioles (2)4–12 mm long. Primary racemes usually all leaf-opposed, c. 2–6 cm long, 3–8-flowered, determinate; supplementary inflorescences 1–2-flowered on a filiform axis 1–2 cm long from near base of short shoots, thus appearing axillary; bracts 1–2(3) mm long. Keel 6–8(10) mm long.

Botswana. SE: Kgatleng Distr., Mochudi, fl. iii.1914, *F.A. Rogers* 6735 (K). **Zimbabwe**. S: Chipinge Distr., Rimai (Remayi), 1.ii.1975, *Biegel, Pope & Russell* 4917 (K; MO; SRGH). **Mozambique**. MS: Machanga Distr., Chiloane, iii.1887, *L. Scott* (K). GI: 26.6 km da estrada Vilankulo (Vilanculos)–Nova Mambone (Mambone) para Mabote, fl. & fr. 28.iii.1952, *Barbosa & Balsinhas* 5060 (K; LMA). M: Umbelúzi–Matutuíne (Bela Vista), fl. & fr. 14.iv.1944, *Torre* 6428 (K; LISC).

Also in South Africa, southwards to the Eastern Cape Province. Woodland and savanna; 0–950 m.

Sometimes cultivated as a green manure, e.g. Zimbabwe, Mazowe (Mazoe), Henderson Research Station, *Wild* 3997 (K; LISC; SRGH).

In its typical form this subspecies occurs within the Flora Zambesiaca area only in coastal parts of Mozambique south of Beira. Where it extends inland, principally in the Limpopo drainage area, the short shoots are more marked, the primary racemes are more developed, the axes of the supplementary inflorescences are shorter and less filiform and the flowers are larger, approaching subsp. *virgulata*. *Pedro & Pedrógão* 948 (LMA) from Maputo, Boane–Impamputo, Movene, comes close to subsp. *virgulata* and some intermediate material from southern Zimbabwe seems better referred to subsp. *virgulata*. *Wild* 7507 (K; SRGH) from the Umkondo Mine in the Save Valley, together with *M.D. Dale* 31 (K; SRGH) from the same place, is from a malachite outcrop and the reduction is probably largely induced physiologically.

135. **Crotalaria misella** Polhill, Crotalaria Africa & Madagascar: 293, fig. 80 (1982). —Lock, Leg. Afr. Check-list: 192 (1989). Type: Mozambique, Mueda Distr., andados 30 km de Chomba para Negomano, ao longo do rio Matiu, *Torre & Paiva* 11873 (COI; EA; K; LISC, holotype; LMU; SRGH).

Prostrate annual with slender thinly puberulous stems to 60 cm long. Leaves all well spread, 3-foliolate; leaflets 8–12 × 7–11 mm, broadly obovate or obcordate, thinly appressed puberulous beneath; petiole c. 8–12 mm long; stipules 1–2 mm long, subulate. Flowers 2 at the top of a slender leaf-opposed peduncle 0.8–2 cm long; bracts 1–1.5 mm long, linear; bracteoles on the pedicel, minute. Calyx 2.5–3 mm long, puberulous; upper lobes rather broadly triangular, acuminate, a little longer than the tube. Standard narrowly elliptic-obovate, yellow, glabrous; wings a little shorter than the keel; keel 4.5–5 mm long, angular, with a narrow twisted beak. Pod subsessile, globose, 4 mm across, minutely puberulous, 2–4-seeded. Seeds not seen.

Mozambique. N: Mueda Distr., andados 30 km de Chomba para Negomano, ao longo do rio Matiu, c. 300 m, fl. & fr. 13.iv.1964, *Torre & Paiva* 11873 (COI; EA; K; LISC; LMU; SRGH).

Known only from the type, collected from riverine woodland on sand; c. 300 m.

136. **Crotalaria kapiriensis** De Wild. in Bull. Jard. Bot. État **5**: 23 (1915). —E.G. Baker, Legum. Trop. Africa: 37 (1926). —Wilczek in F.C.B. **4**: 109 (1953) as "*kapirensis*". —Polhill, Crotalaria Africa & Madagascar: 294 (1982). —Lock, Leg. Afr. Check-list: 184 (1989). Type from Dem. Rep. Congo (Katanga).

Crotalaria spinosa subsp. *aculeata* sensu E.G. Baker in J. Linn. Soc., Bot. **42**: 312 (1914) pro parte.

Crotalaria spinosa var. *macrocarpa* Baker f. in J. Linn. Soc., Bot. **42**: 313 (1914). Type: Zambia, Mumbwa, *Macaulay* 627 (K, holotype).

Crotalaria aculeata sensu Verdoorn in Bothalia **2**: 413 (1928) pro parte.

Similar to *C. aculeata* but up to 2 m tall and larger in all parts. Calyx 7–8 mm long; lobes narrowly attenuate-triangular, 2–2.5 times as long as the tube. Keel 1.5–1.8 cm long. Pod 2–3 × (1)1.2–1.4 cm, almost oblong-ellipsoid, only slightly broadened distally, c. 28–36-seeded.

Zambia. W: Kitwe, fl. & fr. 18.iii.1955, *Fanshawe* 2149 (K; NDO; SRGH). C: Lusaka Distr., Mt. Makulu, fl. & fr. 4.iv.1955, *Exell, Mendonça & Wild* 1399 (BM; LISC; SRGH). S: Choma National Forest (Siamambo Forest Reserve), fl. & fr. 13.iii.1960, *White* 7769 (FHO; K). **Zimbabwe**. W: Hwange Distr., Kazungula, fr. iv.1955, *R.M. Davies* 1117 (K; SRGH). C: Gweru (Gwelo) Kopje, fl. & fr. 5.iii.1967, *Biegel* 1967 (K; SRGH). E: Nyanga Mts., Rhodes Hotel, iii.1969, *Corby* 2099 (K; SRGH).

Also in Dem. Rep. Congo (Katanga). In seasonally damp or disturbed places, probably beginning to spread as a weed; 900–1400 m.

137. **Crotalaria aculeata** De Wild. in Ann. Mus. Congo, Sér. IV, Bot. [Études Fl. Katanga] **1**: 185, t. 46 (1903). —Verdoorn in Bothalia **2**: 413 (1928). —Wilczek in F.C.B. **4**: 107 (1953). —Brenan in Mem. New York Bot. Gard. **8**: 246 (1953) excl. specim. *Brass* 18020. —Torre in C.F.A. **3**: 50 (1962). —Polhill in F.T.E.A., Leguminosae, Pap.: 969 (1971); Crotalaria Africa & Madagascar: 295 (1982). —Lock, Leg. Afr. Check-list: 163 (1989). Type from Dem. Rep. Congo (Katanga).

Erect spiny short-lived perennial, wiry to bushy, 0.3–1.5 m tall; branches subappressed to crisped pubescent. Leaves mostly clustered on short shoots, 3-foliolate; leaflets 0.5–1.5(2.5) × 0.2–0.6(1.2) cm, oblanceolate, truncate or emarginate, appressed pubescent beneath; petiole of main leaves 0.3–1.2(2.5) cm long; stipules 1–2 mm long, linear-subulate, recurved. Flowers mostly solitary on 0.7–2.5(3.5) cm long spine-tipped axes from short shoots all along the branches, very rarely with occasional short racemes on long shoots; bracts 0.7–1(2) mm long, linear; bracteoles on the pedicel, setaceous. Calyx (3)4–5 mm long, appressed pubescent; lobes triangular, acuminate to attenuate, 1–2 times as long as the tube. Standard elliptic, yellow lined and later flushed reddish, densely pubescent outside; wings as long as the keel; keel 0.8–1.2 cm long, angular, with a long narrow twisted beak. Pod subsessile, 1–2.3 × 0.5–0.9 cm, oblong-obovoid to cylindrical, puberulous, c. 10–28-seeded. Seeds 2–2.5 mm long, oblique-cordiform, smooth, brown.

Subsp. **aculeata**

Crotalaria spinosa subsp. *aculeata* (De Wild.) Baker f. in J. Linn. Soc., Bot. **42**: 312 (1914); Legum. Trop. Africa: 37 (1926).

Pod 1–1.5 cm long, 0.5–0.8 cm in dorsi-ventral diameter, 1.5–2.5 times as long as broad, oblong-obovoid to shortly oblong-clavate, distinctly broadened distally.

Zambia. B: Sesheke Distr., Lusu, fl. & fr. 10.vii.1962, *Fanshawe* 6927 (K; NDO). N: Mbala Distr., Kawimbe, fl. 2.v.1957, *Richards* 9513 (K; SRGH). W: Solwezi Distr., R. Mwafwe, fl. & fr. 18.vii.1930, *Milne-Redhead* 730 (K). C: Lusaka, University of Zambia Campus, fl. & fr. 4.ix.1972, *Strid* 2053 (K). S: Mazabuka Distr., west of Mazabuka, Kafue Flats, Nanga Pilot Polder, fl. & fr. 13.v.1957, *Angus* 1590 (K; SRGH). **Malawi**. N: Chitipa Distr., 3 km on road to Nthalire, fr. 18.ix.1975, *Pawek* 10143 (K; MAL; MO; SRGH).

Also west to Angola, northwards into Dem. Rep. Congo (Katanga) and southern and western Tanzania, and in Madagascar.

Subsp. *claessensii* (De Wild.) Polhill, with more elongate pods, extends through east Dem. Rep. Congo, northwest Tanzania, Burundi, Rwanda and Uganda to the Sudan. Grassland, seasonally damp and disturbed places, probably spreading; 900–1700 m.

See note below under *Crotalaria spinosa*.

138. **Crotalaria spinosa** Hochst. ex Benth. in Hooker, London J. Bot. **2**: 576 (1843) excl. *Klotschy* 522. —J.G. Baker in F.T.A. **2**: 17 (1871) pro parte. —E.G. Baker in J. Linn. Soc., Bot. **42**: 312, t. 13B (1914) pro parte. —Wilczek in F.C.B. **4**: 105 (1953) pro majore parte. —Brenan in Mem. New York Bot. Gard. **8**: 246 (1953). —Torre in C.F.A. **3**: 49 (1962). —Polhill in F.T.E.A., Leguminosae, Pap.: 970 (1971); Crotalaria Africa & Madagascar: 296 (1982). —Lock, Leg. Afr. Check-list: 204 (1989). Lectotype, selected by Brenan, loc. cit., from Ethiopia.

Similar to *C. aculeata* but smaller in all parts. Erect or spreading annual or short-lived perennial, 0.2–0.8 m tall. Leaflets mostly 4–15 × 1–4 mm, oblanceolate to obtriangular, truncate or emarginate; petioles c. 2–8 mm long. Calyx 2.5–3 mm long. Keel 4–6 mm long. Pods 7.5–9 × 3.5–5 mm, oblong-obovoid, c. 6–8-seeded.

Zambia. N: Mbala–Tunduma road, Saisi R., fl. 12.iv.1962, *Richards* 16329 (K; SRGH). W: Kitwe, fl. & fr. 28.iii.1967, *Fanshawe* 10008 (K; NDO). C: Lusaka, University Campus, fl. & fr. 12.iii.1972, *Kornaś* 1396 (K). S: east of Chunga, Lochinvar National Park, fl. & fr. 14.iii.1972, *van Lavieren, Sayer & Rees* 734 (K; SRGH). **Zimbabwe**. N: c. 1.5 km from Chinhoyi (Sinoia) on old road from Chinhoyi (Sinoia) to Karoi, fl. 18.ii.1969, *Corby* 2084 (K; SRGH). E: Nyanga, 10.iii.1964, *Corby* 1086 (SRGH).

Principally along and between the Western and Great Rift Valleys northwards to Ethiopia, but also in Angola, and with scattered modern records in other parts of tropical Africa; no records for the Flora Zambesiaca area existed at Kew prior to 1955, but this species is apparently now well established in the Kafue Basin and elsewhere. Grassland and disturbed places, usually in sites with a high water-table or catchment areas of local run-off; 950–1500 m.

C. kapiriensis and *C. aculeata* have been included in this species at times and as pointed out by Brenan (1953) they differ in no essentials other than the size of the parts. Nevertheless where all three grow together, around Lusaka for example, the differences are maintained. The species are probably all spreading here as weeds of disturbed ground and overgrazed places. Where the natural ranges abut in Dem. Rep. Congo (Katanga) and along the Western Rift Valley some slight introgression does seem to occur, but if examined carefully, very few specimens are difficult to place.

139. **Crotalaria eremicola** Baker f. in J. Linn. Soc., Bot. **42**: 311 (1914). —Verdoorn in Bothalia **2**: 414 (1928) as "*erimicola*". —Burtt Davy, Fl. Pl. Ferns Transvaal, pt. 2: 400 (1932). —Schreiber in Merxmüller, Prodr. Fl. SW. Afrika, fam. 60: 23 (1970). —Polhill, Crotalaria Africa & Madagascar: 297 (1982). —Lock, Leg. Afr. Check-list: 177 (1989). Type from Namibia.

Perennial with a number of erect, slender, spiny stems from a taproot or small branched rootstock, 15–40 cm tall; branches strigulose to subappressed or crisped puberulous. Leaves 3-foliolate; leaflets 3–8 × 1–3.5 mm, oblanceolate to obovate, puberulous to strigose beneath; petiole 1–5 mm long; stipules 0.5–1.5 mm long, subulate or linear-lanceolate. Flowers 1–2 on slender patent 0.5–1(1.5) cm long spines all along the branches; bracts 0.5–1 mm long, linear-lanceolate; bracteoles on the pedicel, small. Calyx 3–4 mm long, appressed puberulous; lobes triangular, 1–2 times as long as the tube. Standard obovate to oblate, yellow, subglabrous to pubescent on the upper half outside; wings nearly as long as the keel; keel 5–11 mm long, angular, with a narrow twisted beak. Pod subsessile, 4–7 mm across, subglobose, appressed puberulous, 2–4-seeded. Seeds 2 mm long, oblique-cordiform, rugulose, brown.

Subsp. **parviflora** Polhill, Crotalaria Africa & Madagascar: 298 (1982). —Lock, Leg. Afr. Check-list: 177 (1989). Type from South Africa (Northern Province).

Crotalaria spinosa var. *schlechteri* Baker f. in J. Linn. Soc., Bot. **42**: 312 (1914). Type from South Africa (Gauteng).

Keel 5–6 mm long.

Botswana. SE: 60 km NW of Serowe, fl. & fr. 25.iii.1965, *Wild & Drummond* 7299 (K; SRGH). **Mozambique**. GI: Guijá Distr., entre Chókwe (Guijá) e Nalazi, fl. & fr. 10.v.1944, *Torre* 6611 (LISC).

Also in South Africa, southwards to the Northern Cape Province. Open places, grassland on sand and valley depressions on clay with *Acacia*; 100–900 m.

Subsp. *eremicola*, with keel 8–11 mm long, occurs lower down the Orange R. and in southern Namibia.

140. **Crotalaria prolongata** Baker in F.T.A. **2**: 25 (1871). —E.G. Baker in J. Linn. Soc., Bot. **42**: 284 (1914). —Wilczek in F.C.B. **4**: 132 (1953). —Torre in C.F.A. **3**: 40 (1962). —Polhill in F.T.E.A., Leguminosae, Pap.: 972 (1971); Crotalaria Africa & Madagascar: 308, fig. 83/1–6 (1982). —Lock, Leg. Afr. Check-list: 198 (1989). TAB. 3,7: **40**, fig. A. Type from Angola.

Erect annual, 15–75 cm tall, later developing a few curved-ascending branches from near the base; stem densely covered with longish (0.5–1 mm) slightly spreading hairs. Leaves mostly 3-foliolate, the uppermost 1-foliolate; leaflets 2–7 × 1–3.5 cm, oblong-oblanceolate to oblong-obovate, rounded to a markedly apiculate apex, finely pubescent on both surfaces, with lateral nerves slightly raised beneath; petiole 1.5–4 cm long, channelled; stipules 2–4 mm long, setaceous. Racemes terminal, sessile, 3–15 cm long, ± congested, occasionally branched, also with flowers clustered in axils below; bracts 2–4 mm long, setaceous; bracteoles on the pedicel, 0.7–1 mm long. Calyx 3.5–4.5 mm long, densely silvery appressed pubescent; upper lobes narrowly attenuate-triangular, ± twice as long as the tube. Standard elliptic or ovate, dull yellow, flushed reddish-brown or maroon, extensively pubescent outside; wings brighter yellow, a little shorter than the keel; keel 8–9 mm long, angular, with a long narrow twisted beak incurved a the tip. Pod sessile, 6–8 × 4–5 mm, shortly oblong-obovoid, densely appressed pubescent outside, hairy along suture inside, 1–2-seeded. Seeds 3–3.5 mm long, very obliquely cordiform, smooth, yellowish-brown.

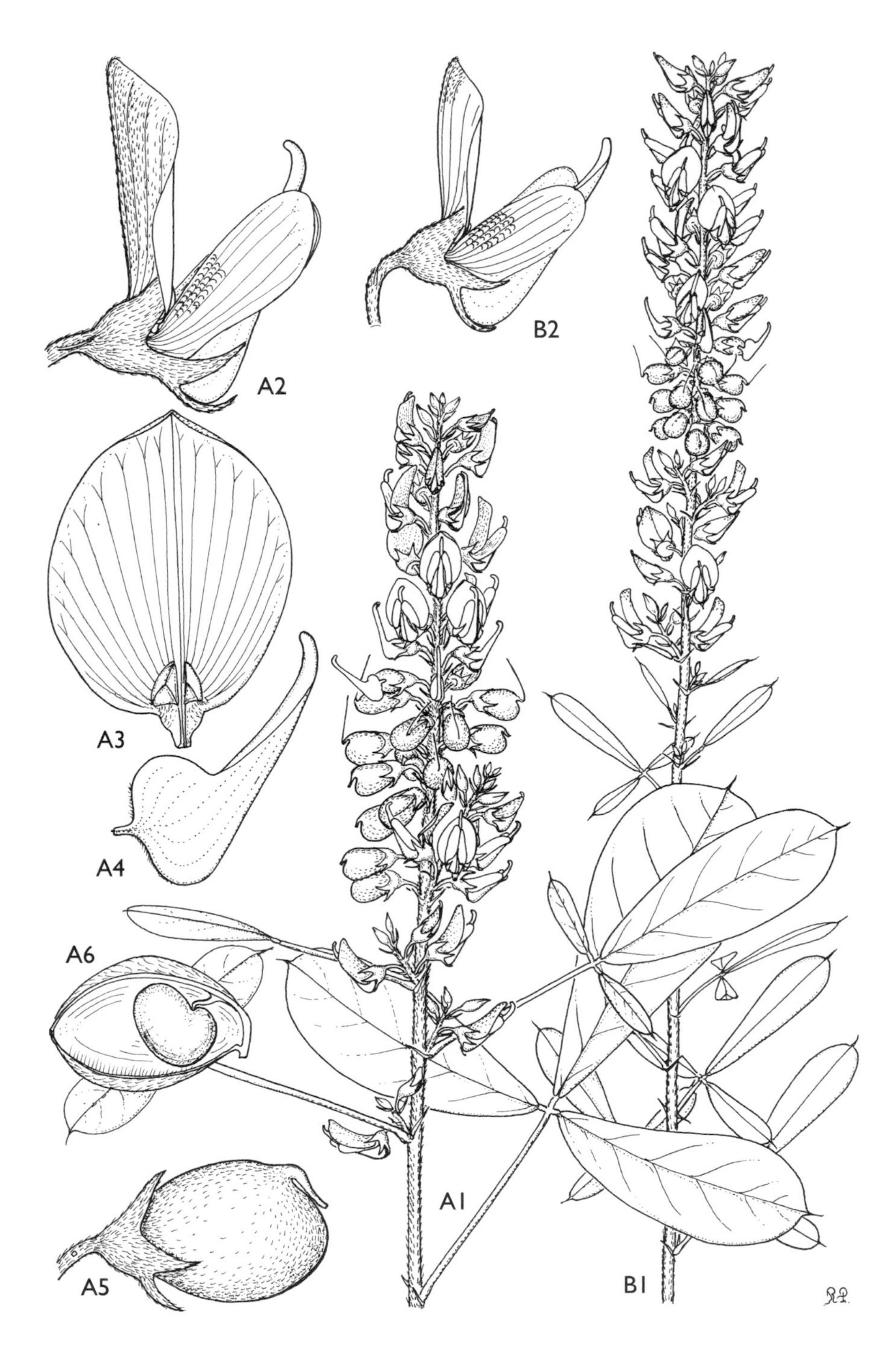

Tab. 3,7: **40**. A. —CROTALARIA PROLONGATA. A1, flowering and fruiting branch, apical part (× 1), from *Fanshawe* 3100; A2, flower (× 4); A3, standard (× 4); A4, keel (× 4), A2–A4 from *E.A. Robinson* 6658; A5, pod (× 4); A6, valve of dehisced pod (× 6), A5 & A6, from *White* 2481. B. —CROTALARIA CUSPIDATA. B1, flowering and fruiting branch, apical part (× 1), from *Richards* 18143; B2, flower (× 4), from *Richards* 5457. Drawn by Roger Polhill. From Crotalaria Africa & Madagascar.

Zambia. B: Zambezi (Balovale) to Kabompo, iii.1961, *Drummond & Rutherford-Smith* 7319 (SRGH). N: Kaputa Distr., Lake Tanganyika, Sumbu, fl. & fr. 5.iv.1957, *Richards* 9036 (K). W: 25 km north of Kabompo Gorge, fl. & fr. 19.iv.1965, *E.A. Robinson* 6658 (K; SRGH). E: 5 km north of Lundazi, fl. & fr. 27.iv.1952, *White* 2481 (K; FHO). **Malawi**. N: Mzimba Distr., 9.5 km north of Mzambazi, fl. & fr. 10.iii.1978, *Pawek* 13958 (K). C: Kasungu National Park (Game Reserve), Lifupa Camp, fl. & fr. 21.vi.1970, *Brummitt* 11616 (K).

Also in Angola, Dem. Rep. Congo (Katanga) and Tanzania (Ufipa). Miombo woodland in higher rainfall areas and damp places by lakes, rivers and marshes; c. 900–1500 m.

141. **Crotalaria cuspidata** Taub. in Engler, Pflanzenw. Ost-Afrikas **C**: 205 (1895). —E.G. Baker in J. Linn. Soc., Bot. **42**: 284 (1914). —Polhill in Kew Bull. **22**: 323 (1968); in F.T.E.A., Leguminosae, Pap.: 972 (1971); Crotalaria Africa & Madagascar: 309, fig. 83/7, 8 (1982). —Lock, Leg. Afr. Check-list: 173 (1989). Type from Dem. Rep. Congo (Lake Albert).

Crotalaria seretii De Wild. in Ann. Mus. Congo, Sér. V, [Études Fl. Bas-Moyen-Congo] **3**: 200 (1910). —Wilczek in F.C.B. **4**: 133 (1953). TAB. 3,7: **40**, fig. B. Type from Dem. Rep. Congo (Orientale).

Erect annual, (0.3)0.5–1.3 m tall, sparingly branched; stem obscurely ribbed, densely silky pilose. Leaves mostly 3-foliolate, the uppermost generally 1-foliolate; leaflets 15–45 × 3–13 mm, linear-oblanceolate to oblong-oblanceolate, obtuse or rounded to the markedly apiculate apex, pilose on both surfaces, more densely beneath; petiole 7–10(15) mm long; stipules 2–4 mm long, setaceous. Racemes terminal, sessile, (5)10–30 cm long, densely many-flowered, with other flowers clustered in the axils below; bracts 2–3 mm long, setaceous; bracteoles on the pedicel, 0.5–1 mm long. Calyx 3.5–4 mm long, densely silky pubescent; upper lobes narrowly attenuate-triangular, 1.5–2.5 times as long as the tube. Standard elliptic, dull yellow, veined and flushed brown outside, densely pubescent apically outside; wings a little shorter than the keel; keel 7–9 mm long, angular, with a long narrow twisted beak incurved at the tip. Pod sessile, 5–6 mm long, shortly oblong-obovoid, densely pubescent outside, hairy along the suture inside, 1–2-seeded. Seeds c. 3 mm long, oblique-cordiform, smooth, yellow to brown.

Zambia. N: Mbala–Mpulungu, Chilongowelo turning, fl. & fr. 2.v.1955, *Richards* 5457 (K; LISC; SRGH). W: Ndola, fl. & fr. 29.vi.1955, *Fanshawe* 2351 (K; NDO; SRGH).

Also in Nigeria, Central African Republic, Angola, Dem. Rep. Congo and Tanzania. Grassy places in woodland, by lakes, rivers and marshes, often on sand; 750–1600 m.

142. **Crotalaria morumbensis** Baker f. in J. Linn. Soc., Bot. **42**: 267 (1914). —Wilczek in F.C.B. **4**: 79 (1953). —Polhill in F.T.E.A., Leguminosae, Pap.: 973 (1971); Crotalaria Africa & Madagascar: 310 (1982). —Lock, Leg. Afr. Check-list: 192 (1989). Type from Dem. Rep. Congo (Katanga).

Erect annual, 0.2–1.5 m tall, developing a few ascending branches only on older plants; stem obscurely ribbed, covered with fine longish (0.5–1 mm) subappressed hairs. Leaves 1-foliolate; basal leaflets 2–4.5 × 0.8–2 cm, obovate-elliptic, elliptic or oblong, grading to linear-lanceolate above, c. 4.5–9 × 0.3–0.7 cm, glabrous to sparsely pubescent above, thinly appressed pubescent beneath; petiole c. 2–3 mm long; stipules 2–4 mm long, setaceous. Racemes terminal, subsessile, 5–25 cm long, subdensely many-flowered, developing clusters of flowers and short racemes from axils of upper reduced leaves; bracts 2–3 mm long, linear; bracteoles on the pedicel, 0.5–1 mm long. Calyx 3.5–4 mm long, densely subappressed pubescent; lobes narrowly attenuate-triangular, 2–3 times as long as the tube. Standard elliptic, yellow, lined reddish, thinly pubescent outside; wings a little shorter than the keel; keel 7–8.5 mm long, angular, with a long narrow twisted beak, slightly incurved at the tip. Pod sessile, 5–6 × 3–4 mm, shortly oblong-obovoid, pubescent outside and along the suture inside, 2-seeded. Seeds c. 3 mm long, oblique-cordiform, smooth, yellow to greyish-green or purplish.

Zambia. W: Copperbelt Distr., Mufulira, fl. & fr. 29.iv.1964, *Fanshawe* 8614 (K; NDO; SRGH). C: Chisamba Forest Reserve, Kamaila Hill, fl. & fr. 23.iv.1972, *Kornaś & Medwecka Kornaś* 1642 (K). E: Chipata Distr., 48 km from Chipata (Fort Jameson) on Nsefu Road, fr. 11.v.1963, *van Rensburg* 2117 (K; SRGH). **Malawi**. C: Lilongwe Distr., Dzalanyama Forest Reserve, below Choulongwe (Chaulongwe) Falls, fr. 26.iv.1970, *Brummitt* 10157 (K). S: Machinga Distr., Munde Hill, fr. 30.iv.1982, *Patel* 882 (K; MAL).

Also in adjacent parts of Dem. Rep. Congo and Tanzania. Miombo woodland; 500–1300 m.

143. **Crotalaria cephalotes** Steud. ex A. Rich., Tent. Fl. Abyss. **1**: 156 (1847). —J.G. Baker in F.T.A. **2**: 23 (1871). —E.G. Baker in J. Linn. Soc., Bot. **42**: 276 (1914). —Eyles in Trans. Roy. Soc. South Africa **5**: 370 (1916). —Verdoorn in Bothalia **2**: 384 (1928). —Hutchinson, Botanist South. Africa: 483 (1946). —Merxmüller in Proc. & Trans. Rhodesia Sci. Assoc. **43**: 18 (1951). —Brenan in Mem. New York Bot. Gard. **8**: 245 (1953). —Wilczek in F.C.B. **4**: 200 (1953). —Hepper in F.W.T.A., ed. 2, **1**: 549 (1958). —Torre in C.F.A. **3**: 34 (1962). —Binns, First Check List Herb. Fl. Malawi: 79 (1968). —Polhill in F.T.E.A., Leguminosae, Pap.: 983, fig. 135 (1971). —Drummond in Kirkia **8**: 217 (1972). —Jacobsen in Kirkia **9**: 160 (1973). —Gonçalves in Garcia de Orta, Sér. Bot. **5**: 66 (1982). —Polhill, Crotalaria Africa & Madagascar: 311, fig. 84 (1982). —Lock, Leg. Afr. Check-list: 170 (1989). TAB. 3,7: **41**. Syntypes from Ethiopia.

Crotalaria cephalotes var. *moeroensis* Baker f. in J. Linn. Soc., Bot. **42**: 277 (1914). —Wilczek in F.C.B. **4**: 201 (1953). Type from Dem. Rep. Congo (Katanga, L. Mweru).

Erect annual 10–70(100) cm tall, developing numerous curved-ascending branches, particularly from near the base, densely covered with long mostly spreading hairs. Leaves almost all 3-foliolate, uppermost sometimes 1-foliolate, crowded and larger just below the inflorescences; leaflets mostly 15–30 × 4–6 mm, upper ones up to 50 × 8 mm, linear-oblanceolate to oblong-oblanceolate, rounded at the apex, long-cuneate to the base, pilose on both surfaces; petioles mostly 5–15 mm long, uppermost up to 2.5 cm, narrowly winged; stipules 0. Flowers in dense subsessile heads or racemes; bracts 1–2 mm long, setaceous; pedicel recurved near apex, without bracteoles. Calyx 3–4 mm long, densely hairy; lobes narrowly triangular, 2–3 times as long as the tube. Standard obovate, pale yellow, marked red at the base inside, veined reddish and medially pubescent outside; wings nearly as long as the keel; keel 4–5.5 mm long, angular, with a narrow twisted beak. Pod sessile, (4)5–6 × 3–5 mm, ovoid-globose to bluntly ellipsoid, densely pilose outside, hairy inside along upper suture behind the placenta, 2-seeded. Seeds 2–2.5 mm long, oblique-cordiform, smooth.

Zambia. B: 5 km Mongu–Kaoma (Mankoya), fl. 2.iv.1964, *Verboom* 1027 (K; SRGH). N: Mbala, Ndundu, fl. & fr. 2.v.1957, *Richards* 9510 (BR; K; SRGH). W: 80 km NE of Kabompo on Chizela (Chizera) road, fl. 27.iii.1961, *Drummond & Rutherford-Smith* 7414 (K; SRGH). C: 24 km NW of Kabwe, Chambesi, fl. & fr. 8.iv.1972, *Kornaś* 1569 (K). E: Chipata Distr., east of Machinje Hills, fl. & fr. 15.v.1965, *Mitchell* 2993 (K; SRGH). S: Choma Distr., 1.5 km south of Masuku Mission, fl. & fr. 27.iii.1952, *White* 2346 (FHO; K). **Zimbabwe**. N: Zvimba Distr., angle of Susuje R. and Kildonan road, fl. & fr. 29.iii.1969, *Pope* 11 (K; SRGH). W: Hwange Distr., Kazuma Range, fl. & fr. 12.v.1972, *Gibbs Russell* 1978 (K; SRGH). C: Chegutu (Hartley), Poole Farm, fl. & fr. 21.iii.1948, *Hornby* 2918 (K; MO; SRGH). E: Mutasa Distr., Honde Valley, Makanga R., fr. 4.iv.1966, *Corby* 1594 (K; SRGH). **Malawi**. N: Rumphi–Mzuzu, 1 km from Engucwini, fl. & fr. 21.v.1970, *Brummitt* 10996 (K; MAL). C: Dedza Distr., Chongoni Forest, fl. & fr. 16.v.1969, *Salubeni* 1342 (K; MAL). S: Zomba, Nschema, fl. 23.iv.1955, *Banda* 85 (K; MAL; SRGH). **Mozambique**. N: Malema, Campo Experimental do C.I.C.A., fl. & fr. 27.iv.1961, *Balsinhas & Marrime* 456 (K; LISC; LMA; LMU). Z: 62.6 km de Mocuba para Milange, fl. & fr. 9.vi.1949, *Barbosa & Carvalho* 2993 (K; LISC; LMA). T: Moatize Distr., Zóbuè, fl. & fr. 12.vii.1942, *Torre* 4402 (K; LISC; MO).

Widespread in savanna areas from Senegal to Ethiopia, south to Angola and Zimbabwe. Miombo woodland and short open grassland, frequently on disturbed often sandy ground; 60–1700 m.

Diminutive plants from around Lake Mweru have been referred to var. *moeroensis*. If considered worthy of formal recognition then *Richards* 9277 (K) from Mporokoso–Nsama road, fl. & fr. 16.iv.1957, should be included.

144. **Crotalaria argyrolobioides** Baker in Bull. Misc. Inform., Kew **1897**: 249 (1897). —E.G. Baker in J. Linn. Soc., Bot. **42**: 384 (1914). —Verdoorn in Bothalia **2**: 401 (1928). —Brenan in Mem. New York Bot. Gard. **8**: 248 (1953). —Binns, First Check List Herb. Fl. Malawi: 78 (1968). —Brummitt in Wye Coll. Malawi Proj. Rep.: 65 (1973). —Polhill in F.T.E.A., Leguminosae, Pap.: 975 (1971); Crotalaria Africa & Madagascar: 312 (1982). —Lock, Leg. Afr. Check-list: 166 (1989). Syntypes: Malawi, Nyika Plateau, *Whyte* 109 and 117 (K, syntypes).

Crotalaria kasikiensis Baker f. in Rev. Zool. Bot. Africaines **21**: 299 (1932). —Wilczek in F.C.B. **4**: 258 (1953). Type from Dem. Rep. Congo (Katanga).

Crotalaria senegalensis sensu Hutchinson, Botanist South. Africa: 508 (1946), non (Pers.) Bacle ex DC.

Tab. 3,7: **41**. CROTALARIA CEPHALOTES. 1, part of young plant (heads become considerably enlarged in older plants) (× 1); 2, flower (× 6); 3, standard (× 6); 4, wing (× 6); 5, keel (× 6); 6, anthers (× 9); 7, section through gynoecium (× 6), 1–7 from *Polhill & Paulo* 1983; 8, pod (× 6); 9, seed (× 6), 8 & 9 from *Newbould & Jefford* 2377. Drawn by Roger Polhill. From F.T.E.A.

Erect well-branched annual or short-lived perennial, 0.3–2 m tall; branches generally with short appressed hairs. Leaves 3-foliolate; leaflets 8–20(30) × 2–9(13) mm, oblanceolate to oblong-elliptic, ± shortly appressed hairy beneath, almost always glabrous above; petiole 3–15(25) mm long; stipules 0. Racemes sometimes dense at first, usually elongating, (1)3–15 cm long, 8–40-flowered, later developed on numerous fairly short flowering branches, sometimes a few flowers also in the axils; bracts 1–2 mm long, generally subulate; bracteoles on the pedicel, setaceous. Calyx 4–6 mm long, appressed pubescent; upper lobes attenuately or acuminately triangular, (1)1.5–2.5 times as long as the tube. Standard elliptic, yellow, lined or flushed reddish, apically pubescent outside; wings shorter than the keel; keel 9–12 mm long, angular, with a narrow twisted beak slightly incurved at the tip. Pods sessile, obliquely inserted, 7–8 × 5–5.5 mm, shortly cylindrical when ripe, appressed pubescent, 6–12-seeded. Seeds 1.5–2 mm long, rounded-cordiform, smooth.

Zambia. N: Isoka Distr., Nyika Plateau, fl. & fr. 24.viii.1962, *Verboom* 641 (K; LISC). W: Chingola Distr., Luano, fl. & fr. 1.xi.1968, *Mutimushi* 2780 (K; NDO). **Malawi**. N: Mzimba Distr., Mzuzu, Marymount, fl. & immat. fr. 6.xiii.1970, *Pawek* 3675 (K). S: Zomba Plateau, fl. & fr. 26.vii.1970, *Brummitt* 12237 (K; LISC; MAL; PRE; SRGH). **Mozambique**. N: Lichinga (Vila Cabral), fl. & fr. 15.vi.1934, *Torre* 176 (BM; COI; K; LISC).

Also in Dem. Rep. Congo (Katanga) and southern Tanzania, principally on high ground along Western Rift Valley from Marungu Mts. to Mulanje, but also descending along rivers and found as far west as Luano. Montane grassland and bushland at forest edges, also higher rainfall miombo woodland and riverine associations; (900)1350–2200 m.

Broadly speaking *C. argyrolobioides* replaces *C. annua* on higher ground and where the ranges approximate there are some indications that intermediates occur, but superimposed on this *C. argyrolobioides* behaves quite differently where it descends to low altitudes around the southern end of Lake Tanganyika. Plants from here, e.g. Mbala Distr., Mwambeshi R., fl. & fr. 5.ix.1960, *Richards* 13196 (K; SRGH); Mwemba Village, fl. & fr. 9.vii.1970, *Sanane* 1254 (K), are luxuriant with large more elliptic leaflets, racemes lax from inception, bracts and indumentum notably short, departing thus from *C. argyrolobioides* and further from *C. annua*. This extreme form is distinctive, but linked to the typical form by other plants such as *Mutimushi* 2780 (cited above) and the type of *C. kasikiensis*.

One plant from Mozambique, Lichinga (Vila Cabral), fl. & immat. fr. 17.v.1948, *Pedro & Pedrógão* 3662 (LMA) is doubtfully referred to this species. It has a rhizomatous rootstock (not preserved) with stems 30–50 cm tall, generally similar to *C. argyrolobioides* but with some hairs on the leaflets above, only 2 ovules and immature ovoid-globose pods 5–6 mm long.

145. **Crotalaria annua** Milne-Redh. in Hooker's Icon. Pl. **33**: t. 3243 (1934). —Polhill in F.T.E.A., Leguminosae, Pap.: 974 (1971); Crotalaria Africa & Madagascar: 313 (1982). —Lock, Leg. Afr. Check-list: 165 (1989). Type: Zambia, Solwezi, *Milne-Redhead* 422 (K, holotype).

Crotalaria adenocarpoides sensu Wilczek in F.C.B. **4**: 264 (1953) pro minore parte.

Crotalaria symoensiana Timp. in Bull. Soc. Roy. Bot. Belgique **91**: 168 (1959). Type from Dem. Rep. Congo (Katanga).

Erect annual, (0.3)0.5–1.8 m tall, with rather stiffly ascending branches above; stem densely spreading pilose. Leaves 3-foliolate; leaflets 8–20 × 3–8 mm, oblanceolate to elliptic-obovate, pilose on both surfaces; petiole 4–15 mm long; stipules 0. Racemes mostly 5–16 cm long, laxly to densely 12–40-flowered, sometimes with a few flowers in the axils below; bracts 2–3(4) mm long, linear-subulate; bracteoles on the pedicel, setaceous. Calyx 6–7 m long, densely pilose; upper lobes narrowly attenuate-triangular, 3–4 times as long as the tube. Standard elliptic, yellow lined red, apically pubescent outside; wings shorter than the keel; keel 1–1.2 cm long, angular, with a narrow twisted beak slightly incurved at the tip. Pods sessile, obliquely but sub-basally inserted, 7–8 × 5–6 mm, ± ovoid-ellipsoid, subappressed pubescent, 8–16-seeded. Seeds 1.5–2.2 mm long, rounded-cordiform, smooth.

Zambia. N: Kasama Distr., Chishimba Falls, fl. & fr. 20.v.1962, *E.A. Robinson* 5186 (BR; K; LISC; SRGH). W: Kitwe, fl. & immat. fr. 28.ii.1954, *Fanshawe* 879 (K; NDO). C: Mkushi Distr., Nkolonga Estates, banks of Chimyamyone, fl. & immat. fr. 15.viii.1965, *Moffat* 17 (K). **Malawi**. N: Mzimba Distr., Mzuzu, Marymount, fl. & fr. 5.v.1969, *Pawek* 2367 (K).

Congo–Zambezi divide from Angola to N Malawi, extending into adjacent parts of Dem. Rep. Congo and Tanzania. Miombo woodland, sometimes by streams or in dambos; 1100–1500 (1600) m.

146. **Crotalaria congesta** Polhill, Crotalaria Africa & Madagascar: 315, fig. 85/19–27 (1982). — Lock, Leg. Afr. Check-list: 172 (1989). Type: Malawi, Viphya Plateau, Vernal Pool, *Pawek* 4755 (K, holotype; MAL).
Crotalaria sp. F of Polhill in F.T.E.A., Leguminosae, Pap.: 986 (1971).

Erect annual, 10–50 cm tall, developing long ascending branches from the base; branches strigulose with white hairs usually marked brown at the base. Leaves 3-foliolate; leaflets 10–20 × 3–6 mm, oblanceolate to oblong-obovate, strigose-pubescent beneath; petiole 3–10 mm long; stipules 0. Flowers 6–20 in terminal sessile heads or crowded above a short racemose part, and with clusters of flowers in the axils below; bracts 1–2 mm long, linear; pedicels 4–8 mm long, with minute bracteoles near the middle. Calyx 4–5 mm long, densely strigose-pubescent with generally brown-based hairs; upper lobes attenuate-triangular, (2)3–4 times as long as the tube. Standard obovate or oblong-obovate, yellow, usually lined reddish-brown, densely strigose-pubescent outside; wings a little shorter than the keel; keel 7–8 mm long, angular, with a narrow twisted beak, slightly incurved at the tip. Pods sessile, 6–8 × 4–5 mm, oblong-ellipsoid to shortly cylindrical, densely strigose-pubescent, 6–10-seeded. Seeds 1.8–2 mm long, oblong-cordiform, smooth, with a well developed aril.

Malawi. N: Mzimba Distr., Viphya Plateau, Chikangawa Forest, fl. & immat. fr. 22.v.1967, *Salubeni* 706 (K; SRGH); Mzimba Distr., Lusangadzi Forest, fl. & fr. 23.vi.1974, *Pawek* 8743 (K; MO).

Also in Dem. Rep. Congo (Kivu), Burundi and south-western Tanzania. Miombo woodland, *Protea* scrub and montane grassland; 1650–1750 m.

147. **Crotalaria nyikensis** Baker in Bull. Misc. Inform., Kew **1897**: 250 (1897). —E.G. Baker in J. Linn. Soc., Bot. **42**: 280 (1914). —Verdoorn in Bothalia **2**: 402 (1928). —Brenan in Mem. New York Bot. Gard. **8**: 245 (1953). —Binns, First Check List Herb. Fl. Malawi: 79 (1968). —Polhill in F.T.E.A., Leguminosae, Pap.: 985 (1971); Crotalaria Africa & Madagascar: 315 (1982). —Lock, Leg. Afr. Check-list: 193 (1989). Type: Malawi, Nyika Plateau, *Whyte* 197 (K, holotype).
Crotalaria kyimbilae Harms in Bot. Jahrb. Syst. **54**: 380 (1917). —E.G. Baker, Legum. Trop. Africa: 29 (1926). Type from Tanzania.

Erect, ultimately bushy annual or short-lived perennial, 50–130 cm tall; branches densely covered with longish mostly spreading hairs. Leaves 3-foliolate; leaflets mostly 8–20 × 2–7 mm, linear-oblanceolate to oblanceolate-oblong, the lowermost sometimes oblong-obovate, glabrous or nearly so above, rather densely appressed pilose beneath; petiole 3–11 mm long; stipules 0. Flowers in sessile to shortly pedunculate dense heads or short racemes 2–5 cm long, rarely in longer dense racemes; bracts (1)2–3 mm long, linear-lanceolate, usually ± as long as the pedicel; bracteoles near top of the pedicel, 1–1.5 mm long, setaceous. Calyx 4–5 mm long, densely often brownish villous; upper lobes attenuate-triangular, ± twice as long as the tube. Standard elliptic-obovate, yellow, veined reddish-brown and densely hairy outside; wings a little shorter than the keel; keel (7)7.5–8.5(9) mm long, angular, with a narrow twisted beak, slightly incurved at the tip. Pods sessile, 4.5–6 × 3–4 mm, ovoid-ellipsoid, somewhat pointed, pilose, 1–2-seeded. Seeds 2 mm long, oblique-cordiform, smooth, with a conspicuous aril.

Zambia. N: Mbala Distr., Ndundu, fl. & fr. 6.v.1959, *Richards* 11369 (K). **Malawi**. N: Chitipa Distr., Misuku Hills, above Misuku Rest House, fl. & fr. 14.vii.1970, *Brummitt* 12122 (K; LISC; MAL; PRE; SRGH). C: Ntchisi Mt., fl. 24.vii.1946, *Brass* 16895 (NY) — fide Brenan, loc. cit. (1953).

Also in southern Tanzania. Montane grassland and bushland at forest edges, descending into miombo woodland principally along the rivers; 1050–2000 m.

The few specimens collected away from the main highland centres – Zambia, Kambole R., *Richards* 8299 (K); Chishimba Falls, *E.A. Robinson* 5205 (K; SRGH); Shiwa Ngandu, *Greenway* 5407 (EA; K) – look somewhat anomalous, with rather weak growth, small flowers and bracts and a tendency to more elongate racemes.

148. **Crotalaria florida** Welw. ex Baker in F.T.A. **2**: 30 (1871). —E.G. Baker in J. Linn. Soc., Bot. **42**: 293 (1914). —Verdoorn in Bothalia **2**: 412 (1928). —Wilczek in F.C.B. **4**: 195 (1953). —Torre in C.F.A. **3**: 44 (1962). —Polhill in F.T.E.A., Leguminosae, Pap.: 976 (1971);

Crotalaria Africa & Madagascar: 319, fig. 86 (1982). —Lock, Leg. Afr. Check-list: 179 (1989). Type from Angola.

Subshrub, several-stemmed from a woody rootstock, (0.1)0.5–1.2 m tall, developing numerous shortish flowering branches above; plants smelling of honey when dried; branches ribbed, densely strigulose with frequently brown-based hairs, sometimes with slightly longer hairs interspersed. Leaves mostly 3-foliolate, uppermost sometimes 1-foliolate; leaflets 5–22 × 2–10 mm, oblong-oblanceolate to oblong-obovate or obovate, slightly coriaceous, densely strigulose or strigose-puberulent beneath; petiole 3–15(20) mm long, channelled; stipules 0. Primary racemes 2–5 cm long, ± laxly 6–10-flowered, with numerous shorter inflorescences developed on lateral branches and in the axils; bracts 1–3 mm long, linear-lanceolate, usually shorter than the pedicel; bracteoles on the pedicel or calyx, 0.5–4 mm long. Calyx 4–5 mm long, densely strigulose or strigose-puberulent; lobes acuminately or attenuately triangular, 0.8–2 times as long as the tube, often slightly involute, ± lanate inside (particularly the lower one). Standard elliptic-obovate to obovate, yellow, sometimes lined brown, glabrous or nearly so outside; wings as long as the keel or a little shorter; keel 8–11 mm long, angular, with a narrow twisted beak, slightly incurved at the tip. Pods subsessile, 6–8 × 4–6 mm, ovoid-globose, pointed, strigose-puberulent, 2-seeded. Seeds 2–2.5 mm long, rounded-cordiform, smooth, dark, with a small aril.

1. Upper calyx lobes evenly attenuate-triangular, 1.5–2 times as long as the tube; bracteoles on the pedicel, usually small · iii) var. *congolensis*
– Upper calyx lobes acuminately triangular, a little shorter to a little longer than the tube · · 2
2. Bracteoles on the pedicel, rarely at the base of the calyx, 0.5–1 mm long · · · i) var. *florida*
– Bracteoles on the calyx, 1.5–4 mm long · ii) var. *monosperma*

i) Var. **florida**. TAB. 3,7: **42**.

Crotalaria sphaerocarpa sensu Hutchinson, Botanist South. Africa: 509 (1946) non Perr. ex DC.

Bracteoles on the pedicel or rarely at the base of the calyx, 0.5–1 mm long. Calyx with short acuminately triangular lobes.

Zambia. B: Kaoma (Mankoya), near Luena R., fl. 20.xi.1959, *Drummond & Cookson* 6681 (K; LISC; SRGH). N: Mbala–Kambole road, above turning to Lunzua Falls, fl. 7.iv.1955, *Richards* 5345 (K; LISC; SRGH). W: Ndola, fl. 25.iii.1969, *Mutimushi* 3006 (K; NDO; SRGH). C: Kabwe (Broken Hill), fl. 14.viii.1963, *Fanshawe* 7934 (K; NDO).

Also in Angola, Dem. Rep. Congo and Tanzania in a broad zone (7–15°S) along the Congo–Zambezi watershed. Miombo woodland; 1000–1500 m.

ii) Var. **monosperma** (De Wild.) R. Wilczek in F.C.B. **4**: 196 (1953). —Polhill in F.T.E.A., Leguminosae, Pap.: 977 (1971); Crotalaria Africa & Madagascar: 320 (1982). Type from Dem. Rep. Congo (Katanga).

Crotalaria monosperma De Wild., Pl. Bequaert. **2**: 503 (1924). —E.G. Baker, Legum. Trop. Africa: 32 (1926).

Crotalaria kipiriensis R. Wilczek in Bull. Jard. Bot. État **23**: 179 (1953); in F.C.B. **4**: 209 (1953). Type from Dem. Rep. Congo (Katanga).

Bracteoles inserted on the calyx, linear, 1.5–4 mm long. Calyx as var. *florida*.

Zambia. N: Mbala Distr., Kanyika, fl. 25.ii.1965, *Richards* 19705 (K; LISC); Kawimbe Rocks, fl. 11.ii.1969, *Sanane* 459 (K).

Also on high ground in Tanzania (Ufipa) and Dem. Rep. Congo (Katanga) (Marungu and Kundelungu). Miombo woodland, at higher altitudes than var. *florida*; 1500–1750 m.

iii) Var. **congolensis** (Baker f.) R. Wilczek in F.C.B. **4**: 196 (1953). —Polhill, Crotalaria Africa & Madagascar: 320 (1982). Type from Dem. Rep. Congo (Katanga).

Crotalaria globifera var. *congolensis* Baker f. in De Wildeman, Pl. Bequaert. **3**: 320 (1925). —E.G. Baker, Legum. Trop. Africa: 875 (1930).

Bracteoles on the pedicel, small. Calyx with the upper lobes attenuately triangular, 1.5–2 times as long as the tube.

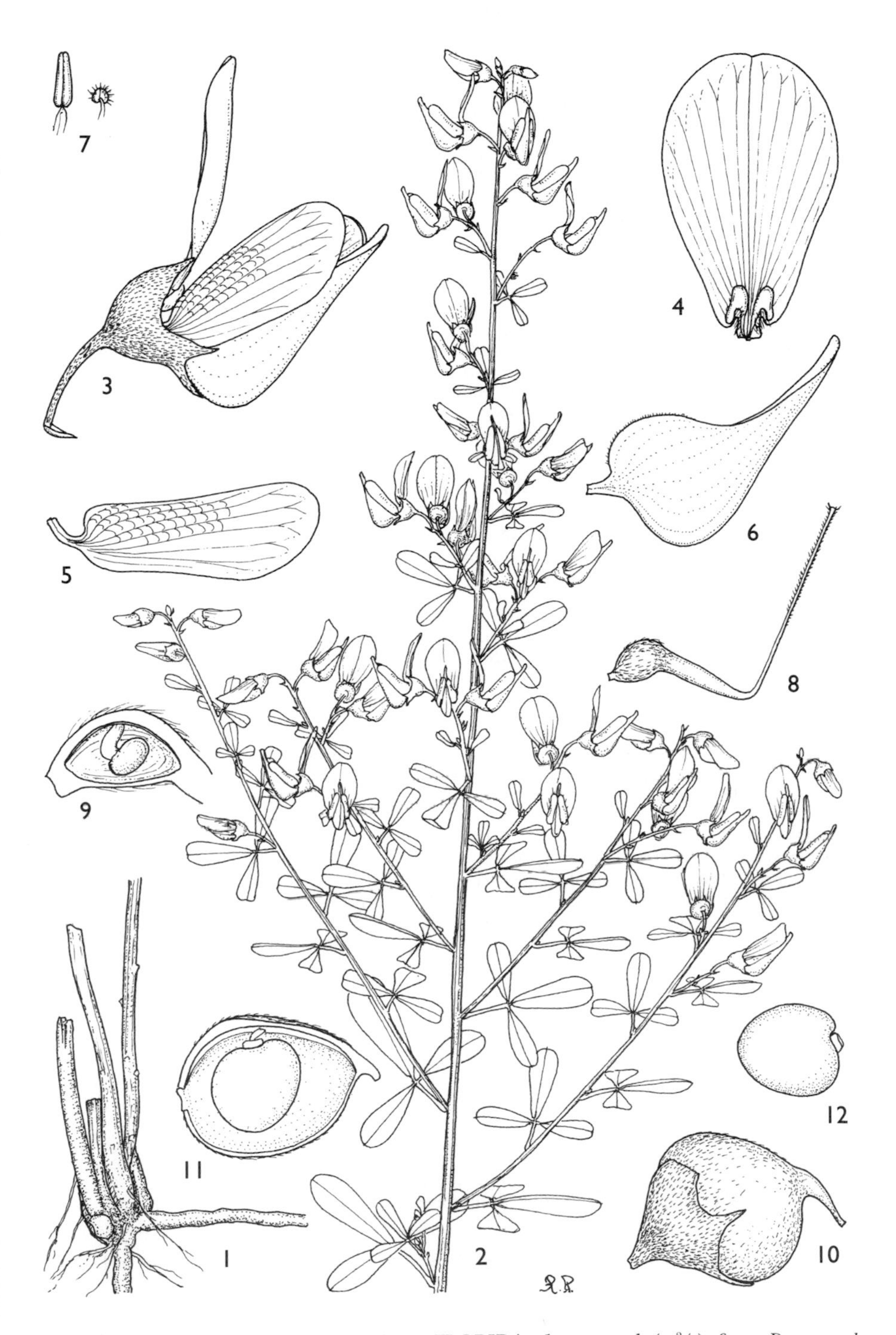

Tab. 3,7: **42**. CROTALARIA FLORIDA var. FLORIDA. 1, rootstock (× $^2/_3$), from *Drummond & Cookson* 6681; 2, flowering shoot (× 1); 3, flower (× 4); 4, standard (× 4); 5, wing (× 4); 6, keel (× 4); 7, anthers (× 6); 8, gynoecium (× 4); 9, section through ovary (× 6), 2–9 from *Richards* 5413; 10, pod (× 4); 11, valve of same (× 4), 10 & 11 from *White* 1842; 12, seed (× 6), from *Richards* 15220. Drawn by Roger Polhill. From Crotalaria Africa & Madagascar.

Zambia. W: Luano, fl. 20.iv.1966, *Fanshawe* 9702 (K; NDO; SRGH); Luano, 15.v.1967, *Mutimushi* 1931 (K; NDO; SRGH).

Also in adjacent parts of Dem. Rep. Congo (Katanga). Miombo woodland; c. 1200 m.

These two specimens are attributed to *C. florida* var. *congolensis* with some hesitation – they are both subprostrate, c. 20 cm high, which is quite uncharacteristic of the species.

The variants have restricted ranges and to some extent different facies, but they do intermingle and intergrade with var. *florida*.

149. **Crotalaria kambolensis** Baker f., Legum. Trop. Africa: 34 (1926). —Polhill, Crotalaria Africa & Madagascar: 320 (1982). —Lock, Leg. Afr. Check-list: 184 (1989). Type: Zambia, Lake Tanganyika, Kambole, *Nutt* (K, holotype).

Crotalaria robynsii R. Wilczek in Bull. Jard. Bot. État **23**: 170 (1953); in F.C.B. **4**: 194 (1953). Type from Dem. Rep. Congo.

Crotalaria pseudoflorida R. Wilczek in Bull. Jard. Bot. État **23**: 171 (1953); in F.C.B. **4**: 197 (1953). Type from Dem. Rep. Congo (Katanga).

Crotalaria piedboeufii R. Wilczek in Bull. Jard. Bot. État **23**: 172 (1953); in F.C.B. **4**: 197 (1953). Type from Dem. Rep. Congo.

Crotalaria florida var. *richardsiana* Torre in Mem. Junta Invest. Ultramar, sér. 2, **19**: 34, t. 12 (1960); in C.F.A. **3**: 44 (1962). Type from Angola.

Similar to *C. florida*, and with the same sweet smell when dry, but the vertical rootstock bright yellow inside, the stems velvety tomentose with obviously spreading hairs, the flowering branches variably developed but generally more contracted and generally no branches below the pseudopanicle, and the petals usually not marked reddish.

Zambia. N: Mbala Distr., Chilongowelo, fl. & fr. 5.v.1955, *Richards* 5540 (K; SRGH). W: Mwinilunga Distr., Chinkobolo (Sinkabolo) Dambo, fl. 29.i.1938, *Milne-Redhead* 4383 (K).

Also in Central African Republic, Dem. Rep. Congo and Angola. *Brachystegia* and *Cryptosepalum* woodland, commonly where the water-table is relatively high, near dambos, streams, etc.; 1200–1600 m.

150. **Crotalaria alemanniana** Torre in Mem. Junta Invest. Ultramar, sér. 2, **19**: 34, t. 13 (1960); in C.F.A. **3**: 45 (1962). —Polhill, Crotalaria Africa & Madagascar: 321 (1982). —Lock, Leg. Afr. Check-list: 164 (1989). Type from Angola.

Stems several from a woody rootstock, erect to 60 cm, developing numerous progressively shortened flowering branches above, ribbed, densely strigulose. Leaves mostly 3-foliolate, uppermost reduced, some 1-foliolate; leaflets 8–15 × 2–8 mm, oblanceolate to obovate, rounded, truncate or emarginate, glabrous or nearly so above, densely white strigulose beneath; petiole 3–10 mm, slender, slightly grooved; stipules 0. Primary racemes short, few-flowered, most of the flowers on lateral branches progressively abbreviated upwards and in the axils, forming a leafy pseudopanicle; bracts 0.5–1.5 mm long, linear-subulate; bracteoles usually on the pedicel, setaceous. Calyx 4–5 mm long, densely strigulose; lobes triangular, somewhat acuminate or attenuate, 1.5–2 times as long as the tube. Standard elliptic-obovate to broadly elliptic, yellow tinged red, sparsely hairy outside; wings equalling the keel or shorter; keel 9–11 mm long, angular, with a narrow twisted beak incurved at the tip. Pod subsessile, 6 × 5 mm, ovoid-globose, 2-seeded (ovules 4–8). Seeds not seen.

Zambia. B: 35 km WSW of Kabompo on road to Zambezi (Balovale), fl. 24.iii.1961, *Drummond & Rutherford-Smith* 7295 (K; LISC; SRGH). W: 90 km SSW of Solwezi on road to Kasempa, fl. 22.iii.1961, *Drummond & Rutherford-Smith* 7187 (K; LISC; SRGH).

Also in Angola. Deciduous woodland, sometimes on Kalahari Sands; c. 900–1200 m.

151. **Crotalaria quangensis** Taub. in Bot. Jahrb. Syst. **23**: 177 (1896). —E.G. Baker in J. Linn. Soc., Bot. **42**: 286 (1914). —Torre in C.F.A. **3**: 41 (1962). —Polhill, Crotalaria Africa & Madagascar: 321 (1982). —Lock, Leg. Afr. Check-list: 200 (1989). Type from Angola.

Stems several to many, erect from a napiform and/or rhizomatous rootstock, 10–70(100) cm tall, simple to sparsely branched, finely ribbed, appressed puberulous to spreading pubescent. Leaves all or almost all 3-foliolate, uppermost occasionally reduced, 1-foliolate; leaflets 8–45 × 2–17 mm, mostly oblanceolate, but varying from linear-oblanceolate to elliptic-obovate, appressed puberulous to pubescent beneath;

petiole 0.2–2.5 cm long, broadly channelled to distinctly winged; stipules 0. Racemes sessile to shortly pedunculate, 1–17 cm long, densely to laxly 15–40-flowered, sometimes with flowers or flowering branches from upper axils; bracts 1.5–4 mm long, linear-lanceolate, shorter to longer than the pedicel; bracteoles inserted on the pedicel, 0.5–1.5 mm long, setaceous. Calyx 3–5 mm long, thinly to densely pubescent with appressed to rather irregularly arranged hairs; lobes narrowly triangular and slightly acuminate, rarely subulate with inrolled margins, (1)1.2–1.5(2) times as long as the tube. Standard elliptic or elliptic-obovate to subcircular, yellow, sometimes lined or flushed reddish, glabrous outside in the Flora Zambesiaca area; wings ± as long as the keel; keel 6–10 mm long, angular, with a narrow twisted beak. Pod sessile, 5–7 × 4–5.5 mm, ovoid-globular, appressed puberulous to glabrous except on upper side, 2–8-seeded. Seeds 2–2.5 mm long, rounded-cordiform, smooth, brown, exarillate.

Var. **quangensis** —Polhill, Crotalaria Africa & Madagascar: 322 (1982).
Crotalaria kutchiensis Baker f. in J. Linn. Soc., Bot. **42**: 279 (1914). —Torre in C.F.A. **3**: 36 (1962). Lectotype, chosen by Torre, loc. cit., from Angola.
Crotalaria malangensis var. *overlaetii* R. Wilczek in Bull. Jard. Bot. État **23**: 174 (1953); in F.C.B. **4**: 203 (1953). Type from Dem. Rep. Congo (Katanga).
Crotalaria francoisiana Duvign. & Timp. in Bull. Soc. Roy. Bot. Belgique **91**: 150 (1959). Type from Dem. Rep. Congo (Katanga).
Crotalaria xassenguensis Torre in Mem. Junta Invest. Ultramar, sér. 2, **19**: 28 (1960); in C.F.A. **3**: 39 (1962). Type from Angola.

Stems 10–30 cm tall, mostly simple, with spreading or less often appressed hairs. Petiole 2–6(10) mm long; leaflets 8–22 × (1.5)3–7(10) mm. Racemes sessile, 1–3(7) cm long, dense to sublax, sometimes with flowers in the axils below. Keel 6–7 mm long. Pod 5–5.5 × 4–4 mm, 2-seeded.

Zambia. W: Kabompo Distr., Kabompo Gorge, fl. 24.x.1966, *Leach & Williamson* 13466 (K; SRGH); Mwinilunga Distr., Matonchi Farm, fl. 5.x.1937, *Milne-Redhead* 2588 (BM; BRLU; K).
Also in Angola and adjacent parts of Dem. Rep. Congo (Katanga). Characteristic of dambos on Kalahari Sands, occasionally burnt places in miombo woodland, but then generally with atypical appressed indumentum (both forms may grow together, see *Milne-Redhead* 901 (K) from 65 km south of Mwinilunga); c. 1200–1400 m. Flowers in the dry season.

Var. **malangensis** (Baker f.) Polhill, Crotalaria Africa & Madagascar: 322 (1982). Lectotype, chosen by Torre (1962), from Angola.
Crotalaria malangensis Baker f. in J. Linn. Soc., Bot. **42**: 287 (1914). —Torre in C.F.A. **3**: 42 (1962).

Stems 30–60 cm tall, occasionally with 1–several flowering branches above, appressed puberulous. Petioles mostly 8–25 mm long; leaflets 20–45 × 5–17 mm. Racemes sessile, 8–16 cm long, sublax to lax. Keel (6)7–8 mm long. Pod 6–7 × 5.5–6 mm, 2(6)-seeded.

Zambia. W: Mwinilunga Distr., Mwinilunga–Kalene road, fl. & fr. 25.v.1969, *Mutimushi* 3486 (K; NDO).
Also in Angola and Dem. Rep. Congo (Katanga). Miombo woodland.
This variant is characteristic of plateau woodland NW of the Flora Zambesiaca area and the specimen cited comes very close to var. *quangensis*. *Milne-Redhead* 2532 (K), from Mwinilunga, is even more aberrant, with 2 ovules as general in var. *malangensis*, but rather large flowers (keel 9 mm long) more characteristic of var. *mullendersii* (R. Wilczek) Polhill, the variety which predominates on the plateaux of the Katanga in Dem. Rep. Congo. The calyx lobes are unusually short and involute for the species as a whole, and the stems are rather more branched than usual, approaching *C. florida* in these respects.

152. **Crotalaria graminicola** Taub. ex Baker f. in J. Linn. Soc., Bot. **42**: 291 (1914). —Hepper in Kew Bull. **11**: 118 (1956); in F.W.T.A., ed. 2, **1**: 549 (1958). —Polhill in F.T.E.A., Leguminosae, Pap.: 977, fig. 134 (1971); Crotalaria Africa & Madagascar: 325, fig. 89 (1982). Lectotype, selected by Polhill (1982), from Nigeria.
Crotalaria diloloensis Baker f. in Rev. Zool. Bot. Africaines **21**: 297 (1932). —Milne-Redhead in Bull. Misc. Inform., Kew **1940**: 52 (1940). —Hutchinson, Botanist South. Africa: 520 (1946). —Wilczek in F.C.B. **4**: 257, t. 15 (1953). —Duvigneaud & Timperman in Bull. Soc. Roy. Bot. Belgique **91**: 151, fig. 6 (1959). —Torre in C.F.A. **3**: 42 (1962). —Lock, Leg. Afr. Check-list: 180 (1989). Type from Dem. Rep. Congo (Katanga).

Crotalaria stenocladon Baker f. in Rev. Zool. Bot. Africaines **21**: 297 (1932). Type from Dem. Rep. Congo (Katanga).

Crotalaria praecox Milne-Redh. in Hooker's Icon. Pl. **33**: t. 3244 (1934). Type: Zambia, Mwinilunga Distr., Matonchi Farm, *Milne-Redhead* 1010 (K, holotype).

Crotalaria diloloensis var. *prostrata* R. Wilczek in Bull. Jard. Bot. État **23**: 215 (1953); in F.C.B. **4**: 258 (1953). Type from Dem. Rep. Congo (Katanga).

Stems numerous, spreading to erect from a vertical and/or rhizomatous rootstock, simple to sparingly branched and 5–20 cm tall at flowering time (subsequent vegetative shoots branching and taller), slender, slightly ribbed, appressed puberulous to subappressed pubescent (hairs 0.1–0.2 mm long). Leaves 3-foliolate; leaflets 5–15 × 2–7 mm, oblanceolate to obovate, rounded to emarginate at the apex, appressed puberulous to pubescent beneath, rarely also above; petiole 2–12 mm long, broadly channelled to slightly winged; stipules 0. Racemes 2–10(18) cm long, sublaxly to laxly 6–20(30)-flowered; bracts 1–3 mm long, linear to linear-lanceolate; bracteoles on the pedicel, setaceous. Calyx (2.5)3–4 mm long, appressed pubescent; upper lobes lanceolate-triangular, 1.5–2 times as long as the tube. Standard obovate, yellow, generally lined and later flushed reddish-purple or brown, pubescent at least along the midvein outside, often more extensively on the upper part; wings a little shorter than the keel; keel 4–5(7) mm long, angular, with a narrow twisted beak at least slightly incurved at the tip. Pod nearly sessile, 3.5–5 mm long, subglobose-ovoid, appressed puberulous, 1–4-seeded (ovules 2–6(8)). Seeds 1.5–2 mm long, rounded-cordiform, smooth, brown, generally exarillate in the Flora Zambesiaca area.

Zambia. N: Mbala, sandpits, fl. & fr. 18.viii.1960, *Richards* 13112 (K; SRGH). W: Solwezi Dambo, fl. 10.ix.1952, *White* 3217 (FHO; K). C: 15 km from University of Zambia campus along Great East Road, fl. & fr. 19.ix.1972, *Strid* 2175 (K; LUS).

Also from Ghana to Central African Republic, Burundi, Tanzania, Dem. Rep. Congo and Angola. Miombo woodland, grassland and dambo edges subject to burning; 1250–1800 m. Flowers in the dry season after fires.

153. **Crotalaria pseudodiloloensis** R. Wilczek in Bull. Jard. Bot. État **23**: 212 (1953); in F.C.B. **4**: 253 (1953). —Polhill in F.T.E.A., Leguminosae, Pap.: 979 (1971); Crotalaria Africa & Madagascar: 326 (1982). —Lock, Leg. Afr. Check-list: 199 (1989). Type from Dem. Rep. Congo (Katanga).

Crotalaria bianoensis Duvign. ex Timp. in Bull. Soc. Roy. Bot. Belgique **91**: 169, fig. 9 (1959). Type from Dem. Rep. Congo (Katanga).

Stems several to many, erect from a slender vertical and rhizomatous rootstock, 8–35 cm tall, slender, simple to sparingly branched, scarcely ribbed, appressed or subappressed pubescent. Leaves often little developed at flowering time, mostly 3-foliolate, uppermost often 1-foliolate; leaflets 4–14 × 0.5–3(4) mm, linear-oblanceolate to oblanceolate, acute to obtuse or rarely rounded at the apex, often folded lengthwise, glabrous or nearly so above, finely pubescent beneath; petiole 1–3(10) mm long, channelled; stipules 0. Racemes (3)7–12 cm long, subdense at first, becoming lax, (6)10–24-flowered, often with flowers in the axils below; bracts 1–3(6) mm long, linear-lanceolate, usually shorter than the pedicel; bracteoles on the pedicel, setaceous. Calyx 3–4 mm long, appressed pubescent; lobes triangular to triangular-lanceolate, 1.2–1.5(2) times as long as the tube. Standard broadly elliptic-obovate, yellow, usually flushed crimson-purple, usually thinly but extensively pubescent outside; wings ± as long as the keel; keel 5–7 mm long, angular, with a narrow twisted beak slightly incurved at the tip. Pod nearly sessile, 5–7 × 3–4 mm, ovoid-globose to shortly oblong-ovoid, appressed pubescent, 4–10-seeded. Seeds c. 2 mm long, rounded-cordiform, smooth, brown, exarillate.

Zambia. N: Mbala Distr., Sunzu Hill, fl. & fr. 7.xi.1958, *Robson & Fanshawe* 505 (BM; K; LISC; SRGH); Great North Road, 16 km north of Mpika, fl. 17.ix.1958, *Lawton* 473 (K; NDO).

Also in Dem. Rep. Congo (Katanga) and southern Tanzania. Miombo woodland subject to burning, often in disturbed places.

154. **Crotalaria luondeensis** R. Wilczek in Bull. Jard. Bot. État **23**: 209 (1953); in F.C.B. **4**: 252 (1953). —Polhill in F.T.E.A., Leguminosae, Pap.: 980 (1971); Crotalaria Africa & Madagascar: 329 (1982). —Lock, Leg. Afr. Check-list: 189 (1989). Type from Dem. Rep. Congo (Katanga).

Stems numerous, erect from a woody rootstock, 15–45 cm tall, sometimes developing short branches above, tomentose with long fine twisted hairs. Leaves 3-foliolate; leaflets 10–20 × 1–4 mm, linear-oblanceolate, pointed, tomentose beneath; petiole 1–5 mm long; stipules 0. Racemes 2–8 cm long, subdensely many-flowered, sometimes a few other flowers in the axils below; bracts 2–4 mm long, linear to linear-lanceolate, a little shorter to a little longer than the pedicel; bracteoles on the pedicel, setaceous. Calyx 3.5–5 mm long, tomentose; upper lobes triangular, 1.5–2 times as long as the tube. Standard obovate to obovate-circular, yellow, turning orange, medially pubescent to extensively subtomentose outside; wings nearly as long as the keel; keel 7–8 mm long, angular, with a straight twisted beak. Pod sessile, 5–6 mm long, ovoid-globose, tomentose, 4–6-seeded. Mature seeds not seen.

Mozambique. N: Mts. east of Lago Niassa (Lake Nyasa), 1900, fl. *W.P. Johnson* s.n. (K).
Also in Dem. Rep. Congo (Marungu Mts.) and southern Tanzania. Upland grassland.

155. **Crotalaria lukafuensis** De Wild. in Ann. Mus. Congo, Sér. IV, Bot. [Études Fl. Katanga] **1**: 184 (1903). —E.G. Baker in J. Linn. Soc., Bot. **42**: 379 (1914). —Wilczek in F.C.B. **4**: 242 (1953). —Polhill, Crotalaria Africa & Madagascar: 329, fig. 90/1–7 (1982). —Lock, Leg. Afr. Check-list: 189 (1989). Type from Dem. Rep. Congo (Katanga).
Crotalaria pseudo-descampsii Baker f. in Rev. Zool. Bot. Africaines **21**: 298 (1932). Type from Dem. Rep. Congo (Katanga).

Perennial, developing a slender rhizomatous rootstock, but flowering from first year of growth; stems erect, 25–70 cm tall, branching above, densely spreading pilose, the hairs 1–2 mm long and twisted when dry. Leaves almost all 3-foliolate, uppermost grading to bracts; leaflets 8–25 × 1–4 mm, linear-oblanceolate to oblanceolate, pointed, sparsely pilose above, silky pilose beneath; petiole 1–3 mm long; stipules 0. Racemes sessile, short, dense, 12–30-flowered; bracts 5–7 mm long, lanceolate; bracteoles on the pedicel, setaceous to linear. Calyx 5–6(7) mm long, silky tomentose; upper lobes broadly triangular, slightly acuminate, 1–1.5 times as long as the tube. Standard broadly obovate, yellow, silky tomentose outside; wings and standard longer than the keel; keel subangular, with a twisted slightly incurved beak, 6–7.5 mm long. Pod sessile, 7 × 4 mm, oblong-ellipsoid, pilose, 6–10-seeded. Seeds 1.5 mm long, oblique-cordiform, smooth.

Zambia. W: 25 km north of Kabompo Gorge, fl. & fr. 19.iv.1965, *E.A. Robinson* 6648 (BR; K); Solwezi, fl. 9.iv.1960, *E.A. Robinson* 3489 (K; SRGH).
Also in Dem. Rep. Congo (Katanga). Miombo woodland; 1200–1350 m.

156. **Crotalaria axilliflora** Baker f. in J. Linn. Soc., Bot. **42**: 300 (1914). —Wilczek in F.C.B. **4**: 248 (1953) pro majore parte. —Polhill, Crotalaria Africa & Madagascar: 329, fig. 90/8–16 (1982). —Lock, Leg. Afr. Check-list: 166 (1989). Lectotype, chosen by Wilczek, loc. cit., from Dem. Rep. Congo (Lake Tanganyika).
Crotalaria mokoroensis R. Wilczek in Bull. Jard. Bot. État **23**: 206 (1953); in F.C.B. **4**: 247 (1953). Type from Dem. Rep. Congo (Lake Mweru).

Erect annual, 30–75 cm tall, developing ascending branches; stem densely silvery silky pubescent to pilose, the hairs 0.5–1(2) mm long. Leaves all or almost all 3-foliolate, uppermost sometimes 1-foliolate; leaflets 6–30 × 1–10 mm, linear-oblanceolate, oblong-oblanceolate or oblong-obovate, obtuse to rounded at the apex, glabrous to sparsely pilose above, thinly appressed pubescent to pilose beneath; petiole 2–6 mm long; stipules 0. Racemes 1–4 cm long, densely 6–20-flowered, with other flowers developing in the axils below and sometimes also on abbreviated lateral branches; bracts 3–7 × 0.1–0.7 mm, linear to narrowly linear-lanceolate; pedicel 3–4(6 in fruit) mm long; bracteoles setaceous or lacking. Calyx 5–6 mm long, silky tomentose; lobes acuminately triangular, 1.5–2 times as long as the tube. Standard broadly obovate, yellow, silky tomentose outside; wings as long as the keel; keel 6–8 mm long, subangular, with a narrow twisted beak hooked inwards at the tip. Pod sessile, 7–8 × 4–5 mm, ovoid-ellipsoid, pilose, 4–8-seeded. Seeds 1.5–2.5 mm long, oblique-cordiform, smooth.

Zambia. N: Mansa (Fort Rosebery), fl. & fr. 8.v.1964, *Fanshawe* 8557 (K; NDO).
Also in Dem. Rep. Congo (Katanga). Miombo woodland; c. 1200 m.

157. **Crotalaria axillifloroides** R. Wilczek in Bull. Jard. Bot. État **23**: 208 (1953); in F.C.B. **4**: 250 (1953). —Polhill, Crotalaria Africa & Madagascar: 331 (1982). —Lock, Leg. Afr. Check-list: 166 (1989). Type from Dem. Rep. Congo (Katanga).
Crotalaria axillifloroides var. *gracilis* R. Wilczek in Bull. Jard. Bot. État **23**: 208 (1953) excl. typo.

Erect annual, (15)30–75 cm tall, developing ascending branches from near the base upwards; stem densely silvery spreading pilose, the hairs 1–2 mm long and twisted when dried. Leaves 3-foliolate; leaflets 8–20 × 2–7 mm, oblong-oblanceolate, obtuse, pilose on both surfaces, ± silky beneath; petiole 2–4 mm long; stipules 0. Racemes sessile, 3–12 cm long, densely 12–60-flowered, with other flowers in the axils below; bracts leaf-like, 6–12 × 2–5 mm, elliptic-lanceolate; pedicels 1–2(3 in fruit) mm long; bracteoles 0. Calyx 6–7 mm long, silky tomentose; lobes triangular-lanceolate, 2–3 times as long as the tube. Standard obovate, yellow, silky tomentose outside; wings as long as the keel; keel 6–8 mm long, subangular, with a narrow twisted beak hooked inwards at the tip. Pod sessile, 6–7 × 4–5 mm, ovoid to ovoid-ellipsoid, silky pilose, 6–10-seeded. Seeds 1.5–1.8 mm long, oblique-cordiform, smooth.

Zambia. N: Kawambwa Distr., track to Ntumbachushi (Timnatushi) Falls, fl. 19.iv.1957, *Richards* 9338 (K; SRGH). W: 25 km north of Kabompo Gorge, fl. & fr. 19.iv.1965, *E.A. Robinson* 6650 (K; SRGH).
Also in Dem. Rep. Congo (Katanga). Miombo woodland; 1200–1700 m.

158. **Crotalaria argenteotomentosa** R. Wilczek in Bull. Jard. Bot. État **23**: 207 (1953); in F.C.B. **4**: 249 (1953). —Polhill, Crotalaria Africa & Madagascar: 332 (1982). —Lock, Leg. Afr. Check-list: 165 (1989). Type from Dem. Rep. Congo (Katanga).

Erect annual 20–100 cm tall, usually with rather slender ascending branches; stem densely silvery silky pilose. Leaves mostly 3-foliolate, uppermost sometimes 1-foliolate; leaflets 8–20 × 1–4 mm, linear-oblanceolate to oblanceolate, ± pointed, pilose beneath and generally thinly so above; petiole 1–4 mm long; stipules 0. Racemes 3–10 cm, long, subdensely to laxly 6–20-flowered, with other flowers developing in the axils below; bracts 3–6 mm long, linear-lanceolate to lanceolate; pedicel 2–5 mm long; bracteoles on the pedicel, setaceous, or obscure. Calyx 5–7 mm long, silky tomentose; lobes triangular lanceolate, 2–2.5 times as long as the tube. Standard broadly elliptic-obovate to subcircular, yellow, finely lined red, silky tomentose outside but hairs rarely obscuring the surface completely; wings longer than the keel; keel 8–9 mm long, subangular, with a twisted beak slightly to strongly incurved at the tip. Pod sessile, 7–9 × 4.5–7 mm, ovoid-ellipsoid to ovoid-globose, densely silky pilose, 6–10-seeded. Seeds 2.5–3 mm long, oblique-cordiform, smooth.

Subsp. **argenteotomentosa**

Beak of the keel strongly incurved (± hooked) at the tip; style hairy on both sides above; standard oblong-obovate; stem hairs generally subappressed and rarely exceeding 1 mm; upper calyx lobes usually broadly triangular-lanceolate and no more than twice as long as the tube.

Zambia. N: Kaputa Distr., Nsama, fl. 4.iv.1957, *Richards* 9017 (K).
Also in Dem. Rep. Congo (eastern parts of the Katanga Province). Miombo woodland; 1200 m.
The specimen cited, the only one from Zambia, is ± intermediate between typical material of the two subspecies.

Subsp. **dolosa** Polhill, Crotalaria Africa & Madagascar: 332 (1982). Type: Zambia, Mbala–Mpulungu, close to Chilongowelo turning, *Richards* 5456 (K, holotype).
Crotalaria sparsifolia sensu Hutchinson, Botanist South. Africa: 520 (1946) non Baker, based on *Hutchinson & Gillett* 3958 (K); the BM duplicate of this number is *C. bemba* R. Wilczek.

Beak of keel often hardened and slightly curved at the tip, often to one side, but not hook-like; style with long hairs above on one edge only; standard subcircular; stem hairs often rather longer and slightly more spreading; calyx lobes rather narrowly triangular-lanceolate, 2–2.5 times as long as the tube.

Zambia. N: Mbala Distr., Lake Tanganyika, Kumbula (Nmbulu) Island, fl. & fr. 11.iv.1955, *Richards* 5397 (BR; K; SRGH); Mbala Distr., Chisungu Estate, fl. & fr. 20.iv.1959, *McCallum-Webster* 831 (K).

Known only from the Mbala District. Miombo woodland; 750–1500 m.

159. **Crotalaria subspicata** Polhill in Kew Bull. **22**: 325, fig. 35/1–9 (1968); in F.T.E.A., Leguminosae, Pap.: 981 (1971); Crotalaria Africa & Madagascar: 332 (1982). —Lock, Leg. Afr. Check-list: 206 (1989). Type from SE Tanzania.

Erect annual to 30 cm, usually unbranched, spreading pilose. Leaves 3-foliolate, uppermost sometimes 1-foliolate; leaflets 10–25 × 3–6 mm, oblong-oblanceolate to oblong-obovate, obtuse or rounded at the apex, sparsely pilose above, thinly so beneath; petiole 2–5 mm long; stipules 0. Racemes sessile, 3–10 cm long, dense, many-flowered, with other flowers in the axils below; bracts 5–10 mm long, narrowly lanceolate, attenuate at either end; bracteoles 0. Calyx 4–5 mm long, densely pilose; lobes narrowly attenuate-triangular, 3 times as long as the tube. Standard elliptic, pale yellow, pilose outside; wings nearly as long as the keel; keel 5–5.5 mm long, angular with a narrow twisted beak. Pod sessile, 4–5 mm long, ovoid-globose, pilose, 4–6-seeded. Seeds rounded cordiform, 1.5 mm across, smooth, exarillate.

Malawi. S: Machinga Distr., 50 km from Zomba towards Namwera, fl. & fr. 13.iii.1964, *Correia* 204 (LISC).

Also in SE Tanzania. Miombo woodland; 500 m.

160. **Crotalaria subtilis** Polhill, Crotalaria Africa & Madagascar: 333 (1982). —Lock, Leg. Afr. Check-list: 206 (1989). Type from Dem. Rep. Congo (Katanga).

Crotalaria axillifloroides var. *gracilis* R. Wilczek in Bull. Jard. Bot. État **23**: 209 (1953) pro parte; in F.C.B. **4**: 251 (1953). Type as above.

Erect annual 10–40 cm tall; stem slender, sometimes with ascending branches, densely pilose. Leaves 3-foliolate; leaflets 5–15 × 1–4 mm, linear-oblanceolate to oblong-oblanceolate, mostly obtuse, thinly pilose on both surfaces; petiole 1–5 mm long; stipules 0. Racemes 3–10 cm long, laxly 8–30-flowered; bracts leaf-like, 5–9 mm long, lanceolate, acuminate, narrowed to a slight stalk, mostly exceeding the flowers and fruits; bracteoles 0. Calyx 3.5–4.5 mm long, pilose; lobes narrowly triangular-lanceolate, 2.5–3 times as long as the tube. Standard elliptic to subcircular, yellow lined reddish-brown, uniformly subdensely pilose outside; wings exceeding the keel; keel 5–6 mm long, angular, with a narrow twisted beak slightly incurved at the tip. Pod sessile, 4 mm long, globose-ellipsoid, 2-seeded (ovules 4–6). Seeds 1.5 mm long, ± cordiform, smooth.

Zambia. N: Mbala Distr., Kambole Escarpment, fl. & fr. 4.vi.1957, *Richards* 9991 (K); halfway to Kasama from Mbala, N'tingila Ranch, fl. & fr. 8.vi.1980, *Hooper & Townsend* 1881 (K; NDO).

Also in adjacent part of Katanga Province (Dem. Rep. Congo). Miombo woodland; 1200–1500 m.

161. **Crotalaria vandenbrandii** R. Wilczek in Bull. Jard. Bot. État **23**: 184 (1953); in F.C.B. **4**: 212 (1953). —Polhill in F.T.E.A., Leguminosae, Pap.: 981 (1971); Crotalaria Africa & Madagascar: 333 (1982). —Lock, Leg. Afr. Check-list: 209 (1989). Type from Dem. Rep. Congo (Katanga).

Erect annual 20–50 cm tall, developing ascending branches, densely long-pilose, the hairs 1–2 mm long, twisted when dried. Leaves 3-foliolate, uppermost occasionally 1-foliolate; leaflets 8–20 × 2–6 mm, oblong-oblanceolate, pointed, thinly pilose on both surfaces; petiole 1.5–6 mm long; stipules 0. Racemes sessile, 3–15 cm long, densely many-flowered, with other flowers developing in the axils below; bracts leaf-like, linear-lanceolate, 4–8 mm long, as long as the flowers and fruits; bracteoles 0. Calyx 4–5 mm long, pilose; lobes attenuate-triangular, 3 times as long as the tube. Standard elliptic to oblong-obovate, pale yellow, densely pilose outside; wings nearly as long as the keel; keel 6–7 mm long, angular, with a narrow ± straight twisted beak. Pod sessile, 4–6 mm long, globose-ellipsoid, densely pilose, 2-seeded. Seeds 1.5–2 mm long, oblique-cordiform, smooth, with a distinct aril.

Zambia. N: Mbala Distr., Ndundu, fl. & fr. 3.iv.1959, *McCallum-Webster* 712 (K); Mbala Distr., Lunzua Agricultural Station, fl. 25.iii.1955, *Richards* 5159 (K; SRGH).
Also in adjacent parts of Dem. Rep. Congo and Tanzania. Miombo woodland; 1500–1800 m.

162. **Crotalaria egregia** Polhill, Crotalaria Africa & Madagascar: 335, fig. 92/1–10 (1982). — Lock, Leg. Afr. Check-list: 177 (1989). TAB. 3,7: **43**, fig. A. Type: Zambia, Mporokoso–Nsama, *Richards* 9286 (K, holotype).

Erect annual 30–50 cm tall, developing ascending branches from near the base, densely pilose. Leaves 1-foliolate; leaflets mostly 20–50 × 4–10 mm, oblong-lanceolate, thinly pilose on both surfaces; petiole 2–3 mm long; stipules 0. Racemes 6–20 cm long, sublaxly 12–40-flowered, with a few flowers developing in the axils below; bracts (3)4–6 mm long, linear-lanceolate, ± as long as the pedicel; bracteoles 0. Calyx 6–7 mm long, densely pilose; upper lobes narrowly acuminate-triangular, 3 times as long as the tube. Standard elliptic, pale yellow, ± marked brownish, silky tomentose outside; wings a little shorter than the keel; keel 10–11 mm long, angular, with a narrow twisted beak. Pod sessile, 8–10 × 6 mm, shortly ovoid-ellipsoid, pilose, 8-seeded. Seeds 2.5 mm long, oblique-cordiform, smooth.

Zambia. N: Mbala Distr., Chilongowelo Escarpment, fl. 6.iv.1962, *Richards* 16259 (K; SRGH); Kawambwa, fl. & fr. 14.iv.1959, *Mutimushi* 25 (K; NDO).
Known only from northern Zambia. Miombo woodland and dambo margins; 1200–1500 m.

163. **Crotalaria kipandensis** Baker f. in J. Linn. Soc., Bot. **42**: 266 (1914). —Verdoorn in Bothalia **2**: 414 (1928). —Wilczek in F.C.B. **4**: 96 (1953). —Polhill in F.T.E.A., Leguminosae, Pap.: 982 (1971). —Drummond in Kirkia **8**: 218 (1972). —Polhill, Crotalaria Africa & Madagascar: 337, fig. 92/11–12 (1982). —Lock, Leg. Afr. Check-list: 185 (1989). TAB. 3,7: **43**, fig. B. Lectotype, chosen by R. Wilczek, loc. cit., from Dem. Rep. Congo (Katanga).
Crotalaria pseudokipandensis R. Wilczek in Bull. Jard. Bot. État **23**: 160 (1953); in F.C.B. **4**: 95 (1953). Type from Dem. Rep. Congo (Katanga).

Erect annual 20–60 cm tall, sometimes developing a few weak branches from near the base, densely pilose. Leaves 1-foliolate; leaflet 2–4(5) × 0.6–1.2(1.5) cm, oblong-lanceolate to oblong-elliptic, pilose on both surfaces; petiole 2–5 mm long; stipules 0. Racemes sessile, 4–20 cm long, densely many-flowered, with a few flowers developing in the axils below; bracts leaf-like, 6–12 mm long, linear-lanceolate; bracteoles 0. Calyx 5–8 mm long, densely silky pilose; upper lobes narrowly triangular-lanceolate, 2.5–4 times as long as the tube. Standard elliptic, pale yellow, red marked, silky tomentose outside; wings ± as long as the keel; keel 6–7 mm long, subangular, with a narrow twisted beak. Pod sessile, 6–8 × 4–5 mm, shortly ovoid-ellipsoid, pilose, 4–8-seeded. Seeds 2–2.5 mm long, oblique-cordiform, smooth, with or without a small aril.

Zambia. W: Chizela Distr., 12 km Chizela (Chizera) to Solwezi, fl. 22.iii.1961, *Drummond & Rutherford-Smith* 7196 (BR; K; LISC; SRGH). C: 9.5 km east of Lusakà, fl. & fr. 13.iii.1956, *King* 341 (K). E: Lundazi Distr., fl. & fr. iii.1962, *Verboom* 611 (K). S: Mumbwa, fl. 20.iii.1963, *van Rensburg* 1747 (K; SRGH). **Zimbabwe**. N: Makonde Distr., Chinhoyi (Sinoia), Kafete road, fl. 18.ii.1969, *Corby* 2085 (K; SRGH). **Malawi**. N: Mzimba Distr., Champhila (Champira) Forest, fl. & fr. 20.iv.1974, *Pawek* 8413 (K; MAL; MO; SRGH). C: Dedza Distr., Chongoni Forestry School, fl. & fr. 3.iv.1967, *Salubeni* 629 (K; MAL; SRGH). S: Ntcheu Distr., Lower Kirk Range, Chipusiri, *Exell, Mendonça & Wild* 950 (BM; LISC; SRGH). **Mozambique**. N: Lichinga (Vila Cabral), fr. 17.v.1948, *Pedro & Pedrógão* 3670 (LMA).
Also in Dem. Rep. Congo (Katanga) and SE Tanzania. Miombo woodland and associated grassland; 900–1700 m.

164. **Crotalaria lepidissima** Baker f. in J. Linn. Soc., Bot. **42**: 281 (1914). —Wilczek in F.C.B. **4**: 185 (1953). —Torre in C.F.A. **3**: 36 (1962). —Polhill, Crotalaria Africa & Madagascar: 337 (1982). —Lock, Leg. Afr. Check-list: 187 (1989). Type from Dem. Rep. Congo (Katanga).

Stems ultimately numerous from a stout woody rhizomatous rootstock, but flowering from first year of growth, erect to 0.8 m, developing a number of flowering branches above, ribbed, densely covered with fine spreading hairs mostly 0.8–1 mm long. Leaves all 3-foliolate or uppermost occasionally 1-foliolate; leaflets 8–35 × 3–10 mm, oblanceolate to oblong-obovate or obovate, pilose beneath and usually also

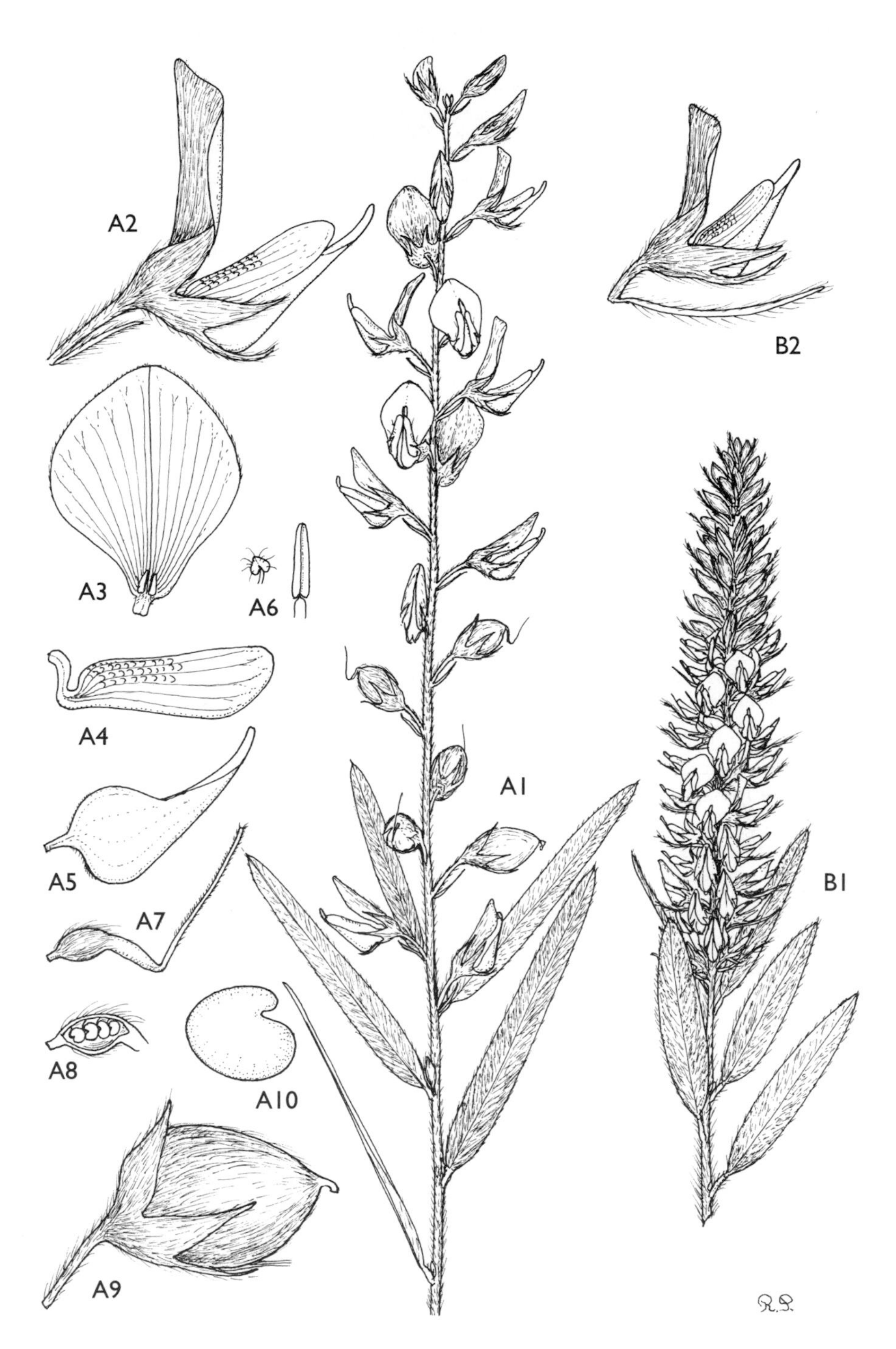

Tab. 3,7: **43**. A. —CROTALARIA EGREGIA. A1, flowering and fruiting branch (× 1); A2, flower (× 3); A3, standard (× 3); A4, wing (× 3); A5, keel (× 3); A6, anthers (× 8); A7, gynoecium (× 3); A8, section through ovary (× 4); A9, pod (× 3); A1–A9 from *Richards* 9286; A10, seed (× 6), from *Richards* 5691. B. —CROTALARIA KIPANDENSIS. B1, flowering branch, apical part (× 1); B2, flower (× 3), B1 & B2 from *Drummond & Rutherford-Smith* 6974. Drawn by Roger Polhill. From Crotalaria Africa & Madagascar.

above; petiole 3–15 mm long, channelled or narrowly winged; stipules 0. Racemes (1.5)3–10(17) cm long, densely many-flowered; bracts 3–6 mm long, linear-lanceolate to lanceolate-caudate, exceeding the pedicel; bracteoles on the pedicel, setaceous. Calyx 5–7 mm long, pilose; lobes attenuate-triangular, 2–3 times as long as the tube. Standard oblong-obovate, clear yellow, sometimes finely reddish lined, subdensely pilose outside; wings as long as the keel; keel 9–11 mm long, angular, with a long narrow twisted beak, conspicuously incurved at the tip. Pod subsessile, 5–6 × 3.5–5 mm, ovoid-globose, pubescent, 1–2-seeded. Seeds 2 mm long, oblique-cordiform, smooth, with a slightly swollen rim-aril.

Zambia. N: Lumangwe, fl. 14.xi.1957, *Fanshawe* 3993 (K; NDO). W: 19 km from Solwezi to Mwinilunga, fl. 20.xi.1972, *Strid* 2508 (K; LUS).

Also in Angola and Dem. Rep. Congo (Katanga). Miombo woodland, often in disturbed or damp places, or grassland of dambos and seepage areas; 1200–1350 m.

165. **Crotalaria kipilaensis** R. Wilczek in Bull. Jard. Bot. État **23**: 175 (1953); in F.C.B. **4**: 203 (1953). —Polhill, Crotalaria Africa & Madagascar: 337 (1982). —Lock, Leg. Afr. Check-list: 185 (1989). Type from Dem. Rep. Congo (Katanga).

Erect ± branching annual 25–80 cm tall; stem densely appressed to slightly spreading pubescent. Leaves 3-foliolate, uppermost occasionally 1-foliolate; leaflets (10)15–35 × (5)7–15 mm, oblong-oblanceolate to oblong-obovate, rounded at the apex, appressed pubescent beneath; petiole 8–20 mm long, slightly winged; stipules 0. Racemes sessile, 3–10 cm long, densely many-flowered, with other flowers in the axils below; bracts 3–4 mm long, linear to linear-lanceolate, exceeding the pedicel; bracteoles on the pedicel, setaceous. Calyx 5–6 mm long, densely appressed pubescent; lobes narrowly triangular-lanceolate, upper ones 4 times as long as the tube. Standard elliptic to elliptic-obovate, yellow, densely pubescent to tomentellous outside; wings shorter than the keel; keel 6–8 mm long, angular, with a long narrow twisted beak incurved at the tip. Pod sessile, 5–6 × 4–4.5 mm, ovoid-globose, appressed pubescent, 2-seeded. Seeds 2–2.5 mm long, oblique-cordiform, smooth, with a pronounced aril.

Zambia. N: about 1.5 km from Mbala on road to Itimbwe (Itembwe) Gorge, fl. & fr. 24.iv.1959, *McCallum-Webster* 911 (K). W: Kitwe, fl. & fr. 6.iv.1957, *Fanshawe* 3148 (K; NDO). C: 3 km north of Kabwe (Broken Hill), fl. & fr. 15.iii.1961, *Drummond & Rutherford-Smith* 6912 (K; LISC; SRGH).

Also in Angola and Dem. Rep. Congo (Katanga). Miombo woodland; 1250–1700 m.

166. **Crotalaria streptorrhyncha** Milne-Redh. in Hooker's Icon. Pl. **33**: t. 3245 (1934). —Polhill, Crotalaria Africa & Madagascar: 338 (1982). —Lock, Leg. Afr. Check-list: 205 (1989). Type: Zambia, Solwezi Distr., Mutanda Bridge, *Milne-Redhead* 569 (K, holotype).

Similar to *C. kipilaensis* but bracts (3)4–7 mm long, calyx lobes ovate-lanceolate, thinly pubescent to glabrous apart from margins, standard oblong-obovate, pubescent only at the apex outside, wings longer than the keel, beak of the keel only slightly incurved and rather shorter, and pod a little larger, glabrous except along the upper suture.

Zambia. W: Mwinilunga Distr., 7 km north of Kalene Hill, fl. & fr. 17.iv.1965, *E.A. Robinson* 6621 (K); 8 km west of Solwezi, fl. 18.iii.1961, *Drummond & Rutherford-Smith* 7005 (K; SRGH).

Known only from western Zambia. Miombo and riverside woodland; c. 1200–1350 m.

167. **Crotalaria ionoptera** Polhill, Crotalaria Africa & Madagascar: 340, fig. 95/1–10 (1982). —Lock, Leg. Afr. Check-list: 183 (1989). Type: Zambia, Mwinilunga Distr., Matonchi, Kalenda Plain, *E.A. Robinson* 3629 (K, holotype; MO).

Slender erect branching annual 20–70 cm tall; stem densely appressed puberulous. Leaves 1-foliolate above, 3-foliolate below; leaflets 8–22 × 0.5–4 mm, mostly linear-oblanceolate, acute, lower ones sometimes oblong-oblanceolate, appressed puberulous beneath; petiole 1–5(7) mm long; stipules 0. Racemes 3–18 cm long, very laxly 4–12-flowered, sometimes with a few flowers in the axils below; bracts 1.5–2.5 mm long, linear, shorter than the pedicel. Calyx 3.5–4.5 mm long, with appressed to irregularly arranged hairs; upper lobes narrowly triangular, slightly acuminate, 1.5–2 times as long as the tube. Standard elliptic, reddish-brown and pubescent outside, violet with yellow basal marks inside; wings blue to violet, shorter than the keel; keel 1–1.2 cm long, angular, with a long narrow reddish-brown beak slightly incurved at the tip. Pod subsessile, 6–9 × 3–6 mm, shortly oblong-ellipsoid to ellipsoid-globose, pubescent, 16–20-seeded. Seeds 2 mm across, oblique-cordiform, smooth.

Zambia. B: Mongu Airport, fr. 23.iii.1964, *Verboom* 1000 (K). W: Mwinilunga Distr., Matonchi Farm, fl. 13.ii.1938, *Milne-Redhead* 4558 (K).

Known only from Zambia along the upper reaches of the Zambezi R. Open grassy places on sand or dried out laterite pavements; 1050–1400 m.

168. **Crotalaria crebra** Polhill, Crotalaria Africa & Madagascar: 341, fig. 96/1–9 (1982). —Lock, Leg. Afr. Check-list: 172 (1989). TAB. 3,7: **44**, fig. A. Type: Zambia, Mwinilunga Distr., 7 km north of Kalene Hill, *E.A. Robinson* 6624 (K, holotype).

Erect annual 50–70 cm tall, with numerous ascending branches, the ultimate flowering branches mostly suppressed to give axillary clusters of flowers and leaves; stem rather densely subappressed pubescent. Leaves mostly 3-foliolate, uppermost reduced, usually 1-foliolate; leaflets 7–15 × 2–5 mm, oblong-oblanceolate, thinly pilose on both surfaces; petiole 2–4 mm long; stipules 0. Flowers 4–10 in terminal sessile heads and on suppressed lateral branches below, also a few in the axils; bracts progressively reduced inwards, outer ones foliaceous, the innermost linear; pedicel 1–3 mm long, sometimes with setaceous bracteoles. Calyx 4–5 mm long, densely pilose; upper lobes narrowly acuminate-triangular, 2–3 times as long as the tube. Standard elliptic-obovate, yellow, finely lined reddish, densely silky pubescent outside; wings ± as long as the keel; keel 8–10 mm long, angular, with a straight twisted beak. Pod sessile, 6 × 4 mm, shortly oblong-ellipsoid, pubescent, 6–10-seeded. Mature seeds not seen.

Zambia. W: Mwinilunga Distr., 7 km north of Kalene Hill, fl. & fr. 17.iv.1965, *E.A. Robinson* 6624 (K).

Known only from the type gathering. Woodland fringe; 1200–1300 m.

169. **Crotalaria pygmaea** Polhill, Crotalaria Africa & Madagascar: 342, fig. 96/10–18 (1982). —Lock, Leg. Afr. Check-list: 200 (1989). TAB. 3,7: **44**, fig. B. Type: Zambia, near Kawambwa, *Bullock* 3813 (BM; BR; K, holotype).

Small annual 2–10 cm tall; stem erect, densely appressed or subappressed pubescent, developing numerous ascending branches from near the base. Leaves mostly 3-foliolate, uppermost sometimes 1-foliolate; leaflets 7–20 × 3–6 mm, oblong-oblanceolate, appressed pilose on both surfaces; petiole (3)5–10 mm long, narrowly winged; stipules 0. Flowers (2)6–16 in sessile terminal heads, also clustered in a few axils below; bracts 2–4 mm long, linear; pedicel 3–5 mm long, without bracteoles. Calyx 5 mm long, silky tomentose; upper lobes narrowly triangular, 2–3 times as long as the tube. Standard elliptic, pointed, pale yellow, tomentellous outside; wings markedly shorter than the keel; keel 10–12 mm long, angular, with a long narrow twisted beak. Pod sessile, 7–8 × 3.5–4 mm, oblong-ellipsoid, appressed pubescent, 20–24-seeded. Mature seeds not seen.

Zambia. N: Kaputa Distr., Nsama, fl. 3.iv.1957, *Richards* 8993 (K); Mporokoso Distr., Kalungwishi R., Lumangwe Falls, fl. 14.iv.1989, *Goyder, Pope & Radcliffe-Smith* 3027 (K; NDO).

Known only from the area east of Lake Mweru. Miombo woodland and grassland, sometimes in damp places; 1050–1200 m.

170. **Crotalaria sparsifolia** Baker in Bull. Misc. Inform., Kew **1897**: 249 (1897). —E.G. Baker in J. Linn. Soc., Bot. **42**: 304 (1914). —Verdoorn in Bothalia **2**: 401 (1928). —Hutchinson, Botanist South. Africa: 496 (1946), but not p. 520. —Binns, First Check List Herb. Fl. Malawi: 80 (1968). —Polhill in F.T.E.A., Leguminosae, Pap.: 988 (1971). —Drummond in Kirkia **8**: 218 (1972). —Polhill, Crotalaria Africa & Madagascar: 343, fig. 82/4 (1982). —Lock, Leg. Afr. Check-list: 204 (1989). TAB. 3,7: **45**, fig. D. Syntypes: Malawi, between Kondowe and Karonga, *Whyte* s.n.; Nyika Plateau, *Whyte* s.n. (K, syntypes).

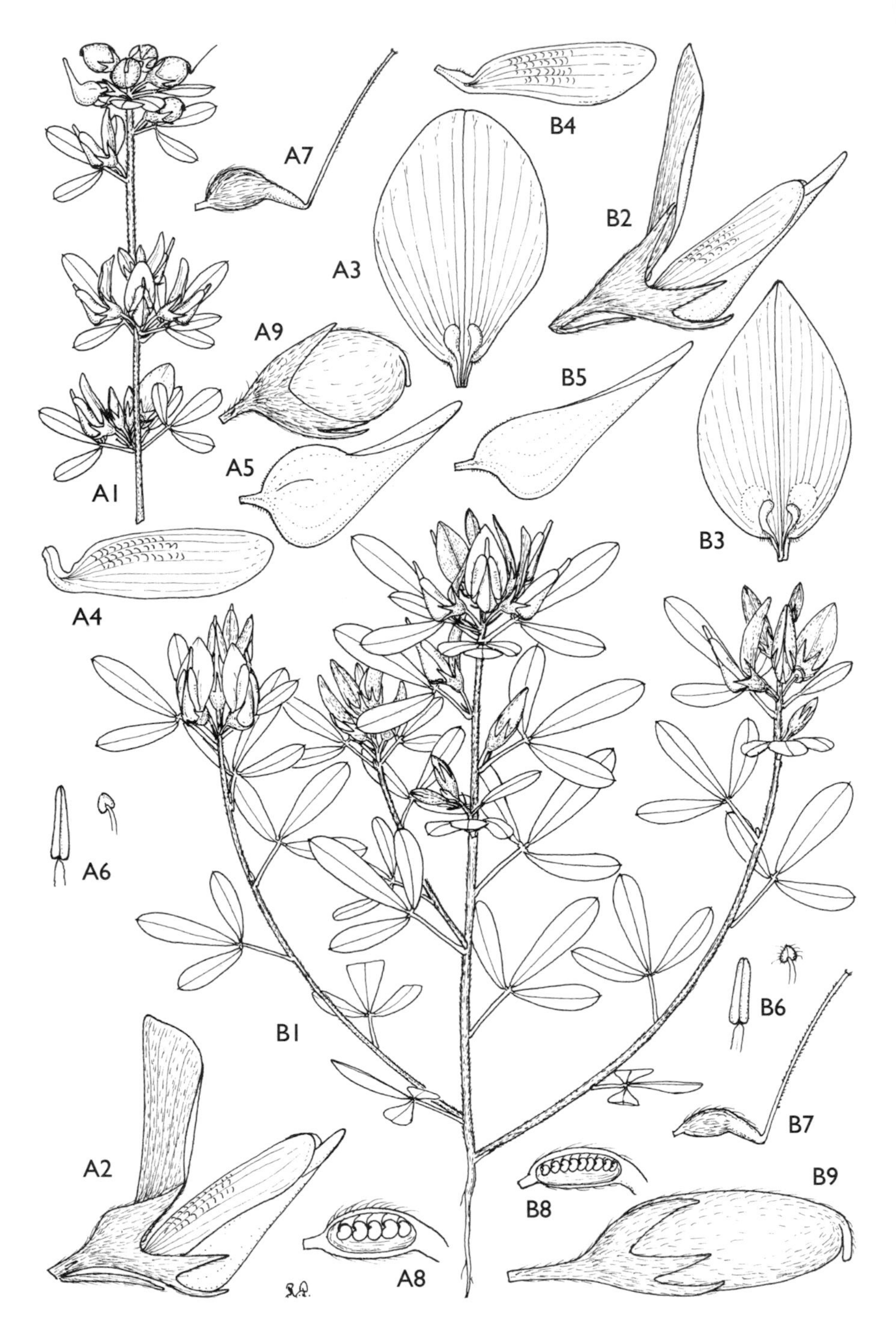

Tab. 3,7: **44**. A. —CROTALARIA CREBRA. A1, flowering and fruiting branch, apical part (× 1); A2, flower (× 4); A3, standard (× 4); A4, wing (× 4); A5, keel (× 4); A6, anthers (× 8); A7, gynoecium (× 4); A8, section through ovary (× 6); A9, pod (× 4); A1–A9 from *E.A. Robinson* 6624. B. —CROTALARIA PYGMAEA. B1, plant habit (× 1), from *Bullock* 3813; B2, flower (× 3); B3, standard (× 3); B4, wing (× 3); B5, keel (× 3); B6, anthers (× 8); B7, gynoecium (× 3); B8, section through ovary (× 6); B9, pod (× 3), B2–B9 from *Richards* 9438. Drawn by Roger Polhill. From Crotalaria Africa & Madagascar.

Erect annual 10–50 cm tall, developing numerous slender ascending branches; stem densely silvery appressed pubescent. Leaves sometimes small or deciduous at flowering time, mostly 3-foliolate, upper ones usually 1-foliolate; leaflets 4–15 × 0.5–4 mm, linear-oblanceolate to oblanceolate, glabrous to sparsely pubescent above, subdensely appressed pubescent beneath; petiole 1–8 mm long; stipules 0. Racemes 3–12 cm long, very laxly 6–12-flowered, with other flowers developing in the axils below; bracts 1–2 mm long, linear-subulate; pedicel 1–2.5 mm long, generally with small setaceous bracteoles. Calyx 3–4 mm long, densely appressed pubescent to tomentellous; upper lobes triangular-acuminate, 1.5–2.5 times as long as the tube. Standard elliptic-obovate, pale yellow, with deeper yellow red-fringed marks at the base inside, reddish lined and pubescent to tomentellous outside; wings a little shorter than the keel; keel (4)5–6 mm long, angular, with a narrow twisted beak slightly incurved at the tip. Pod sessile, 4–5 × 3–4 mm, ovoid-globose, densely pubescent, 3–6-seeded. Seeds 1.5–2 mm across, oblique-cordiform, smooth.

Zambia. N: 25 km north of Mpika, fl. & fr. 25.v.1962, *E.A. Robinson* 5245 (K; SRGH). W: Kitwe, fl. & fr. 20.v.1955, *Fanshawe* 2290 (K; NDO; SRGH). C: Serenje Distr., Kundalila Falls, fl. & fr. 24.viii.1963, *Richards* 18211 (K). E: Katete, fl. & fr. 24.iii.1955, *Exell, Mendonça & Wild* 1160 (BM; LISC; SRGH). S: Mazabuka Distr., Kafue Gorge, fl. & fr. 9.vi.1963, *van Rensburg* 2295 (K; SRGH). **Zimbabwe**. N: Mazowe Distr., Concession, fl. & fr. 4.iv.1969, *Corby* 2129 (K; SRGH). **Malawi**. N: Mzimba Distr., Mzuzu, Marymount, fl. & fr. 2.viii.1974, *Pawek* 8860 (K; MO; SRGH). C: Dedza Distr., Chongoni Forestry School, fl. & fr. 10.v.1967, *Salubeni* 702 (K; SRGH). S: Machinga Distr., Munde Hill, fl. & fr. 30.iv.1982, *Patel* 890 (K; MAL). **Mozambique**. N: Sanga Distr., Macaloge (Mecaloja), fl. & fr. 28.viii.1934, *Torre* 245 (BM; COI; K; LISC). Z: Alto Molócuè, arredores, fr. 5.vii.1943, *Torre* 5671 (LISC).

Also in Dem. Rep. Congo (Katanga) and southern Tanzania. Miombo woodland; 950–2000 m.

Greenway 5433 (EA; K), from Zambia, Shiwa Ngandu, fl. 20.vii.1938, appears to represent a new species characterized by numerous upper leaves 1-foliolate, the standard only thinly pubescent outside and the keel 7 mm long, slightly incurved at the tip.

171. **Crotalaria schliebenii** Polhill in Kew Bull. **22**: 328, fig. 36/1–9 (1968); in F.T.E.A., Leguminosae, Pap.: 989 (1971); Crotalaria Africa & Madagascar: 344 (1982). —Lock, Leg. Afr. Check-list: 202 (1989). Type from Tanzania.

Erect annual or short-lived perennial 0.6–1 m tall, developing numerous ascending flowering branches above, and sometimes a few long ones from near the base; stem densely silvery subappressed pubescent. Leaves 3-foliolate; leaflets 8–20 × 1–3 mm, linear-oblanceolate to oblong-oblanceolate, thinly pilose above, subdensely so beneath; petiole 3–7 mm long; stipules 0. Racemes sessile, 3–14 cm long, densely many-flowered; bracts 2–3.5 mm long, linear-subulate; pedicel 1–2.5 mm long, without bracteoles. Calyx 3.5–4 mm long, silvery tomentose; upper lobes narrowly attenuate-triangular, 2–3 times as long as the tube. Standard obovate, yellow, with darker red-fringed marks at the base inside, brown lined and subdensely pubescent to tomentellous outside; wings nearly as long as the keel; keel 4.5–5.5 mm long, angular, with a nearly straight twisted beak. Pod sessile, 4–5 × 3–3.5 mm, ovoid-globose, pubescent, 2-seeded. Seeds 2–2.5 mm long, ovate-cordiform, smooth, with a prominent aril.

Mozambique. N: 10 km de Nampula para Corrane, próximo do rio Colave, fl. & fr. 14.iv.1961, *Balsinhas & Marrime* 402 (BM; K; LISC; LMA); Mogovolas Distr., entre Corrane e Nametil, fl. & fr. 18.viii.1935, *Torre* 948 (COI; LISC).

Also in SE Tanzania. Riverside grassland and swampy places.

172. **Crotalaria quarrei** Baker f. in Rev. Zool. Bot. Africaines **23**: 181 (1933). —Wilczek in F.C.B. **4**: 246 (1953). —Polhill in Kew Bull. **22**: 335 (1968); in F.T.E.A., Leguminosae, Pap.: 994 (1971); Crotalaria Africa & Madagascar: 344, fig. 82/5 (1982). —Lock, Leg. Afr. Check-list: 200 (1989). Neotype, see Wilczek, loc. cit., from Dem. Rep. Congo (Katanga).

Small annual 5–10 cm tall, with numerous decumbent to shortly ascending branches from the base; stem slender, densely appressed pubescent. Leaves mostly 3-foliolate, the uppermost often 1-foliolate; leaflets 5–16 × 3–8 mm, oblanceolate-elliptic to obovate, subacute to rounded at the apex, thinly pilose above, more densely so beneath; petiole 2–8 mm long; stipules 0. Flowers 6–20 in sessile to

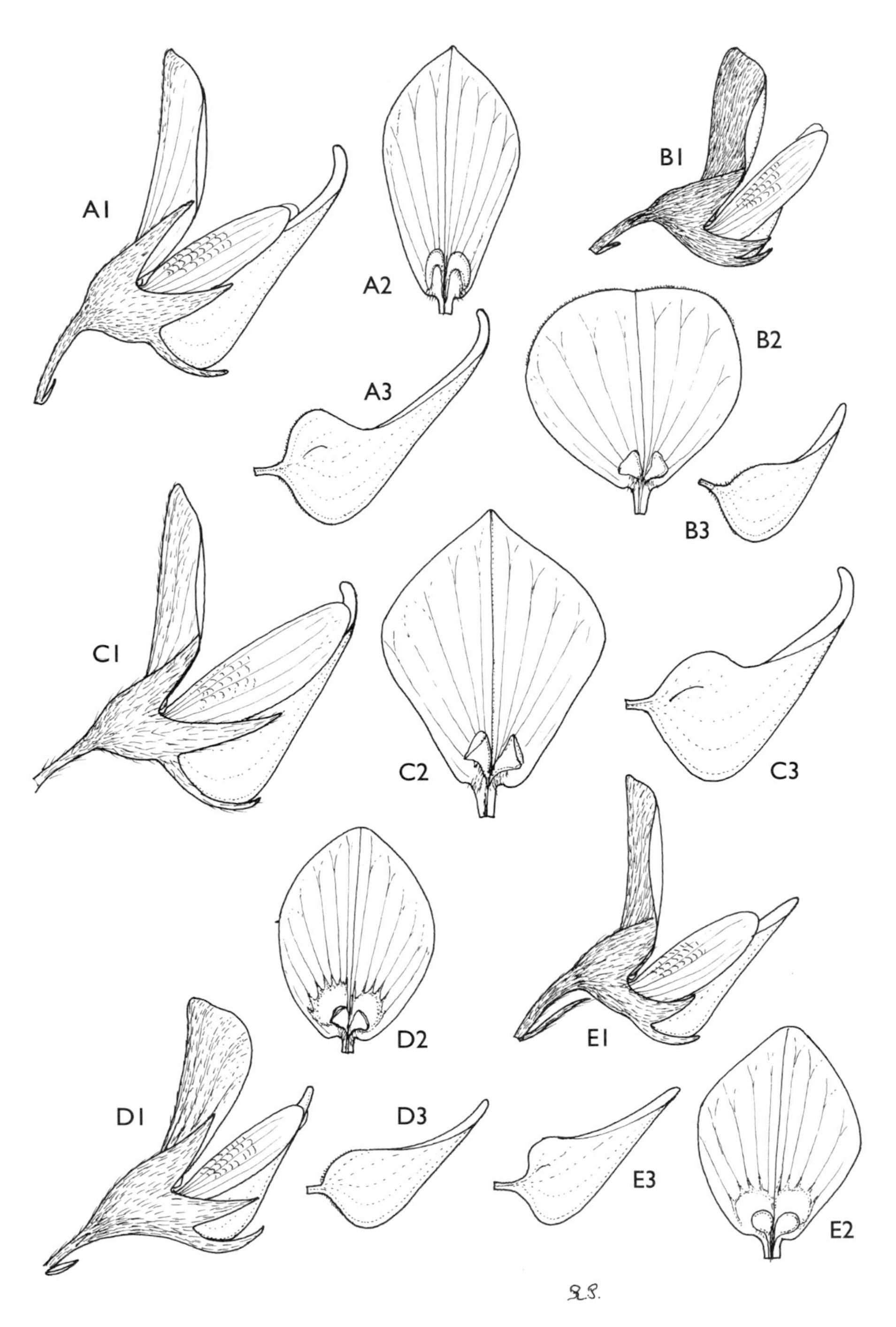

Tab. 3,7: **45**. Flowers of *Crotalaria* sect. *Dispermae*: 1, flower; 2, standard; 3, keel. A. —CROTALARIA ELISABETHAE (× 4), from *Mutimushi* 2425; B. —CROTALARIA FILICAULIS var. FILICAULIS (× 6), from *Pawek* 8502B; C. —CROTALARIA JOHNSTONII (× 6), from *J. Williamson* 62; D. —CROTALARIA SPARSIFOLIA (× 6), from *Pawek* 8492A; E. —CROTALARIA QUARREI var. QUARREI (× 6), from *Richards* 5257. Drawn by Roger Polhill. From Crotalaria Africa & Madagascar.

pedunculate heads or racemes, sometimes with a few flowers in the axils below; bracts 1–3 mm long, linear; pedicel 1.5–2.5 mm long, without obvious bracteoles. Calyx 2.5–3 mm long, densely appressed pubescent; upper lobes triangular, 1–2 times as long as the tube. Standard obovate, bright yellow, with red marks at the base inside, reddish lined and tomentellous outside; wings a little shorter than the keel; keel 3.5–4.5 mm long, angular, with a narrow twisted beak. Pod sessile, 4–5 × 3–3.5 mm, ovoid-globose, pubescent, 12–16-seeded. Seeds 1 mm long, rounded-cordiform, smooth.

Var. **quarrei** TAB. 3,7: **45**, fig. E.

Inflorescences sessile or with a short peduncle 2–6 mm long, subumbelliform, sometimes with a few more laxly inserted flowers below the head (growing out somewhat in fruit); bracts longer than the pedicel.

Zambia. N: Kasama Distr., Chishimba Falls, fl. & fr. 31.iii.1955, *Richards* 5257 (K). W: Kitwe, fl. & fr. 15.iii.1968, *Mutimushi* 2549 (K; NDO; SRGH). S: 32 km from Mumbwa, fl. & fr. 19.iii.1963, *van Rensburg* 1697 (K; SRGH).

Also in Cameroon, Chad, Central African Republic, Burundi, Angola, Dem. Rep. Congo (Katanga) and Tanzania. Miombo woodland and open disturbed places; 1200–1750 m.

Var. **longipes** Polhill in Kew Bull. **22**: 336 (1968); in F.T.E.A., Leguminosae, Pap.: 994 (1971); Crotalaria Africa & Madagascar: 345 (1982). Type from SE Tanzania.

Inflorescences with a peduncle 10–20 mm long, laxly racemose or with most of the flowers ± crowded above but not umbelliform; bracts mostly shorter than the pedicel.

Zambia. N: Mbala Distr., Chilongowelo, fl. & fr. 19.iii.1952, *Richards* 1114 (K); Chisungu, fl. & fr. 16.iv.1962, *Richards* 16333 (K; SRGH).

Also in southern Tanzania. Growing in situations similar to var. *quarrei*; 1250–1450 m.

173. **Crotalaria filicaulis** Welw. ex Baker in F.T.A. **2**: 24 (1871). —E.G. Baker in J. Linn. Soc., Bot. **42**: 291 (1914). —Verdoorn in Bothalia **2**: 401 (1928). —Wilczek in F.C.B. **4**: 207 (1953). —Torre in C.F.A. **3**: 43 (1962). —Jacobsen in Kirkia **9**: 160 (1973). —Polhill, Crotalaria Africa & Madagascar: 345, fig. 82/2 (1982). —Lock, Leg. Afr. Check-list: 178 (1989). Type from Angola.

Erect annual 20–120 cm tall; stem slender, often with numerous ascending branches and shorter flowering branches above, usually densely silvery appressed pubescent, elsewhere sometimes sparsely puberulous to subglabrous. Leaves mostly 3-foliolate, uppermost sometimes 1-foliolate; leaflets 10–20(30) × 0.5–2.5(4) mm, linear-oblanceolate, appressed puberulous beneath; petiole 1–7(12) mm long; stipules 0. Terminal racemes 6–25 cm long, laxly 10–40-flowered, shorter on lateral branches, sometimes developing a few flowers in the axils below; bracts 1–4 mm long, shortly linear-elliptic to linear-subulate; pedicel 1.5–4 mm long, with setaceous bracteoles. Calyx 3–3.5 mm long, usually silvery tomentellous, elsewhere sometimes subglabrous; upper lobes rather broadly triangular, often slightly acuminate, ± twice as long as the tube. Standard obovate to oblate, pale yellow, marked red at the base inside, lined red outside, tomentellous outside or elsewhere sometimes puberulous only near the apex; wings longer than the keel; keel 4–5 mm long, subangular, with a tapered twisted beak. Pod sessile, 4–5 × 3.5–4 mm, ovoid-globose, puberulous, 2-seeded. Seeds 2.5–3 mm long, rounded-cordiform.

Var. **filicaulis**. TAB. 3,7: **45**, fig. B.

Standard 3–5 mm wide, obovate, tomentellous outside; bracts 1–1.5 mm long; petioles 1–5(7) mm long.

Zambia. B: Kaoma (Mankoya), fl. & fr. 5.iv.1966, *E.A. Robinson* 6918 (K; SRGH). N: Mbala Distr., near Sumbawanga road, fl. & fr. 21.iv.1963, *Richards* 18117 (K; SRGH). W: Ndola, fl. & fr. 28.iii.1957, *Fanshawe* 3101 (K; NDO). C: Serenje, fl. & fr. 28.iii.1984, *Brummitt, Chisumpa & Nshingo* 16945 (K; NDO; SRGH). E: Petauke Distr., 4 km west of Kachalola on Great East Road, fl. 17.iii.1959, *Robson* 1735 (K). **Zimbabwe**. N: Bindura Distr., 27 km from Bindura towards

Mount Darwin, fl. & fr. 2.iii.1971, *Corby* 2192 (K; SRGH). C: Chegutu Distr., Poole Farm, fl. & fr. 14.iii.1948, *Hornby* 2860 (K; LISC; MO; SRGH). E: Mutare Distr., Odzi, fl. & fr. 14.vi.1936, *Eyles* 8601 (K; SRGH). **Malawi**. N: Mzimba Distr., Mbawa Experimental Station, fl. & fr. 5.iv.1955, *Jackson* 1586 (BR; K; SRGH). C: 48 km from Kasungu on M1, fl. & fr. 7.iv.1978, *Pawek* 14350 (K; MAL; MO; SRGH). **Mozambique**. N: 16 km NE of Mandimba Border Post, fl. & fr. 3.v.1960, *Leach & J. Brunton* 9908 (K; SRGH).

Also in Angola, Dem. Rep. Congo (Katanga) and southern Tanzania. Miombo woodland, often in disturbed places on sandy or gravelly soils; 500–1800 m.

Var. **grandiflora** Polhill, Crotalaria Africa & Madagascar: 345 (1982). Type from Dem. Rep. Congo (Katanga).

Crotalaria filicauloides R. Wilczek in Bull. Jard. Bot. État **23**: 177 (1953); in F.C.B. **4**: 206 (1953). Type as for var. *grandiflora*.

Standard 6–8 mm wide, rounded-obovate to oblate, tomentellous outside; bracts 2–4 mm long; petioles mostly 3–12 mm long.

Zambia. N: Mbala Distr., Kambole Escarpment, fl. & fr. 21.iv.1969, *Richards* 24498 (K); Mweru Wantipa, road from Mporokoso to Nsama, fl. 16.iv.1957, *Richards* 9284A (K).

Also in Katanga as far west as the Kundelungu Plateau. Miombo and its fringes on gritty soils; 1200–1500 m.

A very distinctive large-flowered form of the species, but apparently intergrading with var. *filicaulis* at the top of the escarpment south of Lake Tanganyika, e.g. Kambole Escarpment, late fl. & fr. 14.vi.1961, *Richards* 15278 (K).

Var. *decaulescens* (R. Wilczek) Polhill, from the Luapula drainage area in Katanga between Lubumbashi and Lake Mweru, is similar to var. *filicaulis*, but is altogether much less hairy, the calyx subglabrous and the standard puberulous only towards the apex.

174. **Crotalaria kuiririensis** Baker f. in J. Linn. Soc., Bot. **42**: 300 (1914). —Torre in C.F.A. **3**: 48 (1962). —Drummond in Kirkia **8**: 218 (1972). —Polhill, Crotalaria Africa & Madagascar: 346 (1982). —Lock, Leg. Afr. Check-list: 185 (1989). Type from Angola.

Erect annual 30–65 cm tall, with slender ascending densely silvery appressed pubescent branches. Leaves 1-foliolate above, 3-foliolate below; leaflets 10–35 × 1–11 mm, linear-oblanceolate to oblong-lanceolate or oblong-elliptic, densely silky pubescent beneath; petiole 2–15 mm long; stipules 0. Racemes short, few-flowered (often seeming longer by suppression of upper leaves), with flowers clustered in many of the axils below; bracts 2–3 mm long, linear; pedicel 1–2 mm long, with small setaceous bracteoles on the upper part. Calyx 4–4.5 mm long, tomentellous; upper lobes broadly triangular-acuminate, 1.5–2 times as long as the tube. Standard elliptic-obovate, yellow, lined reddish-brown, tomentellous outside; wings nearly as long as the keel; keel 5.5–6 mm long, angular, with a narrow twisted beak sharply incurved at the tip. Pod sessile, 4.5–5 × 4 mm, ovoid-globose, silky pubescent, 2-seeded. Seeds 2.5 mm across, oblique-cordiform.

Zambia. B: Zambezi Distr., Chavuma, fl. 20.iv.1954, *Gilges* 344 (K; SRGH). W: Kitwe, fr. 20.v.1955, *Fanshawe* 2284 (K; NDO). C: 19 km north of Kapiri Mposhi, fl. & fr. 8.vi.1963, *Drummond* 8272 (K; SRGH). **Zimbabwe**. N: Gokwe South Distr., Charama Plateau, fl. & fr. 11.iv.1963, *Bingham* 621 (K; LISC; SRGH).

Also in Dem. Rep. Congo (Katanga) and southern Angola. Deciduous woodland, principally on Kalahari Sands; 1050–1250 m.

175. **Crotalaria duboisii** R. Wilczek in Bull. Jard. Bot. État **23**: 199, t. 5 (1953); in F.C.B. **4**: 238, t. 13 (1953). —Polhill, Crotalaria Africa & Madagascar: 346 (1982). —Lock, Leg. Afr. Check-list: 176 (1989). Type from Dem. Rep. Congo (Katanga).

Subsp. **mutica** Polhill, Crotalaria Africa & Madagascar: 346 (1982). —Lock, Leg. Afr. Check-list: 176 (1989). Type: Zambia, Mbala Distr., Simanwe Farm, *Richards* 4969 (BR; K, holotype; SRGH).

Erect annual (subsp. *duboisii* can be perennial), 0.5–1 m tall, with ascending branches at least above in older plants; stem slightly ribbed, densely strigulose with brown-based hairs. Leaves 3-foliolate; leaflets 8–20 × 1.5–5 mm, linear-oblanceolate to oblong-oblanceolate, strigulose on both surfaces; petiole 4–14 mm long; stipules

0. Racemes 9–25 cm long, laxly 12–30-flowered, sometimes with a few flowers in the axils below; bracts 1–2 mm long, linear-lanceolate; bracteoles on the pedicel, setaceous, 0.5–1 mm long. Calyx 4–5 mm long, brownish strigulose with some longer more flexuous hairs; lobes irregularly oblong-triangular, truncate, involute, lanate inside, shorter than the tube. Standard elliptic-obovate, yellow, usually flushed brown, glabrous outside; wings shorter than the keel; keel 1.1–1.3 cm long, angular with a narrow twisted beak, incurved at the tip. Pods subsessile, 9–10 × 5–6 mm, oblong-ellipsoid, densely strigose-puberulent, 8–14-seeded. Seeds 2–2.5 mm long, oblique-cordiform, smooth, with a white aril.

Zambia. N: Mbala Distr., Senga Hill, fl. & fr. 21.v.1962, *E.A. Robinson* 5208 (K; SRGH); Chinsali Distr., c. 60 km SW of Isoka, fl. & immat. fr. 15.iii.1975, *Hooper & Townsend* 781 (K; SRGH).
Known only from northern Zambia. Miombo woodland; 1500–1800 m.
Subsp. *duboisii*, with triangular acute calyx lobes, longer bracts and bracteoles, occurs on the higher plateaux of the Katanga Province (Dem. Rep. Congo). *Jeff* 31 (BM) from Luwingu, fl. iv.1922, has more pointed calyx lobes than usual in subsp. *mutica*, but the bracts and bracteoles are typical of the Zambian plants.

176. **Crotalaria bemba** R. Wilczek in Bull. Jard. Bot. État **23**: 139 (1953); in F.C.B. **4**: 116 (1953). —Polhill, Crotalaria Africa & Madagascar: 348 (1982). —Lock, Leg. Afr. Check-list: 168 (1989). Type from Dem. Rep. Congo (Katanga).

Erect annual herb, (10)20–60 cm tall, with 0–numerous ascending branches, appressed pubescent. Leaves mostly 3-foliolate, uppermost often 1-foliolate; leaflets 8–30 × 2–8 mm, linear-oblanceolate to oblong-oblanceolate, appressed pubescent beneath; petiole 0–2 mm; stipules 0. Racemes (3)5–12 cm long, laxly 4–30-flowered, developing other racemes and single flowers in the axils below; bracts 1.5–5 mm long, linear-lanceolate, shorter than the pedicel; bracteoles minute or lacking. Calyx (4)6–7 mm long, densely appressed pubescent; upper lobes broadly triangular-lanceolate, acuminate, ± twice as long as the tube. Standard oblate, yellow, lined brown, tomentellous outside; wings longer than the keel; keel 7–9 mm long, subangular with a twisted beak incurved at the tip. Pod sessile, 7–9 × 5–7 mm, ovoid-globose, appressed pubescent, 6–12-seeded. Seeds 1.5–2.5 mm long, oblique-cordiform, smooth.

Zambia. N: Muchinga Escarpment, 15 km north of Mporokoso, fl. 15.iv.1989, *Goyder, Pope & Radcliffe-Smith* 3038 (K; NDO); Mbala Distr., Chilongowelo Escarpment, fl. & fr. 6.iv.1962, *Richards* 16263 (K; SRGH). W: Kalulushi Distr., Chati, fl. 28.ii.1964, *Mutimushi* 639 (K; NDO).
Also in Dem. Rep. Congo (Katanga). Miombo woodland; 1200–1800 m.

177. **Crotalaria blanda** Polhill, Crotalaria Africa & Madagascar: 348, fig. 98/1–10 (1982). — Lock, Leg. Afr. Check-list: 168 (1989). Type: Zambia, Mansa (Fort Rosebery), *Fanshawe* 8574 (K, holotype; LISC; NDO).

Erect annual 0.8–1.5 m tall, developing numerous slender flowering branches above, densely silvery appressed pubescent. Leaves mostly 3-foliolate, upper ones 1-foliolate; leaflets 10–20 × 0.7–2 mm, linear-oblanceolate, pubescent beneath; petiole 2–4 mm long; stipules 0. Racemes 8–20 cm long, laxly 10–30-flowered, with other flowers on shorter racemes or in the axils below; bracts 1.5–2.5 mm long, linear-subulate to linear-lanceolate; pedicel 4–7 mm long, with setaceous bracteoles. Calyx 3.5–4 mm long, densely covered with fine irregularly arranged hairs; upper lobes triangular, 1.5–2.5 times as long as the tube. Standard broadly elliptic, yellow, later flushed reddish, silvery tomentellous outside; wings a little shorter than the keel; keel 1–1.2 cm long, angular, with a long narrow twisted beak hooked inwards at the tip. Pod sessile, 5–7 × 4–5 mm, shortly ovoid-ellipsoid, densely pubescent, 10–12-seeded. Seeds 1.5 mm long, oblique-cordiform, smooth.

Zambia. N: Mansa (Fort Rosebery), fl. & fr. 5.v.1964, *Fanshawe* 8574 (K; LISC; NDO).
Known only from the type. Miombo scarp woodland; c. 1200 m.

178. **Crotalaria arcuata** Polhill, Crotalaria Africa & Madagascar: 349, fig. 99/1–9 (1982). — Lock, Leg. Afr. Check-list: 165 (1989). TAB. 3,7: **46**, fig. A. Type: Zambia, Mweru Wantipa, Muzombwe, *Whellan* 1387 (K, holotype; SRGH).

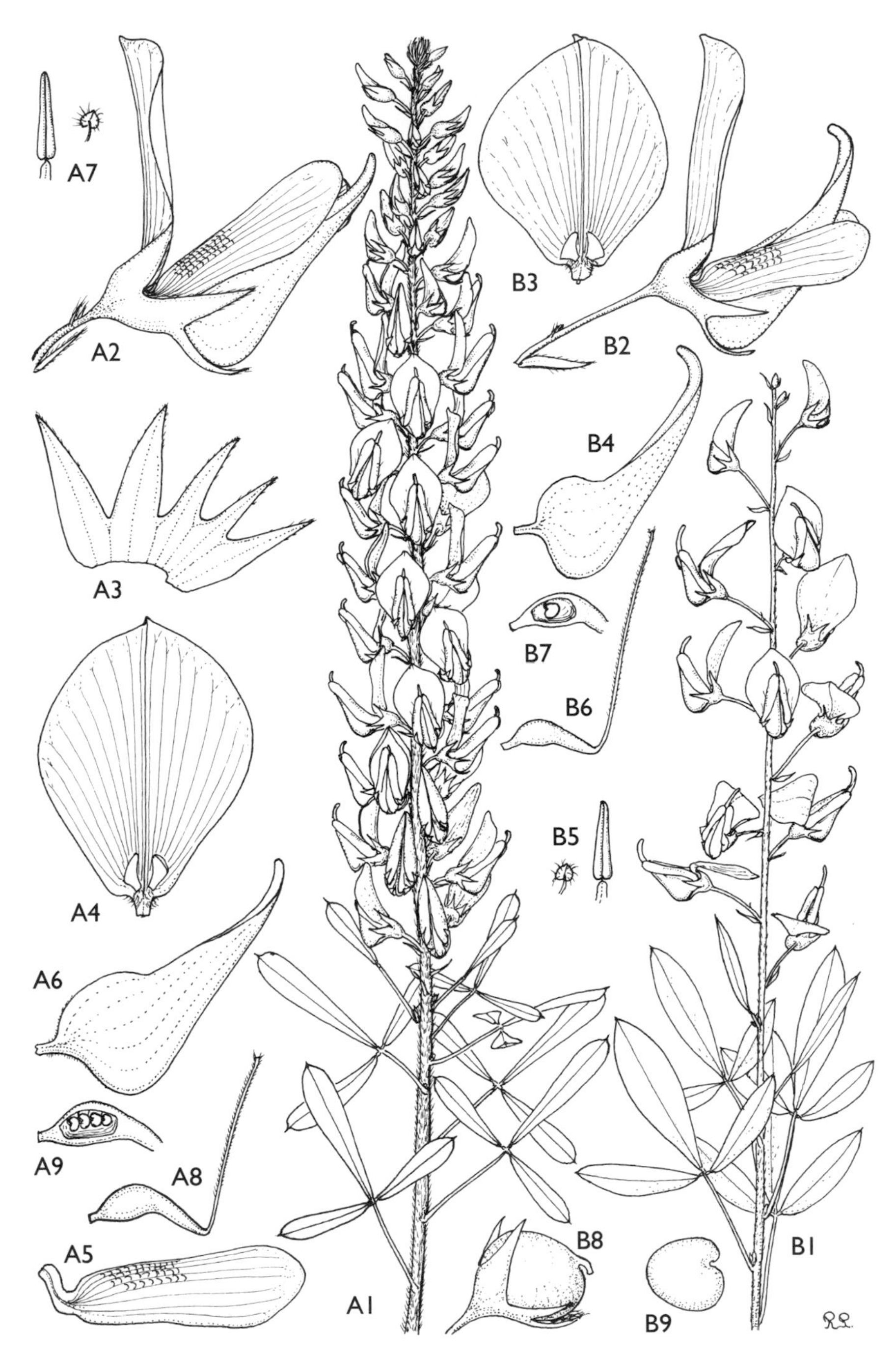

Tab. 3,7: **46**. A. —CROTALARIA ARCUATA. A1, apical portion of flowering branch (× 1); A2, flower (× 3); A3, calyx, opened out (× 3); A4, standard (× 3); A5, wing (× 3); A6, keel (× 3); A7, anthers (× $4^1/_2$); A8, gynoecium (× 3); A9, section through ovary (× $4^1/_2$), A1–A9 from *Whellan* 1387. B. —CROTALARIA PEREGRINA. B1, apical portion of flowering branch (× 1); B2, flower (× 3); B3, standard (× 3); B4, keel (× 3); B5, anthers (× $4^1/_2$); B6, gynoecium (× 3); B7, section through ovary (× $4^1/_2$), B1–B7 from *Richards* 5454; B8, pod (× 3); B9, seed (× $4^1/_2$), B8 & B9 from *Richards* 5834. Drawn by Roger Polhill. From Crotalaria Africa & Madagascar.

Strictly erect annual to 90 cm, sparsely branched below the flowering part, strigose-pubescent to pilose. Leaves mostly 3-foliolate, upper ones 1-foliolate; leaflets 12–35 × 3–8 mm, linear-oblanceolate to oblong-oblanceolate, obtuse or rounded to the apiculate tip, appressed pubescent beneath; petiole 7–20 mm long; stipules 0. Racemes (3)15–40 cm long, subdensely many-flowered, with other flowers developing in the axils below, and sometimes with numerous short flowering branches; bracts (1)3–5 mm long, linear-lanceolate; pedicel 5–7 mm long, with small bracteoles. Calyx 4–7 mm long, usually glabrous except on the lobe margins, occasionally strigose-pubescent with brown-based hairs; upper lobes narrowly triangular, 2–3 times as long as the tube. Standard elliptic-obovate, yellow, often lined to flushed reddish, glabrous to medially pubescent outside; wings a little shorter than the keel; keel 1.1–1.3 cm long, angular, with a long narrow twisted beak curved inwards at the tip. Young pod appressed pubescent, (2)6–8-ovulate. Seeds not seen.

Zambia. N: Kaputa Distr., Mweru Wantipa, Nsama, fl. & immat. fr. 3.iv.1957, *Richards* 8991 (K); 40 km north of Mporokoso, fl. 15.iv.1989, *Goyder, Pope & Radcliffe-Smith* 3044 (K; NDO).

Known only from around Lakes Mweru and Mweru Wantipa in northern Zambia and adjacent Dem. Rep. Congo. Miombo woodland, sometimes in damp sandy places; 900–1200 m.

179. **Crotalaria peregrina** Polhill, Crotalaria Africa & Madagascar: 350, fig. 99/10–18 (1982). — Lock, Leg. Afr. Check-list: 196 (1989). TAB. 3,7: **46**, fig. B. Type: Zambia, Mbala, Chilongowelo Drive, *Richards* 5454 (BR; K, holotype; LISC; SRGH).

Erect annual 40–70 cm tall, developing a number of ascending branches from near the base, appressed pubescent. Leaves mostly 3-foliolate, upper ones 1-foliolate; leaflets 15–45 × 2–12 mm, linear to narrowly elliptic, pointed and apiculate, thinly appressed puberulous beneath; petiole 0.5–2.5 cm long; stipules 0. Racemes 12–30 cm long, laxly many-flowered, sometimes with short flowering branches and flowers clustered in the axils below; bracts leaf-like, 3–7 mm long, lanceolate-caudate; pedicel 7–11 mm long, bearing small bracteoles. Calyx 4–5 mm long, glabrous except for the lobe margins; upper lobes narrowly triangular, 2.5–3 times as long as the tube. Standard elliptic, yellow with brown lines and a few hairs outside near the apex; wings a little shorter than the keel; keel 10–11 mm long, angular, with a long narrow twisted beak hooked inwards at the tip. Pod sessile, 5 × 4 mm, ovoid-globose, glabrous, 2-seeded. Seeds 2.5–3 mm long, oblique-cordiform, smooth.

Zambia. N: Mbala Distr., Inona (Inono) R., fl. & immat. fr. 8.iv.1959, *Richards* 11092 (K); Mbala–Mpulungu road, near Mukoma turning, fl. & fr. 29.v.1961, *Richards* 15161 (BR; K; SRGH).

Known only from the Mbala District. Miombo woodland in disturbed, sandy, sometimes damp places; 1050–1450 m.

180. **Crotalaria eurycalyx** Polhill, Crotalaria Africa & Madagascar: 351, fig. 98/11–14 (1982). — Lock, Leg. Afr. Check-list: 177 (1989). Type: Zambia, Kawambwa Distr., road to Nchelenge, *Richards* 9377 (BR; K, holotype).

Similar to *C. arcuata* but bracts leaf-like, lanceolate to linear-elliptic, narrowed at both ends, 5–8 mm long, pedicels 2–4 mm long, calyx 6–8 mm long with broad, ovate lobes 3–5 times as long as the tube and generally thinly pubescent. The pod is not known but the ovary is hairy only on the upper side and 8-ovulate.

Zambia. N: Kawambwa Distr., road to Nchelenge, fl. 20.iv.1957, *Richards* 9377 (BR; K).

Known only from the type. Miombo woodland; 1200 m.

181. **Crotalaria elisabethae** Baker f. in J. Linn. Soc., Bot. **42**: 301, t. 10 (1914). —Milne-Redhead in Bull. Misc. Inform., Kew **1940**: 52 (1940). —Wilczek in F.C.B. **4**: 213 (1953). —Torre in C.F.A. **3**: 48 (1962). —Polhill, Crotalaria Africa & Madagascar: 351, fig. 82/1 (1982). — Lock, Leg. Afr. Check-list: 177 (1989). TAB. 3,7: **45**, fig. A. Lectotype, chosen by Milne-Redhead, loc. cit., from Dem. Rep. Congo (Katanga).

Crotalaria acuminatissima Baker f. in J. Linn. Soc., Bot. **42**: 293 (1914). —Torre in C.F.A. **3**: 45 (1962). Type from Angola (Bié).

Crotalaria mumbwae Baker f. in J. Linn. Soc., Bot. **42**: 300 (1914). —Verdoorn in Bothalia **2**: 402 (1928). Type: Zambia, near Mumbwa, *Macaulay* 399 (K, holotype).

Crotalaria kamatinii R. Wilczek in Bull. Jard. Bot. État **23**: 182 (1953); in F.C.B. **4**: 210 (1953). Type from Dem. Rep. Congo (Katanga).

Crotalaria subumbellata Torre in Mem. Junta Invest. Ultramar, sér. 2, **19**: 32, t. 10 (1960); in C.F.A. **3**: 41 (1962). Types from Angola (Moxico).

Erect branching annual (0.1)0.3–1 m tall; stem densely appressed pubescent. Leaves mostly 3-foliolate, upper ones often 1-foliolate; leaflets 4–25 × 1–5(7) mm, mostly linear-oblanceolate, oblanceolate or oblong-oblanceolate, lower ones sometimes oblong-obovate, rounded to retuse at the apex, appressed pubescent beneath, occasionally also above; petiole 2–12 mm long; stipules 0. Racemes variably developed, 1–10 cm long, subumbellate to sublaxly many-flowered, with many other flowers on shorter racemes or heads developing on ± abbreviated lateral branches and in the axils; bracts (1)1.5–2.5 mm long, linear-subulate to linear-lanceolate; pedicel 2–5 mm long, with setaceous bracteoles on the pedicel. Calyx 3.5–5 mm long, appressed pubescent; lobes narrowly triangular, 2–3 times as long as the tube. Standard elliptic to oblong-obovate, yellow marked purplish-brown, medially and apically pubescent outside; wings markedly shorter than the keel; keel 7–9.5 mm long, angular, with a long narrow beak incurved at the tip. Pod sessile, 4.5–5 × 3.5–4 mm, ovoid-globose, pubescent, 2-seeded. Seeds 1.5–2 mm long, oblique-cordiform, smooth, with a prominent aril.

Zambia. N: 25 km west of Kasama, fl. & fr. 7.v.1961, *E.A. Robinson* 4639 (K; SRGH). W: Kitwe, fl. 27.iv.1966, *Mutimushi* 1397 (K; NDO; SRGH). C: Serenje Distr., Lake Lusiwasi, fl. & fr. 5.iv.1961, *Richards* 14943 (K; SRGH).

Also in Dem. Rep. Congo (Katanga) and Angola. Miombo woodland; 1200–1500 m.

Plants with short racemes and small leaves, the upper ones less frequently 1-foliolate, were formerly referred to *C. acuminatissima*, but the differences mainly seem to be due to growth conditions, though some local variation patterns may be superimposed.

182. **Crotalaria phyllostachys** Baker in Bull. Misc. Inform., Kew **1897**: 250 (1897). —E.G. Baker in J. Linn. Soc., Bot. **42**: 301 (1914). —Verdoorn in Bothalia **2**: 402 (1928). —Polhill, Crotalaria Africa & Madagascar: 352 (1982). —Lock, Leg. Afr. Check-list: 197 (1989). Syntypes: Malawi, Nyika Plateau, *Whyte* s.n.; Mpata to commencement of the Tanganyika Plateau, *Whyte* s.n. (K, syntypes).

Annual, erect to 1 m, ultimately shortly branched above; stem densely silky pubescent. Leaves 3-foliolate, sometimes with smaller 1-foliolate leaves subtending axillary flower clusters; leaflets 7–20 × 1–3 mm, progressively smaller above, linear-oblanceolate, pointed, silky pubescent to pilose beneath; petiole flattened, 3–10 mm long; stipules 0. Racemes 4–10 cm long, lax, but most of the flowers in axillary clusters well down the stem; bracts of terminal racemes 2–2.5 mm long, a little shorter than the pedicel, linear-lanceolate; bracteoles on the pedicel, setaceous. Calyx 4–5 mm long, densely subappressed to spreading pubescent; upper lobes attenuate-triangular, 2–3 times as long as the tube. Standard elliptic, yellow lined red, medially pubescent outside; wings shorter than the keel; keel 9–11 mm long, angular, with a narrow twisted beak, slightly hooked at the tip. Pod sessile, 4–5 mm long, ovoid-globose, pubescent, 1–2-seeded. Seeds 2 mm long, rounded-cordiform, smooth, with a small aril.

Malawi. N: Karonga Distr., descending to Karonga from crossroads, fl. & fr. 8.vii.1973, *Pawek* 7165A (K; MAL; MO); Kyelekera, 32 km WSW of Karonga, fl. 5.vi.1989, *Brummitt* 18364 (K; MAL).

Known only from northern Malawi. Miombo woodland; 750–1800 m.

This species is known only from the syntypes and the specimens cited above. It flowers in the dry season and may have been overlooked as a consequence, but it does appear to be very local in its distribution.

183. **Crotalaria gamwelliae** Baker f. in J. Bot. **76**: 20 (1938). —Polhill in F.T.E.A., Leguminosae, Pap.: 985 (1971); Crotalaria Africa & Madagascar: 353 (1982). —Lock, Leg. Afr. Check-list: 179 (1989). Type: Zambia, Mbala Distr., *Gamwell* 216 (BM, holotype; SRGH).

Erect annual, 30–70 cm tall, developing ascending branches from the base and ultimately many short flowering branches near the top; stem slender, appressed pubescent. Leaves mostly 3-foliolate, the upper ones 1-foliolate; leaflets 10–30 × 2–10 mm, mostly linear-oblanceolate to oblong-oblanceolate, the lower ones sometimes more obovate, glabrous or nearly so above, finely appressed pubescent beneath; petiole 4–20 mm long; stipules 0. Racemes 6–15(30) cm long, lax or upper part dense, main ones 16–40-flowered, other short racemes and clusters of flowers in the axils below; bracts 2–3 mm long, linear, shorter than the pedicel; bracteoles on upper part of the pedicel, setaceous. Calyx 4–5 mm long, appressed pubescent; upper lobes attenuate-triangular, c. 2–3 times as long as the tube. Standard elliptic-obovate, yellow, often veined reddish-brown, pubescent medially and apically outside; wings shorter than the keel; keel 8–11 mm long, angular, with a narrow twisted beak, slightly incurved at the tip. Pods subsessile, 4–6 × 4 mm, subglobose-ovoid, slightly pointed, appressed pubescent, 2-seeded. Seeds 2–3 mm long, rounded-cordiform, smooth, with a prominent aril.

Zambia. N: Mbala, Chilongowelo Drive, fl. 3.v.1955, *Richards* 5487 (K; SRGH). C: Great North Road, 5 km from Serenje turning, fl. 5.iv.1961, *Richards* 14939 (K).
Also in southern Tanzania. Miombo woodland; 1000–1500 m.

184. **Crotalaria johnstonii** Baker in Bull. Misc. Inform., Kew **1897**: 250 (1897). —E.G. Baker in J. Linn. Soc., Bot. **42**: 285 (1914). —Verdoorn in Bothalia **2**: 403 (1928). —Polhill, Crotalaria Africa & Madagascar: 353 (1982). —Lock, Leg. Afr. Check-list: 183 (1989). TAB. 3,7: **45**, fig. C. Type: Malawi, Chitipa (Fort Hill), *Whyte* s.n. (K, holotype).
Crotalaria leucotricha Baker in Bull. Misc. Inform., Kew **1897**: 251 (1897). —E.G. Baker in J. Linn. Soc., Bot. **42**: 285 (1914). Type: Malawi, Chitipa (Fort Hill), *Whyte* s.n. (K, holotype).

Erect branching annual 15–40 cm tall; branches slender, ascending, often mostly from near the base of the plant, densely pubescent with appressed to slightly spreading hairs at first, somewhat glabrescent. Leaves mostly 3-foliolate, uppermost occasionally 1-foliolate; leaflets 8–20 × 2–7 mm, oblanceolate to elliptic-oblanceolate, pubescent on both surfaces; petiole 4–10 mm long; stipules 0. Racemes 4–13 cm long, subdense at top, lax below, 10–30-flowered, developing other flowers in the axils below; bracts 2–3 mm long, linear, subequalling mature pedicels; bracteoles 0. Calyx 4–5 mm long, densely pubescent; upper lobes narrowly triangular, 2–3 times as long as the tube. Standard obovate, yellow marked brown, ± densely or at least extensively pubescent outside; wings nearly as long as the keel; keel (5)5.5–6.5 mm long, angular, with a narrow twisted beak incurved at the tip. Pod subsessile, globose, 3.5–4 mm across, 1–2-seeded. Seeds 2.5 mm long, oblique-cordiform, smooth, arillate.

Malawi. N: NW of Karonga, Ngerenge, fl. & fr. 7.viii.1952, *J. Williamson* 62 (BM); 32 km west of Karonga, fl. 26.iv.1977, *Pawek* 12723 (K; MAL; MO).
Known only from a small area of northern Malawi around Karonga. Miombo woodland; 1050–1200 m.

185. **Crotalaria bredoi** R. Wilczek in Bull. Jard. Bot. État **23**: 183 (1953); in F.C.B. **4**: 212 (1953). —Polhill, Crotalaria Africa & Madagascar: 354 (1982). —Lock, Leg. Afr. Check-list: 169 (1989). Type from Dem. Rep. Congo (Katanga).

Erect branching annual 4–9 cm tall, spreading pilose. Leaves mostly 3-foliolate, uppermost sometimes 1-foliolate; leaflets 6–18 × 1–4 mm, oblanceolate, pilose on both surfaces; petiole 3–10 mm long; stipules 0. Racemes 5–30 cm long, many-flowered, subdense to lax below, rarely with a few other flowers in the axils below; bracts 2–4 mm long, linear, a little shorter than the mature pedicel; bracteoles on the pedicel, setaceous. Calyx 3–4 mm long, finely pubescent with irregularly arranged hairs; lobes acuminately triangular, 1–1.5 times as long as the tube. Standard elliptic to oblong-obovate, yellow, ± finely lined brown, pubescent outside; wings shorter than the keel; keel 7–8 mm long, angular, with a narrow twisted beak slightly incurved at the tip. Pod sessile, 4–4.5 × 3–3.5 mm, ovoid-globose, pubescent, 2-seeded. Seeds 2–2.5 mm long, oblique-cordiform, smooth, with a pronounced aril.

Zambia. N: Kaputa Distr., Mweru Wantipa, Muzombwe, fl. & fr. 16.xii.1960, *Richards* 13729 (K; LISC; SRGH); Kaputa Distr., Mkupa Katandula, fl. 10.vi.1950, *Bullock* 2933 (K).

Also in adjacent parts of the Katanga Province of Dem. Rep. Congo. Wet grassland around swamps and dambos; 900–1050 m.

186. **Crotalaria laxiflora** Baker in Bull. Misc. Inform., Kew **1895**: 64 (1895). —Taubert in Engler, Pflanzenw. Ost-Afrikas **C**: 423 (1895). —E.G. Baker in J. Linn. Soc., Bot. **42**: 296 (1914). —Polhill in F.T.E.A., Leguminosae, Pap.: 986 (1971); Crotalaria Africa & Madagascar: 355 (1982). —Lock, Leg. Afr. Check-list: 187 (1989). Type: Zambia, Lake Tanganyika, Fwambo, *Carson* 28 (K, holotype).

Slender erect branching herb 25–70 cm tall, glabrous. Leaves 3-foliolate; leaflets 7–20 × 1–2.5 mm, obcuneiform to linear-oblanceolate; petiole 2–10 mm long; stipules 0. Racemes 2–10 cm long, laxly 4–16-flowered; bracts 0.5–1(1.5) mm long, shortly linear-lanceolate; bracteoles 0. Calyx 3.5–5 mm long; upper lobes attenuate-triangular, 2–3 times as long as the tube. Standard elliptic, pale yellow lined brown; wings much shorter than the keel; keel 7–9 mm long, angular, with a long narrow twisted beak. Pod subsessile, 5–6 × 4–4.5 mm, ovoid-globose, 1–2-seeded. Seeds 2.5–3 mm across, rounded-cordiform, smooth, with a small aril.

Var. **laxiflora** —Polhill, Crotalaria Africa & Madagascar: 355 (1982).

Leaflets rounded to truncate at the apex. Keel 7–8 mm long, with a straight beak only slightly incurved at the very tip.

Zambia. N: Mbala Distr., Mukoma Escarpment, fl. 7.iv.1962, *Richards* 16292 (K; SRGH); Mbala Distr., road to Itimbwe (Itembwe) Gap, fl. & fr. 24.iv.1959, *Richards* 11354 (BR; K; LISC).

Escarpments around the southern end of Lake Tanganyika, just extending into Tanzania. Miombo woodland, generally in disturbed places; 1200–1750 m.

Var. **acuta** Polhill, Crotalaria Africa & Madagascar: 355 (1982). Type: Zambia, Mbala Distr., Sunzu Hill, *Richards* 15074A (K, holotype; SRGH).

Leaflets acute. Keel 8–9 mm long, distinctly incurved at the tip.

Zambia. N: Mbala Distr., Kilema Farm, by Manyesi River, fl. & fr. 11.v.1962, *Richards* 16453 (K; SRGH); 40 km from Mbala (Abercorn) to Tunduma, fl. & fr. 12.iv.1962, *Richards* 16316 (K; SRGH).

Known only from between Mbala and the Tanzanian border, largely eastwards of var. *laxiflora* but both varieties known from around the township itself. Miombo woodland, generally on rocky hills and in disturbed places; 1500–1950 m.

187. **Crotalaria nuda** Polhill, Crotalaria Africa & Madagascar: 355, fig. 101/1–9 (1982). —Lock, Leg. Afr. Check-list: 193 (1989). TAB. 3,7: **47**, fig. A. Type: Zambia, 55 km SW of Lusaka, Sala Reserve, *E.A. Robinson* 4990 (K, holotype; SRGH).

Crotalaria sphaerocarpa sensu Hutchinson, Botanist South. Africa: 501 (1946) non Perr. ex DC.

Generally a slender erect branching annual 30–90 cm tall, but sometimes a tufted suffrutex, glabrous. Leaves mostly 3-foliolate, uppermost generally 1-foliolate; leaflets 5–18 × 1–3.5 mm, linear-oblanceolate, obtuse to truncate; petiole 0.5–2(4) mm long; stipules 0. Racemes 3–12 cm long, 12–30-flowered, generally lax, upper

---

Tab. 3,7: **47**. A. —CROTALARIA NUDA. A1, flowering branch (× 1); A2, flower (× 4); A3, standard (× 4); A4, keel (× 4); A5, anthers (× 8); A6, gynoecium (× 4); A7, section through ovary (× 6), A1–A7 from *E.A. Robinson* 4990; A8, pod (× 4), from *van Rensburg* 2894; A9, seed (× 6), from *Hutchinson & Gillett* 3708. B. —CROTALARIA LIMOSA. B1, flowering branch (× 1); B2, flower (× 4); B3, standard (× 4); B4, keel (× 4); B5, anthers (× 8); B6, gynoecium (× 4); B7, section through ovary (× 6), B1–B7 from *Milne-Redhead* 4318; B8, pod (× 4); B9, seed (× 6), B8 & B9 from *E.A. Robinson* 3606. C. —CROTALARIA DEBILIS. C1, habit (× 1); C2, flower (× 6); C3, standard (× 6); C4, keel (× 6); C5, anthers (× 12); C6, gynoecium (× 6); C7, section through ovary (× 6); C8, pod (× 6), C1–C8 from *E.A. Robinson* 3707. Drawn by Roger Polhill. From Crotalaria Africa & Madagascar.

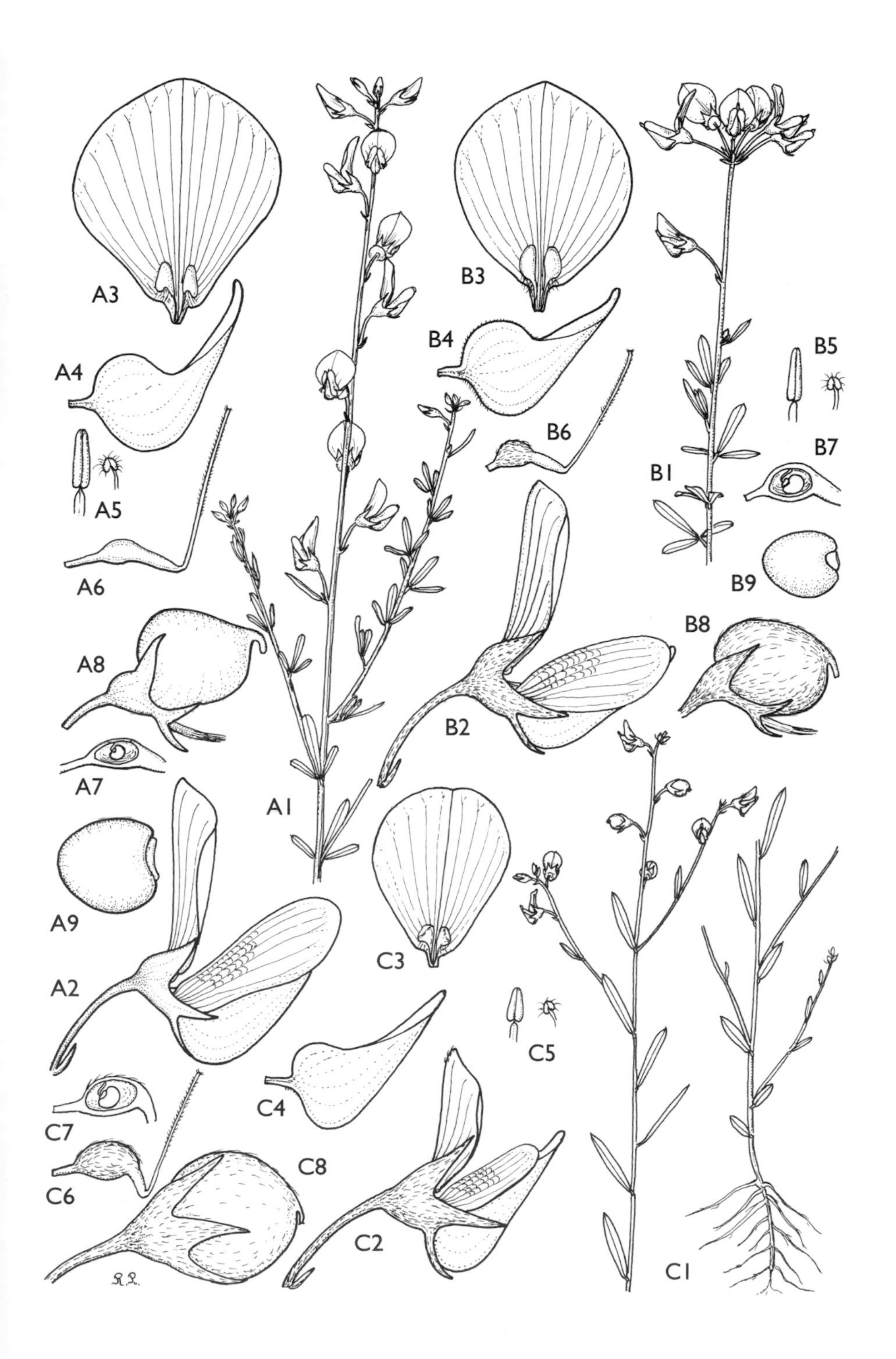
A3
B3
B4
A4
B5
B6
B7
A5
B1
A6
B9
B8
A8
B2
A7
A1
A9
C3
A2
C5
C4
C7
C8
C6
C2
C1

part occasionally congested; bracts 1–2.5 mm long, linear-lanceolate to narrowly elliptic; pedicel 3–6 mm long, sometimes with minute bracteoles medially. Calyx 3.5–4 mm long; upper lobes narrowly triangular-acuminate, 1.5–2 times as long as the tube. Standard obovate-circular, yellow lined reddish; wings ± as long as the keel; keel (4)5–7 mm long, angular, with a narrow twisted beak slightly incurved at the tip. Pod with a stipe up to 1 mm long, (3)5–6 × 3–3.5 mm, ovoid-globose, 2-seeded. Seeds 2.5 mm across, rounded cordiform, smooth, with a small aril.

Zambia. N: 32 km SSW of Mpika, by road at summit of Luangwa Escarpment, fl. 30.iii.1984, *Brummitt, Chisumpa & Nshingo* 17017 (K; NDO). W: Ndola Distr., Lake Ishiku, fl. & fr. 16.iv.1957, *Fanshawe* 3188 (K; NDO). C: 13 km from Lusaka on old road to Mumbwa, fl. & fr. 3.iv.1964, *van Rensburg* 2894 (K; SRGH).

Not known elsewhere. Generally in seasonally damp places that have dried out, occasionally in woodland, quite commonly on dolomite limestones; 1150–1550 m.

188. **Crotalaria limosa** Polhill, Crotalaria Africa & Madagascar: 356, fig. 101/10–18 (1982). —Lock, Leg. Afr. Check-list: 188 (1989). TAB. 3,7: **47**, fig. B. Type: Zambia, Mwinilunga Distr., Matonchi Farm, *Milne-Redhead* 4318 (BR; K, holotype).

Erect branching annual 30–40 cm tall; branches ascending, slender, appressed puberulous, glabrescent. Leaves mostly 3-foliolate, the uppermost sometimes 1-foliolate; leaflets 5–12 × 1–3 mm, narrowly oblong-oblanceolate, mostly obtuse, thinly puberulous beneath; petiole 0.5–1.5 mm long; stipules 0. Racemes 2–4 cm long, subumbellate with a few laxly inserted flowers below, 6–12-flowered; bracts 1–2 mm long, linear; pedicel 3–5 mm long, sometimes with setaceous bracteoles. Calyx 3–3.5 mm long, puberulous; lobes attenuate-triangular, twice as long as the tube. Standard obovate-circular, bright yellow, glabrous or nearly so; wings nearly as long as the keel; keel 6 mm long, angular, with a narrow twisted beak only slightly incurved at the tip. Pod sessile, 5 × 3 mm, ovoid-ellipsoid or shortly oblong-ellipsoid, puberulous, 1–2-seeded. Seeds 2 mm across, rounded-cordiform, smooth, arillate.

Zambia. W: Mwinilunga Distr., Matonchi, Kalenda Plain, fl. & fr. 16.iv.1960, *E.A. Robinson* 3606 (K; SRGH).

Known only from Matonchi, west of Mwinilunga, in the more swampy parts of the laterite dambo, in shallow water or on the dried out surface; c. 1400 m.

189. **Crotalaria tenuipedicellata** Baker f. in J. Linn. Soc., Bot. **42**: 266 (1914). —Wilczek in F.C.B. **4**: 78 (1953). —Polhill, Crotalaria Africa & Madagascar: 357 (1982). —Lock, Leg. Afr. Check-list: 207 (1989). Lectotype, chosen by Wilczek, loc. cit., from Dem. Rep. Congo (Lake Mweru).

Erect annual, 0.5–1.2 m tall, branching above; branches thinly appressed puberulous. Leaves 1-foliolate; leaflets 20–40 × 1–5 mm, linear-oblanceolate to narrowly oblanceolate-oblong, lowermost oblanceolate, thinly puberulous to minutely strigulose beneath; petiole 2–6 mm long; stipules 0. Racemes 2–10 cm long, 6–16-flowered, lax but uppermost flowers subumbelliform, also a few flowers developing in the axils below; bracts 1–1.5 mm long, linear to shortly linear-lanceolate; pedicels 5–7 mm long, slender, without bracteoles. Calyx 3–3.5 mm long, appressed puberulous; upper lobes broadly triangular, slightly acuminate, 1.5 times as long as the tube. Standard oblong-obovate, pale yellow lined reddish, thinly puberulous on upper part outside; wings markedly shorter than the keel; keel 8–9 mm long, angular, with a long narrow twisted beak strongly incurved at the tip. Pod sessile, 4 × 3.5 mm, ovoid-globose, puberulous, 2-seeded. Mature seeds not seen.

Zambia. N: Kaputa Distr., Mporokoso–Nsama road, fl. & fr. 16.iv.1957, *Richards* 9285 (K). W: Chingola, fl. 19.iii.1968, *Mutimushi* 2558 (K; NDO).

Also in adjoining parts of the Katanga Province of Dem. Rep. Congo. Miombo woodland; 1200–1350 m.

190. **Crotalaria campestris** Polhill, Crotalaria Africa & Madagascar: 357, fig. 95/11–19 (1982). —Lock, Leg. Afr. Check-list: 170 (1989). Type: Zambia, 45 km east of Mwinilunga, *E.A. Robinson* 3561 (K, holotype; MO; SRGH).

Slender erect annual 30–60 cm tall, branching above, glabrous. Leaves subsessile, 1-foliolate; leaflets 5–13 × 0.7–1.5 mm, linear-oblanceolate, acute; stipules 0. Racemes 2–7 cm long, very laxly 3–8-flowered, slightly flexuous; bracts 1–2 mm long, linear-lanceolate; pedicel 4–8 mm long, with small bracteoles near the base. Calyx 3–4 mm long; upper lobes attenuate-triangular, 2–2.5 times as long as the tube. Standard elliptic, pale yellow, later flushed pink; wings much shorter than the keel; keel 11–12 mm long, angular, with a long straight narrow twisted beak. Pod with a stipe 3–4 mm long, 8–11 (including stipe) × 5–6 mm, rounded ellipsoid, many-seeded (ovules 40). Seeds 1 mm in diameter.

Zambia. W: Mwinilunga, fl. & fr. 16.v.1969, *Mutimushi* 3398 (K; NDO); Mwinilunga Distr., Matonchi, Kalenda Plain, fl. & fr. 16.iv.1960, *E.A. Robinson* 3629A (K).
Known only from the Mwinilunga District and adjacent parts of the Katanga Province of Dem. Rep. Congo. Grassy plains on Kalahari Sands and laterite pavements; c. 1400 m.

191. **Crotalaria basipeta** R. Wilczek in Bull. Jard. Bot. État **23**: 141 (1953); in F.C.B. **4**: 80 (1953). —Polhill, Crotalaria Africa & Madagascar: 358 (1982). —Lock, Leg. Afr. Check-list: 167 (1989). Type: Zambia, Mweru Wantipa, Kabwe Marsh, *Bredo* 3154 (BR, holotype; K).
Crotalaria mabobo R. Wilczek in Bull. Jard. Bot. État **23**: 132 (1953); in F.C.B. **4**: 94 (1953). Type from Dem. Rep. Congo (Lubumbashi).

Erect annual 50–120 cm tall, branching above; stem appressed puberulous, glabrescent, lower part sometimes spongy and with adventitious roots. Leaves 1-foliolate or sometimes mostly simple; leaflets 10–35 × 1–3 mm, linear-oblanceolate to narrowly oblong, acute and apiculate, glabrous to thinly appressed puberulous beneath; petiole 0–4 mm long; stipules 0. Racemes 6–20 cm long, laxly many-flowered, with other flowers developing in the axils below; bracts 1–2 mm long, linear-subulate; pedicel 3–6 mm long, sometimes with minute bracteoles. Calyx (2.5)3–5 mm long, appressed puberulous to subglabrous; upper lobes usually narrowly triangular-acuminate, (1.5)2–2.5 times as long as the tube. Standard elliptic-obovate, orange-yellow, lined or flushed reddish, glabrous to thinly puberulous outside; wings distinctly shorter than the keel; keel (6)7–9 mm long, angular, with a long narrow twisted beak slightly incurved at the tip. Pod sessile, 5–6 × 3.5–4 mm, ovoid-globose, puberulous, 2-seeded. Seeds 2.5 mm long, oblique-cordiform, smooth, with a pronounced aril.

Zambia. N: Kaputa Distr., Mweru Wantipa, Selemani, fl. 17.iv.1961, *Phipps & Vesey-FitzGerald* 3271 (K; LISC; SRGH). S: Shishamba R., fl. & fr. 10.iv.1963, *Mitchell* 21/1 (BR; LISC; SRGH).
Also in adjacent parts of the Katanga Province of Dem. Rep. Congo. Wet grassland and open places in woodland and thicket; 900–1150 m.

192. **Crotalaria ringoetii** Baker f. in De Wildeman, Pl. Bequaert. **2**: 509 (1924); Legum. Trop. Africa: 26 (1926). —Wilczek in F.C.B. **4**: 78 (1953). —Binns, First Check List Herb. Fl. Malawi: 79 (1968). —Polhill in F.T.E.A., Leguminosae, Pap.: 987 (1971); Crotalaria Africa & Madagascar: 359 (1982). —Lock, Leg. Afr. Check-list: 201 (1989). Type from Dem. Rep. Congo (Katanga).

Erect laxly branching annual 0.2–1 m tall; stem slender, appressed puberulous. Leaves 1-foliolate; leaflets 10–45 × 1–8 mm, linear-oblanceolate above to oblong-elliptic below, upper ones acute, appressed puberulous beneath; petiole 1–3(4) mm long; stipules 0. Racemes 5–25 cm long, laxly many-flowered, with a few other flowers developing in the axils below; bracts 0.5–1.5 mm long, linear to shortly linear-lanceolate; pedicel 2–3(5) mm long, without bracteoles. Calyx 2.5–3.5 mm long, appressed puberulous with generally brown-based hairs; upper lobes attenuately triangular-lanceolate, 2–2.5 times as long as the tube. Standard elliptic-obovate, pale yellow, with reddish lines and medially puberulous outside, and with red fringed basal patches inside; wings shorter than the keel; keel 4–5 mm long, angular, with a narrow twisted slightly incurved beak. Pod sessile, 3–3.5 × 3 mm, ovoid-globose, puberulous, 2-seeded. Seeds 2–2.7 mm long, oblique-cordiform, smooth, with a small aril.

Zambia. N: Mbala Distr., Mwambeshi Estate, fl. & fr. 20.iv.1959, *Richards* 11296 (K; LISC; SRGH). W: Luanshya, fr. 24.iv.1955, *Fanshawe* 2252 (K; NDO; SRGH). C: 10 km south of Kapiri

Mposhi, fl. & fr. 27.iii.1955, *Exell, Mendonça & Wild* 1217 (BM; LISC; SRGH). E: Chipata Distr., Ngoni area, fl. ii.1962, *Verboom* 464 (K; LISC; SRGH). **Malawi**. N: Rumphi Distr., 1.5 km along road to the Nyika Plateau, fl. & fr. 26.iv.1973, *Pawek* 6573A (K); Kasungu Distr., Chimaliro Forest, fl. & fr. 23.iv.1974, *Pawek* 8491 (K; MO; SRGH). S: Machinga Distr., Nyambi, near Chipunga R., fl. & fr. 26.iv.1955, *Jackson* 1639 (K; SRGH). **Mozambique**. N: Ngauma Distr., 8 km south of Massangulo, fl. & fr. 26.v.1961, *Leach & Rutherford-Smith* 11017 (K; SRGH).

Also in adjacent parts of Tanzania and Dem. Rep. Congo. Miombo woodland, persisting in disturbed places; 1050–1650 m.

193. **Crotalaria passerinoides** Taub. in Engler, Pflanzenw. Ost-Afrikas **C**: 204 (1895). —E.G. Baker in J. Linn. Soc., Bot. **42**: 257 (1914). —Polhill in F.T.E.A., Leguminosae, Pap.: 987 (1971); Crotalaria Africa & Madagascar: 359 (1982). —Lock, Leg. Afr. Check-list: 195 (1989). Type from Tanzania.

Similar to *C. ringoetii* but stems more leafy, the branches, if any, more crowded, the racemes subdense, the linear bracts as long as the pedicels, the standard becoming reddish lined and sometimes splotched but without red fringed dark yellow patches at the base inside, and the seeds with a prominent aril exceeding the hilar sinus.

Zambia. N: Mbala Distr., Saisi R., fl. 15.iv.1959, *Richards* 11225 (BR; K; SRGH); 105 km east of Kasama, Chambeshi Flats, fl. & fr. 22.iv.1961, *E.A. Robinson* 4615 (K; SRGH).

Also in Dem. Rep. Congo (Katanga) and Tanzania. Wet grassland; 1100–1750 m.

194. **Crotalaria debilis** Polhill, Crotalaria Africa & Madagascar: 359, fig. 101/19–26 (1982). —Lock, Leg. Afr. Check-list: 173 (1989). TAB. 3,7: **47**, fig. C. Type: Zambia, banks of Kafue R., 11 km north of Chingola, *E.A. Robinson* 3707 (K, holotype).

Diminutive erect branching annual 15–25 cm tall; stem wiry, subglabrous. Leaves simple, subsessile, without stipules; blades 4–12 × 1–2.5 mm, linear-lanceolate and acute above to oblanceolate-oblong and obtuse below, shortly rounded to the base, sparsely puberulous beneath. Racemes 2–4 cm long, laxly 3–6-flowered, with a few other flowers developing in the axils below; bracts 1–2 mm long, linear; pedicels 2–4 mm long, without bracteoles. Calyx 2–3 mm long, thinly appressed puberulous; upper lobes attenuate-triangular, 1.5 times as long as the tube. Standard obovate, yellow, lined and later flushed red, subglabrous; wings shorter than the keel; keel 4 mm long, angular, with a narrow twisted only slightly incurved beak. Pod sessile, 3 mm long, ovoid-globose, minutely puberulous, 1–2-seeded. Seeds 1.5 mm long, oblique-cordiform, smooth, with a prominent aril.

Zambia. W: Mwinilunga Distr., Matonchi, Kalenda Plain, fl. 16.iv.1960, *E.A. Robinson* 3629B (K).

Known only from western Zambia. Short wet grassland by rivers and drying laterite pavements; 1350–1400 m.

195. **Crotalaria carsonii** Baker f. in J. Linn. Soc., Bot. **42**: 290 (1914). —Polhill, Crotalaria Africa & Madagascar: 360 (1982). —Lock, Leg. Afr. Check-list: 170 (1989). Type: Zambia, Mbala Distr., Namkolo Mission (Niomkolo), *Carson* s.n. (K, holotype).

Erect annual 20–85 cm tall; branches numerous, slender, ascending, often most from near the base, with shorter flowering branches above, subdensely shortly pubescent. Leaves mostly 3-foliolate, upper ones 1-foliolate; leaflets 5–12 × 0.5–2 mm, linear-oblanceolate, glabrous to sparsely puberulous above, appressed puberulous beneath; petiole 0.5–4 mm long; stipules 0. Racemes 2.5–12 cm long, laxly to subdensely 8–30-flowered, developing other flowers in the axils below; bract 1–3 mm long, linear-subulate; pedicel 2–4 mm long, sometimes with small setaceous bracteoles. Calyx 2.5–3 mm long, subappressed pubescent; upper lobes narrowly triangular, attenuate or slightly acuminate, 1.5–2.5 times as long as the tube. Standard elliptic to elliptic-obovate, yellow, sometimes lined brown, pubescent outside; wings slightly shorter than the keel; keel 4.5–5 mm long, angular, with a narrow twisted beak slightly incurved at the tip. Pod sessile, 3–4 × 2–3 mm, ovoid-globose, pubescent, 2-seeded. Seeds 1.5–2 mm long, oblique-cordiform, smooth, arillate.

Zambia. N: Mbala Distr., Kapata–Sondwe (Sondwa), fl. & fr. 18.iv.1950, *Bullock* 2881 (K); Kawambwa, fl. & fr. 17.iv.1957, *Richards* 9330 (K). W: Mwinilunga Distr., 7 km north of Kalene

Hill, fl. & fr. 16.iv.1965, *E.A. Robinson* 6576 (K; SRGH).
Known only from Zambia. Damp places at edges of rivers and swamps, also shallow soil on rock outcrops and laterite pans; 780–1350 m.
At present represented only by three discrete populations, which look rather different but seem to be conspecific.

Typical form. Leaflets mostly obtuse to truncate. Racemes lax; bracts shorter than the pedicel; bracteoles present. Upper calyx lobes narrowly attenuate-triangular, 2–2.5 times as long as the tube. Standard elliptic, pointed. Pod 3 × 2–2.5 mm. Seeds with a prominent aril much exceeding the hilar sinus.
At the southern end of Lake Tanganyika, e.g. *Bullock* 2881, cited above.

Kawambwa form. Leaflets obtuse to truncate. Racemes subdense; bracts longer than the pedicel; bracteoles lacking. Calyx as typical form. Standard elliptic-obovate, rounded. Pod 3.5 × 3 mm. Seeds with a prominent aril.
Kawambwa area south of Lake Mweru, e.g. *Richards* 9330, cited above.

Zambezi form. Leaflets acute. Racemes with uppermost flowers crowded, lax below; bracts shorter than the pedicel; bracteoles absent. Upper calyx lobes narrowly triangular, slightly acuminate, 1.5 times as long as the tube. Standard elliptic. Pod 3.5–4 × 3 mm. Seeds with only a small aril.
Zambezi rapids north of Kalene Hill and thereabouts, e g. *E.A. Robinson* 6576, cited above.

196. **Crotalaria confertiflora** Polhill, Crotalaria Africa & Madagascar: 360, fig. 102/12–22 (1982). —Lock, Leg. Afr. Check-list: 172 (1989). Type: Zambia, 50 km south of Kasama, Chibutubutu, *E.A. Robinson* 4598 (BR; K, holotype; SRGH).

Erect annual 20–70 cm tall, developing relatively short flowering branches above; stem shortly appressed pubescent. Leaves 3-foliolate, uppermost reduced, occasionally 1-foliolate; leaflets 3–12 × 0.5–2 mm, linear-oblanceolate or oblanceolate, appressed puberulous beneath; petiole 1–3 mm long; stipules 0. Flowers 6–12 in sessile terminal heads; bracts 1–1.5 mm long, linear-lanceolate; pedicel 1.5–2 mm long, without bracteoles. Calyx 2.5 mm long, densely pubescent with irregularly arranged hairs; upper lobes narrowly triangular-acuminate, 1.5–2 times as long as the tube. Standard oblong-obovate, yellow, tomentellous outside; wings almost as long as the keel; keel 4.5–5 mm long, angular, with a very narrow twisted beak, slightly incurved at the tip. Pod sessile, 3.5–4 × 3 mm, ovoid-globose, pubescent, 2-seeded. Seeds 2 mm long, oblong-cordiform, smooth, with a prominent aril.

Zambia. N: Kasama Distr., Chibutubutu, fr. 18.vi.1961, *E.A. Robinson* 4703 (K; SRGH); same place and date, *E.A. Robinson* 4704 (K).
Known only from the type locality. Seasonally damp flood-plain.

197. **Crotalaria microthamnus** Robyns ex R. Wilczek in Bull. Jard. Bot. État **23**: 181 (1953); in F.C.B. **4**: 210 (1953). —Polhill, Crotalaria Africa & Madagascar: 362 (1982). —Lock, Leg. Afr. Check-list: 191 (1989). Type from Dem. Rep. Congo (Lake Mweru).

Bushy annual 35–80 cm tall; branches slender, appressed puberulous. Leaves mostly 3-foliolate, uppermost on main and suppressed lateral branches sometimes 1-foliolate; leaflets 5–10 × 1–3 mm, obcuneiform to oblanceolate, appressed puberulous beneath; petiole 2–6 mm long; stipules 0. Racemes short, few-flowered, most of the flowers solitary or clustered in axils all along the branches; bracts 1 mm long, linear to shortly linear-lanceolate. Calyx 3–4 mm long, appressed puberulous; upper lobes narrowly attenuate-triangular, 2–3 times as long as the tube. Standard elliptic to oblong-obovate, yellow, usually lined brown, glabrous to thinly puberulous on the upper part outside; wings nearly as long as the keel; keel 6–7 mm long, angular, with a straight narrow twisted beak. Pod sessile, 3–4 mm across, subglobose, puberulous, 1–2-seeded. Mature seeds not seen.

Zambia. N: Kaputa Distr., Mweru Wantipa, road to Bulaya, fl. 4.iv.1957, *Richards* 9010 (K); Nchelenge Distr., Lake Mweru, Kafulwe, fl. 26.iv.1957, *Richards* 9426 (K).
Known only from around Lakes Mweru and Mweru Wantipa. Wet grassland; c. 950–1050 m.
E.A. Robinson 4672 (EA; K) from miombo woodland 53 km west of Mporokoso, Chiwala, fr. 21.v.1961, represents an imperfectly known species, apparently related to *C. kibaraensis* R. Wilczek. It occurs in drier habitats and is easily distinguished from *C. microthamnus* by the

smaller flowers in clusters in the axils accompanied by few reduced branches and the larger leaflets puberulous above.

198. **Crotalaria pudica** Polhill, Crotalaria Africa & Madagascar: 363 (1982). —Lock, Leg. Afr. Check-list: 199 (1989). Type from Tanzania.

Stems numerous, ascending from a small woody rootstock, 15–30 cm tall, slender, slightly ribbed, strigulose-puberulous with brown-based hairs. Leaves 3-foliolate; leaflets 8–15 × 2–5 mm, oblanceolate, rather densely appressed puberulous beneath; petiole 2–7 mm long, broadly grooved; stipules 0. Racemes 2–4 cm long, subdensely many-flowered; bracts 2–4 mm long, linear-lanceolate, 0.7–1.2 times as long as the pedicel; bracteoles on the pedicel, 1.5–2 mm long. Calyx 3–4 mm long, tomentellous; lobes triangular, 1.2–1.5 times as long as the tube. Standard obovate, yellow lined reddish-brown, densely puberulous over upper two-thirds outside; wings ± as long as the keel; keel 5–6 mm long, angular, with a straight twisted beak. Ovary hairy, 2-ovulate. Pod unknown.

Malawi. N: Rumphi Distr., Nyika National Park, near Thazima Gate, fl. 5.iv.1981, *Salubeni* 2975 (MO). C: Dedza Distr., Chongoni Forest School, fl. 12.iii.1963, *Salubeni* 1 (K; SRGH); same locality, 13.iii.1967, *Salubeni* 588 (K; SRGH).

Also in southern Tanzania. Miombo woodland; c. 1650 m.

199. **Crotalaria pseudotenuirama** Torre in Mem. Junta Invest. Ultramar, sér. 2, **19**: 36 (1960); in C.F.A. **3**: 47 (1962). —Polhill in Kew Bull. **22**: 331 (1968); in F.T.E.A., Leguminosae, Pap.: 989 (1971); Crotalaria Africa & Madagascar: 363 (1982). —Lock, Leg. Afr. Check-list: 199 (1989). Type from Angola.

Erect annual 20–90 cm tall, with generally numerous slender ascending branches; stem appressed pubescent. Leaves 3-foliolate; leaflets 8–20 × 2–5 mm, linear-oblanceolate to oblong-oblanceolate, mostly rounded to truncate at the apex, appressed pubescent beneath; petiole 2–9 mm long; stipules 0. Racemes 3–13 cm long, subdensely to laxly 10–30-flowered; bracts 1–2 mm long, linear to linear-lanceolate; pedicel 2–4 mm long, sometimes with minute bracteoles. Calyx 2.5–3.5 mm long, brownish tomentose; upper lobes triangular, 1–1.5 times as long as the tube. Standard obovate, pale yellow, veined brown, golden-brown tomentose outside; wings ± as long as the keel; keel 4–6 mm long, angular, with a narrow twisted beak slightly incurved at the tip. Pod sessile, 5–6 mm long, ovoid-globose, pubescent, 1–2-seeded. Seeds 2 mm long, ovate-cordiform, smooth, brown to purplish, with a prominent aril.

Zambia. B: Kaoma Distr., Luena Flats, Lizulu, fl. & fr. 14.vi.1964, *Verboom* 1061 (K; SRGH). N: Mbala Distr., Lunzua Agriculture Station (Lunzuwa), fl. & fr. 20.iv.1962, *E.A. Robinson* 5091 (B; K; LISC; SRGH). W: Kitwe, fl. & fr. 21.iii.1954, *Fanshawe* 989 (BR; K; NDO; SRGH). C: Mkushi Distr., 43 km east of Kabwe on Old Mkushi Road, Kantwite River, fl. 20.iv.1972, *Kornaś* 1631 (K). E: Chimutengo/Petauke, fl. & fr. 3.x.1966, *Mutimushi* 1563 (K; NDO). S: 19 km north of Choma, fl. & fr. 22.iii.1957, *E.A. Robinson* 2161 (K; SRGH). **Malawi**. N: Chitipa Distr., 13 km east of Chinunkha junction, fl. 18.iv.1976, *Pawek* 11103 (K; MO). C: Kasungu National Park (Game Reserve), Lingadzi road, fr. 21.vi.1970, *Brummitt* 11602 (K; MAL).

With a wide but rather scattered distribution in tropical Africa, from Senegal to Ethiopia, south to Angola, Dem. Rep. Congo (Katanga) and the Flora Zambesiaca area. Wet grassland along rivers, edges of swamps and dambos; 1000–1750 m.

Mutimushi 84, 1366 and 2596 (all K; NDO) from Kitwe seem to represent a distinct local form in which the racemes are unusually contracted.

200. **Crotalaria hyssopifolia** Klotzsch in Peters, Naturw. Reise Mossambique **6**, pt. 1: 55 (1861). —J.G. Baker in F.T.A. **2**: 24 (1871) pro parte. —Taubert in Engler, Pflanzenw. Ost-Afrikas **C**: 205 (1895). —E.G. Baker in J. Linn. Soc., Bot. **42**: 286 (1914) pro parte. —Verdoorn in Bothalia **2**: 403 (1928). —Hepper in F.W.T.A., ed. 2, **1**: 549 (1958). —Binns, First Check List Herb. Fl. Malawi: 79 (1968) excl. syn. —Polhill in F.T.E.A., Leguminosae, Pap.: 991 (1971). —Drummond in Kirkia **8**: 218 (1972). —Polhill, Crotalaria Africa & Madagascar: 363 (1982). —Lock, Leg. Afr. Check-list: 182 (1989). Type: Mozambique, Querimba Island, *Peters* s.n. (B†, holotype).

Crotalaria gracillima Klotzsch in Peters, Naturw. Reise Mossambique **6**, pt. 1: 55 (1861). Type: Mozambique, Cabaceira Peninsula, *Peters* s.n. (B†, holotype).

Annual or short-lived perennial 10–70 cm tall, developing numerous ascending or spreading branches; stem slender, appressed puberulous. Leaves 3-foliolate; leaflets 8–25 × 2–8(12) mm, linear-oblanceolate to oblong-oblanceolate or obovate, rounded, truncate or emarginate at the apex, appressed pubescent beneath; petiole 2–12(18) mm long; stipules 0. Racemes 2–11(20) cm long, subdensely to sublaxly (10)20–40-flowered, sometimes with a few other flowers in the axils below; bracts 0.7–1.5 mm long, linear to shortly linear-lanceolate; pedicel 1.5–4 mm long, with minute bracteoles. Calyx 2–3.5 mm long, appressed pubescent; upper lobes narrowly triangular, 2–3 times as long as the tube. Standard elliptic-obovate, yellow, reddish lined and pubescent outside; wings a little shorter than the keel; keel 3.5–5 mm long, angular, with a narrow twisted beak slightly incurved at the tip. Pod sessile, 3–4(4.5) mm long, ovoid-globose, appressed pubescent, 1–2-seeded. Seeds 1.5–2 mm long, subcordiform, smooth, with a generally prominent aril.

Zimbabwe. E: Mutare (Umtali), Circular Drive, fl. & fr. 7.v.1967, *Corby* 1921 (K; SRGH). **Malawi**. N: Chitipa (Fort Hill), fl. & fr. vii.1896, *Whyte* s.n. (K). S: Mt. Mulanje, Likhubula (Likabula) Valley, fl. & fr. 4.iv.1988, *Chapman* 9023 (K; MAL; MO). **Mozambique**. N: 26 km east of Camuana, fl. & fr. 21.v.1961, *Leach & Rutherford-Smith* 10961 (K; LISC; SRGH). Z: 19 km from Maganja da Costa towards Régulo Ingive, fl. & fr. 26.ix.1949, *Barbosa & Carvalho* 4191 (K; LISC; LMA). MS: Gondola Distr., Matsinho (Chimoio), fl. & fr. 27.iii.1948, *Garcia* 774 (BM; LISC).

Widespread in tropical Africa from Senegal to Ethiopia, south through East Africa to the eastern part of the Flora Zambesiaca area. Miombo woodland and grassland, in open, disturbed, often damp (rarely swampy) places; 0–1200 m.

Also recorded from Mozambique, Tete, by Gonçalves in Garcia de Orta, Sér. Bot. **5**: 67 (1982).

See note under *C. tenuirama* below.

201. **Crotalaria tenuirama** Welw. ex Baker in F.T.A. **2**: 24 (1871). —E.G. Baker in J. Linn. Soc., Bot. **42**: 297 (1914) pro majore parte. —Wilczek in F.C.B. **4**: 216 (1953) pro minore parte. —Torre in C.F.A. **3**: 46 (1962). —Polhill, Crotalaria Africa & Madagascar: 365 (1982). —Lock, Leg. Afr. Check-list: 207 (1989). Type from Angola.

Very similar to *C. hyssopifolia* but leaves and flowers more laxly inserted, branches all fairly short, slender and rather stiffly ascending-spreading, the indumentum everywhere shorter and sparser, reduced to a small area of apical puberulence on the standard outside (in Angola sometimes more hairy).

Zambia N: Mporokoso Distr., 53 km west of Mporokoso, Chiwala, fr. 21.v.1961, *E.A. Robinson* 4670 (EA; K; SRGH). W: Chingola, fl. 19.iii.1968, *Mutimushi* 2561 (K; NDO; SRGH).

Also in Angola and Dem. Rep. Congo (Katanga). Miombo woodland; 1300–1400 m.

Fanshawe 3114 (K; NDO), from Luanshya, has flowers infested with caterpillars and, like plants of *C. hyssopifolia* similarly attacked, has unusually congested racemes.

The status of *C. tenuirama* as a species distinct from the more widespread and variable *C. hyssopifolia* is rather uncertain, but they are geographically separated and no intermediates have yet been found in the area where the ranges of the two species approximate between Lakes Tanganyika and Malawi.

202. **Crotalaria parvula** Welw. ex Baker in F.T.A. **2**: 25 (1871). —E.G. Baker in J. Linn. Soc., Bot. **42**: 276 (1914). —Wilczek in F.C.B. **4**: 203 (1953). —Torre in C.F.A. **3**: 34 (1962). —Binns, First Check List Herb. Fl. Malawi: 79 (1968). —Polhill in F.T.E.A., Leguminosae, Pap.: 991 (1971); Crotalaria Africa & Madagascar: 366 (1982). —Lock, Leg. Afr. Check-list: 195 (1989). Type from Angola.

Annual, with slender erect stem, 5–40 cm tall, developing numerous ascending or spreading stems from the base, often also short flowering branches at the top; stem rather densely appressed puberulous. Leaves 3-foliolate; leaflets 5–12 × 1.5–4 mm, linear-oblanceolate to oblanceolate, rounded to truncate or retuse, appressed puberulous beneath; petiole 2–5 mm long; stipules 0. Racemes or heads sessile, dense, 1–3(6 in fruit) cm long, many-flowered; bracts 1–1.5 mm long, linear; pedicel 1–2 mm long, sometimes with minute bracteoles. Calyx 2–3 mm long, ± densely appressed puberulous; upper lobes narrowly attenuate-triangular, 2–3 times as long as the tube. Standard oblong-obovate, pale yellow, reddish marked basally inside, red lined outside and medially puberulous to tomentellous, glabrous near the edges; wings shorter than the keel; keel 3–4 mm long, angular, with a narrow twisted beak.

Pod sessile, 2–3.5 mm long, ovoid-globose, puberulous, 2-seeded. Seeds 1–1.5 mm long, ovate-cordiform, smooth, with a prominent aril.

Zambia. N: Mbala Distr., Chisungu, fl. & fr. 16.iv.1962, *Richards* 16334 (K; SRGH). W: Kafue R., 11 km north of Chingola, fl. & fr. 4.v.1960, *E.A. Robinson* 3699 (K; SRGH). **Malawi**. N: Chitipa Distr., Misuku, fl. & fr. 28.v.1958, *S.T. Hoyle* 1 (K). C: Dedza Distr., Bembeke Mission, 25.iv.1971, *Pawek* 4690 (K).

Also in Nigeria, Cameroon, Central African Republic, Dem. Rep. Congo, Burundi, Angola and southern Tanzania. Short damp grassland in open disturbed places, sometimes on rock pavements; 1350–1800 m.

203. **Crotalaria alexandri** Baker f. in J. Linn. Soc., Bot. **42**: 299 (1914) as *'alexanderi'*. —Wilczek in F.C.B. **4**: 261 (1953). —Binns, First Check List Herb. Fl. Malawi: 78 (1968). —Polhill in F.T.E.A., Leguminosae, Pap.: 992 (1971). —Drummond in Kirkia **8**: 217 (1972). —Polhill, Crotalaria Africa & Madagascar: 367, fig. 103/1–10 (1982). —Lock, Leg. Afr. Check-list: 164 (1989). TAB. 3,7: **48**, fig. A. Type from Kenya.

Crotalaria buchananii Baker f., Legum. Trop. Africa: 33 (1926). —Verdoorn in Bothalia **2**: 400 (1928). Type: Malawi, Shire Highlands, Blantyre, *Buchanan* 62 (K, holotype).

Crotalaria randii Baker f. in J. Bot. **64**: 302 (1926); Legum. Trop. Africa: 875 (1930). Type: Zimbabwe, Mwami (Miami), *Rand* 13 (BM, holotype).

Erect annual 5–40 cm tall, developing ascending branches from near the base; stem slender, densely ± spreading pilose. Leaves 3-foliolate; leaflets 5–15 × 1–5 mm, linear-oblanceolate to oblong-oblanceolate, acute to rounded and apiculate, almost always pilose on both surfaces; petiole 3–6 mm long; stipules 0. Racemes up to 1(3) cm long, with most of the flowers clustered in many of the axils below; bracts 2–4 mm long, linear; bracteoles 0. Calyx 2.5–5 mm long, densely hairy; lobes narrowly triangular, 2–2.5 times as long as the tube. Standard oblong-obovate, pale yellow, marked red at base inside, reddish veined and densely pubescent outside; wings a little shorter than the keel; keel 4.5–6 mm long, angular, with a narrow twisted beak. Pod sessile, 4–5 × 3–3.5 mm, shortly oblong-ellipsoid, pilose, 7–12-seeded. Seeds 1–1.5 mm across.

Zambia. N: Mbala–Tunduma, Saisi Bridge, fl. & fr. 12.iv.1962, *Richards* 16325 (K; LISC; SRGH). W: Ndola Distr., Misaka Forest Reserve, fr. 7.v.1969, *Mutimushi* 3071 (K; NDO). C: 58 km north of Kabwe (Broken Hill), fl. & fr. 15.iii.1961, *Drummond & Rutherford-Smith* 6917 (BR; K; LISC; SRGH). E: 74 km Rumphi to Chitipa, fl. 9.iv.1969, *Pawek* 2041C (MAL). S: 16 km north of Choma, fl. & fr. 15.iv.1963, *van Rensburg* 1937 (K; SRGH). **Zimbabwe**. N: west of Mutoko (Mtoko), fl. & fr. 17.ii.1966, *Corby* 1463 (K; SRGH). C: Harare, Hatfield, fl. & fr. 7.iv.1957, *Whellan* 1217 (K; SRGH). E: Mutare (Umtali), Circular Drive, fl. & fr. 19.iv.1961, *Chase* 7466 (K; LISC; MO; P; SRGH). **Malawi**. N: Mzimba Distr., Mzuzu, fl. & fr. 16.vii.1973, *Pawek* 7200 (K; SRGH). C: Dedza Plateau, 38.5 km south of Dedza, fl. & fr. 29.iii.1978, *Pawek* 14151A (K; MO). S: Ntcheu Distr., Kirk Range, Dzonze (Zonze) Hill, fl. 17.iii.1955, *Exell, Mendonça & Wild* 991 (BM; LISC; SRGH). **Mozambique**. MS: Manica Distr., Garuso, fl. & fr. iv.1935, *Gilliland* Q.1834 (K).

Also in Rwanda, Burundi, S Ethiopia, Uganda, Kenya and Tanzania. Disturbed places in open miombo woodland or short grassland; 500–1750 m.

204. **Crotalaria bequaertii** Baker f. in J. Linn. Soc., Bot. **42**: 299 (1914). —Wilczek in F.C.B. **4**: 215 (1953). —Torre in C.F.A. **3**: 47 (1962). —Binns, First Check List Herb. Fl. Malawi: 78 (1968). —Polhill in F.T.E.A., Leguminosae, Pap.: 993 (1971). —Drummond in Kirkia **8**: 217 (1972). —Polhill, Crotalaria Africa & Madagascar: 368 (1982). —Lock, Leg. Afr. Check-list: 168 (1989). Type from Dem. Rep. Congo (Katanga).

Crotalaria bequaertii var. *pubescens* R. Wilczek in Bull. Jard. Bot. État **23**: 186 (1953); in F.C.B. **4**: 215 (1953). Type from Dem. Rep. Congo (Katanga).

Similar to *C. alexandri*, but with generally shorter hairs (0.1–0.3 mm long), the leaflets variable but often more wedge-shaped, broadest near the apex, subacute to rounded or often truncate, glabrous or shortly pubescent above, bracts much shorter than the pedicel, 0.5–1.5 mm long, and the pod 2-seeded. Seeds sometimes (Mbala region) with an aril.

Zambia. B: Mongu Distr., 48 km Mongu–Kaoma (Mankoya), fl. & fr. 2.iv.1964, *Verboom* 1026 (K; SRGH). N: Mporokoso Distr., Chiwala, 53 km west of Mporokoso, fl. & fr. 21.v.1961, *E.A. Robinson* 4674 (K; SRGH). W: Kitwe, fl. 10.iv.1957, *Fanshawe* 3158 (K; NDO). C: Lusaka Distr.,

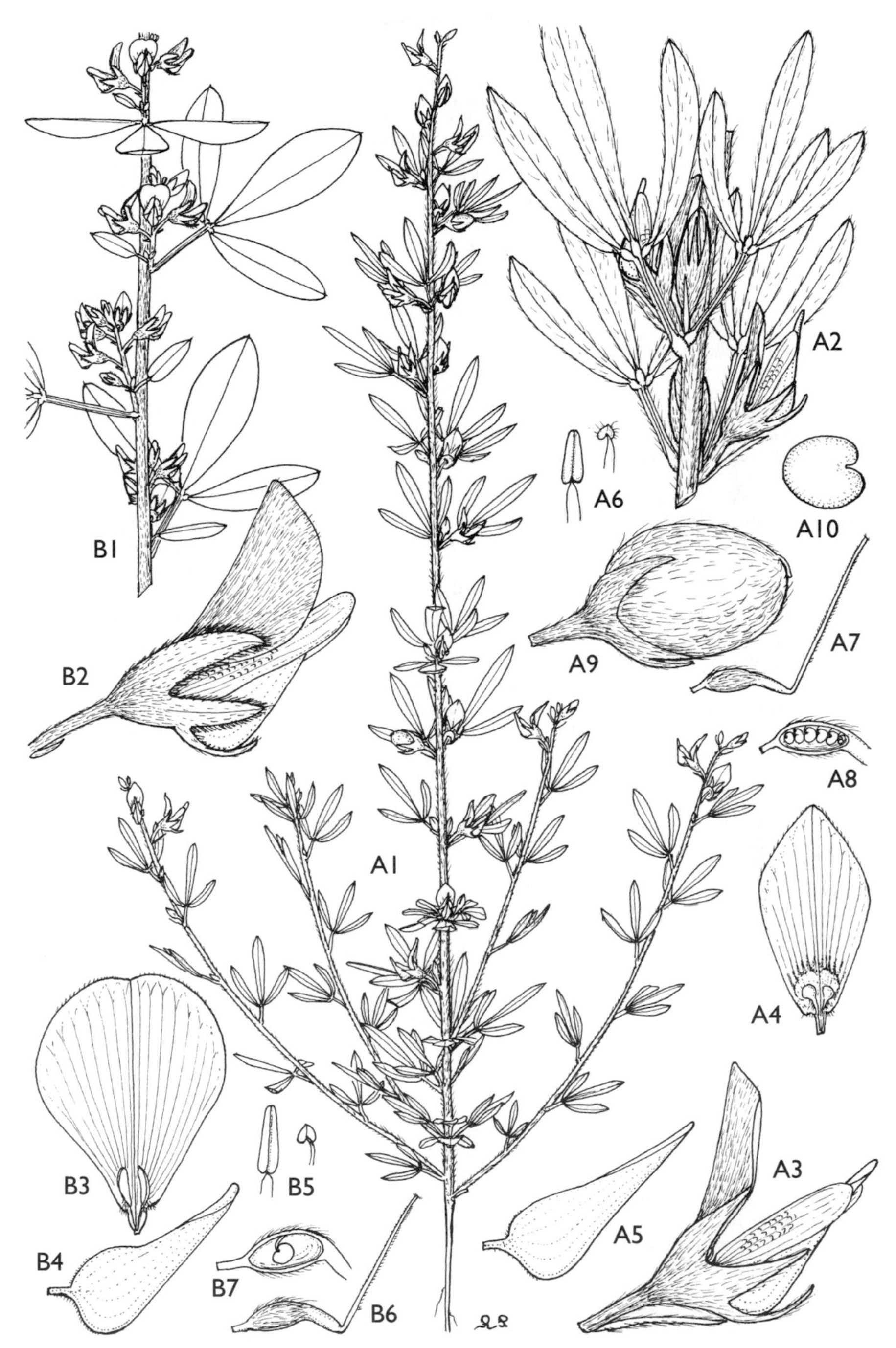

Tab. 3,7: **48**. A. —CROTALARIA ALEXANDRI. A1, plant habit (× 1); A2, detail of flowering nodes (× 3); A3, flower (× 6); A4, standard (× 6); A5, keel (× 6); A6, anthers (× 12); A7, gynoecium (× 6); A8, section through ovary (× 9), A1–A8 from *Richards* 16231; A9, pod (× 6), from *Richards* 16325; A10, seed (× 9), from *E.A. Robinson* 2863. B. —CROTALARIA PRAETEXTA. B1, part of a flowering stem (× 1); B2, flower (× 6); B3, standard (× 6); B4, keel (× 6); B5, anthers (× 12); B6, gynoecium (× 6); B7, section through ovary (× 9), B1–B7 from *Hooper & Townsend* 644. Drawn by Roger Polhill. From Crotalaria Africa & Madagascar.

Great East Road between Undaunda and Rufunsa, fl. & fr. 6.iv.1972, *Kornaś* 1550 (K). S: 8 km east of Choma, fl. & fr. 25.iii.1955, *E.A. Robinson* 1143 (K; SRGH). **Zimbabwe**. N: 27 km from Bindura to Mount Darwin, fl. & fr. 2.v.1971, *Corby* 2200 (K; SRGH). C: Chegutu Distr., Msengezi, fl. & fr. iv.1955, *Conradie* 1 (K; SRGH). **Malawi**. N: Rumphi–Mzuzu road, 1 km from Engucwini, fl. & fr. 2.v.1970, *Brummitt* 11009 (K; LISC; MAL; SRGH). C: Kazungu Distr., Lisasadzi, fl. & fr. 9.iv.1955, *Jackson* 1621 (K; SRGH). **Mozambique**. N: Niassa, fr. 25.v.1961, *Leach & Rutherford-Smith* 10999 (SRGH, not seen).

Also in Dem. Rep. Congo (Katanga), Tanzania and Angola. Open often disturbed places in woodland and short grassland, often on sand; 950–1700 m.

As indicated by R. Wilczek in F.C.B. **4**: 215 (1953) this species is rather variable in the shape of the leaflets, the length of the hairs and whether hairs occur on the upper surface of the leaflets. Plants from around Mbala are exceptional in sometimes having distinctly arillate seeds, e.g. Mbala, Golf Course, fl. & fr. 12.iv.1959, *McCallum-Webster* 708 (BR; K); Mbala, Sandpits, fl. & fr. 17.iv.1961, *Richards* 13787 (K; SRGH).

205. **Crotalaria praetexta** Polhill, Crotalaria Africa & Madagascar: 368, fig. 103/11–17 (1982). —Lock, Leg. Afr. Check-list: 198 (1989). TAB. 3,7: **48**, fig. B. Type: Zambia, 104 km Solwezi to Mwinilunga, *Drummond & Rutherford-Smith* 7179 (BR; K, holotype; LISC; NDO; P; SRGH).

Erect annual 10–30 cm tall, sometimes with a few weak branches from the base, with numerous mostly suppressed flowering branches above; stem densely covered with short stiff appressed to slightly spreading hairs. Leaves mostly 3-foliolate, but with reduced and 1-foliolate leaves on abbreviated flowering branches; leaflets 7–35 × 2–13 mm, oblong-lanceolate to elliptic-oblong or obovate, appressed puberulous beneath; petiole 3–18 mm long, narrowly winged; stipules 0. Flowers in clusters or short racemes, terminal and on much reduced 1–2 leaved branches in the axils below, also in the axils below that; bracts 0.5–1 mm long, linear; pedicel 1.5–3 mm long, without bracteoles. Calyx 2–3 mm long, strigulose-puberulous, densely so on the tube, less densely on the lobe surfaces but very densely in a fringe on the lobe margins; upper lobes obliquely triangular-lanceolate, curved on the upper side, 2–3 times as long as the tube. Standard elliptic-obovate, yellow lined reddish-brown, tomentellous outside; wings ± as long as the keel; keel 4–5 mm long, angular, with a twisted beak. Pod sessile, 3.5–4 mm long, ovoid-globose, appressed puberulous, 2-seeded. Seeds 1.5–2 mm long, oblique-cordiform, smooth.

Zambia. N: near Kawambwa, fl. & fr. 26.iv.1957, *Richards* 9440 (K). W: Luanshya, fl. 26.iii.1955, *Fanshawe* 2235 (K; NDO). C: 10 km south of Kapiri Mposhi, fl. & fr. 27.iii.1955, *Exell, Mendonça & Wild* 1221 (BM; LISC; SRGH).

Known only from the northern and western parts of Zambia. Miombo woodland; 1200–1400 m.

206. **Crotalaria ephemera** Polhill in Kew Bull. **22**: 334, fig. 37 (1968); Crotalaria Africa & Madagascar: 369, fig. 104 (1982). —Lock, Leg. Afr. Check-list: 177 (1989). TAB. 3,7: **49**. Type from Dem. Rep. Congo (Katanga).

Small spreading annual; stem 5–10(15) cm tall, branching from the base, appressed pubescent. Leaves all or almost all 3-foliolate, uppermost occasionally 1-foliolate; leaflets 5–20 × 2–10 mm, oblanceolate to oblong-elliptic or obovate, appressed pubescent beneath, sometimes above also; petiole 4–10 mm long, narrowly winged; stipules 0. Flowers in small terminal heads or short racemes, also with a few flowers in the axils below; bracts 0.5–1.5 mm long, linear; pedicel 2–4 mm long, without bracteoles. Calyx 2.5–3 mm long, appressed puberulous; upper lobes triangular, 2–3 times as long as the tube. Standard elliptic or oblong-obovate, yellow, lined reddish, densely puberulous outside; wings nearly as long as the keel; keel 4–5 mm long, angular, with a narrow twisted beak. Pod sessile, 3–3.5 mm long, ovoid-globose, appressed puberulous, 2-seeded. Seeds 1.5 mm long, rounded-cordiform, smooth.

Zambia. N: Mporokoso Distr., Chiwala, 53 km west of Mporokoso, fl. & fr. 21.v.1961, *E.A. Robinson* 4671 (EA; K; SRGH). W: Luano, fl. & fr. 11.v.1967, *Mutimushi* 1906 (K; NDO; SRGH).

Also in adjoining parts of Katanga (Dem. Rep. Congo). Miombo woodland, in open places; 1200–1400 m.

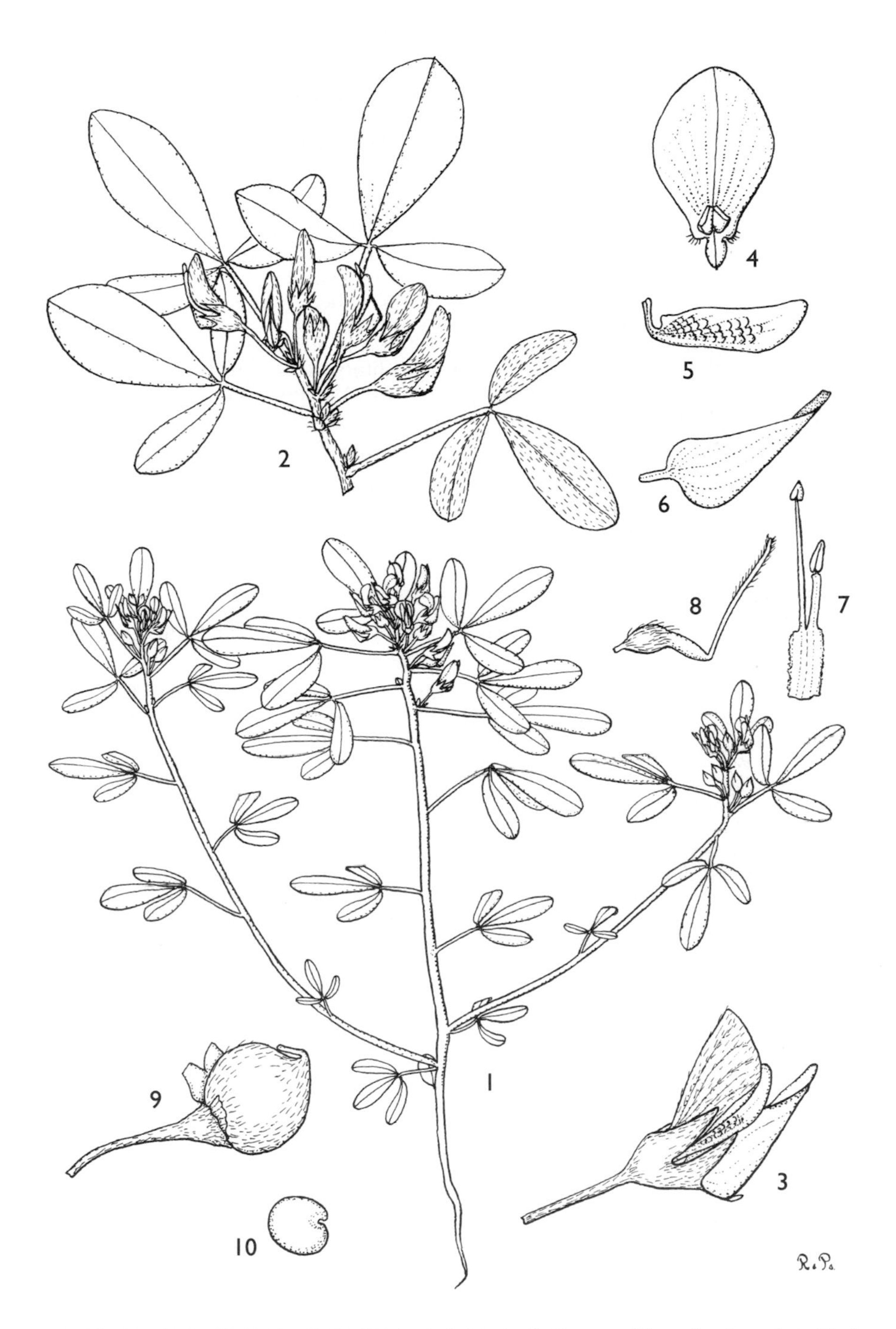

Tab. 3,7: **49**. CROTALARIA EPHEMERA. 1, habit (× 1); 2, part of flowering branch (× 2); 3, flower (× 6); 4, standard (× 6); 5, wing (× 6); 6, keel (× 6); 7, two stamens (× 6); 8, gynoecium (× 6), 1–8 from *de Witte* 207; 9, pod (× 6); 10, seed (× 6), 9 & 10 from *Fanshawe* 2241. Drawn by Roger Polhill. From Crotalaria Africa & Madagascar.

207. **Crotalaria sertulifera** Taub. in Bot. Jahrb. Syst. **23**: 178 (1896). —E.G. Baker in J. Linn. Soc., Bot. **42**: 384 (1914). —Wilczek in F.C.B. **4**: 230, fig. 15 (1953). —Polhill in F.T.E.A., Leguminosae, Pap.: 993 (1971); Crotalaria Africa & Madagascar: 369 (1982). —Lock, Leg. Afr. Check-list: 203 (1989). Type from Dem. Rep. Congo (Katanga).

Small annual, with a very short stem and relatively long radiating procumbent branches from the base; branches slender, 5–10 cm long, appressed puberulous. Leaves 3-foliolate (uppermost often ± suppressed); leaflets 6–14 × 5–13 mm, obovate-circular, rounded or retuse at the apex, appressed puberulous beneath; petiole 3–12 mm long; stipules 0. Racemes subumbelliform, with 3–8 flowers at the top of the slender peduncle, 2–6.5 cm long; bracts 0.7–1 mm long, subulate; pedicel 2–5 mm long, without bracteoles. Calyx c. 3 mm long, appressed puberulous; upper lobes narrowly triangular, 1.5–2 times as long as the tube. Standard obovate, yellow, sometimes lined reddish-brown, puberulous outside at least towards the apex and margins; wings shorter than the keel; keel 4–5 mm long, angular, with a narrow twisted beak. Pod sessile, 7–9 × 3–3.5 mm, subcylindrical, slightly broadened basally, slightly curved along the upper suture, puberulous, 12–16-seeded. Seeds 1 mm long, broadly cordiform, smooth.

Zambia. N: 6 km NE of Mansa (Fort Rosebery) on old disused road to Luwingu, fl. & fr. 11.iv.1961, *Angus* 2816 (FHO; K). W: Ndola, fl. 7.iii.1954, *Fanshawe* 945 (K; NDO). S: Mumbwa Distr., Kawena Protected Forest, Namintondwa Hill, fl. & fr. 11.iii.1972, *Kornaś* 1370 (K).
Also in Dem. Rep. Congo (Katanga) and western Tanzania. Short grassland and bare places in miombo woodland; 1200–1400 m.

208. **Crotalaria baumii** Harms in Warburg, Kunene-Samb. Exped. Baum: 253 (1903). —E.G. Baker in J. Linn. Soc., Bot. **42**: 277 (1914). —Torre in C.F.A. **3**: 35 (1962). —Polhill, Crotalaria Africa & Madagascar: 370 (1982). —Lock, Leg. Afr. Check-list: 167 (1989). Type from S Angola.

Prostrate annual; stems numerous, radiating, slender, 10–40 cm long, well branched, appressed puberulous to finely ± spreading pubescent, glabrescent. Leaves 3-foliolate, uppermost reduced, sometimes 1-foliolate; leaflets 5–13 × 3–11 mm, obovate to obovate-circular, rarely oblanceolate, rounded or emarginate at the apex, punctate above, appressed puberulous to finely pubescent beneath, rarely also above; petiole 1–5 mm long; stipules 0. Racemes 1–3 cm long, subumbelliform, sometimes with 1–few flowers below the main head, 6–many-flowered; bracts 1.5–3.5 mm long, linear to linear-spathulate; pedicel 2–3 mm long, sometimes with setaceous bracteoles. Calyx 2.5–4 mm long, densely pubescent; upper lobes triangular-acuminate, 1.2–2 times as long as the tube. Standard obovate, yellow, ± densely pubescent on the upper half outside; wings a little shorter than the keel; keel 5–6 mm long, angular, with a narrow twisted beak. Pod sessile, 3–4 × 2.5–3 mm, ovoid-globose, appressed pubescent, 2-seeded. Seeds 1.5–2 mm long, broadly cordiform, smooth.

Zambia. B: Kabompo Distr., 21 km WSW of Kabompo on Zambezi (Balovale) road, fl. 24.iii.1961, *Drummond & Rutherford-Smith* 7269 (K; SRGH); Kaoma Distr., Ndanda Dispensary, fl. & fr. 14.iv.1964, *Verboom* 1060 (K; SRGH).
Also in southern Angola and western Dem. Rep. Congo (Katanga). Woodland on Kalahari Sands, pans or disturbed places in short grassland; c. 1100–1150 m.

103. BOLUSIA

By B.-E. van Wyk

Bolusia Benth. in Hooker's Icon. Pl. **12**: t. 1163 (1873). —Hutchinson, Gen. Fl. Pl. **1**: 360 (1964). —Polhill in Bot. Syst. **1**: 323 (1976).

Erect to spreading perennial herbs or small suffrutices (one species sometimes annual?). Leaves digitately 3-foliolate, rarely 1-foliolate; stipules lanceolate to asymmetrically ovate, more or less truncate at the base, obscurely or markedly dentate, with 1–5 teeth along the outer edge. Inflorescences leaf-opposed, short, racemose, with 1–5 flowers; bracts and bracteoles linear to lanceolate, persistent. Flowers yellow to whitish-cream or flushed with pink or purple. Calyx 5-lobed, slightly zygomorphic, the upper pair of lobes slightly broader than the lower 3 lobes. Standard rounded to broadly ovate, cucullate, slightly to markedly emarginate, with two callosities at point of attachment of the claw; wing petals obovate-falcate, shorter than the standard, with prominent auricles at the base; keel petals narrowly linear, helically coiled through several (3–4) turns. Vexillary stamen part of the sheath or sometimes free; 5 larger anthers basifixed, alternating with 5 shorter subbasifixed anthers. Ovary several-ovuled; style long, filiform, glabrous, helically coiled; stigma small and terminal. Pods shortly stipitate, oblong-ellipsoid, often somewhat clavate, markedly inflated, glabrous, smooth, dehiscent. Seeds obliquely cordiform to nearly horseshoe-shaped, with a deep hilar sinus; surface smooth to verrucose; rim aril inconspicuous.

A small genus of five named species endemic to tropical Africa south of the equator. All except *B. amboensis* are relatively rare and poorly represented in herbarium collections. Only one named species, *B. ervoides* (Welw. ex Baker) Torre from south-western Angola, does not occur in the Flora Zambesiaca area (a further imperfectly known species occurs in Tanzania, see Polhill in F.T.E.A., Leguminosae, Pap.: 995, fig. 136 (1971)).

Bolusia is remarkable for the peculiar keel that forms a spirally coiled beak, a feature not present in the closely related *Crotalaria*. The diagnostic characters of *Bolusia*, namely the dentate stipules, coiled keels, glabrous styles and chromosome base number of x=9 are taken at face value to keep the genus separate from *Crotalaria*, where the stipules are not dentate, the keel never coiled, the style almost always with one or two lines of hairs along the upper half and where the chromosome number is x=7 or 8. The following four species occur in the Flora Zambesiaca area, one of which is described here as new:

1. Leaflets glabrous beneath (at most with a few hairs along the midribs); stems glabrous or almost so · 1. *amboensis*
– Leaflets distinctly hairy beneath; stems sparsely to densely silky · · · · · · · · · · · · 2
2. Leaves predominantly 1-foliolate, basal ones sometimes 3-foliolate · · · · · · · · 3. *acuminata*
– Leaves all 3-foliolate · 3
3. Stipules entire to obscurely dentate, 3–7(10) × 3–6(8) mm; flowers 8–10 mm long; pods 1.7–2.7(3.5) cm long · 2. *resupinata*
– Stipules (at least the larger ones) markedly dentate, with up to 5 distinct teeth, (6)10–15(18) × (3)6–9(11) mm; flowers 12–16 mm long; pods c. 4.8 cm long · · · · · · · · · · · · 4. *grandis*

1. **Bolusia amboensis** (Schinz) Harms in Bull. Herb. Boissier, sér. 2, **1**: 876 (1901). —Harms in Warburg, Kunene-Samb.-Exped. Baum: 260 (1903). —E.G. Baker, Legum. Trop. Africa: 169 (1926). —Schreiber in Merxmüller, Prodr. Fl. SW. Afrika, fam. 60: 15 (1970). —Lock, Leg. Afr. Check-list: 161 (1989). TAB. 3,7: **50**, fig. 1. Type from Namibia.
Phaseolus amboensis Schinz in Bull. Herb. Boissier **7**: 36 (1899).
Bolusia rhodesiana Corbishley in Bull. Misc. Inform., Kew **1920**: 329 (1920). —E.G. Baker, Legum. Trop. Africa: 169 (1926). —Drummond in Kirkia **8**: 217 (1972). —Lock, Leg. Afr. Check-list: 161 (1989). Type: Zimbabwe, Rusape, *Hislop* 26 (K, holotype and isotype).

Erect to procumbent perennial herb (sometimes annual?) 0.2–0.5 m high. Leaves digitately 3-foliolate; leaflets (5)13–22(36) × 3–9(14) mm, variable in shape and texture, linear, narrowly elliptic to broadly obovate, the upper ones often markedly narrower than the lower ones, glabrous above, glabrous beneath or with hairs along the midribs only; petiole (4)6–12(17) mm long; stipules 3–7(10) × 3–5(7) mm, asymmetrically ovate, ± truncate at the base, obscurely to markedly dentate, with one or two teeth along the outer edge. Inflorescences leaf-opposed or terminal, short, racemose, 1–4-flowered; rhachis (5)15–40(50) mm long; peduncle 5–20 mm long; bracts small, 2–4 mm long, linear; bracteoles small, c. 0.5 mm long, linear. Flowers 8–10 mm long, white, whitish-cream to yellow and flushed pink or purple. Calyx 5–8 mm long; tube slightly shorter than the lobes. Pods c. 25 × 8 mm, smooth, dehiscent. Seeds brown, c. 3 × 2 mm, smooth.

Caprivi Strip. Popa Falls in the Okavango R., fl. & fr. 19.i.1956, *de Winter* 4348 (K; WIND). **Botswana**. N: Ngamiland Distr., Shakawe, north of airstrip, fl. & fr. 24.iv.1975, *Biegel, Müller & Gibbs Russell* 4993 (K; MO; PRE; SRGH). **Zambia**. B: Senanga Distr., Sioma, 80 km south of Senanga, west bank of the Zambezi R., fl. & fr. 1.ii.1975, *Brummitt, Chisumpa & Polhill* 14211 (K; MO; PRE; SRGH). N: Chinsali Distr., Shiwa Ngandu, fl. & fr. 2.vi.1956, *E.A. Robinson* 1558 (K; SRGH). **Zimbabwe**. C: Marondera (Marandellas), fl. 29.ix.1966, *Corby* 1640 (K; PRE); Chegutu Distr., c. 18 km SE of Norton, fl. & fr. 1.iv.1973, *Pope* 915 (MO; PRE; SRGH).

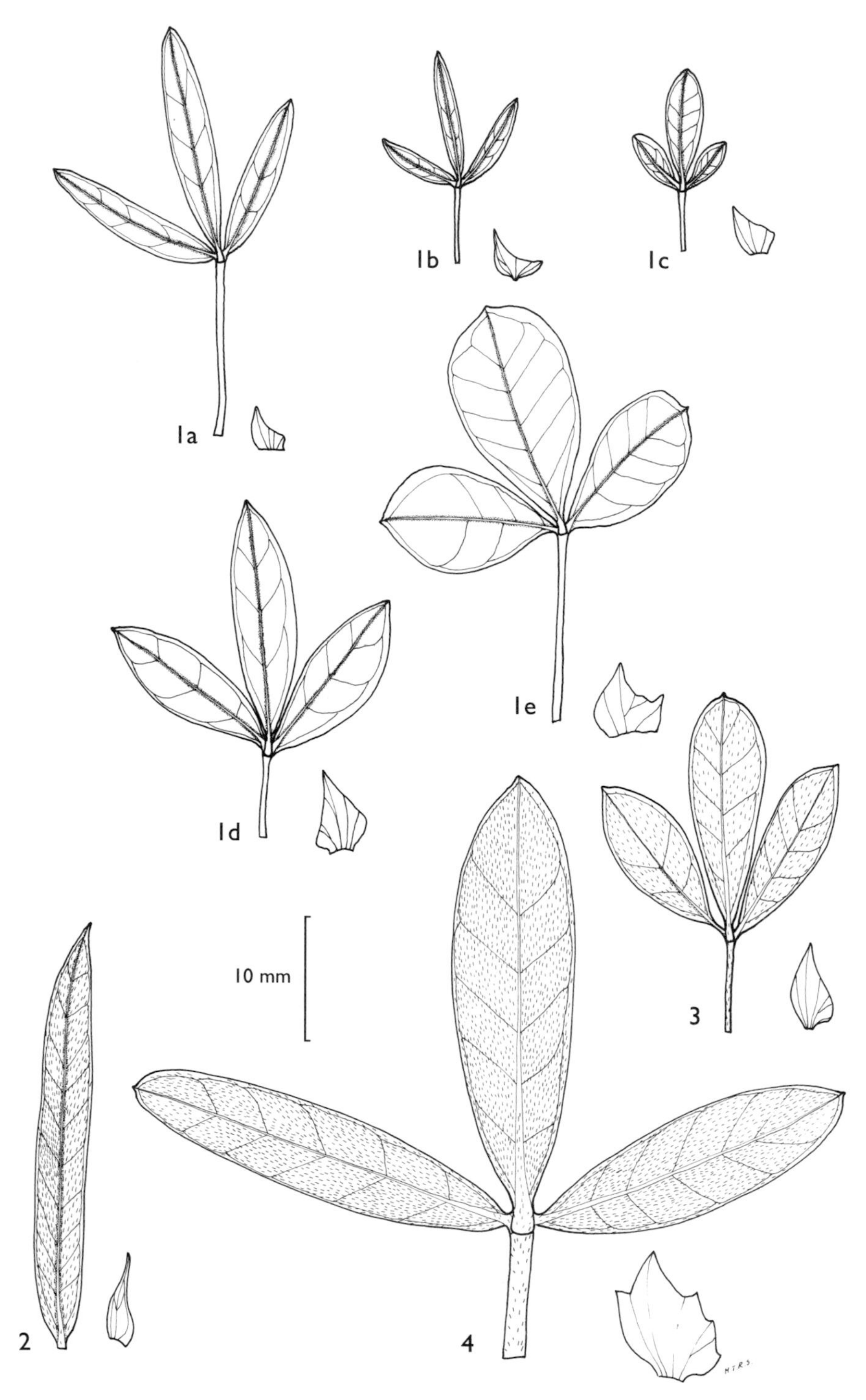

Tab. 3,7: **50**. 1. Leaves and stipules of *Bolusia* species. 1, *B. amboensis,* 1a from *P.A. Smith* 1242; 1b from *Loeb* 117; 1c from *Biegel* 385; 1d from *Eyles* 3861; 1e from *Dept. Agric. (Zambia)* 4; 2, *B. acuminata*, from *Acocks* 12481; 3, *B. resupinata*, from *St. Clair-Thompson* 1283; 4, *B. grandis*, from *Lawton* 1521. Drawn by Mahalia Schubert.

Also in Dem. Rep. Congo (Katanga), south-eastern Angola and north-eastern Namibia. Open grassland, woodland, lake margins, river banks and disturbed areas, in flat sandy areas, often on Kalahari Sand; c. 1000–1500 m.

This is the only widely distributed species and it is quite variable in the size and shape of the leaflets. Some of this variation is illustrated in TAB. 3,7: **50**, fig. 1a–1e. In addition to regional variation, there is also considerable variation within a plant – lower leaflets are usually markedly broader than the upper ones. A robust form from Zimbabwe, known as *B. rhodesiana*, is doubtfully distinct and is here included in *B. amboensis*. The flowers are said to be white flushed with purple, but sometimes also as dirty white or yellow tinged with purple. Plants are often described as annuals but this may be based on seedling recruitment after rains, as most specimens are woody towards their bases and at least some of them show clear signs of having resprouted after fire.

2. **Bolusia resupinata** Milne-Redh. in Hooker's Icon. Pl. **33**: t. 3246 (1934). —Lock, Leg. Afr. Check-list: 161 (1989). TAB. 3,7: **50**, fig. 3. Type: Zambia, Serenje Distr., Musha Hills near Kanona on the Mpika–Kabwe (Broken Hill) road, *St. Clair-Thompson* 1283 (K, holotype and isotype).

Erect to procumbent perennial herb 0.2–0.4 m high. Leaves digitately 3-foliolate; leaflets (8)10–15(20) × 3–6(10) mm, narrowly elliptic to obovate, the upper ones often markedly narrower than the lower ones, sparsely silky hairy above and beneath, ± glabrescent above; petiole 6–9(15) mm long; stipules 3–7(10) × 3–6(8) mm, asymmetrically ovate, not dentate or obscurely dentate. Inflorescences leaf-opposed or terminal, short, racemose, 1–4-flowered; rhachis 10–20 mm long; peduncle 5–15 mm long; bracts small, 2–4 mm long, linear, silky hairy; bracteoles small, c. 0.5 mm long, linear. Flowers 8–10 mm long, pale yellow flushed purple. Calyx 5–7 mm long; tube slightly shorter than the lobes. Pods 17–27(35) × 6–11 mm, smooth, dehiscent. Seeds brown, c. 3 × 2 mm, smooth to slightly verrucose.

Zambia. N: Mpika Distr., fl. vi.1961, *Verboom* 226 (K; SRGH). C: Kapiri Mposhi (Kapiri), fl. 6.viii.1957, *Fanshawe* 3428 (K). **Malawi**. C: Lilongwe Distr., Dzalanyama Forest, 13 km from Sinyala Gate on road to Choulongwe (Chaulongwe) Falls, fl. 26.vi.1970, *Brummitt* 11698 (K).

Not known elsewhere. Open grassland, sandy soil; 1050–1700 m.

A poorly known species, almost indistinct from *B. amboensis* except for the silky hairs on the stems and leaves.

3. **Bolusia acuminata** (DC.) Polhill, Crotalaria Africa & Madagascar: 375 (1982). TAB. 3,7: **50**, fig. 2. Type from South Africa (Northern Cape Province).

Crotalaria acuminata DC., Prodr. **2**: 128 (1825).

Bolusia capensis Benth. in Hooker's Icon. Pl. **12**: t. 1163 (1873). —Lock, Leg. Afr. Check-list: 161 (1989). Type from South Africa (Northern Cape Province).

Erect to procumbent perennial herb, 0.2–0.3 m high. Leaves predominantly 1-foliolate, basal ones sometimes digitately 3-foliolate; leaflets (15)17–45(55) × 3–11(13) mm, variable in shape, usually linear to narrowly elliptic, usually acute to acuminate at the apex, sometimes the basal ones broadly obovate to broadly oblong with rounded apices, glabrous above, densely silky hairy beneath; petiole 5–10(12) mm long, densely silky hairy; stipules (2)4–7 × 2–4 mm, lanceolate to narrowly ovate, acuminate, not dentate or rarely with 1–2 small teeth. Inflorescences leaf-opposed or terminal, short, racemose, 1–3(5)-flowered; rhachis (15)20–35 mm long; peduncle 10–20 mm long; bracts small, 2–4 mm long, linear, silky hairy; bracteoles small, 0.5–1 mm long, linear, silky hairy. Flowers 8–10 mm long, yellow. Calyx 5–8 mm long; tube markedly shorter than the lobes. Pods 18–24 × 5–8 mm, smooth, dehiscent. Seeds brown, c. 3 × 2.5 mm, distinctly verrucose.

Botswana. SE: Kweneng Distr., 8 km to Ngware from turn-off from Molepolole–Letlhakeng road, fl. 21.ix.1977, *Hansen* 3191 (K; PRE; SRGH).

Also in South Africa (Northern Cape, North-West and Free State Provinces). Open grassland and Kalahari sandveld, sandy areas; c. 1400 m.

The only species with predominantly 1-foliolate leaves and can also be distinguished by the linear, silky leaflets and lanceolate stipules. It has a fairly limited distribution area.

4. **Bolusia grandis** B.-E. van Wyk, sp. nov.* TAB. 3,7: **50**, fig. 4; TAB. 3,7: **51**. Type: Zambia, Mbala (Abercorn)–Mpulungu road, not far from Mukoma turning, *Richards* 5308 (K, holotype and isotype; SRGH).

Procumbent, trailing, perennial herb, up to 0.5 m across. Leaves digitately 3-foliolate; leaflets 25–65(75) × 4–11(20) mm, variable in shape, upper ones usually linear, lower ones narrowly elliptic, both with rounded apices, glabrous above, densely silky hairy beneath; petiole 10–17 mm long, silky hairy; stipules foliaceous,

---

*__Bolusia grandis__ B.-E. van Wyk, sp. nov., inter species generis partibus omnibus 1.5–2-plo majoribus differt, foliolis 25–65(75) × 4–11(13) mm, stipulis foliaceis 6–15(18) × 6–11 mm, floribus 12–16 mm longis, stipulis etiam manifeste ad marginem externum dentatis dentibus 3–5 magnis (non 1–2(3) parvis). Typus: Zambia, Mbala (Abercorn)–Mpulungu road, not far from Mukoma turning, *Richards* 5308 (K, holotypus et isotypus; SRGH).

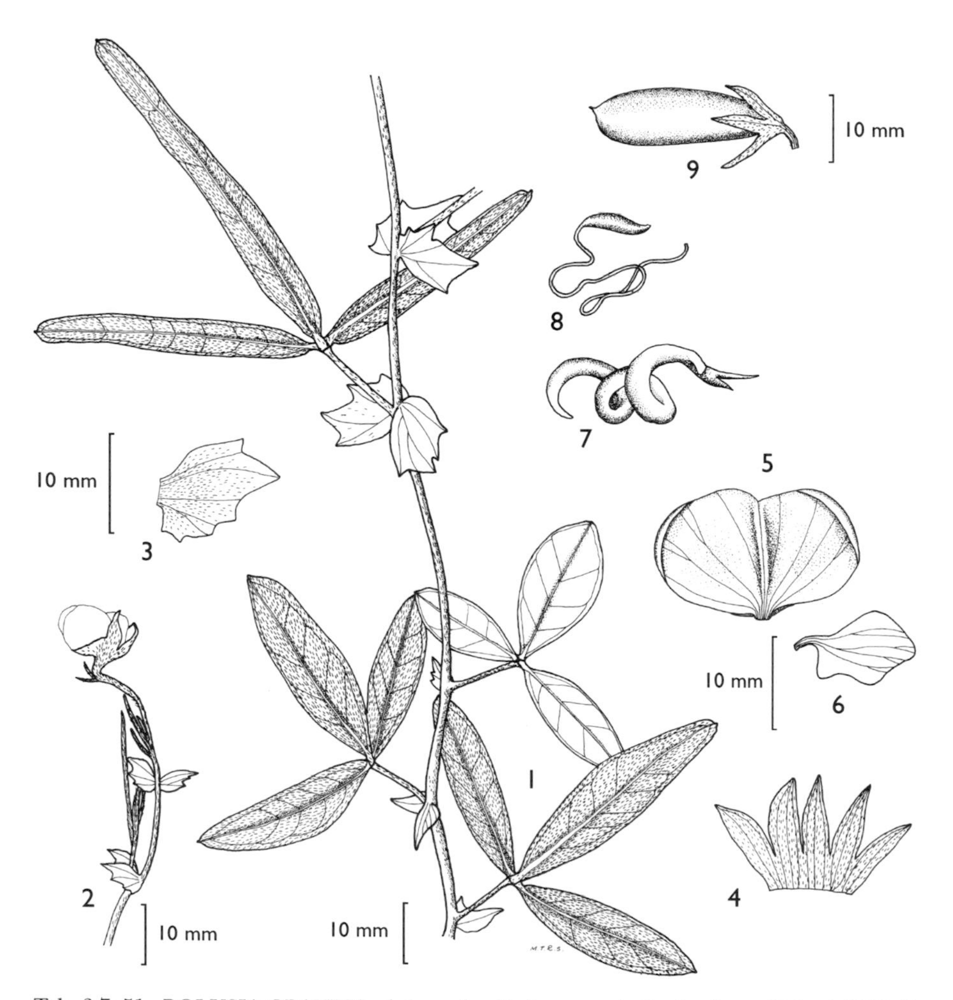

Tab. 3,7: **51**. BOLUSIA GRANDIS. 1, branch with leaves and stipules, from *Richards* 5308; 2, inflorescence; 3, stipule; 4, calyx opened out with the upper lobes to the left; 5, standard; 6, wing; 7, keel; 8, gynoecium; 9, young pod, 2–9 from *Lawton* 1521. Drawn by Mahalia Schubert.

very large, (6)10–15(18) × (3)6–9(11) mm, ovate to broadly ovate-auriculate, acuminate, usually distinctly dentate with 3–5 prominent teeth. Inflorescences leaf-opposed or terminal, short, racemose, 1–3-flowered; rhachis (15)20–35 mm long; peduncle 5–15 mm long; bracts 2–4(6) mm long, linear, silky hairy; bracteoles 1–2(3) mm long, linear, silky hairy. Flowers 12–16 mm long, cream coloured. Calyx 8–12 mm long; tube markedly shorter than the lobes. Pods very large, c. 48 × 10–14 mm, smooth, dehiscent. Seeds brown, c. 4 × 3 mm, smooth.

Zambia. N: Mbala (Abercorn) to Mpulungu road, near Mukoma, fl. & fr. 29.v.1961, *Richards* 15165 (K; SRGH); Isoka airstrip, fl. 8.ix.1968, *Lawton* 1521 (K).
Also in Dem. Rep. Congo (Katanga). Woodland, in sandy soil; 1050–1350 m.
Easily recognized by the large size of the leaves, flowers and pods, as well as the large, dentate stipules and silky pubescence.

104. LOTONONIS (DC.) Eckl. & Zeyh.

By R.M. Polhill

Lotononis (DC.) Eckl. & Zeyh., Enum. Pl. Afr.: 176 (1836) *nom. conserv.* —van Wyk, A synopsis of the genus *Lotononis* in Contrib. Bolus Herb., No.14: 1–292 (1991).
Ononis sect. *Lotononis* DC., Prodr. **2**: 522 (1825).
Amphinomia DC., Prodr. **2**: 522 (1825) *nom. rejic.*
Listia E. Mey., Comment. Pl. Afr. Austr.: 80 (1836).

Herbs. Leaves in Flora Zambesiaca area digitately 3-foliolate or lower ones occasionally 1-foliolate, elsewhere sometimes all 1-foliolate or up to 8-foliolate; stipules small or foliaceous, usually single at each node. Inflorescences terminal or leaf-opposed, the flowers in racemes, heads, clusters or single; bract small to foliaceous, sometimes inserted on the pedicel; bracteoles small or more often lacking. Calyx usually with lateral and upper lobes joined higher on either side, sometimes (sect. *Oxydium*) with 5 subequal lobes. Standard predominantly yellow, or elsewhere sometimes white, pink or blue, hairy or glabrous outside; wings shorter or longer than the keel, often hairy, generally sculptured; keel with apex rounded, pointed or somewhat beaked, often hairy. Stamens in a sheath open on the upper side; anthers markedly dimorphic, 4 oblong to linear and basifixed, 5 ovate and dorsifixed, the carinal one intermediate. Ovary sessile to stipitate; style tapered, with a small stigma. Pods linear, oblong or ovate, flat or turgid, rarely folded and twisted, usually dehiscent, few–many-seeded. Seeds rounded to generally oblique-cordiform, smooth to finely tuberculate.

A genus of about 150 species, almost all in southern Africa, only a few extending into eastern tropical Africa, the Mediterranean region and Pakistan.
Five sections occur in the Flora Zambesiaca area, as described in more detail by van Wyk, loc. cit. (1991).
Sect. **Listia** (E. Mey.) B.-E. van Wyk. Perennials, prostrate or procumbent, glabrous or nearly so, rooting at the nodes. Stipules paired. Flowers in pedunculate racemes or subumbels, many-flowered; bracteoles present. Calyx with upper and lateral lobes fused higher. Standard broadly rounded, glabrous outside; keel half oblong-elliptic, obtuse or rounded. Pods oblong to linear or folded like a concertina, many-seeded. *Spp. 1–3.*
Sect. **Lipozygis** (E. Mey.) Benth. Perennials, erect from a woody rootstock, resprouting after burning, often long-hairy. Stipules single, often foliaceous. Flowers in terminal subumbels; bracts inserted some way up on well developed pedicels; bracteoles generally absent. Calyx with upper and lateral lobes fused higher. Standard oblong to oblong-ovate, subequalling the keel, pubescent outside; keel half oblong-elliptic, obtuse or rounded, usually hairy. Pods scarcely longer than the calyx, ovoid, dehiscent, c. 8-seeded. *Sp. 4.*
Sect. **Leptis** (E. Mey. ex Eckl. & Zeyh.) Benth. Perennials, prostrate or procumbent. Stipules generally single, paired in *L. carinata.* Flowers single or clustered opposite alternate leaves or on short shoots; bracteoles present or absent. Calyx with upper and lateral lobes fused higher. Standard as long to slightly longer than the keel, with hairs at least medially outside; keel half oblong-elliptic to narrowly oblong, rounded or obtuse, usually hairy. Pods ovoid to oblong, dehiscent, usually c. 10-seeded. *Spp. 5–9.*
Sect. **Leobordea** (Del.) Benth. Annuals or rarely short-lived perennials. Stipules generally single. Flowers 1–numerous at nodes of opposite leaves; bracteoles absent. Calyx with upper and lateral lobes fused higher. Standard usually shorter than the keel, hairy outside; keel half

oblong-elliptic to narrowly oblong, obtuse to somewhat pointed, usually hairy. Pods little longer than calyx, dehiscent, usually c. 10-seeded. *Spp.10, 11.*
Sect. **Oxydium** Benth. Perennials, sometimes short-lived and flowering in first season; hairs biramous. Stipules single. Flowers 1–few opposite leaves and sometimes on extended stalks; bracteoles absent. Calyx subequally lobed. Standard with a well developed claw dilated at the base, ± as long as the keel, usually glabrous except for hairs along the midvein outside; keel half oblong-elliptic, often curved, pointed and often beaked. Pods various, 2–many-seeded. *Spp. 12–15.*

1. Stipules paired (one sometimes smaller than the other); flowers numerous in pedunculate racemes or subumbels, with recurving pedicels and elongate or folded pods; perennial herbs, often with prostrate stems rooting at the nodes (sect. *Listia*) ··············· 2
– Stipules single at each node except *L. carinata*; flowers 1–few or if numerous then on short peduncles with erect pedicels and short pods; habit various, rarely rooting at the nodes ·· 4
2. Calyx lateral lobes narrowly triangular, attenuate, little shorter than the tube; keel 6–7 mm long; pods (12)15–18 mm long, narrowly oblong; terminal leaflets mostly 2–3 times as long as the lateral leaflets (unless plants very stunted) ···················· 1. *angolensis*
– Calyx lateral lobes short and broadly triangular, obtuse, about half as long as the tube; keel 8–9 mm long; pods 7–10 mm long, straight or plicate; terminal leaflet not much longer than the lateral leaflets ·· 3
3. Pods narrowly oblong, not plicate, pubescent; flowers mostly subumbelliform, with no more than 1–2 below those crowded at the top; calyx generally pubescent (least so in Mozambique) ·· 2. *bainesii*
– Pods crimped and laterally folded like a concertina, glabrous except along the upper suture; flowers in short racemes; calyx glabrous or nearly so ················ 3. *listii*
4. Hairs basifixed, fine, mostly rather irregularly arranged and sometimes lacking from upper surface of leaflets; keel obtuse to rounded at the apex, usually hairy (glabrous or almost so in *L. mucronata*); upper and lateral lobes of calyx joined higher on either side ······· 5
– Hairs slightly biramous, silvery, appressed and mostly unidirectional on both sides of the leaflets; keel shortly to distinctly beaked, glabrous; calyx with sinuses at ± same level (sect. *Oxydium*) ·· 12
5. Flowers 6–24 in shortly pedunculate subumbels terminal on erect pilose shoots; bracts inserted on the erect pedicel (easiest to see in fruit when pedicels extend up to 7 mm); stipules resembling small leaflets (sect. *Lipozygis*) ····················· 4. *eriantha*
– Flowers fewer, leaf-opposed along the branches (sometimes on short shoots); bract at the base of the pedicel except sometimes in *L. mucronata* and *L. stolzii*, but then stems prostrate; stipules not resembling leaflets in shape (though enlarged in *L. stipulosa* and *L. carinata*) ········ 6
6. Leaves all alternate; mostly perennial herbs with a slender branched rootstock, short-lived in *L. calycina*, see couplet 8 (sect. *Leptis*) ························· 7
– Leaves opposite at flowering nodes (alternate elsewhere); annuals, but exceptionally becoming a short-lived perennial in *L. platycarpa* (sect. *Leobordea*) ··············· 11
7. Stipules paired at the nodes; plant bushy with stems mostly ascending ······ 8. *carinata*
– Stipules single at the nodes; stems prostrate ································· 8
8. Wings shorter than the keel, scarcely longer than the calyx; clusters of flowers strictly leaf-opposed; short-lived perennial, flowering in the first year, and developing numerous branches from the top of a thickened, undivided taproot; pods shortly exserted from calyx, pubescent ··· 9. *calycina*
– Wings as long or nearly as long as the keel and distinctly longer than the calyx; clusters of flowers mostly terminal on short shoots; perennials with slender branched rootstocks; pods much longer than calyx, glabrous except near apex and upper margin (unknown in *L. stolzii*) ·· 9
9. Keel 7–8 mm long, glabrous or nearly so in Flora Zambesiaca area ······· 6. *mucronata*
– Keel 9–11 mm long, extensively pubescent ································ 10
10. Pedicels 4–7 mm long, much exceeding the small bracts (which are sometimes borne up to halfway up the longer pedicels); calyx 5.5–6 mm long; hairs on branchlets mostly shorter than the diameter of the branchlet; leaflets usually emarginate ············· 5. *stolzii*
– Pedicels 1–2(3) mm long, not much longer than the bract; calyx 7–8 mm long; hairs on branchlets longer than the diameter of the branchlet; leaflets rounded at the apex ····· ······································· 7. *decumbens* subsp. *rehmannii*
11. Stipules linear to linear-lanceolate or linear-oblanceolate, 1-nerved ······ 10. *platycarpa*
– Stipules broadly ovate to oblate, several-nerved from the base ············ 11. *stipulosa*
12. Flowers or flower clusters sessile or nearly so; calyx ± as long as the corolla; pods 3–5 mm long, 1–6-seeded ·· 13

– Flowers supported on a well developed pedicel and often also a peduncle; calyx shorter than the corolla; pods 10–20 mm long, 10–20-seeded 14
13. Flowers 1(2) per node; annual or short-lived perennial, developing numerous branches from the top of an undivided taproot .... 12. *brachyantha*
– Flowers 3–7 per node; perennial with numerous stems from a slender branched rootstock .... 13. *crumanina*
14. Peduncle, if present, only slightly more slender than subtending internode and shorter than the opposed petiole; perennial with numerous stems from a slender branched rootstock .... 14. *laxa*
– Peduncles filiform, markedly more slender than subtending internode and much longer than the opposed petiole; annual or short-lived perennial, developing numerous branches from the top of an undivided taproot .... 15. *serpentinicola*

1. **Lotononis angolensis** Welw. ex Baker in F.T.A. **2**: 6 (1871). —Hiern, Cat. Afr. Pl. Welw. **1**: 195 (1896). —Dummer in Trans. Roy. Soc. South Africa **3**: 301 (1913). —E.G. Baker, Legum. Trop. Africa: 16 (1926). —Wilczek in F.C.B. **4**: 274, t. 18 (1953). —White, F.F.N.R.: 159 (1962). —Torre in C.F.A. **3**: 3 (1962). —Milne-Redhead in F.T.E.A., Leguminosae, Pap.: 814 (1971). —Lock, Leg. Afr. Check-list: 215 (1989). —van Wyk in Contrib. Bolus Herb., No. 14: 102, fig. 27 (1991). TAB. 3,7: **52**. Lectotype selected by van Wyk, loc. cit., from Angola.

Argyrolobium deflexiflorum Baker in Bull. Misc. Inform., Kew **1897**: 253 (1897). Type: Malawi, Chitipa (Fort Hill), *Whyte* s.n. (K, holotype).

Stems numerous, prostrate, radiating from a strong perennial taproot, forming a mat, (5)20–60(100) cm long, rooting at the nodes; whole plant, apart from pods, glabrous or nearly so. Leaves 3-foliolate; terminal leaflet much longer than the lateral leaflets, 15–55 × 4–10 mm, narrowly elliptic-oblong, narrowed to the shortly rounded tip; petiole shorter than the terminal leaflet; stipules paired, one smaller than the other, the larger 4–10 mm long, linear-lanceolate. Inflorescences leaf-opposed, erect, subumbelliform to shortly racemose, 2–10(16) cm long, 12–30-flowered; bracts 2–3 mm long, linear to linear-lanceolate; bracteoles a little smaller. Calyx 3.5–4 mm long; lateral lobes triangular-acuminate, shorter than the tube. Standard yellow veined purple, a little shorter than the keel; keel 6–7 mm long, obtuse. Pods deflexed, (12)15–18 × 3–3.5 mm, narrowly oblong, slightly curved, pointed, finely appressed pubescent, 16–24-seeded. Seeds 1–1.5 mm across, oblique-cordiform.

Zambia. N: Mbala Distr., road to Uningi Pans, fl. & fr. 17.iv.1963, *Richards* 18105 (K). W: Solwezi, fl. 26.vii.1964, *Fanshawe* 8850 (K; NDO). C: Lusaka, fl. & fr. 16.iii.1957, *Noak* 153 (K; SRGH). **Malawi**. N: Chitipa Distr., 16 km east of Chitipa, Kaseye Mission, fl. 26.iv.1977, *Pawek* 12674 (K; MAL; SRGH). C: Ntcheu Distr., west foot of Chirobwe Mt. near Ntcheu–Dedza road, fl. & fr. 25.iii.1977, *Brummitt, Seyani & Patel* 14900 (K; MAL).

Also in Cameroon, Burundi, Uganda, Kenya, Tanzania, Dem. Rep. Congo (Katanga) and Angola. Grassy places, commonly in damp sandy or rocky spots along tracks, in ditches, pans and at dambo edges, becoming a weed of cultivation; 1000–1750 m.

Also grown in Zimbabwe at Marondera, Grasslands Research Station, fl. & fr. 27.v.1961, *Corby* 1013 (K; SRGH).

2. **Lotononis bainesii** Baker in F.T.A. **2**: 6 (1871). —Dummer in Trans. Roy. Soc. South Africa **3**: 301 (1913). —E.G. Baker, Legum. Trop. Africa: 16 (1926). —Burtt Davy, Fl. Pl. Ferns Transvaal, pt. 2: 387 (1932). —Schreiber in Merxmüller, Prodr. Fl. SW. Afrika, fam. 60: 81 (1971). —Lock, Leg. Afr. Check-list: 215 (1989). —van Wyk in Contrib. Bolus Herb., No. 14: 104, fig. 27 (1991). Type from South Africa (probably Northern Province).

Lotononis bainesii var. *pottiae* Burtt Davy, Fl. Pl. Ferns Transvaal, pt. 2: 387 (1932). Type from South Africa (Northern Province).

Stems numerous, prostrate, radiating from a strong perennial taproot, forming a mat, 10–80 cm long, rooting at the nodes, finely pubescent, glabrescent. Leaves 3-foliolate; terminal leaflet somewhat larger than the laterals, 5–55 × 2–10 mm, narrowly elliptic-oblong to elliptic or slightly obovate, obtuse, rounded or emarginate at the tip, subglabrous to thinly appressed pubescent, glabrescent; petiole a little shorter to longer than the terminal leaflet; stipules paired, one smaller than the other, the larger 3–10 mm long, lanceolate, attenuate. Inflorescences leaf-opposed, often on short shoots, erect, mostly subumbelliform, sometimes 1–2 flowers below the others,

Tab. 3,7: **52**. LOTONONIS ANGOLENSIS. 1, part of flowering branch (× 1); 2, flower (× 4); 3, calyx, opened out (× 4); 4, standard (× 4); 5, wing (× 4); 6, keel (× 4); 7, androecium and gynoecium (× 4); 8, androecium, opened out (× 6); 9, gynoecium (× 4); 10, fruits (× 1); 11, pod (× 2); 12, seed (× 6), 1–12 from *Haarer* 2406. Drawn by D.R. Thompson.

2–23 cm long, 6–20-flowered; bracts 2–3 mm long, linear to linear-oblanceolate; bracteoles smaller. Calyx 3.5–4 mm long; lateral lobes broadly triangular, ± half as long as the tube, at least thinly pubescent. Standard yellow, often pale, a little shorter than the keel; keel 8–9 mm long, obtuse. Pods becoming deflexed, 8–10 × 2 mm, narrowly oblong, slightly curved, pubescent, c. 8–10-seeded. Seeds c. 1 mm across, oblique-cordiform.

Botswana. SW: Ghanzi Distr., Okwa Valley, 1 km NE of Namibian border, fl. 15.xii.1975, *Skarpe* S–0 (K). SE: between Manyana and Ramotswa, fl. 31.iii.1977, *Hansen* 3107 (C; GAB; K; PRE). **Mozambique**. GI: Chibuto Distr., Maniquenique, Mondiane, fl. 19.vi.1960, *Lemos & Balsinhas* 150 (K; LISC). M: between Santaca and Quinta da Pedra, 90 km south of Maputo, fl. 3.vi.1948, *Gomes e Sousa* 3741 (K).

Also in Namibia and South Africa (Northern Cape, North West, Northern, Gauteng and Mpumalanga). Damp grassy places in valleys, disturbed places along channels, dams and quarries, also in cultivated areas; up to c. 1000 m.

Tried at a number of pasture research stations as a forage and cover plant.

See also note under *Lotononis listii.*

3. **Lotononis listii** Polhill in Bot. Syst. **1**: 324 (1976). —Lock, Leg. Afr. Check-list: 219 (1989). —van Wyk in Contrib. Bolus Herb., No. 14: 104, fig. 27 (1991). Syntypes from South Africa (Eastern Cape).

Listia heterophylla E. Mey., Comment. Pl. Afr. Austr.: 81 (1836), non *Lotononis heterophylla* Eckl. & Zeyh. —Bentham in Hooker, London J. Bot. **3**: 338 (1844). —Harvey in F.C. **2**: 66 (1862). —Eyles in Trans. Roy. Soc. South Africa **5**: 369 (1916). —E.G. Baker, Legum. Trop. Africa: 20 (1926). —Burtt Davy, Fl. Pl. Ferns Transvaal, pt. 2: 389 (1932). —Hutchinson, Botanist South. Africa: 462 (1946). —White, F.F.N.R.: 158 (1962). —Schreiber in Merxmüller, Prodr. Fl. SW. Afrika, fam. 60: 74 (1970). —Drummond in Kirkia **8**: 223 (1972). Types as for species.

Stems numerous, prostrate to weakly ascending, radiating from a strong perennial taproot, usually forming a mat, 10–120 cm long, rooting at the nodes, finely pubescent, glabrescent. Leaves 3-foliolate; terminal leaflet somewhat larger than the laterals, 8–40 × 3–12 mm, narrowly elliptic-oblong to slightly obovate, pointed or rounded at the apex, subglabrous; petiole shorter to longer than the terminal leaflet; stipules paired, one smaller than the other, the larger 3–5 mm long, linear-lanceolate to ovate-lanceolate. Inflorescences leaf-opposed, often on short shoots, erect, at least shortly racemose, 2–10 cm long, 8–24-flowered; bracts 1.5–3 mm long, linear to linear-oblanceolate; bracteoles filiform. Calyx 3–4 mm long; lateral lobes broadly triangular, less than half as long as the tube, glabrous. Standard bright yellow to orange, often lined or flushed reddish to purplish-brown, a little shorter than the keel; keel 8–9 mm long, obtuse. Pods 7–10 × 2.5–4 mm, plicate, with 3–5 lateral folds, glabrous except along the upper suture, maturing rather few seeds. Seeds 1–1.5 mm long, oblique-cordiform.

Botswana. N: Tsao (Tsau)–Nxainxai (Xai Xai), fl. 19.iv.1980, *P.A. Smith* 3354 (K; SRGH). SE: Gaborone, Broadhurst, Segoditshane R., fl. & fr. 4.viii.1977, *Hansen* 3148 (C; GAB; K; PRE; SRGH; UPS; WAG). **Zambia**. B: Sesheke Distr., Machili, fl. & fr. 21.ix.1969, *Mutimushi* 3789 (K; NDO). S: Choma Distr., Mapanza Mission, fl. & fr. 19.xii.1952, *E.A. Robinson* 15 (K). **Zimbabwe**. N: Hurungwe Distr., Mwami (Miami), K.34 Experimental Farm, fl. 4.x.1946, *Wild* 1296 (K; SRGH). W: Bulawayo Distr., near Heany, fl. & fr. 31.viii.1956, *Whellan* 1124 (K; SRGH). C: Harare, University Campus, fl. & fr. 12.ix.1974, *Pope & Biegel* 1363 (K; SRGH). E: Mutare Distr., Chimedza R., c. 3 km from its junction with Sungwizi (Tsungwesi), fl. & fr. 8.ix.1955, *Drummond* 4863 (K; SRGH). S: Beitbridge Distr., c. 88 km east of Beitbridge, Chiturupadzi Dip, Camp area, fl. 18.iii.1967, *Mavi* 243 (K; SRGH).

Also in Namibia, South Africa (southwards to the Eastern Cape Province), Swaziland and Lesotho. Grassy places along streams and flood plains, often in damp sandy spots, along roadsides and in cultivated ground; 900–1450 m.

Despite the peculiar fruits, on the basis of which the separate genus *Listia* was founded, *Lotonins heterophylla* is very similar to *L. bainesii* and easily confused. B.-E. van Wyk (loc. cit.) notes that aberrations occur with *L. heterophylla* having straight fruits and *L. bainesii* plicate ones, but no such have been noted in the Flora Zambesiaca area, where the ranges do not seem to overlap except in Botswana. Material seen from Mozambique labelled as *Listia, Lotononis listii* or *L. angolensis* all seems to be *L. bainesii,* but, as noted by van Wyk, loc. cit., the calyx is not as hairy as is normal for that species. *L. listii* grows close to the Mozambique border however, and may well occur there.

Also cultivated in Zimbabwe at Marondera, Grasslands Research Station, fl. 13.vi.1961, *Corby* 1020 (K; SRGH) and 18.iv.1967, *Corby* 1917 (K; SRGH).

4. **Lotononis eriantha** Benth. in Hooker, London J. Bot. **2**: 605 (1843). —Harvey in F.C. **2**: 59 (1862). —Dummer in Trans. Roy. Soc. South Africa **3**: 308 (1913). —Burtt Davy, Fl. Pl. Ferns Transvaal, pt. 2: 387 (1932). —Drummond in Kirkia **8**: 223 (1972). —Lock, Leg. Afr. Check-list: 218 (1989). —van Wyk in Contrib. Bolus Herb., No. 14: 118, fig. 34 (1991). Type from South Africa (Gauteng).

Stems numerous, erect from a woody rootstock, simple or sparingly branched above, 8–25 cm tall, resprouting after burning, pilose. Leaves 3-foliolate, shortly petiolate; leaflets 8–20 × 3–7 mm, oblanceolate-elliptic to obovate-elliptic, pilose, especially beneath; stipules single at each node, resembling a small leaflet, shortly stalked, 8–12 × 2–3 mm. Inflorescences terminal, shortly pedunculate, subumbelliform, 6–16(24)-flowered; bract inserted on the pedicel, 2–3 mm long, filiform; pedicel 3–7 mm long, elongating in fruit (when bract is easier to see); bracteoles lacking. Calyx 5–6 mm long, pilose; upper and lateral lobes inserted higher, c. 2 mm long, subulate-triangular. Standard yellow, as long as the keel, c. 4 mm wide, narrowly obovate-elliptic, pubescent outside at least medially; keel 8–10 mm long, rounded at the tip, pubescent. Pods spreading, shortly exserted, 8–10 × 3–3.5 mm, elliptic-oblong, glabrescent, few-seeded. Seeds 1.5–2 mm across, rounded oblique-cordiform.

Zimbabwe. E: Nyanga Distr., Pungwe R., fl. & fr. 6.xi.1930, *T.C.E. Fries, Norlindh & Weimarck* 2737 (K); Mutasa Distr., Nusa Plateau, fl. x.1934, *Gilliland* 887 (K).

Also in South Africa (Gauteng, Northern Province, Mpumalanga and KwaZulu-Natal). Pyrophyte of montane grassland and firebreaks; c. 1800 m.

B.-E. van Wyk, loc. cit., compares the two Zimbabwe specimens cited above with material of the species from the Drakensberg, where the species seems to intergrade to some extent with *Lotononis corymbosa* (E. Mey.) Benth. The highveld material that includes the type of *Lotononis eriantha* tends to have characteristically smaller narrow leaflets.

5. **Lotononis stolzii** Harms in Bot. Jahrb. Syst. **54**: 379 (1917). —E.G. Baker, Legum. Trop. Africa: 16 (1926). —Milne-Redhead in F.T.E.A., Leguminosae, Pap.: 816 (1971). —Lock, Leg. Afr. Check-list: 224 (1989). —van Wyk in Contrib. Bolus Herb., No. 14: 137, fig. 38 (1991). Lectotype, selected by van Wyk, loc. cit., from Tanzania.

Stems numerous, prostrate, radiating from a slender branched perennial rootstock, forming mats, 15–40 cm long, slender, pubescent. Leaves 3-foliolate, shortly petiolate; stipules single at a node, 1–6 mm long, shortly stalked, linear to narrowly oblong-elliptic or oblanceolate; leaflets 5–15 × 3–7 mm, obovate, slightly emarginate, pubescent beneath, in Tanzania sometimes also above. Flowers 1–4 at the ends of short shoots; bracts 1–2 mm long, linear to linear-elliptic or linear-oblanceolate, lower ones recaulescent, emerging from the longer pedicel; pedicels 4–7 mm long; bracteoles c. 1 mm long, filiform. Calyx c. 5.5–6 mm long, pubescent; upper and lateral lobes joined higher, 2 mm long, narrowly triangular-subulate. Standard yellow, with a brown midvein outside, longer than the keel, pubescent outside; keel 9–10 mm long, shortly rounded at the tip, pubescent outside. Pods and seeds not seen.

Malawi. N: Rumphi Distr., Nyika Plateau, Kasaramba Area, fl. 26.v.1967, *Banda* 949 (K; MAL); Chitipa Distr., Domwe Peak to Nganda Peak road, fl. 16.iv.1975, *Pawek* 9271 (K; MAL; MO; SRGH).

Also in the Southern Highlands of Tanzania. Montane grassland; 2100–2350 m.

6. **Lotononis mucronata** Conrath in Bull. Misc. Inform., Kew **1908**: 222 (1908). —Dummer in Trans. Roy. Soc. South Africa **3**: 325 (1913). —Burtt Davy, Fl. Pl. Ferns Transvaal, pt. 2: 388 (1932). —Lock, Leg. Afr. Check-list: 221 (1989). —van Wyk in Contrib. Bolus Herb., No. 14: 135, fig. 39 (1991). Type from South Africa (Gauteng/Mpumalanga).

Lotononis gerrardii Dummer in Trans. Roy. Soc. South Africa **3**: 325 (1913). Type from South Africa (KwaZulu-Natal).

Lotononis gerrardii var. *transvaalensis* Dummer in Trans. Roy. Soc. South Africa **3**: 326 (1913). Type from South Africa (Northern Province).

Lotononis ornata Dummer in Trans. Roy. Soc. South Africa **3**: 327 (1913). Type from Lesotho.

Lotononis affinis Burtt Davy, Fl. Pl. Ferns Transvaal, pt. 2: 389 (1932) non Eckl. & Zeyh. (1836), *nom. illegit.* Type from South Africa (Mpumalanga).

Lotononis sp. 2 (*Wild* 4454) of Drummond in Kirkia **8**: 223 (1972).

Stems numerous, prostrate, radiating from a slender branched perennial rootstock, forming mats, 10–30 cm long, slender, pubescent. Leaves 3-foliolate, shortly petiolate; leaflets 4–8 × 2–5 mm, obovate, slightly emarginate, minutely mucronate, pubescent beneath; stipules single at a node, 1–3 mm long, almost linear to very narrowly elliptic-oblong. Flowers 1–3 terminal and leaf-opposed, mostly at the ends of rather short shoots; bracts at the base of the pedicel or up to halfway up on longer pedicels, 1 mm long, shorter than the pedicel, linear; pedicels 2–9 mm long; bracteoles absent. Calyx 4–5 mm long, pubescent; upper and lateral lobes joined higher, 1–2 mm long, narrowly triangular-subulate. Standard bright yellow, as long as the keel, pubescent medially outside; keel 7–8 mm long, shortly rounded at the tip, glabrous or pubescent towards the apex outside. Pods 7–12 × 3–4 mm, narrowly oblong-obovate, glabrous except along upper suture and near apex. Seeds 1–1.5 mm long, oblique-cordiform.

Zimbabwe. E: Nyanga Distr., Nyangani (Inyangani) Mt., fl. 16.iii.1982, *Pope, Müller & Polhill* 2031 (K; SRGH) — not typical (see note); Mutasa Distr., Stapleford, fl. ix.1948, *Munch* 136 (K; SRGH); Mutare Distr., Himalayas, Engwa, fl. 2.iii.1954, *Wild* 4454 (K; LISC; SRGH). **Mozambique**. MS: Sussundenga Distr., Tsetserra, fl. 7.ii.1955, *Exell, Mendonça & Wild* 242 (LISC).

Also in South Africa (southwards to the Free State and KwaZulu-Natal), Swaziland and Lesotho. Short montane grassland, among rocks and on bare ground; 1800–2550 m.

The specimens from Nyanga, around Stapleford at about 1800 m, agree well with some of the more luxuriant material from South Africa, but the several specimens from Mt. Nyangani, from 1900 to 2550 m, have markedly elongate pedicels, with the bracts attached, mostly 6–9 mm long, rather than 2–4 mm, and the flowers are rather different, with the keel shorter than the wings and quite glabrous. The single gathering from Engwa on the Himalayas, at 2100 m, cited above, is somewhat intermediate, with partly elongate pedicels, but the keel longer than the wings and with just a few minute hairs. B.-E. van Wyk, loc. cit., has drawn attention to the specimens with long pedicels and suggested they may be distinct at infraspecific level.

7. **Lotononis decumbens** (Thunb.) B.-E. van Wyk in Contrib. Bolus Herb., No. 14: 138, fig. 39 (1991). Type from South Africa (Eastern Cape Province).
Ononis decumbens Thunb., Prodr. Pl. Cap., part 2: 129 (1800).

Subsp. **rehmannii** (Dummer) B.-E. van Wyk in Contrib. Bolus Herb., No. 14: 139, fig. 39 (1991). Type from South Africa (Mpumalanga).
Lotononis rehmannii Dummer in Trans. Roy. Soc. South Africa **3**: 326 (1913). —Burtt Davy, Fl. Pl. Ferns Transvaal, pt. 2: 389 (1932). —Lock, Leg. Afr. Check-list: 223 (1989).

Stems numerous, prostrate, radiating from a slender branched perennial rootstock, forming mats, 10–30 cm long, slender, pubescent to pilose. Leaves 3-foliolate, shortly petiolate; leaflets 4–10 × 2–7 mm, obovate, apically rounded, apiculate, pubescent beneath; stipules single at a node, 1–4 mm long, almost linear to very narrowly elliptic-oblong or oblanceolate, sometimes stalked. Flowers 1–4 terminal on short shoots; bracts at or near the base of the pedicel, 1–1.5 mm long, linear; pedicels 1–2(3) mm long; bracteoles minute or elsewhere lacking. Calyx 7–8 mm long, pubescent; upper and lateral lobes joined higher, 3–3.5 mm long, narrowly attenuate-triangular. Standard yellow, as long as the keel, pubescent outside, elsewhere sometimes sparsely so; keel 10–11 mm long, shortly rounded and purple at the tip, pubescent outside, elsewhere sometimes sparsely so. Pods 11–12 × 3–3.5 mm, narrowly oblong-obovate, glabrous except near the apex.

Zimbabwe. E: Nyanga Distr., Sanyatwe, fl. 13.iii.1965, *Corby* 1258 (K; LISC; SRGH).

Also in South Africa (Northern Province, Mpumalanga and Free State) and Lesotho. Habitat not recorded, but probably short grassland above 1800 m.

The single record from Zimbabwe is somewhat anomalous, having bracteoles, otherwise unknown in the species, and altogether more hairy than usual. The habit and structure of the inflorescence is, however, a good match with the South African material.

Subsp. *decumbens*, with longer narrower flowers and narrower leaflets, occurs in the Eastern Cape Province of South Africa.

8. **Lotononis carinata** (E. Mey.) Benth. in Hooker, London J. Bot. **2**: 609 (1843). —Harvey in F.C. **2**: 64 (1862). —Dummer in Trans. Roy. Soc. South Africa **3**: 323 (1913). —Lock, Leg. Afr. Check-list: 216 (1989). —van Wyk in Contrib. Bolus Herb., No. 14: 139, fig. 38 (1991). Lectotype, selected by van Wyk, loc. cit., from South Africa (Eastern Cape Province).

Lipozygis carinata E. Mey., Comment. Pl. Afr. Austr.: 80 (1836).
Lotononis transvaalensis Dummer in Trans. Roy. Soc. South Africa **3**: 292 (1913). —Burtt Davy, Fl. Pl. Ferns Transvaal, pt. 2: 387 (1932). —Lock, Leg. Afr. Check-list: 225 (1989). Types from South Africa (Mpumalanga).
Lotononis pauciflora Dummer in Trans. Roy. Soc. South Africa **3**: 293 (1913). —Lock, Leg. Afr. Check-list: 222 (1989). Type from South Africa (KwaZulu-Natal).
Lotononis florifera Dummer in Trans. Roy. Soc. South Africa **3**: 323 (1913). —Burtt Davy, Fl. Pl. Ferns Transvaal, pt. 2: 388 (1932). —Lock, Leg. Afr. Check-list: 218 (1989). Type from South Africa (KwaZulu-Natal).
Lotononis florifera var. *major* Burtt Davy, Fl. Pl. Ferns Transvaal, pt. 2: 388 (1932). Type from South Africa (Mpumalanga).

Stems numerous, erect and spreading from a branched perennial rootstock, 10–30 cm tall, slender, pubescent to pilose on younger parts. Leaves 3-foliolate, shortly petiolate; leaflets 8–15 × 2–5 mm, linear-elliptic to oblanceolate, pointed, mucronate, appressed pubescent on both surfaces; stipules paired at the nodes, 3–8 mm long, linear-elliptic. Flowers single or in clusters of up to 7, leaf-opposed; bracts 1–2 mm long, nearly as long as the pedicel, filiform or subulate; bracteoles lacking. Calyx 6–7 mm long, pubescent; upper and lateral lobes joined higher, 3–4 mm long, subulate. Standard yellow, longer than the keel, extensively pubescent outside; wings a little shorter than the keel; keel 8–10 mm long, obtuse, extensively pubescent. Pods 8–10 × 3 mm, oblong, glabrous except for a few hairs along the upper suture, c. 10-seeded. Mature seeds not seen.

Mozambique. M: Catuane, Empresa Agro-Pecuária de Catuane, viveiros piscícolas, c. 500 m em direcção Foz do Rio, fl. 29.ix.1983, *Zunguze, Boane & Dungo* 578 (K; LMU).
Also in South Africa as far as southern KwaZulu-Natal. Short grassland, the single gathering from Mozambique collected near a river at 500 m, but extends into montane areas further south.

9. **Lotononis calycina** (E. Mey.) Benth. in Hooker, London J. Bot. **2**: 611 (1843). —Dummer in Trans. Roy. Soc. South Africa **3**: 323 (1913). —Burtt Davy, Fl. Pl. Ferns Transvaal, pt. 2: 388 (1932). —Schreiber in Merxmüller, Prodr. Fl. SW. Afrika, fam. 60: 81 (1970). —Lock, Leg. Afr. Check-list: 216 (1989). —van Wyk in Contrib. Bolus Herb., No. 14: 142, fig. 39 (1991). Lectotype, selected by van Wyk, loc. cit., from South Africa (Eastern Cape Province).
Lipozygis calycina E. Mey., Comment. Pl. Afr. Austr.: 78 (1836) excl. var. *b.*
Lotononis tenella var. *calycina* (E. Mey.) Harv. in F.C. **2**: 65 (1862).
Lotononis tenella var. *hirsutissima* Harv. in F.C. **2**: 65 (1862). Lectotype, selected by van Wyk, loc. cit., from South Africa (Gauteng).
Crotalaria erisemoides Ficalho & Hiern in Trans. Linn. Soc. London, Bot. **2**: 17 (1881). —E.G. Baker in J. Linn. Soc., Bot. **42**: 414 (1914); Legum. Trop. Africa: 59 (1926). Type from Angola (Moxico).
Lotononis calycina var. *hirsutissima* (Harv.) Dummer in Trans. Roy. Soc. South Africa **3**: 328 (1913). —Burtt Davy, Fl. Pl. Ferns Transvaal, pt. 2: 388 (1932).
Lotononis calycina var. *acuta* Dummer in Trans. Roy. Soc. South Africa **3**: 328 (1913). —Burtt Davy, Fl. Pl. Ferns Transvaal, pt. 2: 388 (1932). Lectotype, selected by van Wyk, loc. cit., from South Africa (Mpumalanga).
Lotononis erisemoides (Ficalho & Hiern) Torre in Mem. Junta Invest. Ultramar, sér. 2, **19**: 23 (1960); in C.F.A. **3**: 5 (1962). —Lock, Leg. Afr. Check-list: 218 (1989).
Lotononis sp. 1 (*Chase* 3544) of Drummond in Kirkia **8**: 223 (1972).

Short-lived perennial, developing numerous prostrate radiating branches from the top of an undivided taproot, flowering from first year of growth, ultimately (at least in South Africa) rootstock sometimes branched; lower branches numerous, prostrate, forming mats, 5–20 cm long, slender, pilose. Leaves 3-foliolate; leaflets 4–12 × 2–7 mm, elliptic-obovate, apiculate, sparsely pilose to glabrous above, pilose beneath; petiole ± as long as the leaflets or a little shorter; stipules single at a node, 2–10 mm long, linear-elliptic to oblanceolate. Flowers 1–3, leaf-opposed; bracts 1–2 mm long, ± as long as the pedicel, linear; bracteoles absent. Calyx 4–6 mm long, pilose; upper and lateral lobes joined slightly higher, 3–4 mm long, subulate. Standard bright yellow, shorter than or equalling the keel, narrowly elliptic-obovate, extensively pubescent outside; wings distinctly shorter than the keel; keel 8–10 mm long, narrow with a blunt reddish-purple tip, extensively pubescent. Pods 6–8 × 2.5–3 mm, narrowly oblong-obovate, pubescent, c. 8-seeded. Seeds not seen.

Zimbabwe. E: Nyanga, fl. 11.v.1930, *T.C.E. Fries, Norlindh & Weimarck* 2631 (K); edge of Nyanga (Inyanga) road cutting, fl. 10.iii.1964, *Corby* 1084 (K; SRGH); Mutare Distr., Zimunya Communal Land (Zimunya's Reserve), fl. & fr. 31.xii.1950, *Chase* 3544 (K; LISC; SRGH).

Also in Angola, Namibia and widespread in the eastern half of South Africa. Short, montane grassland, sometimes by rivers; c. 1700 m.

The above description is based on the specimens seen from around Nyanga and Mutare, and some material from the northern Drakensberg of South Africa. B.-E. van Wyk has annotated the Nyanga specimens cited above with the comment that they are not typical and they do not key out unambiguously in the key given in his synopsis of the genus. The calyx is distinctly shorter than the keel and the leaflets may be hairy above. However, the species shows considerable variation throughout its range and he did not consider any subdivision practicable at this stage.

10. **Lotononis platycarpa** (Viv.) Pic. Serm. in Webbia **7**: 331 (1950). —Torre in C.F.A. **3**: 4 (1962). —Schreiber in Merxmüller, Prodr. Fl. SW. Afrika, fam. 60: 83 (1970). —Milne-Redhead in F.T.E.A., Leguminosae, Pap.: 813, fig. 118/11–20 (1971). —Drummond in Kirkia **8**: 223 (1972). —Lock, Leg. Afr. Check-list: 222 (1989). —van Wyk in Contrib. Bolus Herb., No. 14: 151, fig. 44 (1991). TAB. 3,7: **53**, fig. A. Type from Egypt.

Lotus platycarpos Viv., Pl. Aegypt., Dec. IV: 14, t. 2/9 (1830).

Leobordea lotoidea Del., Fragm. Fl. Arab. Pétrée: 23 (1830). Type from Arabia.

Lotononis leobordea Benth. in Hooker, London J. Bot. **2**: 607 (1843). —Harvey in F.C. **2**: 61 (1862). —J.G. Baker in F.T.A. **2**: 5 (1871). —Dummer in Trans. Roy. Soc. South Africa **3**: 311 (1913). —Eyles in Trans. Roy. Soc. South Africa **5**: 369 (1916). —E.G. Baker, Legum. Trop. Africa: 17 (1926) *nom. superfl.*, based on *Leobordea lotoidea.*

Lotononis clandestina sensu J.G. Baker in F.T.A. **2**: 8 (1871).

Lotononis clandestina var. *steingroeveriana* Schinz in Verh. Bot. Vereins Prov. Brandenburg **30**: 157 (1888). Type from South Africa (Northern Cape).

Lotononis lotoidea (Del.) Batt., Fl. Alg. **1**: 209 (1889).

Lotononis dinteri Schinz in Vierteljahrsschr. Naturf. Ges. Zürich **52**: 423 (1907). —Dummer in Trans. Roy. Soc. South Africa **3**: 317 (1913). —E.G. Baker, Legum. Trop. Africa: 18 (1926). —Torre in C.F.A. **3**: 4 (1962). Type from Namibia.

Lotononis steingroeveriana (Schinz) Dummer in Trans. Roy. Soc. South Africa **3**: 311 (1913). —Lock, Leg. Afr. Check-list: 224 (1989).

Amphinomia platycarpa (Viv.) Cufod. in Bull. Jard. Bot. État **25**, Suppl. [Enum. Pl. Aethiop. Sperm.]: 22 (1955); Zweit Nachtr.: 23 (1969).

Amphinomia lotoidea (Del.) Maire, Fl. Afr. Nord.: 80 (1987).

Lotononis sp. 3 (*Miller* 2269) & ? *sp. 4* (*Davies* 1224) of Drummond in Kirkia **8**: 223 (1972).

Annual, with a very short stem, soon developing numerous slender radiating prostrate branches up to 5–50 cm long, pilose, exceptionally perennating with other stems from the top of an undivided taproot. Leaves opposite on flowering shoots, 3-foliolate; leaflets 4–8(20) × 1.5–3(4) mm, oblanceolate, slightly apiculate, glabrous to pilose above, pilose beneath; petiole ± as long as the leaflets or a little shorter; stipules single at a node, 2–4 mm long, linear to linear-lanceolate or linear-oblanceolate. Flowers 1–several clustered at the nodes, sessile; bracts 1–2.5 mm long, linear; bracteoles absent. Calyx 3.5–4.5(6) mm long, pilose; upper and lateral lobes joined slightly higher, 1.5–2.5(3) mm long, subulate. Standard pale yellow, shorter than the keel, narrowly elliptic-ovate, extensively pubescent outside; wings distinctly shorter than the keel; keel 6–7(9) mm long, narrow, slightly upcurved to the bluntly pointed reddish tip, extensively pubescent. Pods 4–7(9) × 1.5–2.5 mm, narrowly oblong, often somewhat tapered to the base, slightly downcurved at the tip, pubescent, sometimes glabrescent, c. 10-seeded. Seeds 1–1.5 mm long, oblique-cordiform, smooth, brown.

Botswana. N: Ngamiland Distr., Aha Hills, near the Nxainxai (Xai-Xai) to Qangwa road, fl. 22.iv.1980, *P.A. Smith* 3380 (K; SRGH). SW: 43 km from Ghanzi on Maun road, fl. & fr. 3.viii.1955, *Story* 5091 (K; PRE). SE: Kweneng Distr., Mantswabese Ranch, fl. & fr. 17.iii.1977, *Hansen* 3080 (GAB; K: PRE; SRGH). **Zimbabwe**. N: Murehwa Distr., without precise locality, fl. & fr. 4.ix.1931, *Rattray* 408 (K; SRGH). W: Bulawayo, road to Gwanda, near "Drive In" cinema, fl. 15.v.1958, *Drummond* 5817 (K; SRGH). S: Mberengwa Distr., Belingwe Reserve, fl. & fr. v.1958, *Davies* 2468 (K; SRGH).

Widely distributed in the drier parts of Africa and west Asia, from the Cape Verde Is., through North Africa and the Sahel to W Pakistan, and from Arabia and the Horn of Africa to northern Tanzania, then disjunctly south to southern Angola, the Flora Zambesiaca region, Namibia and the westerly part of South Africa. Open ground, roadsides, cultivated ground and other disturbed places or in short grassland on sand; 900–1400 m.

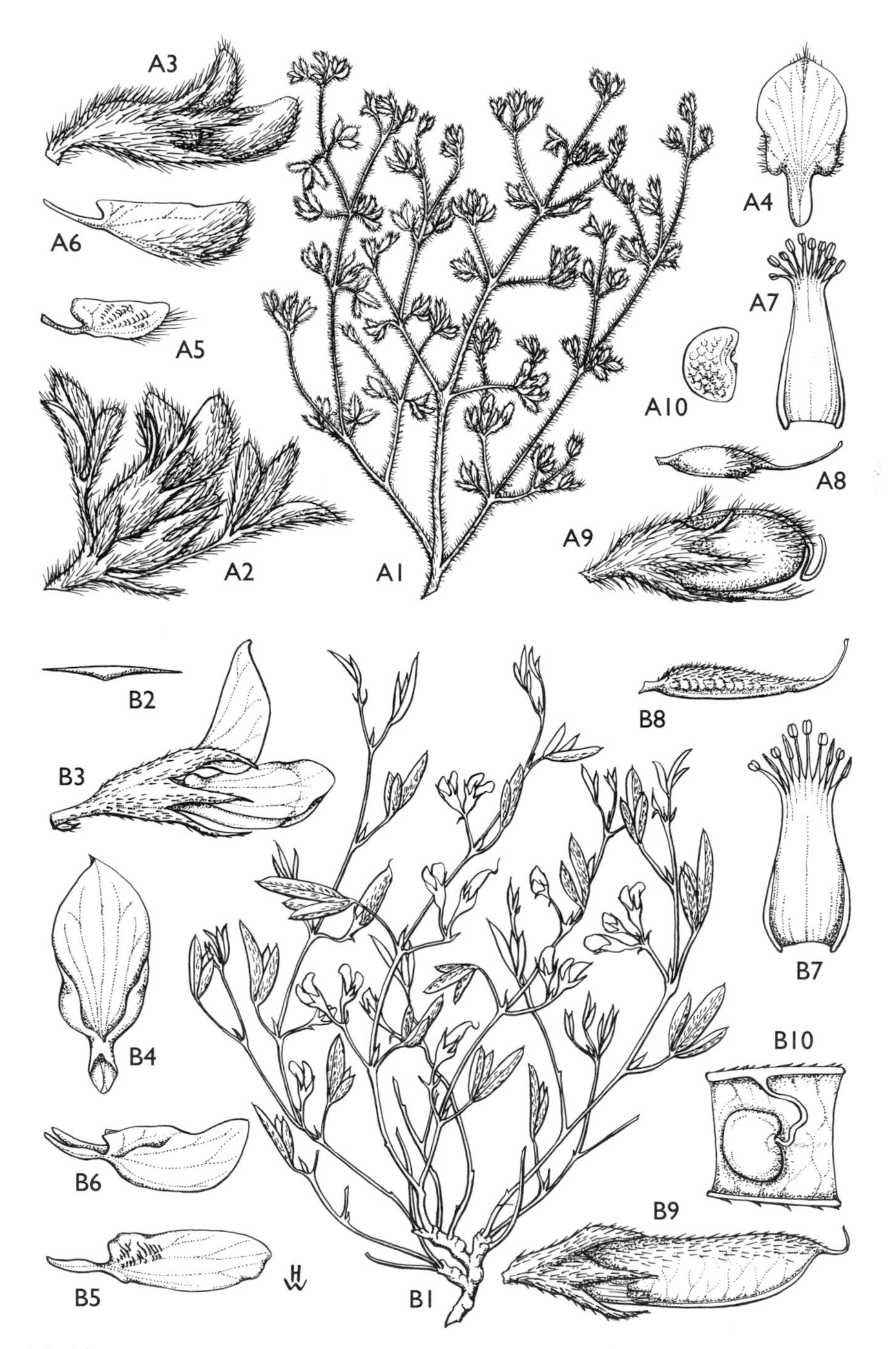

Tab. 3,7: **53**. A. —LOTONONIS PLATYCARPA. A1, habit (× $^2/_3$); A2, flowering node (× 4); A3, flower (× 5); A4, standard (× 5); A5, wing (× 5); A6, keel (× 5); A7, androecium, opened out (× 5); A8, gynoecium (× 5), A1–A8 from *Verdcourt* 1794; A9, pod (× 5); A10, seed (× 7), A9 & A10 from *Bally* 5464. B. —LOTONONIS LAXA. B1, plant habit (× $^2/_3$); B2, hair (× 40); B3, flower (× $3^1/_2$); B4, standard (× $3^1/_2$); B5, wing (× $3^1/_2$); B6, keel (× $3^1/_2$); B7, androecium, opened out (× $3^1/_2$); B8, gynoecium (× $3^1/_2$), B1–B8 from *Newbould* 3407 and *A.S. Thomas* 2155; B9, pod (× 3); B10, portion of pod valve showing one seed (× 5), B9 & B10 from *Merxmüller* 107. Drawn by Heather Wood. From F.T.E.A.

Notably variable over its wide range and particularly so in the western parts of southern Africa, as discussed by van Wyk (loc. cit.). The gatherings from SE and SW Botswana are larger in all parts, the figures given in brackets in the description above. The specimens named as this species from the mountainous Eastern Districts of Zimbabwe belong to the atypical form of *L. calycina*, described above, but are deceptively similar. In the Flora Zambesiaca area, *L. platycarpa* only very exceptionally perennates (as in *Hansen* 3080, cited above), having a short erect stem and many radiating branches from an undivided taproot, the leaves on the flowering shoots are more strictly opposite and the flowers of the normal Zimbabwean form of the species are smaller and probably generally paler in colour.

R.M. Davies 1224 (SRGH), the specimen cited for *Lotononis sp. 4* (*Davies* 1224) of Drummond in Kirkia **8**: 223 (1972), has not been seen, but as an annual from the Southern Province of Zimbabwe is most likely to be this species or the next.

11. **Lotononis stipulosa** Baker f., Legum. Trop. Africa: 18 (1926). —Schreiber in Merxmuller, Prodr. Fl. SW. Afrika, fam. 60: 85 (1970). —Lock, Leg. Afr. Check-list: 224 (1989). —van Wyk in Contrib. Bolus Herb., No. 14: 156, fig. 44 (1991). Type: Zimbabwe, Macheke, *Eyles* 2020 (K, holotype).

Annual, with a short stem, soon developing numerous slender radiating procumbent to prostrate branches up to 10–50 cm long, pilose. Leaves opposite at flowering nodes, otherwise alternate, 3-foliolate, often larger on initial stem and its more distal branches; leaflets 8–25(35) × 1.5–4(6) mm, oblanceolate to narrowly elliptic-oblanceolate, slightly apiculate, subglabrous to pilose above, pilose beneath; petiole shorter to markedly longer than the leaflets; stipules single at a node, 5–10 × 3–10 mm, broadly ovate to oblate, often somewhat acuminate, several-nerved from the base. Flowers several to numerous, produced successively in a cluster at the nodes, sessile, subtended by the conspicuous leaf-like stipules; bracts 3–4 mm long, linear; bracteoles absent. Calyx 5–7 mm long, pilose; upper and lateral lobes joined slightly higher, c. 3 mm long, subulate. Standard yellow, shorter than the keel and scarcely exceeding the calyx, narrowly elliptic-ovate, extensively pubescent outside; wings a little shorter than the standard; keel 6–7 mm long, narrow, slightly upcurved to the obtuse reddish tip, extensively pubescent. Pods 3–6 × 1.5–2 mm, narrowly oblong-obovoid to oblong-elliptic, often somewhat tapered to the base, slightly downcurved at the tip, pubescent distally, c. 8-seeded, but some seeds often aborted. Seeds 1 mm long, oblique-cordiform, smooth, pale brown.

Zambia. S: vicinity of Livingstone, see map in van Wyk, loc. cit., specimen not located. **Zimbabwe**. W: Matobo Distr., Besna Kobila Farm, fl. & fr. iii.1948, *O.B. Miller* 5176 (K; SRGH). C: Chikomba Distr., 1.5 km north of Chivhu (Enkeldoorn), fl. & fr. 20.iv.1969, *Biegel* 2923 (K; SRGH). E: 8 km north of Nyanga (Inyanga) Village, c. 1.5 km NE of main Nyanga North Road, fl. & fr. 28.iv.1967, *Rushworth* 902 (K; LISC; SRGH). S: Masvingo Distr., Makaholi Office, fl. & fr. 30.iv.1969, *Cleghorn* 1975 (K; SRGH).

Also in South Africa (Northern Province and Mpumalanga). Open sandy places on disturbed ground, often along roads and in fallow fields; 900–1750 m.

Easily recognized by the enlarged stipules, which conspicuously subtend the flower clusters. A closely related species in Namibia, formerly confused with *Lotononis stipulosa*, has been segregated recently as *L. bracteosa* B.-E. van Wyk. In that species the enlarged subtending structures are bracts, not stipules.

12. **Lotononis brachyantha** Harms in Repert. Spec. Nov. Regni Veg. **16**: 360 (1920). —E.G. Baker, Legum. Trop. Africa: 17 (1926). —Schreiber in Merxmüller, Prodr. Fl. SW. Afrika, fam. 60: 81 (1970). —Lock, Leg. Afr. Check-list: 216 (1989). —van Wyk in Contrib. Bolus Herb., No. 14: 184, fig. 53 (1991). Type from Namibia.

Annual or short-lived perennial, with a short stem, soon developing numerous slender radiating procumbent to prostrate branches from the top of an undivided taproot; branches up to 10–30 cm long, densely strigose-pubescent with silvery appressed slightly biramous hairs. Leaves 3-foliolate; leaflets 4–20 × 1.5–8 mm, oblanceolate to narrowly elliptic-oblanceolate, ± bluntly pointed, densely strigose-pubescent on both surfaces; petiole mostly shorter than the leaflets, but long on initial stem and luxuriant growth; stipules single at a node, 2–8 mm long, narrowly and obliquely linear-elliptic. Flowers 1(2), leaf-opposed at the nodes, with a short peduncle when paired; bracts 2–3 mm long, linear, slightly exceeding the pedicel; bracteoles absent. Calyx 5–6 mm long, densely strigose-pubescent; upper and lateral lobes joined at ± the same level, c. 3 mm long, triangular-subulate. Standard yellow, shorter than the keel and scarcely exceeding the calyx, trullate above a well developed basally dilated claw, medially pubescent outside; wings a little shorter than the keel; keel c. 6 mm long, narrow, with a short upwardly directed purplish beak, glabrous. Pods c. 5 × 2 mm, almost oblong, slightly widened medially, somewhat turgid, with a distinct rib along the upper suture, finely pubescent, c. 6-seeded. Seeds 1 mm long, oblique-cordiform, minutely roughened, brown.

Botswana. N: Ngamiland Distr., Khardoum Valley, 45 km east of Namibia border, fl. & fr. 15.iii.1965, *Wild & Drummond* 7061 (K; LISC; SRGH). SW: Kgalagadi Distr., Mahuditlhake (Mahudutlake) Pan, c. 335 km west of Kanye, 22–27.v.1967, *Cox* 385 (K). SE: mapped from Sefophe area by van Wyk, loc. cit., but specimen not seen. **Zimbabwe**. S: mapped from Gwanda area by van Wyk, loc. cit., but specimen not seen.

Also in Namibia. Pans and valleys in short grassland; c. 900–1000 m.

Closely related to the following species.

13. **Lotononis crumanina** Burch. ex Benth. in Hooker, London J. Bot. **2**: 612 (1843). —Harvey in F.C. **2**: 62 (1862). —Dummer in Trans. Roy. Soc. South Africa **3**: 313 (1913). —Burtt Davy, Fl. Pl. Ferns Transvaal, pt. 2: 388 (1932). —Schreiber in Merxmüller, Prodr. Fl. SW. Afrika, fam. 60: 82 (1970). —Lock, Leg. Afr. Check-list: 217 (1989). —van Wyk in Contrib. Bolus Herb., No. 14: 184, fig. 53 (1991). Type from South Africa (Northern Cape).

Stems numerous from a slender branched perennial rootstock, prostrate to weakly ascending, up to 10–35 cm long, densely strigose-pubescent with slightly biramous hairs, the shorter ones closely appressed, the longer ones slightly spreading. Leaves 3-foliolate; leaflets 8–20 × 2–5 mm, oblanceolate to narrowly elliptic-oblanceolate, ± bluntly pointed, densely silky strigose on both surfaces with mostly longish straight slender silvery hairs; petiole shorter to longer than leaflets, longer on young and ascending shoots; stipules single at a node, 2–6 mm long, narrowly linear-elliptic. Flowers mostly 3–7 in nearly sessile leaf-opposed clusters; bracts 3–4 mm long, linear, exceeding the negligible pedicel; bracteoles absent. Calyx 4–5.5 mm long, densely hairy; upper and lateral lobes joined to ± the same level, c. 2.5–3.5 mm long, triangular-subulate. Standard yellow, ± as long as the calyx, ovate-trullate above a well developed basally dilated claw, pubescent along the midvein outside; wings a little shorter than the keel; keel 4–5 mm long, falcate into a distinct tapering beak, glabrous. Pods 3–4 × 2–2.5 mm, almost half-elliptic to shortly oblong-ellipsoid, pubescent, 1–2-seeded. Seeds 1.5 mm long, oblique-cordiform, minutely roughened, brown or mottled.

Botswana. SE: Lobatse on the way to Kanye, fl. & fr. ix.1967, *Lambrecht* 342 (K; SRGH).

Also in Namibia and the central parts of South Africa (North-West Province, Northern Cape Province and the Free State). Habitat not recorded for Botswana, but generally in short grassland in pans and valleys on ± sandy calcareous soils; c. 900–1000 m.

14. **Lotononis laxa** Eckl. & Zeyh., Enum. Pl. Afr.: 177 (Jan. 1836). —Bentham in Hooker, London J. Bot. **2**: 612 (1843). —Harvey in F.C. **2**: 63 (1862). —Dummer in Trans. Roy. Soc. South Africa **3**: 314 (1913). —Burtt Davy, Fl. Pl. Ferns Transvaal, pt. 2: 387 (1932). —Brenan in Mem. New York Bot. Gard. **8**: 245 (1953). —Milne-Redhead in F.T.E.A., Leguminosae, Pap.: 816, fig. 118/1–10 (1971). —Drummond in Kirkia **8**: 223 (1972). —Lock, Leg. Afr. Check-list: 219 (1989). —van Wyk in Contrib. Bolus Herb., No. 14: 188, fig. 54 (1991). TAB. 3,7: **53**, fig. B. Lectotype, selected by van Wyk, loc. cit., from South Africa (Eastern Cape).

Crotalaria diversifolia E. Mey., Comment. Pl. Afr. Austr.: 27 (Feb. or later in 1836). Lectotype, selected by van Wyk, loc. cit., from South Africa (Eastern Cape).

Lotononis diversifolia (E. Mey.) Benth. in Hooker, London J. Bot. **2**: 611 (1843).

Lotononis woodii Bolus in J. Bot. **34**: 19 (1896). —Dummer in Trans. Roy. Soc. South Africa **3**: 314 (1913). —Burtt Davy, Fl. Pl. Ferns Transvaal, pt. 2: 388 (1932). Lectotype, selected by van Wyk, loc. cit., from South Africa (KwaZulu-Natal).

Lotononis laxa var. *multiflora* Dummer in Trans. Roy. Soc. South Africa **3**: 315 (1913). —Burtt Davy, Fl. Pl. Ferns Transvaal, pt. 2: 388 (1932). —Brenan in Mem. New York Bot. Gard. **8**: 245 (1953). Lectotype, selected by van Wyk, loc. cit., from South Africa (Mpumalanga).

Stems numerous from a stout branched perennial rootstock, decumbent to shortly ascending, resprouting after burning, up to 10–50 cm long, densely silvery strigulose or strigulose-pubescent with slightly biramous hairs. Leaves mostly 3-foliolate, lower ones sometimes 1-foliolate; leaflets 10–25 × 2–5 mm, mostly linear-elliptic to oblanceolate, obtuse or rounded to a mucronate tip, strigulose-pubescent on both surfaces; petiole shorter than the leaflets; stipules single at a node, resembling small leaflets. Flowers leaf-opposed, solitary or subumbellately 2–5; peduncle 0–16 mm long; bracts 2–3 mm long, linear to linear-oblanceolate; pedicel 3–5 mm long; bracteoles absent. Calyx 5–6 mm long, densely strigulose-pubescent; upper and lateral lobes joined at nearly the same level, narrowly triangular-lanceolate, c. 3 mm long. Standard yellow, fading reddish, ± as long as the other petals, narrowly ovate-elliptic above a well developed basally dilated claw, pubescent along the midvein outside; keel 6–7 mm long, oblong-falcate, with a small pointed beak, glabrous. Pods (12)15–20 × 4–4.5 mm, linear-oblong, slightly tapered towards the apex, strigose-pubescent, 10–20-seeded. Seeds c. 2 mm long, rounded, slightly oblique-cordiform, matt, pale brown or mottled.

Zimbabwe. C: Harare, Greystone Park, fl. & fr. 14.xi.1964, *West* 6138 (K; SRGH). E: Nyanga (Inyanga), fl. & fr. 11.v.1930, *T.C.E. Fries, Norlindh & Weimarck* 2554 (K). **Malawi**. C: Ntchisi Distr., Chenga Hill, fl. 9.ix.1946, *Brass* 17589 (K); Dedza Distr., Chongoni Forest, fl. & fr. 17.iii.1969, *Salubeni* 1274 (K; MAL; SRGH).

Extends from Ethiopia to South Africa (Eastern Cape Province), mostly along the Great Rift Valley in tropical Africa and tending to occur in small disjunct populations, but perhaps overlooked to some extent. Grassland, often most apparent after burning; 1500–1700 m.

van Wyk, loc. cit., suggests that plants with more developed inflorescences, described as var. *multiflora*, are produced during active regrowth after fires.

15. **Lotononis serpentinicola** Wild in Kirkia **5**: 75 (1965). —Drummond in Kirkia **8**: 223 (1972). —Lock, Leg. Afr. Check-list: 224 (1989). —van Wyk in Contrib. Bolus Herb., No. 14: 189, fig. 54 (1991). Type: Zimbabwe, Great Dyke, Nyarasuswe, *Wild* 5743 (K; M; MO; PRE; SRGH, holotype).

Annual or short-lived perennial, with a short stem, soon developing numerous slender radiating prostrate to ascending branches from the top of an undivided taproot; branches up to 10–50 cm long, densely silvery strigose-pubescent with slightly biramous hairs. Leaves 3-foliolate; leaflets 8–25 × 2.5–8 mm, mostly narrowly elliptic-oblanceolate, lower ones more obovate, pointed to rounded to a slightly mucronate tip, strigose-pubescent on both surfaces; petioles shorter than the leaflets; stipules single at a node, resembling a small leaflet. Flowers 1–2, leaf-opposed, with a short rhachis when paired; peduncle 8–20 mm long, filiform; bracts 1–2 mm long, linear; pedicel 3–3.5 mm long; bracteoles absent. Calyx 4–5 mm long, densely strigulose-pubescent; upper and lateral lobes joined to ± same level, 2.4–3 mm long, narrowly and acuminately triangular. Standard bright yellow inside, paler and lined red outside, as long as the keel, elliptic-ovate above a well developed basally dilated claw, pubescent along the midvein outside; wings a little shorter than the keel; keel 5–7 mm long, oblong-falcate with a small purplish beak, glabrous. Pods 10–12 × 3–3.5 mm, narrowly oblong, a little tapered to the apex, strigulose-pubescent, 16–20-seeded. Seeds c. 2 mm across, rounded, only slightly oblique-cordiform, matt, brown.

Zimbabwe. N: Mazowe Distr., Birkdale Pass, 35 km north of Kildonan, fl. & fr. 10.ii.1982, *Brummitt & Drummond* 15836 (K; SRGH). C: 24 km south of Shurugwi on Great Dyke, fl. 16.iii.1964, *Wild* 6364 (K; LISC; SRGH). S: Masvingo Distr., Mashava (Mashaba), on serpentine hill quarry, fl. & fr. 19.ii.1972, *Wild* 7887 (K; LISC; SRGH).

Known only from Zimbabwe, along the Great Dyke and in the Mashava area further south. Grassland and disturbed places by roads and quarries on serpentine or norite soils; 1000–1500 m.

105. LEBECKIA Thunb.

By R.M. Polhill

Lebeckia Thunb., Nov. Gen. Pl.: 139 (1800).

Shrubs or subshrubs. Leaves 1–3-foliolate with a distinct petiole, or apparently simple, pulvinate; leaflets needle-like to flat and broad, sometimes lacking; stipules rarely present, linear. Inflorescences terminal sometimes one-sided racemes, or flowers 1–few on short shoots; bracts and bracteoles small. Calyx with subequal often small lobes. Standard yellow, elliptic to oblate, sometimes with slight thickenings at the top of the claw, glabrous or hairy; keel often a little longer than the standard, curved, apically rounded to shortly and bluntly beaked. Stamens in a sheath open on the upper side; anthers alternately basifixed and shorter, dorsifixed (the carinal one somewhat intermediate). Ovary subsessile to stipitate, with fairly numerous ovules; style curved upwards, slender above the often somewhat thickened lower part; stigma terminal, small. Pod membranous to coriaceous, subsessile to long-stipitate, flattened or less often inflated, sometimes impressed between the seeds, linear to oblong-ellipsoid, sometimes narrowly winged on the upper margin, 3- or more seeded, dehiscent or not. Seeds oblong-reniform to oblique-cordiform, with a small hilum.

A genus of about 35 species in Namibia, Botswana and South Africa (Cape Provinces).

Lebeckia linearifolia E. Mey., Comment. Pl. Afr. Austr.: 33 (1836). —Harvey in F.C. **2**: 86 (1862). —Leistner in Koedoe **2**: 139 & 163 (1959). —Schreiber in Merxmüller, Prodr. Fl. SW. Afrika, fam. 60: 67 (1970). —Lock, Leg. Afr. Check-list: 212 (1989). Type from South Africa (Northern Cape Province).

Shrub 1–2 m tall, virgate, sometimes becoming rounded and rather spreading; whole plant densely covered in very short, appressed to slightly spreading silvery-grey hairs. Leaves simple, sessile (apart from pulvinus), estipulate; leaflet 10–35 × 1–2.5 mm, linear to linear-oblanceolate, tapered to the apex, ± involute. Racemes 2–10 cm long, terminal on mostly relatively short lateral branches, lax, 4–12-flowered; bracts 1–2 mm long, linear-subulate, evanescent; pedicel 1–2 mm long; bracteoles on the pedicel, slightly smaller than the bract. Hypanthium 2–3 mm long, obconic; calyx 4–5 mm long; lobes 1–1.5 mm long, triangular-deltoid, all similar. Standard yellow lined purplish, pubescent outside; wings shorter than the keel; keel 10–12 mm long, oblong-falcate, shortly hairy. Pod 20–30 × 3 mm, linear-oblong, pointed, continuous over the seeds (slightly turgid), 6–8-seeded. Seeds 3 × 1.5 mm, oblong-reniform.

Botswana. SW: Kgalagadi Distr., Kalahari Gemsbok National Park, Twee Rivieren, 16.v.1956, *Story* 5481 (PRE, not seen).
Also in Namibia and South Africa (Northern Cape Province). Calcareous sands in dry river beds and pans.
Leistner in Koedoe **2**: 139–140 (1959) records this species as forming dense stands northwards from Twee Rivieren for c. 6.5 km in the dry river bed of the Nossob on the border of South Africa and Botswana and c. 13 km along the Auob river bed in South Africa, as well as in a pan further west.

Tribe 19. **GENISTEAE***

Genisteae (Bronn) Dumort, Fl. Belg.: 98 (1827). —Polhill in Bot. Syst. **1**: 328 (1976). —Bisby in Adv. Leg. Syst. **1**: 409 (1981).
Sect. *Genisteae* Adans., Fam. Pl. **2**: 320 (1763) *nom. invalid.*
Celeb *Cytiseae* Bercht. & J. Presl, Přir. Rostlin: 229 (1820) *nom invalid.*
Subtribe *Genisteae* Bronn, Pl. Leg.: tab. anal. and p. 132 (1822).
Cytiseae Hutch., Gen. Fl. Pl. **1**: 350 (1964).
Laburneae Hutch., Gen. Fl. Pl. **1**: 355 (1964).
Lupineae Hutch., Gen. Fl. Pl. **1**: 363 (1964).

Shrubs, herbs or small trees. Leaves alternate, opposite or crowded, usually digitately 3(11)-foliolate, sometimes simple or reduced to scale-like or spinous phyllodes; leaflets entire; stipules free or ± joined with the petiole or lacking; stipels absent. Inflorescences terminal or leaf-opposed, rarely axillary. Flowers usually yellow (variegated in *Lupinus*). Calyx usually 2-lipped, occasionally spathaceous. Standard generally without evident auricles or appendages; wings usually sculptured; keel

---

**Argyrolobium* and *Adenocarpus* by E.S. Martins. *Dichilus*, *Melolobium* and *Lupinus* by R.M. Polhill.

oblong-falcate, sometimes beaked or narrowly oblong. Stamens usually all joined into a closed tube, sometimes (in *Argyrolobium, Dichilus* and *Melolobium*) split above; anthers dimorphic, alternately long, basifixed and short, dorsifixed. Style glabrous in African genera, sometimes hairy elsewhere. Pods usually linear or oblong, dehiscent, rarely ellipsoid to globular and indehiscent. Seeds rounded to oblong-ellipsoid, the small hilum on the short side (sometimes arillate outside the Flora Zambesiaca area).

A tribe of 25 genera, mostly in the Mediterranean region, extending into northern Europe and western Asia, with a small secondary centre in southern Africa, represented thinly through the tropical African highlands, Madagascar and the Canary Islands, with *Lupinus* widespread also in the Americas.

Several species are recorded infrequently as ornamentals.

Genista aetnensis (Rafin. ex Biv.) DC., native of southern Europe, has been grown in Harare, e.g. 22.xi.1973, *Biegel* 4382 (K; SRGH). It is a virgate shrub or small tree, with pendent striate branches, scale leaves and yellow flowers with the narrowly oblong keel longer than the standard and silky pubescent; pods shortly oblong, dehiscent.

Laburnum anagyroides Medik., native of SE Europe, is mentioned by Biegel, Check List Ornam. Pl. Rhod. Parks & Gard.: 68 (1977). Small tree with pendent racemes of yellow flowers and leathery, pointed, elastically dehiscent pods, compressed between the few seeds.

Retama monosperma (L.) Boiss., native of the Mediterranean region, has been grown in Jardim Tunduru in Maputo, e.g. Jardim Vasco de Gama, fl. 6.vii.1973, *Balsinhas* 2564 (K). Virgate shrub with drooping branches and fugacious simple leaves; flowers yellow and similar to *Genista aetnensis* but the petals are subequal in length and both the wings and slightly more falcate keel are pubescent; fruit drupe-like, indehiscent, c. 1 cm long.

Spartium junceum L., Spanish Broom, native of southern Europe, is mentioned by Biegel, loc. cit.: 99 (1977). It is a robust virgate shrub with sparse simple leaves, lax racemes of large golden-yellow flowers, with a spathaceous calyx and beaked keel; pods pointing upwards, linear, beaked, with numerous seeds.

Ulex minor Roth and *U. gallii* Planch., both from western Europe, are also mentioned by Biegel, loc. cit.: 106 (1977). *U. minor* is a spiny shrublet to 1 m, with clear yellow petals exserted less than 2 mm from the tubular calyx, which is usually less than 1 cm long. *U. gallii* is similar but the petals are deep golden-yellow and exceeding the calyx by 2.5–3.5 mm; calyx at least 1 cm long.

1. Leaves simple or reduced to spine-tipped phyllodes; cultivated ornamentals · · · · · · · · · 2
– Leaves 3–11-foliolate · 5
2. Spiny shrublets; calyx tubular, nearly as long as the corolla · · · · · · · · · **Ulex** (cultivated)
– Virgate shrubs or small trees with striate branches, unarmed in Flora Zambesiaca area; calyx much shorter than the corolla · 3
3. Calyx spathaceous, deeply split on upper side, the teeth very short; corolla 2–2.5 cm long; fruit a linear dehiscent pod · **Spartium** (cultivated)
– Calyx shortly 2-lipped; corolla c. 1 cm long; fruits as next couplet · · · · · · · · · · · · · · · · 4
4. Fruit a shortly oblong, dehiscent pod; keel-petals longer than the standard · **Genista** (cultivated)
– Fruit drupe-like, indehiscent; keel-petals ± as long as the standard · · **Retama** (cultivated)
5. Leaflets 5–11; keel beaked · 110. **Lupinus**
– Leaflets 3; keel obtuse · 6
6. Small cultivated tree, with pendent racemes of medium-sized flowers (standard 1.5–2 cm long) and leathery dehiscent pods · **Laburnum** (cultivated)
– Herbs or small shrubs; native · 7
7. Stipules semi-sagittate or semi-cordate at the base, with a narrow point of attachment; glandular tubercles often present · 107. **Melolobium**
– Stipules, if present, not lobed; glandular tubercles absent except in *Adenocarpus* · · · · · · 8
8. Stipules inconspicuous or absent, less than 1 mm long if present; leaves well spaced; flowers with wings much shorter than the keel and distinctly spurred · · · · · · · · · · · 106. **Dichilus**
– Stipules conspicuous, usually with a wide point of attachment (often circumcauline) or if inconspicuous (in *Adenocarpus*) then leaves mostly crowded on short shoots; wings as long as or longer than the keel, auriculate but not spurred · 9
9. Pods without glandular tubercles; leaves usually not crowded on short shoots; lip formed by upper calyx lobes more than twice as long as the tube, as long as the lower lip · 108. **Argyrolobium**
– Pods with conspicuous glandular tubercles; leaves mostly crowded on short shoots; lip formed by upper calyx lobes ± as long as the tube, shorter than the lower lip · 109. **Adenocarpus**

106. DICHILUS DC.

By R.M. Polhill

Dichilus DC., Prodr. **2**: 136 (1825). —Schutte & van Wyk in S. Afr. J. Bot. **54**: 182 (1988); & **56**: 244 (1990).

Small slender-stemmed shrubs. Leaves digitately 3-foliolate, petiolate; stipules inconspicuous or absent, up to 1 mm long and caducous if present. Flowers small, in short terminal or leaf-opposed racemes; bracts and bracteoles small. Calyx shorter than the corolla, 2-lipped, the upper lip bifid, the lower trifid. Standard pale yellow, oblong to subcircular, sharply reflexed, auriculate; claw well developed, with calluses at the apex; wings ± two-thirds as long as the keel, with distinct elongate auricles; keel slightly shorter than the standard, rounded at the apex. Stamens all joined in a sheath open on the upper side, with longer basifixed anthers alternating with shorter dorsifixed anthers, the carinal anther intermediate. Style tapered, curved up to a small capitate terminal stigma. Pods pointing upwards, narrowly oblong to linear, laterally compressed between the seeds, dehiscent, 1–8-seeded. Seeds rounded to very shortly oblong, with a small hilar sinus on the short side.

A genus of 5 species in southern Africa.

Racemes terminal on lateral branches, rarely leaf-opposed, laxly 2–4(11)-flowered; pods 4.5–5.5 mm broad · 1. *reflexus*
Racemes leaf-opposed on lateral branches, 1–2(4)-flowered; pods 3–4 mm broad · 2. *lebeckioides*

1. **Dichilus reflexus** (N.E. Br.) A.L. Schutte in S. Afr. J. Bot. **54**: 184 (1988). TAB. 3,7: **54**. Type from South Africa (KwaZulu-Natal).
Argyrolobium reflexum N.E. Br. in Bull. Misc. Inform., Kew **1906**: 18 (1906).
Dichilus lebeckioides sensu Eyles in Trans. Roy. Soc. South Africa **5**: 370 (1916). —sensu Drummond in Kirkia **8**: 219 (1972) non DC.

Small bushy shrub 0.3–1.3 m tall, the thin stems sometimes forming clumps; branches appressed pubescent. Leaflets 5–18 × 2–7 mm, oblanceolate to obovate-oblanceolate, rounded to slightly emarginate at the apex, minutely apiculate, appressed pubescent beneath; petiole shorter than the leaflets; stipules usually absent, if present then only at insertion of leaf-opposed racemes. Racemes terminal on lateral branches, rarely leaf-opposed, 1–4 cm long, laxly 2–4(11)-flowered; bracts 2–3 mm long, linear ± as long as the pedicel; bracteoles on the pedicel, a little smaller. Calyx 4.5–7 mm long, appressed pubescent; lips longer than the tube, shortly lobed. Standard pale yellow, with a few hairs outside medially and apically; wings shorter than the keel; keel 8–10 mm long. Pods 25–35 × 4.5–5.5 mm, narrowly oblong, pubescent only along the sutures, particularly the upper one, (3)4–6-seeded. Seeds 2.5–3 mm across, dark green to almost black.

Zimbabwe. C: Shurugwi (Selukwe), fl. 13.i.1963, *Loveridge* 571 (K; SRGH); Shurugwi, McKinnon Kop, fr. 7.vi.1967, *Biegel* 2188 (K; SRGH). E: Chimanimani Distr., Cashel, fr. 19.iii.1964, *Corby* 1114 (K; SRGH). **Mozambique**. M: Namaacha Distr., Libombos, Mt. M'Ponduine (Mpondium), 800 m, fl. & fr. 22.ii.1955, *Exell, Mendonça & Wild* 499 (LISC).
Also in South Africa (Northern Province, Mpumalanga and KwaZulu-Natal). Disturbed places along streams, on hillsides and by roads; 800–1500 m.
Formerly confused with *D. lebeckioides*.

2. **Dichilus lebeckioides** DC., Prodr. **2**: 136 (1825). —Harvey in F.C. **2**: 77 (1862). —E.G. Baker, Legum. Trop. Africa: 26 (1926). —Schutte & van Wyk in S. Afr. J. Bot. **54**: 185 (1988). —Lock, Leg. Afr. Check-list: 211 (1989). Type from South Africa (Northern Cape Province).

Habit as *D. reflexus*. Leaflets 6–16 × 2–3 mm, narrowly oblanceolate, pointed to shortly rounded and minutely apiculate, appressed pubescent beneath; stipules minute and inconspicuous but generally present. Racemes leaf-opposed on short lateral branches, up to 1 cm long, 1–2(4)-flowered. Flowers similar to *D. reflexus*, but calyx sometimes a little longer, 5–8 mm long. Pods 30–45 × 3–4 mm, pubescent

Tab. 3,7: **54**. DICHILUS REFLEXUS. 1, habit (× 1), from *Pole Evans* 877; 2, flower (× 4); 3, calyx, opened out (× 4); 4, standard (× 4); 5, wing (× 4); 6, keel (× 4); 7, androecium, opened out (× 6); 8, gynoecium (× 4), 2–8 from *Galpin* 13036; 9, fruits (× 1), from *Biegel* 2188; 10, pod (× 2); 11, seed (× 4), 10 & 11 from *Corby* 1114. Drawn by Victoria Friis.

along the sutures or more generally, (3)6–8(9)-seeded. Seeds 2.5–3 mm long, round to oblong in lateral view, brown, often mottled black.

Botswana. N and SE: mapped in Fig. 2 in Schutte & van Wyk in S. Afr. J. Bot. **54**: 185 (1988). **Zimbabwe**. W: Bulawayo, fl. xi.1957, *O.B. Miller* 4620 (K; SRGH).
Also in Namibia and South Africa (Northern and North-West Provinces, Gauteng, Mpumalanga, Free State, Northern Cape).

107. MELOLOBIUM Eckl. & Zeyh.

By R.M. Polhill

Melolobium Eckl. & Zeyh., Enum. Pl. Afr.: 188 (Jan. 1836).
Sphingium E. Mey., Comment. Pl. Afr. Austr.: 65 (Feb. 1836).

Small shrubs or perennial herbs, often sticky with exudate from small surface glands; branches sometimes spine-tipped. Leaves digitately 3-foliolate, sometimes crowded on short shoots; leaflets usually small and linear-oblanceolate; stipules semi-sagittate or semi-cordate with a lateral attachment. Flowers in terminal often spine-tipped racemes, sometimes 1-flowered at the base of the spine; bracts narrow to broad; bracteoles similar and a little smaller. Calyx nearly as long as the corolla, 2-lipped; upper lobes separate or somewhat united; lower lip 3-fid. Standard spathulate to subcircular with a moderately developed claw; keel shortly half oblong-elliptic, apically rounded, the claw more than half as long as the blade. Stamens in a sheath open on the upper side, with longer basifixed anthers alternating with shorter dorsifixed anthers, the carinal anther somewhat intermediate. Style slender, straight or upcurved to a small stigma. Pod subsessile, narrowly oblong to ovate-oblong, often slightly curved, the longer pods impressed between the seeds, dehiscent, 1–8-seeded. Seeds rounded, with a small hilum.

1. Racemes extended, 3–11 cm long, lax, not spine-tipped; bracts cordate · · · · 1. *macrocalyx*
– Racemes abbreviated, mostly less than 3 cm long, with few flowers at the base of the spine-tipped axis (in young luxuriant growth sometimes longer and with more flowers but still with a soft spine at the tip); bracts lanceolate to ovate, rounded at base · · · · · · · · · · · · 2
2. Leaves mostly crowded on short shoots; whole plant villous with hairs c. 1 mm long; wings only slightly exceeding the calyx; pods 7–9 mm long, 2–3-seeded; bracts longer than the calyx tube · 2. *calycinum*
– Leaves not crowded on short shoots; plants variously hairy and glandular, but not villous; wings longer than other petals and well exceeding the calyx; pods 12–15 mm long, 4–6-seeded · 3. *microphyllum*

1. **Melolobium macrocalyx** Dummer in Bull. Misc. Inform., Kew **1912**: 227 (1912). —E.G. Baker, Legum. Trop. Africa: 22 (1926). —Schreiber in Mitt. Bot. Staatssamml. München **2**: 297 (1957); in Merxmüller, Prodr. Fl. SW. Afrika, fam. 60: 89 (1970). —Lock, Leg. Afr. Check-list: 227 (1989). TAB. 3,7: **55**. Lectotype, see Schreiber (1957), from South Africa (Northern Cape Province).

Melolobium macrocalyx var. *longifolium* Dummer in Bull. Misc. Inform., Kew **1912**: 227 (1912) as "*longifolia*". —Leistner in Koedoe **2**: 163 (1959). Type from South Africa (Northern Cape Province).

Melolobium psammophilum Harms in Repert. Spec. Nov. Regni Veg. **11**: 85 (1912). —E.G. Baker, Legum. Trop. Africa: 22 (1926). Type from Namibia.

Melolobium brachycarpum Harms in Repert. Spec. Nov. Regni Veg. **11**: 86 (1912). Type from Namibia.

Melolobium stenophyllum Harms in Repert. Spec. Nov. Regni Veg. **11**: 86 (1912). Syntypes from Namibia.

Melolobium macrocalyx var. *brachycarpum* (Harms) A. Schreib. in Mitt. Bot. Staatssamml. München **2**: 297 (1957).

Melolobium macrocalyx var. *psammophilum* (Harms) A. Schreib. in Mitt. Bot. Staatssamml. München **2**: 297 (1957).

Erect, shrubby somewhat woody plant 0.3–1 m high from a deep taproot, silvery tomentose with fine matted hairs, sometimes with longer wispy hairs interspersed, elsewhere sometimes almost glabrous and conspicuously glutinous. Leaflets 5–12 ×

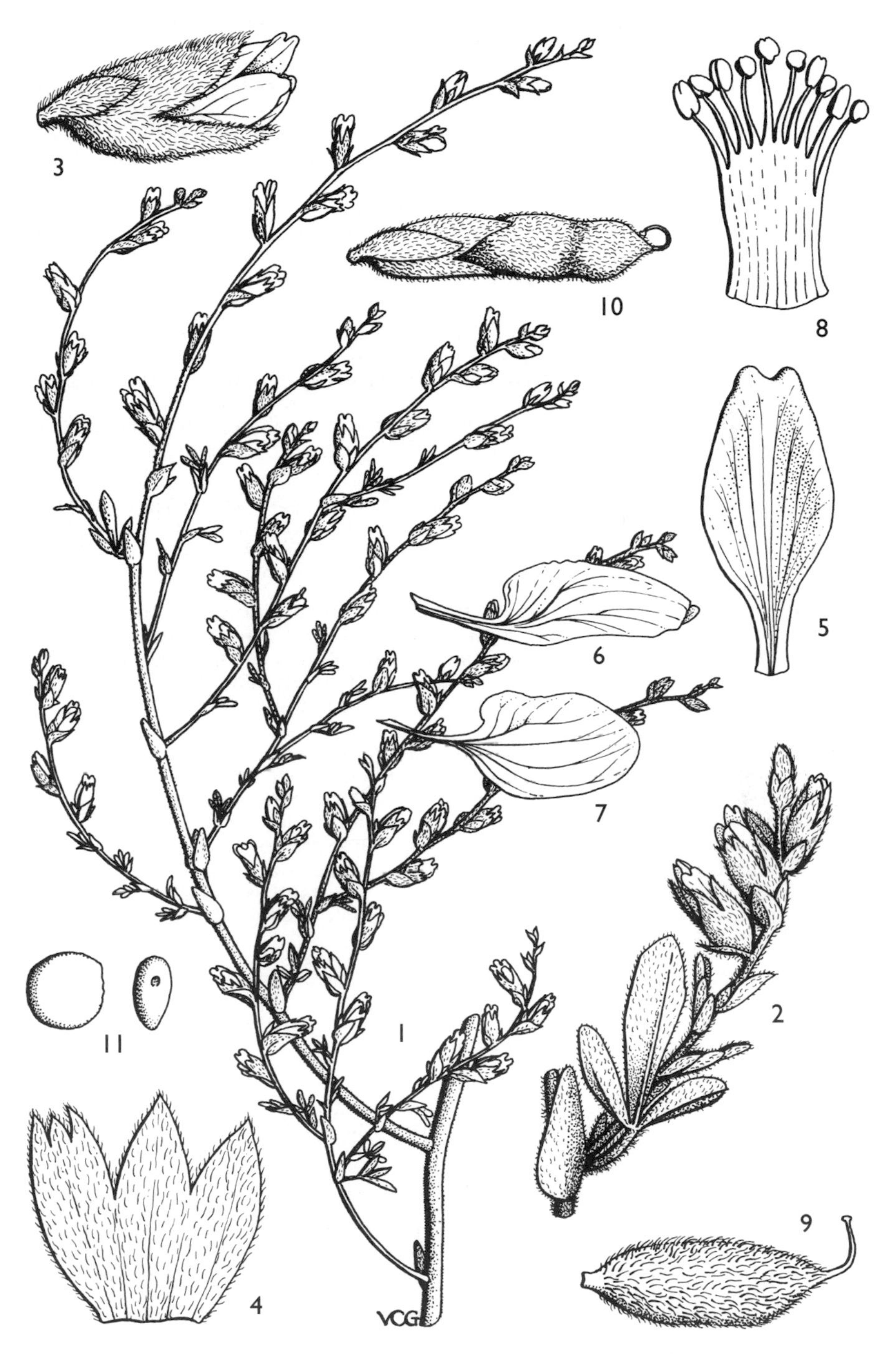

Tab. 3,7: **55**. MELOLOBIUM MACROCALYX. 1, habit (× 1); 2, flowering branch (× 2$^{1}/_{2}$); 3, flower (× 4); 4, calyx, opened out (× 4); 5, standard (× 6); 6, wing (× 6); 7, keel (× 6); 8, androecium, opened out (× 6); 9, gynoecium (× 6), 1–9 from *Story* 4586; 10, pod (× 3$^{1}/_{4}$); 11, seed (× 4); 10 & 11 from *Acocks* 2423. Drawn by Victoria Gordon.

1.5–3 mm, oblanceolate, obtuse to rounded at the apex, densely silvery appressed pubescent to tomentellous, slightly thickened, without evident veins; petiole 1–3 mm long; stipules (2)4–8 × (1)2–4 mm, semi-sagittate, attached laterally, basal part rounded. Racemes terminal and leaf-opposed, 3–11 cm long, lax, 6–16-flowered; bracts 3–7 mm long, cordate; pedicel up to 1 mm long; bracteoles at the top of the pedicel, 3–4 mm long, lanceolate-elliptic. Calyx 6–7 mm long, tomentellous, divided halfway into lips, the upper lobes broadly triangular and ± as long as the joined part, lower lip trifid. Standard yellow, slightly exceeding the calyx, ovate-elliptic above a rather short claw, emarginate, glabrous; wings and keel subequal in length and slightly shorter than the standard; wings without sculpturing; keel obtuse. Pod 8–12 × 3–3.5 mm, shortly oblong, pointed, flattened, compressed between the 2–3 seeds. Seeds 2 mm across, rounded to slightly oblong-ellipsoid, pale brown mottled darker.

Botswana. SW: Ghanzi Distr., 4 km north of Dondong Borehole, fl. 21.vii.1977, *Skarpe* S–185 (K); Kgalagadi Distr., Mahuditlhake (Mahudutlake) Pan, fl. bud, 22–27.v.1967, *Cox* 370 (K). SE: 95 km NW of Molepolole, fl. 14.vi.1955, *Story* 4884 (K; PRE).

Also in Namibia and South Africa (Free State, Northern Cape Province). Mixed woodland and bushland on Kalahari Sands.

Several varieties have been described, but none seem worth retaining with the material now available.

2. **Melolobium calycinum** Benth. in Hooker, London J. Bot. **3**: 350 (1844). —Harvey in F.C. **2**: 78 (1862). —Lock, Leg. Afr. Check-list: 226 (1989). Type from South Africa (Free State).

Melolobium villosum Harms in Repert. Spec. Nov. Regni Veg. **11**: 87 (1912). —Schreiber in Merxmüller, Prodr. Fl. SW. Afrika, fam. 60: 90 (1970). Type from Namibia.

Shrublet, much branched, spiny, 20–60 cm tall, from a deep taproot; branches densely villous with fine spreading hairs c. 1 mm long. Leaves mostly crowded on short shoots; leaflets 2–7 × 1–1.5 mm, oblanceolate, obtuse, often folded lengthwise, villous; stipules 2–5 × 1–2.5 mm, semi-sagittate, attached laterally, basal part rounded. Flowers 2–3 on the lower part of a stiff, straight, spine-tipped, glabrescent axis 1–2 cm long, which is leaf-opposed or terminal on the short shoots, sometimes with several smaller sterile bracts above the flowers; bracts 3–4 mm long, lanceolate to ovate, rounded to subcordate at the base, acuminate, upper sterile ones 0.5–1.5 mm long, lanceolate; pedicel up to 1 mm long; bracteoles at top of the pedicel, 3–5 mm long, lanceolate. Calyx 7–8 mm long, densely villous, divided more than halfway into 2 lips, the upper lip with triangular lobes longer than the joined part, the lower lip trifid. Standard yellow, probably sometimes flushed red, ± as long as the calyx, ovate above a fairly short claw; wings nearly as long as the standard; keel a little shorter, rounded at the apex. Pods 7–9 × 3–3.5 mm, shortly oblong, densely villous, 2–3-seeded. Seeds 2.5–3 mm across, rounded, brown.

Botswana. SW: Kgalagadi Distr., Mahuditlhake (Mahudutlake) Pan, fl. bud, 22–27.v.1967, *Cox* 371 (K).

Also in Namibia and the drier parts of South Africa. Growing with *M. macrocalyx* at Mahudutlake, presumably on Kalahari Sands.

3. **Melolobium microphyllum** (L.f.) Eckl. & Zeyh., Enum. Pl. Afr.: 189 (1836). —Bentham in Hooker, London J. Bot. **3**: 351 (1844). —Harvey in F.C. **2**: 79 (1862). —Schreiber in Merxmüller, Prodr. Fl. SW. Afrika, fam. 60: 89 (1970). —Lock, Leg. Afr. Check-list: 227 (1989). Type from South Africa (Eastern Cape Province).

Ononis microphylla L.f., Suppl. Pl.: 324 (1781). —Thunberg in Fl. Cap., ed. 2, **2**: 585 (1823).

Dichilus candicans E. Mey. in Linnaea **7**: 154 (1832). Type from South Africa (Eastern Cape Province).

Sphingium lamprolobium E. Mey., Comment. Pl. Afr. Austr.: 66 (1836). Type from South Africa (Western Cape).

Sphingium canescens E. Mey., Comment. Pl. Afr. Austr.: 67 (1836). Type from South Africa (Free State).

Sphingium decumbens E. Mey., Comment. Pl. Afr. Austr.: 67 (1836). —Burtt Davy, Fl. Pl. Ferns Transvaal, pt. 2: 390 (1932). Syntypes from South Africa (Eastern Cape Province).

Melolobium candicans (E. Mey.) Eckl. & Zeyh., Enum. Pl. Afr.: 189 (1836). —Bentham in Hooker, London J. Bot. **3**: 351 (1844). —Harvey in F.C. **2**: 78 (1862). —E.G. Baker, Legum. Trop. Africa: 22 (1926). —Burtt Davy, Fl. Pl. Ferns Transvaal, pt. 2: 390 (1932). —Schreiber

in Merxmüller, Prodr. Fl. SW. Afrika, fam. 60: 88 (1970).
Melolobium collinum Eckl. & Zeyh., Enum. Pl. Afr.: 189 (1836). —Bentham in Hooker, London J. Bot. **3**: 351 (1844). Type from South Africa (Western Cape Province).
Melolobium canescens (E. Mey.) Benth. in Hooker, London J. Bot. **3**: 351 (1844). —Harvey in F.C. **2**: 79 (1862).
Melolobium parviflorum Benth. in Hooker, London J. Bot. **3**: 351 (1844). Type from South Africa (Western Cape Province).
Melolobium decumbens (E. Mey.) Benth. in Hooker, London J. Bot. **3**: 352 (1844). —Burtt Davy, Fl. Pl. Ferns Transvaal, pt. 2: 390 (1932). —Schreiber in Merxmüller, Prodr. Fl. SW. Afrika, fam. 60: 88 (1970).
Melolobium microphyllum var. *decumbens* (E. Mey.) Harv. in F.C. **2**: 79 (1862). —E.G. Baker, Legum. Trop. Africa: 22 (1926).
Melolobium accedens Burtt Davy, Fl. Pl. Ferns Transvaal, pt. 2: 390 (1932). Type from South Africa (Mpumalanga).

Shrublet, much branched, spiny and sticky to varying degrees, 20–50 cm tall, from a deep taproot; branchlets pubescent to silvery-white tomentellous, with glands apparent beneath at least on older parts. Leaflets 2–5(10) × 1–3(4) mm, oblanceolate to oblanceolate-obtriangular or somewhat obovate, rounded to emarginate at the apex, often folded lengthwise, usually sparsely pubescent and conspicuously covered with glands which may be sessile, clavate or stalked and capitate; stipules 1.5–3 mm long, semi-sagittate, attached laterally, basal part rounded. Flowers usually 2–4 on the lower part of a stiff, straight, terminal or leaf-opposed, spine-tipped axis 1–2(3) cm long, sometimes with several smaller sterile bracts above the flowers, sometimes in young plants with longer axes to 6 cm long and with up to 8 flowers and the spine apparent but not conspicuous or hardened; bracts 2–3 mm long, lanceolate or ovate, rounded at the base, acuminate, upper sterile ones smaller; pedicel 1–2 mm long; bracteoles on the pedicel 1.5–2 mm long, linear-elliptic. Calyx 5–6 mm long, sparsely to conspicuously pubescent and usually conspicuously covered with sessile or stalked glands, divided ± halfway into 2 lips, the upper lip with triangular-lanceolate lobes largely separate, the lower lip trifid. Standard bright yellow, turning orange or brownish, longer than the calyx, spathulate with a claw about a third as long as the blade; wings at least as long as the standard; keel shorter, obtuse at the apex, 5–6 mm long. Pods 12–15 × 3–3.5 mm, oblong, usually slightly curved, hairy to varying degrees, but hairs longest on sutures especially the placental one, and often conspicuously glandular, 4–6-seeded. Seeds mottled, 2 mm across, nearly straight along the radicular side, rounded on the opposite side, with a very small hilum on the shorter side.

Botswana. N: Boteti (Botletle) R. at Toromoja, fl. & fr. 27.iv.1975, *Ngoni* 456 (K; SRGH) — not typical. SW: Ghanzi Village, fl. & fr. 26.i.1977, *Skarpe* S-117 (K); Kgalagadi Distr., Kang, fl. & fr. 19.vii.1976, *Mott* 990 (K; SRGH). SE: Kweneng Distr., c. 5 km from turn-off towards Ngware from the Molepolole–Letlhakeng road, fl. & fr. 9.viii.1978, *Hansen* 3427 (GAB; K; PRE; SRGH).

Also widespread in Namibia and over much of the drier parts of South Africa from the west to the Eastern Cape Province. Savanna and grassland on Kalahari Sands, persisting in disturbed and grazed places except where trampling is heaviest; 900–1050 m.

The two gatherings seen from northern Botswana, *Ngoni* 456, cited above, and *Richards* 14648 (K), from Makarikari Pan, look significantly different, with luxuriant foliage (leaflets up to 10 mm long) and extended racemes up to 6 cm long and up to 8-flowered, but the axes are softly spine-tipped. The flowers and fruits exactly match other material of this species and they probably represent no more than exceptional growth forms in temporarily moist sites.

108. ARGYROLOBIUM Eckl. & Zeyh.

By E.S. Martins

Argyrolobium Eckl. & Zeyh., Enum. Pl. Afr.: 184 (1836) *nom. conserv.* —Polhill in Kew Bull. **22**: 145–168 (1968).

Herbs, undershrubs or small shrubs, often with a soft indumentum. Leaves, digitately 3-foliolate; stipules small to large, free or sometimes united to each other or to the petiole. Flowers in terminal or leaf-opposed racemes, the inflorescence sometimes contracted and subumbelliform, flowers rarely solitary; cleistogamous flowers with reduced parts and producing small pods often present. Calyx nearly as long as the corolla, deeply divided, the 2 upper lobes free or shortly united into a upper lip, the 3 lower ones united into a 3-toothed lower lip. Corolla yellow; standard obovate to subcircular, shortly clawed, not appendaged; wings often with rows of folds between the veins; keel slightly incurved, obtuse. Stamens usually united into a closed tube, rarely split above, with longer basifixed anthers alternating with shorter dorsifixed anthers. Ovary sessile or subsessile, many-ovulate, tapering upward to the upcurved style; stigma small, terminal. Pod linear-oblong, laterally compressed, sometimes constricted between the seeds, dehiscent. Seeds usually oblong-ovate, compressed with a small sunken hilum.

A genus of about 86 species mostly in South Africa, 15 distributed over the highlands of tropical Africa, 3 in Madagascar and c. 12 in North Africa, southern Europe and from Turkey and the Arabian Peninsula to India.

1. Leaflets of the middle leaves narrowly elliptic to linear · 2
– Leaflets of the middle leaves ovate-lanceolate to broadly obovate · · · · · · · · · · · · · · · · · 3
2. Leaves with petiole 4–12 mm long; leaflets of the upper leaves linear-lanceolate to linear, up to 40 × 2.3 mm; inflorescence pedunculate, 2–9-flowered · · · · · · · · · · · · ·1. *tuberosum*
– Leaves sessile or with petiole up to 1.5 mm long; leaflets of the upper leaves linear-oblanceolate, up to 27 × 3 mm; inflorescence subsessile, 1–2-flowered · · · · 2. *harveianum*
3. Wing petals with rows of crescent-shaped folds between the upper veins towards the base; keel petals usually with a crest on the lateral surface towards the base · · · · · · · · · · · · · 4
– Wing petals without rows of crescent-shaped folds between the veins; keel petals without a crest on the lateral surface · 9
4. Stipules united at first to above the middle along the leaf-opposed margin, soon becoming free, sometimes only the stipules of the apices of the branches united; peduncle 8–16(27) cm long (TAB. 3,7: **55**, fig. B) · 7. *stolzii*
– Stipules free or united only at the very base; peduncle up to 8(12) cm long · · · · · · · · · 5
5. Terminal leaflet of all but the basal leaves 9–7 times as long as the petiole · · · · · · · · · · 6
– Terminal leaflet of all but the basal leaves 1.5–6 times as long as the petiole · · · · · · · · · 7
6. Petiole 3–10 mm long but in lower leaves up to 17 mm long; stipules lanceolate to ovate, up to 18 × 8 mm; inflorescence contracted with 6–12 (or more) flowers; bracts up to 17 mm long · 3. *aequinoctiale*
– Petiole 2–5 mm long but in lower leaves up to 10 mm long; stipules narrowly triangular to linear-lanceolate, up to 12 × 3 mm (in specimens from Angola up to 16 × 9 mm); inflorescence subumbelliform with 2–6(12) flowers; bracts 4–7 mm long · · 4. *macrophyllum*
7. Inflorescence with 5–20(40) flowers; subshrub 30–70(200) cm high with stems usually more than 2 mm in diameter at the base · 5. *fischeri*
– Inflorescence with 1–5(6) flowers; subshrub up to 30 cm high with stems usually less than 2 mm in diameter at the base or slender diffuse perennials with stems prostrate or ascending · 8
8. Slender diffuse perennial with stems usually less than 1 mm (rarely up to 2 mm) in diameter at the base; inflorescence 1–5(6)-flowered; standard 9–13 × 8–12 mm · · · · · · · · 6. *rupestre*
– Subshrub erect to 30 cm high with stems usually up to 2 mm (rarely up to 3 mm) in diameter at the base; inflorescence usually 1-flowered, rarely 2–3-flowered; standard 13–17 × 11–15 mm · 8. *eylesii*
9. Stipules free or only united at the very base, up to 8 × 1 mm, filiform or linear-lanceolate, persistent or somewhat caducous but leaving only a line (TAB. 3,7: **55**, fig. C) · 9. *tomentosum*
– Stipules united along the leaf-opposed margin almost to the apices, up to 14 × 3 mm, lanceolate-ovate, soon caducous but with a small persistent ring 0.5–1.0 mm high · 10. *vaginiferum*

1. **Argyrolobium tuberosum** Eckl. & Zeyh., Enum. Pl. Afr.: 188 (1836). —Bentham in Hooker, London J. Bot. **3**: 341 (1844). —Harvey in F.C. **2**: 69 (1862). —Burtt Davy, Fl. Pl. Ferns Transvaal, pt. 2: 393 (1932). —Polhill in Kew Bull. **22**: 152, t. 1, fig. 23 (1968). —Jacot Guillarmod, Fl. Lesotho: 192 (1971). —Ross, Fl. Natal: 199 (1972). —Drummond in Kirkia **8**: 217 (1972). —Corby in Kirkia **9**: 318 (1974). —Compton, Fl. Swaziland: 257 (1976). —Lock, Leg. Afr. Check-list: 274 (1989). —Retief & Herman in Strelitzia **6**: 441 (1997). TAB. 3,7: **56**, fig. D. Type from South Africa (Eastern Cape Province).

Argyrolobium angustifolium Eckl. & Zeyh., Enum. Pl. Afr.: 188 (1836). Type from South Africa (Eastern Cape Province).

Chasmone tuberosa (Eckl. & Zeyh.) Meisn. in Hooker, London J. Bot. **2**: 73 (1843).

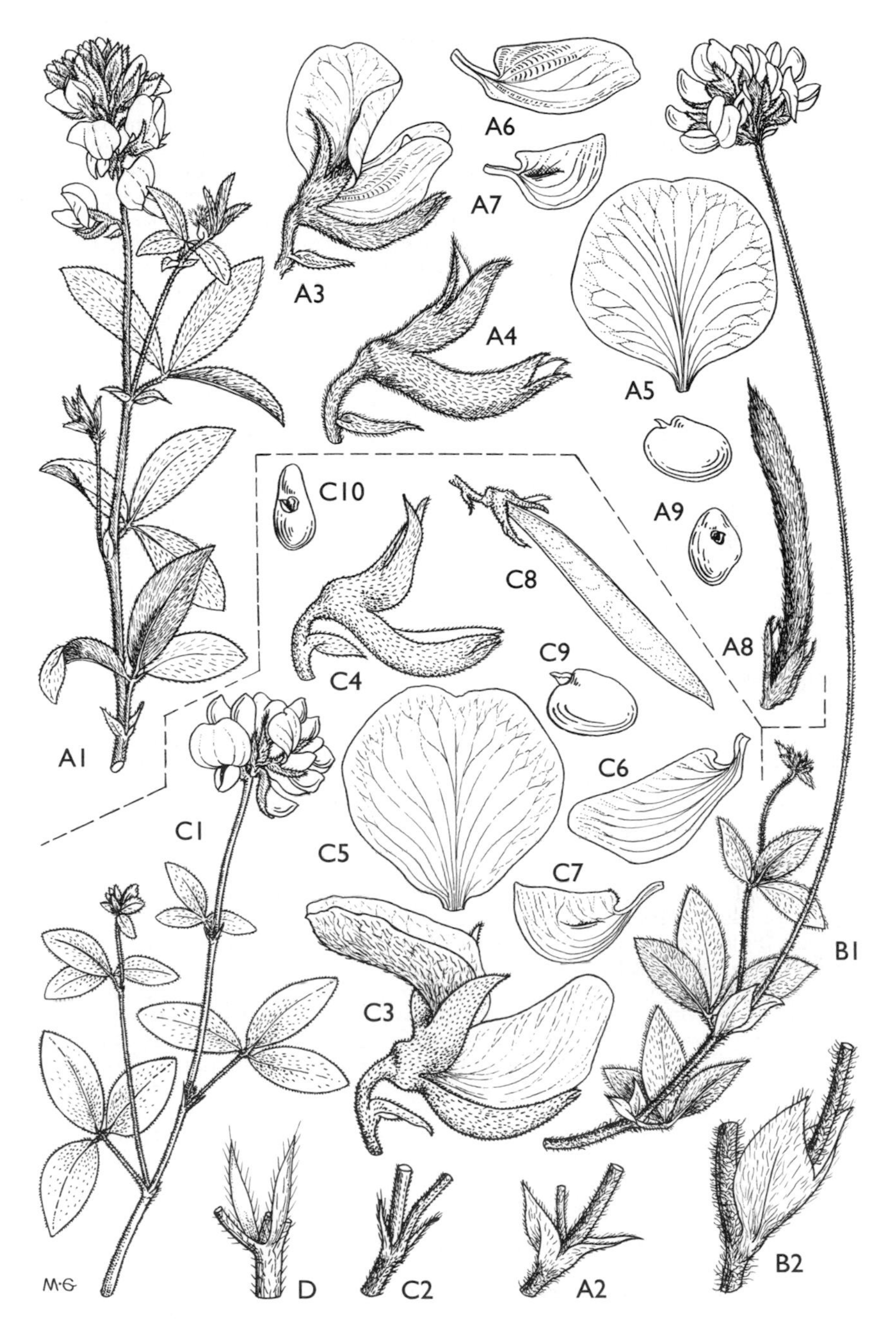

Tab. 3,7: **56**. A. —ARGYROLOBIUM FISCHERI. A1, flowering branch (× $^{2}/_{3}$); A2, stipules (× 2); A3, flower (× 2); A4, calyx (× 2); A5, standard (× 2); A6, wing (× 2); A7, keel (× 2), A1–A7 from *Richards* 15836; A8, pod (× 1); A9, seed, two views (× 4), A8 & A9 from *Peter* 49741. B. —ARGYROLOBIUM STOLZII. B1, flowering branch (× $^{2}/_{3}$); B2, stipules from young shoot (× 2), B1 & B2 from *Richards* 14275. C. —ARGYROLOBIUM TOMENTOSUM. C1, flowering branch (× $^{2}/_{3}$); C2, stipules (× 2), C1 & C2 from *Polhill & Paulo* 1810; C3, flower (× 2); C4, calyx (× 2); C5, standard (× 2); C6, wing (× 2); C7, keel (× 2), C3–C7 from *Milne-Redhead & Taylor* 10425; C8, pod (× 1); C9 & C10, seed, two views (× 4), C8–C10 from *Kerfoot* 1671. D. —ARGYROLOBIUM TUBEROSUM, stipules (× 2), from *Chase* 680. Drawn by Mary Grierson. From F.T.E.A.

A slender herb from a small tuberous rootstock. Stems few-branched, erect or ascending, up to 60 cm high and 0.4–1.5 mm in diameter, angular, subglabrous. Leaves distant, lower ones often with leaflets narrowly elliptic, up to 16 × 4 mm, the upper ones with leaflets linear-lanceolate to linear, 11–80 times as long as broad, 20–40 × 0.5–2.3 mm, acute, sericeous to subglabrous; petiole 4–12 mm long, subglabrous; stipules 6–12 mm long, subulate or lanceolate-setaceous, slightly longer or shorter than the petiole, decurrent to the stem as a ridge. Inflorescence terminal, pedunculate, racemose (1)2–9(12)-flowered. Calyx 7–8 mm long, silky hairy; upper lip 5–6 mm long, 2-lobed to below the middle; lower lip 6–6.5 mm long, 3-toothed, the teeth 2–2.5 mm long, acuminate. Corolla pubescent outside, dark yellow or orange, usually marked with reddish or purple, dark brown or black when dried; standard shortly clawed, 8–9 mm long, ovate to subcircular; wings as long as the standard, narrowly obovate, with narrow claws, asymmetrical at the base and auriculate on one side, with rows of folds between the upper veins; keel 6–6.5 mm long, incurved, obtuse at the apex. Stamens 5–6 mm long, united into a sheath split above; anthers glabrous, the long ones 0.8 mm long and the short ones 0.4 mm long. Ovary densely silky hairy; style c. 3.5 mm long, arcuate, glabrous; stigma capitate and shortly penicillate. Pod not seen, but from Harvey's description c. 5 cm long and 2 mm wide, straight or slightly curved.

Zimbabwe. E: Nyanga, "ad dejectum fluminis Pungwe", 1700 m, fl. 18.xii.1930, *Fries, Norlindh & Weimarck* 3810 (K; LISU; SRGH).

Also in highlands of South Africa and Lesotho. In montane grassland, along streams and rivers; c. 1700 m.

2. **Argyrolobium harveianum** Oliv. in Hooker's Icon. Pl. **16**: sub t. 1525 (1886). —Burtt Davy, Fl. Pl. Ferns Transvaal, pt. 2: 393 (1932) as "*harveyanum*". —Polhill in Kew Bull. **22**: 152 (1968) as "*harveyanum*". —Jacot Guillarmod, Fl. Lesotho: 191 (1971) as "*harveyanum*". —Ross, Fl. Natal: 199 (1972) as "*harveyanum*". —Drummond in Kirkia **8**: 216 (1972) as "*harveyanum*". —Compton, Fl. Swaziland: 257 (1976) as "*harveyanum*". —Lock, Leg. Afr. Check-list: 269 (1989) as "*harveyanum*". —Retief & Herman in Strelitzia **6**: 441 (1997) as "*harveyanum*". Syntypes from South Africa (KwaZulu-Natal and Eastern Cape Province).

Argyrolobium uniflorum Harv. in F.C. **2**: 72 (1862) non Jaub. & Spach (1843) *nom. illegit.*

Slender herb, from a small tuberous rootstock or slender rhizome. Stems simple or few-branched, erect or ascending, up to c. 40 cm high and 0.6–0.8 mm in diameter, glabrous or silky hairy to the apex. Leaves distant, sessile or petiolate; leaflets appressed pubescent, those of the lower leaves 8–20 × 2.5–4 mm, narrowly elliptic, those of the upper leaves 15–27 × 1–3 mm, oblanceolate to linear; petiole up to 1.5 mm long; stipules 2–4.5(6) mm long, subulate, decurrent on the stem as a ridge. Inflorescence leaf-opposed, sessile or very shortly pedunculate, 1–2-flowered. Calyx 9–10 mm long, sericeous; upper lip 6.5–8 mm long, 2-lobed to near base; lower lip as long as the upper one, 3-lobed, the lobes c. 3.5 mm long, acuminate. Corolla bright yellow; standard 11–12 mm long, obovate, obtuse and deeply emarginate at apex, with wavy or crisped margins, sericeous outside; wings c. 8 mm long, obovate, obtuse, asymmetrical at the base, with rows of folds between the upper veins, glabrous; keel as long as the wings, curved, obtuse, glabrous. Stamens united into a sheath c. 8 mm long, split above; anthers glabrous, the longer ones 1.1 mm long. Ovary sericeous; style c. 1 mm long, sericeous. Pod c. 35 × 3.5 mm, compressed, pubescent. Seeds not seen.

Zimbabwe. E: Nyanga Distr., foot of Mt. Nyangani (Inyangani), 1980 m, fl. & fr. immat. 23.x.1946, *Wild* 1545 (K).

Also in South Africa. Montane grassland; c. 1980–2100 m.

3. **Argyrolobium aequinoctiale** Welw. ex Baker in F.T.A. **2**: 46 (1871). —E.G. Baker, Legum. Trop. Africa: 67 (1926); in J. Bot. **70**, Suppl. Polypet.: 228 (1932). —Torre in C.F.A. **3**: 78, t. 6 (1962). —Polhill in Kew Bull. **22**: 153 (1968). —Lock, Leg. Afr. Check-list: 267 (1989). Type from Angola.

Tephrothamnus aequinoctialis (Welw. ex Baker) Hiern, Cat. Afr. Pl. Welw. **1**: 204 (1896) *nom. illegit.*

Suffrutex up to 50 cm high, with a few stems from a small woody rootstock. Stems simple or branching only toward the apex; indumentum yellowish-brown to greyish-brown, of subappressed spreading hairs c. 2.5 mm long. Leaves petiolate; leaflets obovate-oblanceolate to ovate-lanceolate, acute to obtuse and apiculate at the apex, cuneate at the base, tawny to yellowish-grey hairy on both surfaces; terminal leaflet 9–17 times as long as the petiole, 30–75 × 10–30 mm; lateral leaflets smaller; petioles mostly 3–10(17) mm long, sometimes a little longer in the lower leaves; stipules 7–18(30) × (1.2)2–8(13) mm, lanceolate to lanceolate-oblong or ovate, acuminate, joined at the base or free, erect, ± caducous. Inflorescence terminal, sometimes leaf-opposed, usually contracted and subumbelliform, 6–12-flowered or more; peduncle 5–45 mm long; bracts up to 17 mm long, entire or 3-partite; bracteoles up to 10 × 2 mm, linear-lanceolate or lanceolate. Calyx 11–14 mm long, sericeous; upper lip divided nearly to the base, the lobes lanceolate; lower lip with teeth 5–6 mm long, narrowly triangular. Corolla bright yellow; standard 10–13 × 10–13 mm, obovate, subcircular or ± rhomboidal, sericeous outside; wings 10–11 × 4.5 mm, elliptic to obovate-oblong, with rows of folds between the upper veins and with some brown hairs near apex outside; keel little shorter than the wings, curved, the petals half-ovate, obtuse, appendiculate and with a crest on the lateral surface towards the base, claw 2.5–3 mm long. Pod up to 28 × 6 mm, narrowly oblong, flat, silky pubescent, c. 5–9-seeded. Seeds 2–2.5 mm long, elliptic or subreniform to slightly quadrangular in outline, light brown.

Zambia. N: Mporokoso Distr., near Kapatu, 4 km from Mporokoso–Kasama road on Senga Hill road, fr. 2.iv.1984, *Brummitt, Chisumpa & Nshingo* 17108 (K).
Also in Nigeria, Cameroon and Angola. In grassland and open woodland; 1550–1600 m.

4. **Argyrolobium macrophyllum** Harms in Ber. Deutsch. Bot. Ges. **35**: 186 (1917). —E.G. Baker, Legum. Trop. Africa: 185 (1926). —Torre in C.F.A. **3**: 78 (1962). —Polhill in Kew Bull. **22**: 153 (1968); in F.T.E.A., Leguminosae, Pap.: 1002 (1971). —Lock, Leg. Afr. Check-list: 270 (1989). Type from Angola.
Argyrolobium macrophyllum var. *mendesii* Torre in Mem. Junta Invest. Ultramar, sér. 2, **19**: 49 (1960); in C.F.A. **3**: 78 (1962). Type from Angola.

Low spreading shrub or suffrutex up to 70 cm high, with a few simple or laxly branched stems from a small woody rootstock. Stems and branches usually slender, ± densely subappressed pubescent with some longer spreading hairs. Leaves petiolate; leaflets elliptic or elliptic-lanceolate to ovate, sometimes obovate in lowermost leaves, apiculate, cuneate or obtuse at the base, hairy or upper surface glabrous; terminal leaflet (7)10–16 times as long as the petiole, 25–68 × 12–25 mm; lateral leaflets smaller; petioles mostly 2–5 mm long, up to 10(12) mm long in the lowermost leaves; stipules 6–12 × 1.5–3 mm, narrowly triangular or linear-lanceolate, united only at the base (up to 16 × 9 mm, ovate to ovate-lanceolate and caudate in Angolan material), ± spreading or reflexed. Inflorescence terminal or sometimes leaf-opposed, racemose or contracted and subumbelliform, 2–6-flowered or rarely up to 12-flowered; peduncle 3–50(115) mm long; bracts 4–7 mm long, linear-lanceolate. Calyx 8–13 mm long, densely appressed hairy; upper lip divided to near the base with the lobes lanceolate. Corolla bright yellow; standard 11–14 mm long, widely obovate to subcircular, abruptly contracted into a very short claw, undulate at the margins, silvery hairy outside; wings 8–10 mm long, oblong, with rows of folds between the upper veins; keel ± as long as the wings, curved, auriculate with a crest on the lateral surface, claw 2.5–3 mm long. Stamens c. 8.5 mm long; anthers alternating 0.9 and c. 0.5 mm in length. Pod up to 55 × 5 mm, linear-oblong, flat, appressed silky pubescent, 10–18-seeded. Seeds 2.5–3 mm long, ovoid, subreniform-orbicular or slightly quadrangular in outline, dark brown.

Zambia. N: Mbala Distr., Chilongowelo, 1450 m, 21.v.1955, *Richards* 5811 (K). W: Mwinilunga Distr., just north of Matonchi Farm, 29.i.1938, *Milne-Redhead* 4400 (K). **Malawi**. N: Rumphi, 1.6 km on Nyika road, 1460 m, fl. 24.iv.1977, *Pawek* 12628 (K).
Also in Tanzania and Angola. In grassland and grassy places in *Brachystegia* woodland; 1000–1500 m.

5. **Argyrolobium fischeri** Taub. in Engler, Pflanzenw. Ost-Afrikas **C**: 207 (1895). —E.G. Baker, Legum. Trop. Africa: 66 (1926). —Brenan, Check-list For. Trees Shrubs Tang. Terr.: 408 (1949). —Polhill in Kew Bull. **22**: 154, fig. 1/1–10 (1968); in F.T.E.A., Leguminosae, Pap.: 1002, fig. 138/1–10 (1971). —Lock, Leg. Afr. Check-list: 269 (1989). TAB. 3,7: **56**, fig. A. Syntypes from East Africa.

Argyrolobium leucophyllum Baker in Bull. Misc. Inform., Kew **1897**: 253 (1897). —E.G. Baker, Legum. Trop. Africa: 66 (1926). —Brenan, Check-list For. Trees Shrubs Tang. Terr.: 408 (1949). Type: Malawi, Nyika Plateau, *Whyte* 251 (K, holotype).

Argyrolobium mildbraedii Harms in Ber. Deutsch. Bot. Ges. **27**: 93 (1909); in Mildbraed, Deutsch. Zentr.-Afr. Exped. 1907–08, **2**: 247 (1911). Type from Rwanda.

Argyrolobium helenae Buscal. & Muschl. in Bot. Jahrb. Syst. **49**: 471 (1913). Type from Dem. Rep. Congo (Katanga).

Argyrolobium dekindtii Harms in Ber. Deutsch. Bot. Ges. **35**: 186 (1917). —E.G. Baker, Legum. Trop. Africa: 67 (1926). —Torre in C.F.A. **3**: 78 (1962). Type from Angola.

Argyrolobium rufopilosum De Wild. in Bull. Jard. Bot. État **8**: 127 (1923). Type from Dem. Rep. Congo (Kivu).

Argyrolobium aequinoctiale sensu De Wild., Pl. Bequaert. **2**: 517 (1924). —sensu Wilczek in F.C.B. **4**: 281 (1953) non Welw. ex Baker.

Argyrolobium bequaertii De Wild., Pl. Bequaert. **2**: 520 (1924). Type from Dem. Rep. Congo (Kivu).

Argyrolobium shirense sensu De Wild., Pl. Bequaert. **2**: 524 (1924); Pl. Bequaert. **3**: 321 (1925) non Taub.

Argyrolobium aequinoctiale forma *elongatum* R. Wilczek in Bull. Jard. Bot. État **23**: 220 (1953); in F.C.B. **4**: 282 (1953). Type from Dem. Rep. Congo (Kivu).

Bushy woody herb or subshrub 30–70(200) cm high, with several simple or few-branched stems from a woody rootstock. Stems usually more than 2 mm in diameter at the base, densely covered with appressed or ± spreading silvery-grey or golden-brown hairs. Leaves petiolate; leaflets elliptic or ovate-lanceolate, occasionally lanceolate, subacute to round and apiculate at the apex, cuneate to obtuse at the base, sparsely to very densely covered with golden-brown appressed or ± spreading hairs above and beneath, more densely so beneath; terminal leaflet usually 3–6 (very rarely only 2.5–3) times as long as the petiole, 23–51(60) × 10–18(26) mm; petioles mostly 6–16 mm long but in the lowermost leaves sometimes up to 30(40) mm long; stipules 5–12 × 2–6 mm, ovate to ovate-lanceolate, sometimes cordate, acuminate, free or united only at the base, occasionally one of the pair bifid, persistent or very late caducous. Inflorescence terminal or leaf-opposed, racemose and elongate or subumbelliform, racemes 5–20-flowered rarely up to 40-flowered; peduncle 10–80(120) mm long; bracts up to 6 mm long, lanceolate, decreasing and narrowing up the rhachis. Calyx 8–12 mm long, sericeous; upper lip divided by up to c. three quarters of the calyx length, the lobes oblong-lanceolate. Corolla bright yellow, sometimes the petals becoming orange or reddish with age; standard 10–13 × 8–10 mm, widely obovate to subcircular, undulate at the margins, densely silky pubescent outside; wings c. 11 × 2.5–3 mm, oblong and sometimes slightly curved, narrowed at base into a c. 2 mm long claw, with rows of folds between the upper veins, glabrous or sometimes with brown hairs near the apex outside; keel little shorter than the wings, curved, the petals not or only slightly auriculate, with a crest on the lateral surface towards the base and a claw c. 2.5 mm long, glabrous. Pod 30–50 × 4.5–6 mm, narrowly oblong with the valves flat, silky pubescent, c. 12–18-seeded. Seeds 2.5–3 mm long, oblong-ovoid, dark brown.

Zambia. E: Chama Distr., Nyika Plateau, road to Rest House, 2100 m, fl. 2.i.1959, *Richards* 10396 (K). **Malawi**. N: Nkhata Bay Distr., Viphya Plateau, c. 37 km SW of Mzuzu, 1650 m, fl. 12.xi.1972, *Pawek* 5946 (K; MAL; SRGH). S: Blantyre Distr., Soche Mt., 1100 m, fl. 9.i.1938, *Lawrence* 595 (K). **Mozambique**. Z: Milange Distr., Metolola (Metalola), serra de Chiperone, encosta sul, 900 m, fl. 8.ii.1972, *Correia & Marques* 2548 (LMU).

Also in southern Sudan, Ethiopia, Uganda, Kenya, Tanzania, eastern Dem. Rep. Congo and Angola. Submontane woodlands with *Brachystegia boehmii* and *Julbernardia globiflora*, and grassland with *Protea* and other open grasslands subject to annual burning; 900–2100 m.

6. **Argyrolobium rupestre** (E. Mey.) Walp. in Linnaea **13**: 508 (1839). —Bentham in Hooker, London J. Bot. **3**: 345 (1844). —Harvey in F.C. **2**: 73 (1862). —E.G. Baker, Legum. Trop. Africa: 63 (1926). —Burtt Davy, Fl. Pl. Ferns Transvaal, pt. 2: 394 (1932). —Polhill in Kew Bull. **22**: 157, map 1 (1968); in F.T.E.A., Leguminosae, Pap.: 1005 (1971). —Jacot Guillarmod, Fl. Lesotho: 192 (1971). —Ross, Fl. Natal: 199 (1972). —Drummond in Kirkia **8**: 216 (1972). —Corby in Kirkia **9**: 318 (1974). —Lock, Leg. Afr. Check-list: 272 (1989). —Agnew, Upland Kenya Wild Fl., ed. 2: 147 (1994). —Retief & Herman in Strelitzia **6**: 441 (1997). Type from South Africa (Eastern Cape Province).

Low diffuse perennial with a woody taproot. Stems numerous, prostrate, ascending or shortly erect, slender, up to 40 cm long and mostly less than 1 mm in diameter, up to 2.5 mm in diameter at the base, subappressed or spreading golden-brown or greyish hairy. Leaves usually well spaced; leaflets elliptic-lanceolate to obovate-elliptic, sparsely pilose to silky tomentose above and beneath, the terminal leaflets 12–30 × 5–13 mm, mostly 2.5–5 times as long as the petiole; petiole 3–10(15) mm long; stipules 4–10(12) × 1.5–5 mm, lanceolate to ovate, acuminate, free or joined only at the base. Inflorescence leaf-opposed, pedunculate, racemose or subumbelliform, 1–5(6)-flowered; peduncle 0.8–8(10) cm long; bracts 2–8 mm long, lanceolate to linear. Calyx 5–9 mm long, pilose to silky tomentose outside; the upper lip 2-lobed to near the base with the lobes lanceolate. Corolla yellow, often becoming brownish-orange; standard 9–13 × 8–12 mm, widely obovate to suborbicular, sparsely to densely silky pubescent or tomentose outside; wings 8–10 × 2.5–5 mm, oblong to obovate-oblong, with rows of folds between the upper veins, usually with some brown hairs near the apex; keel a little shorter than the wings, curved, the petals with a claw shorter than the blade width, blade semi-orbicular with a crest near base, glabrous or with some bristles along the lower margin. Pod 25–40 × 3.5–5.5 mm, oblong, flat, spreading or subappressed pubescent, c. 8–15-seeded. Seeds 1.5–2 × 1.1–1.5 mm, compressed, oblong-ovoid to oblong-cordate, dark brown.

Inflorescences 1–2(3)-flowered; peduncles 0.8–4(8) cm long; calyx indumentum usually dense and obscuring the surface; standard densely pubescent or tomentose outside; wings 3.5–5 mm wide · subsp. *rupestre*

Inflorescences 2–5(6)-flowered; peduncles (1.5)2.5–9 cm long; calyx indumentum sparse, not obscuring the surface; standard silky pubescent outside but subglabrous towards the margins and in the lower half; wings 2.5–3 mm wide · · · · · · · · · · · · · subsp. *aberdaricum*

Subsp. **rupestre**

Chasmone rupestre E. Mey., Comment. Pl. Afr. Austr.: 74 (1836).

Argyrolobium collinum sensu Eyles in Trans. Roy. Soc. South Africa **5**: 372 (1916) non Eckl. & Zeyh.

Argyrolobium rhodesicum Baker f., Legum. Trop. Africa: 64 (1926). —Goodier & Phipps in Kirkia **1**: 56 (1961) as "*rhodesianum*". Type: Zimbabwe, Mutare (Umtali), *F.A. Rogers* 4029 (BM; K, holotype).

Inflorescences usually 1–2-flowered, with short weak peduncles. Calyx usually with a dense spreading indumentum obscuring the surface. Standard densely pubescent or tomentose outside; wings more than 3 mm wide.

Zimbabwe. E: Nyanga Distr., 56 km east of Rusape, fl. & fr. xi.1957, *O.B. Miller* 4740 (SRGH); prope pagum Inyanga, in colle Nyamoka, 1700 m, fl. 13.i.1931, *Norlindh & Weimarck* 4278 (LISU; PRE). S: Bikita Distr., Old Bikita, 1300 m, fl. & fr. 16.xii.1953, *Wild* 4415 (LISC; PRE). **Mozambique**. MS: Gorongosa Mt., Gogogo summit area, 1700–1868 m, fl. & fr. 12.iii.1972, *Tinley* 2440 (K; LISC; PRE; SRGH), atypical.

Also in South Africa and Lesotho. In submontane and open grassland and in miombo woodland; 1300–1900 m.

The gatherings *Tinley* 2323 and 2440, both from Mozambique, Gorongosa Mt., Gogogo summit area, show a few stipules joined to about the middle, the peduncles up to 8 cm long and calyx 7–10 mm long, in these respects approaching *A. stolzii.*

Subsp. **aberdaricum** (Harms) Polhill in Kew Bull. **22**: 159, map 1 (1968); in F.T.E.A., Leguminosae, Pap.: 1006 (1971). —Lock, Leg. Afr. Check-list: 272 (1989). Type from Kenya.

Argyrolobium aberdaricum Harms in Notizbl. Bot. Gart. Berlin-Dahlem **10**: 75 (1927). —Robyns, Fl. Sperm. Parc Nat. Alb. **1**: 282 (1948).

Argyrolobium virgatum sensu Hutchinson & Bruce in Bull. Misc. Inform., Kew **1941**: 120 (1941). —sensu Wilczek in F.C.B. **4**: 280 (1953) non Baker.

Inflorescences usually 2–5-flowered with long peduncles. Calyx with a sparse to dense appressed or spreading indumentum not obscuring the surface. Standard silky pubescent outside but glabrescent towards the margins and in the lower half; wings 2.5–3 mm wide.

Zambia. N: Isoka Distr., Mafinga Hills, 2000 m, fl. 12.iii.1961, *E.A. Robinson* 4456 (K; SRGH). E: Chama Distr., Nyika Plateau, 2280 m, fr. 7.vi.1962, *Verboom* 640 (K) **Malawi**. N: Nkhata Bay Distr., Viphya Plateau, 56 km SW of Mzuzu, Mpamphala (Mpalampala), 1830 m, fl. 23.ii.1974,

Pawek 8128 (K; MAL; MO; PRE; SRGH). S: Mulanje Distr., Sombani Hut to Fort Lister, NE slope of Mulanje Mt., fl. 13.xi.1979, *Blackmore & Kupicha* 1034 (K).

Also in Ethiopia, Uganda, Rwanda, Kenya and Tanzania. In open montane grassland; 1800–2300 m.

Hilliard & Burtt 4412 and *Richards* 22683 from Nyika Plateau, and *J.D. & E.G. Chapman* 8285, *Hilliard & Burtt* 4546, and 6206, from Mulanje Mt. in southern Malawi, show some features, such as a more dense indumentum, longer petioles, longer and narrower stipules, which approach the circumscription of *A. stolzii*. The same occurs in plants from the highlands of southern Tanzania.

7. **Argyrolobium stolzii** Harms in Ber. Deutsch. Bot. Ges. **35**: 185 (1917). —E.G. Baker, Legum. Trop. Africa: 65 (1926). —Brenan, Check-list For. Trees Shrubs Tang. Terr.: 408 (1949). —Polhill in Kew Bull. **22**: 161, fig. 1/11, 12 (1968); in F.T.E.A., Leguminosae, Pap.: 1006, fig. 138/11, 12 (1971). —Lock, Leg. Afr. Check-list: 273 (1989). TAB. 3,7: **56**, fig. B. Type from Tanzania.

Perennial herb with several trailing and decumbent or ascending stems from a woody taproot. Stems up to 45 cm long, slender, laxly branched, rusty pilose. Leaves petiolate; leaflets lanceolate to widely elliptic or obovate, acute to obtuse at apex, cuneate at base, thinly pilose on both surfaces; terminal leaflets 1.5–3 times as long as the petiole, 17–35(44) × 6–17 mm; petioles 4–13 mm long, pilose; stipules 6–13 × 1.5–4 mm, ovate to narrowly lanceolate, acuminate, some of them joined at first to above the middle along the leaf-opposed margin, usually becoming free. Inflorescence terminal, usually a subumbelliform raceme, (1)2–6-flowered (usually 8–24-flowered in the Southern Highlands of Tanzania); peduncle 8–16 cm long; bracts 3–9 × 1–2 mm, linear-lanceolate, entire. Calyx (7)8–10 mm long, densely to sparsely silky pilose; upper lip divided to ± the middle, the lobes lanceolate; lower lip with teeth 5–6 mm long, narrowly triangular, acute. Corolla bright yellow; standard 8–10 × 7–8 mm, obovate, silky pilose outside; wings 8–10 × 2.5–4.4 mm, with rows of folds between the upper veins; keel as long as the wings, the petals with a crest on the lateral surface towards the base, not appendiculate, with claw c. 2 mm long. Pod up to 40 × 5 mm, narrowly oblong, flat, densely subappressed pilose, with c. 10–14 seeds. Seeds c. 2.5 × 2 mm, light brown.

Malawi. N: Nyika Plateau, below Chelinda Hill opposite Broken rib, 1800 m, fl. 2.ii.1976, *E. Phillips* 1122 (K). S: Zomba Distr., Mt. Malosa, 1200–1800 m, fl. xi. & xii.1896, *Whyte* s.n. (K).

Also in the Southern Highlands of Tanzania. Montane grassland beside swampy areas and near rivers; 1200–2000 m.

A. stolzii is very variable in the Flora Zambesiaca area. Only a few of the specimens seen show the typical joined stipules and the long-peduncled many-flowered inflorescences. Many specimens have all the stipules free, peduncles up to 10 cm long (15–27 cm long in Tanzanian material), and inflorescences usually less than 5-flowered (8–24-flowered in Tanzanian material), e.g., from northern Malawi: *E. Phillips* 458 and 479, from the Nyika Plateau, and *Jackson* 1288, from Mzimba, and from Zomba District: *Balaka & Kaunda* 386, *Banda* 210, *Brummitt* 9147, 9944 , *Exell, Mendonça & Wild* 761, and *Jackson* 1491. These plants approach those of *A. rupestre* subsp. *aberdaricum* which occurs in the same area.

8. **Argyrolobium eylesii** Baker f., Legum. Trop. Africa: 64 (1926). —Wild in Kirkia **5**: 58 (1965). —Polhill in Kew Bull. **22**: 161 (1968). —Drummond in Kirkia **8**: 216 (1972). —Corby in Kirkia **9**: 318 (1974). —Lock, Leg. Afr. Check-list: 268 (1989). Type: Zimbabwe, west of Matopo Hills, Fuller's Hotel, *Eyles* 1175 (BM, holotype).

Erect subshrub up to 30 cm tall, with few to many stems from a woody taproot. Stems slender, usually less than 2 mm in diameter, up to c. 3 mm in diameter at the perennial base, grey or yellowish villous or sericeous. Leaves usually well spaced; leaflets elliptic to widely obovate, subacute to rounded and apiculate at the apex, cuneate to obtuse at the base, sparsely pilose to whitish, greyish or yellowish tomentose or sericeous above and beneath; the terminal leaflets mostly 1.5–3 times as long as the petiole, 17–33 × 9–18 mm; petiole 5–15 mm long; stipules sometimes 3 at each node, 4–6 × 1–1.5 mm, linear to narrowly triangular, free, sericeous. Inflorescence leaf-opposed, pedunculate, racemose but usually 1-flowered, sometimes 2–3-flowered; peduncle 5–70 mm long; bracts 3–6 mm long, linear. Calyx 12–14 mm long, pilose to sericeous outside; the upper lip 2-lobed to near the base, the lobes lanceolate. Corolla yellow, turning reddish with age; standard 13–17 × 11–15 mm, obovate, sparsely to densely silky pubescent or tomentose outside on the upper half, glabrous or subglabrous to the base; wings 11–13 × 4–5 mm, obovate-oblong, with rows of folds between the upper veins, whitish pilose near apex; keel as long as the wings, the petals with a well developed crest towards the base, glabrous. Pod 25–50 × 4–5 mm, oblong, flat, sericeous. Seeds not seen.

Zimbabwe. W: Matobo Distr., Bulawayo–Esigodini (Essexvale), Hope Fountain Mission, 1400 m, fl. 3.ii.1974, *Norrgrann* 496 (K; SRGH). C: Mashaba (Mhlaba) Hills, near Windsor Chrome Mine, fl. 16.i.1962, *Wild* 5599 (K; LISC; PRE; SRGH). S: Masvingo Distr., Danson Claims, Brakfontein, fl. & fr. 14.xii.1966, *Wild* 7584 (K; LISC; PRE; SRGH).

Only known from the serpentine soils of the Great Dyke and from aluminium and chrome soils of western and southern Zimbabwe; 1000–1300 m.

9. **Argyrolobium tomentosum** (Andrews) Druce in Bot. Exch. Club Brit. Isles **1916**: 605 (1917). —Hutchinson, Botanist South. Africa: 455 (1946). —Pole Evans in Bot. Surv. S. Afr. Mem. **22**: 268 (1948). —Robyns, Fl. Sperm. Parc Nat. Alb. **1**: 283 (1948). —Wilczek in F.C.B. **4**: 277 (1953). —Compton, Ann. Check List Fl. Swaziland: 47 (1966). —Polhill in Kew Bull. **22**: 162, fig. 1/13–22 (1968); in F.T.E.A., Leguminosae, Pap.: 1007, fig. 138/13–22 (1971). —Ross, Fl. Natal: 199 (1972). —Drummond in Kirkia **8**: 217 (1972). —Brummitt in Wye Coll. Malawi Proj. Rep.: 64 (1973). —Corby in Kirkia **9**: 318 (1974). —Drummond in Kirkia **10**: 245 (1975). —Compton, Fl. Swaziland: 257 (1976). —Lock, Leg. Afr. Check-list: 273 (1989). —Retief & Herman in Strelitzia **6**: 441 (1997). TAB. 3,7: **56**, fig. C. Type: plate 237 in Andrews, The Botanist's Repository **4** (1802), drawn from a plant cultivated in London from seeds collected in South Africa (Cape Province).

Cytisus tomentosus Andrews, Bot. Repos. **4**, t. 237 (1802).

Chasmone andrewsiana E. Mey., Comment. Pl. Afr. Austr.: 74 (1836) *nom. illegit.* Type as for *Cytisus tomentosus.*

Argyrolobium andrewsianum (E. Mey.) Steud., Nom. Bot., ed. 2, **1**: 129 (1840). —Bentham in Hooker, London J. Bot. **3**: 348 (1844). —Harvey in F.C. **2**: 75 (1862) as "*andrewsiana*". —Eyles in Trans. Roy. Soc. South Africa **5**: 372 (1916). —Harms in Ber. Deutsch. Bot. Ges. **35**: 178 (1917). —E.G. Baker, Legum. Trop. Africa: 64 (1926). —Staner in Rev. Zool. Bot. Africaines **23**: 217 (1933).

Argyrolobium shirense Taub. in Engler, Pflanzenw. Ost-Afrikas **C**: 207 (1895). —E.G. Baker, Legum. Trop. Africa: 68 (1926). —Brenan, Check-list For. Trees Shrubs Tang. Terr.: 408 (1949); in Mem. New York Bot. Gard. **8**: 249 (1953). —Wilczek in F.C.B. **4**: 279 (1953). —Goodier & Phipps in Kirkia **1**: 56 (1961). Types: Malawi, Shire Highlands, *Buchanan* 466, 481 and *Last* s.n. (B†, syntypes; K).

Argyrolobium stuhlmannii Taub. in Engler, Pflanzenw. Ost-Afrikas **C**: 207 (1895). —E.G. Baker, Legum. Trop. Africa: 65 (1926). —Brenan, Check-list For. Trees Shrubs Tang. Terr.: 408 (1949). —Wilczek in F.C.B. **4**: 278 (1953). Type from Rwanda/Uganda border.

Argyrolobium angustistipulatum De Wild., Pl. Bequaert. **2**: 518 (1924). —E.G. Baker, Legum. Trop. Africa: 67 (1926). Type from Dem. Rep. Congo (Kivu).

Erect or scrambling subshrub 70–200 cm tall. Stems up to 7 mm in diameter at the base; branches thinly to densely appressed or spreading fulvous to greyish pubescent, sometimes the hairs grouped, glabrescent. Leaves usually well spaced; leaflets elliptic-lanceolate to elliptic, acute to subacute and apiculate at apex, cuneate at base, usually very sparsely pilose above, more densely so beneath; terminal leaflets mostly 2–3 times as long as the petiole, 20–70 × 9–35 mm; petiole 9–40 mm long; stipules 3–7 × 0.2–1 mm, filiform or linear-lanceolate, free, subglabrous to pilose, persistent or somewhat caducous. Inflorescence leaf-opposed, pedunculate, racemose, sometimes long and lax, sometimes contracted and subumbelliform, usually 8–32-flowered; peduncle 2–10 cm long; bracts 2–4 mm long, linear-oblanceolate. Calyx 9–12 mm long, sometimes reddish, sparsely pubescent or pilose; the upper lip 2-lobed to below the middle with the lobes oblong-lanceolate. Corolla yellow, turning orange or reddish with age; standard 11–13 × 10–14 mm, circular to oblate, sparsely to densely pubescent outside; wings as long as the standard, broad, glabrous or sometimes with some hairs along the midline or to the apex, without any folds between the veins; keel shorter than the wings, the petals 7–9 mm long, without or sometimes with a prominent crest, ciliolate on the upper margin near base. Pod 35–60 × 5–6 mm, oblong, slightly curved, narrowed to the extremity, fulvous appressed pubescent or strigose. Seeds 12–18 per pod, 3–3.5 mm long, dark brown, smooth.

Zambia. N: Mbala Distr., track in Itimbwe (Itembwe) Gorge, 1520 m, fl. & fr. 24.iv.1959, *McCallum-Webster* 912 (K; LISC; SRGH). W: Kitwe, fl. & fr. 18.viii.1967, *Mutimushi* 2017 (K; LISC). E: Nyika Plateau, fr. 24.viii.1962, *Verboom* 664 (K; LISC) **Zimbabwe**. E: Chimanimani Distr., Cashel road, c. 10 km east of Chimanimani (Melsetter Village), 1500 m, fl. & fr. 21.ix.1960, *Rutherford-Smith* 151 (LMU; SRGH). **Malawi**. N: Mzimba Distr., Perekezi Forest Reserve, 30 km from Mzimba, margin of dambo, 1510 m, fl. & fr. 6.vii.1970, *Brummitt* 11827 (K; LISC; PRE; SRGH). C: Dedza Distr., Dedza Mt., 1800 m, fl. 5.iv.1978, *Pawek* 14291 (K; MAL). S: Blantyre Distr., Ndirande Mt., rocky slopes with *Brachystegia* cover, 1430–1520 m, fl. 2.v.1970, *Brummitt* 10314 (K; LISC; SRGH) **Mozambique**. N: Lichinga (Vila Cabral), fl. 8.vi.1934, *Torre* 262 (K; LISC); Ribáuè Distr., south face of Ribáuè Mt., 975 m, fl. 19.vii.1962, *Leach & Schelpe* 11413 (K; SRGH). Z: Gurué, próximo do Pico Namuli, 1500 m, fl. & fr. 29.vi.1943, *Torre* 5603 (BM; LISC). T: Angónia Distr., Monte Domué, 1500 m, fl. 8.vi.1980, *Stefanesco & Nyongani* 551 (LISC; SRGH). MS: Barué Distr., Serra de Chôa, 12 km de Catandica (Vila Gouveia) para Chôa, 1300 m, fl. & fr. 26.v.1971, *Torre & Correia* 18667 (K; LISC; LMU; PRE). M: Namaacha, abaixo das quedas, margem esquerda do R. Impamputo, fl. & fr. 20.vii.1967, *Marques* 2068 (LMU).

Also in north-eastern Dem. Rep. Congo and Uganda, southwards to southern Tanzania and South Africa (Eastern Cape Province). *Brachystegia* woodland on rocky slopes, gully forest and evergreen forest margins, swamp forest margins, in moist grassland on river and dambo margins, rarely growing in water; 600–1800 m.

10. **Argyrolobium vaginiferum** Harms in Bot. Jahrb. Syst. **30**: 323 (1901). —E.G. Baker, Legum. Trop. Africa: 66 (1926) pro parte excl. specim. *Elliot*. —Brenan, Check-list For. Trees Shrubs Tang. Terr.: 408 (1949). —Polhill in Kew Bull. **22**: 164 (1968); in F.T.E.A., Leguminosae, Pap.: 1008 (1971). —Lock, Leg. Afr. Check-list: 274 (1989). Type from Tanzania.

Argyrolobium lejeunei R. Wilczek in Bull. Jard. Bot. État **23**: 219 (1953); in F.C.B. **4**: 277 (1953). Type from Burundi.

Erect or scrambling subshrub 30–100 cm tall. Stems and branches sparsely to densely subappressed greyish pubescent or pilose, sometimes with some spreading longer hairs interspersed, glabrescent. Leaves well spaced; leaflets lanceolate to broadly elliptic, subacute at the apex, subacute to obtuse at the base, sparsely appressed pubescent above and beneath; terminal leaflets 1.5–3 times as long as the petiole, 30–50 × 10–33 mm; petiole 10–20 mm long; stipules 5–14 × 1.5–3 mm, united along the leaf-opposed margin almost to the apices, caducous but persisting as a small ring c. 1 mm high. Inflorescence leaf-opposed, pedunculate, racemose, sometimes long and lax, sometimes short and dense, many-flowered; peduncle 3–8(15) cm long; bracts 3–7 mm long, linear to filiform, often caducous. Calyx 9–13 mm long, sericeous outside; upper lip 2-lobed to below the middle with the lobes oblong-lanceolate. Corolla yellow; standard 10–13 × 10–13 mm, obovate to subcircular, densely pubescent to tomentellous outside; wings as long as the standard, broad, without any folds between the veins; keel shorter than the wings, 7–9 mm long, the petals without any crest, ciliolate on the upper margin. Pod 30–35 × 5–6 mm, oblong, densely silky pubescent. Seeds 8–12 per pod, c. 4 mm long, dark brown or blackish, smooth.

Malawi. N: Rumphi Distr., Nyika Plateau, Chelinda Camp, by chalet no. 3, 2285 m, fl. & fr. 19.v.1970, *Brummitt* 10910 (LISC; PRE; SRGH); Nyika Plateau, Chelinda Bridge, fl. 10.i.1967, *Hilliard & Burtt* 4354 (MAL).

Also known from Burundi and Tanzania. Margins of rainforest and plantations and streamsides; c. 2285 m.

109. ADENOCARPUS DC.

By E.S. Martins

Adenocarpus DC. in Lamarck & De Candolle, Fl. Fr. ed. 3, **5** (Suppl.): 549 (1815). —Gibbs in Bol. Soc. Brot., sér. 2, **41**: 67 (1967).

Much-branched shrubs or small trees, with silky or villous indumentum; branching alternate. Leaves petiolate, digitately 3-foliolate; stipules usually small and caducous or absent. Inflorescence terminal, racemose or capitate. Calyx valvate in bud, 2-lipped, the upper lip deeply bifid, the lower 3-lobed or 3-dentate. Corolla usually yellow, with petals shortly clawed; standard subcircular, as long as or longer than the wings; wings obovate or oblong; keel incurved. Stamens all joined into a closed tube; anthers alternately long and short. Ovary sessile, multi-ovulate. Pod oblong, compressed, dehiscent, glandular-tuberculate or muricate, usually villous. Seeds ovate-oblong in outline, smooth with a small aril.

A genus of about 15 species in southern and western Europe, northern Africa, Madeira and the Canary Islands, with one in tropical Africa.

Adenocarpus mannii (Hook.f.) Hook.f. in J. Linn. Soc. **7**: 189 (1864). —J.G. Baker in F.T.A. **2**: 47 (1871); in Trans. Linn. Soc. London, Bot. **4**: 9 (1894). —Taubert in Engler, Pflanzenw. Ost-Afrikas **C**: 208 (1895). —E.G. Baker, Legum. Trop. Africa: 69 (1926). —Brenan, Check-list For. Trees Shrubs Tang. Terr.: 404 (1949); in Mem. New York Bot. Gard. **8**: 249 (1953). —Wilczek in F.C.B. **4**: 284 (1953). —Hepper in F.W.T.A., ed. 2, **1**: 552 (1958). —White, F.F.N.R.: 138 (1962). —Torre in C.F.A. **3**: 79 (1962). —Gibbs in Bol. Soc. Brot., sér. 2, **41**: 98, fig. 5 (1967). —Milne-Redhead in F.T.E.A., Leguminosae, Pap.: 1009, t. 139 (1971). —Lock, Leg. Afr. Check-list: 267 (1989). —White, Dowsett-Lemaire & Chapman, Evergreen For. Fl. Malawi: 322 (2001). TAB. 3,7: **57**. Type from Bioko (Fernando Po).
Cytisus mannii Hook.f. in J. Linn. Soc., Bot. **6**: 8 (1861).
Adenocarpus benguellensis Welw. ex Baker in F.T.A. **2**: 47 (1871). Type from Angola.

Shrub up to 4.5 m high, erect. Stems much branched, leafy, ± densely pilose, ultimately with the cortex flaking off longitudinally. Leaves usually densely arranged on short shoots, more spaced along the young shoots; leaflets subequal, up to 8(12) × 4(5) mm, oblanceolate to elliptic-obovate, acute or subacute at apex, cuneate at base, often with the edges incurved, usually densely sericeous beneath, sparsely so above; petiole shorter than the leaflets; stipules small. Inflorescence terminal, contracted, racemose, few-flowered. Calyx silky-hirsute; upper lip c. 6 mm long, deeply 2-lobed with the lobes triangular-ovate; lower lip longer than the upper one, deeply 3-lobed with the lobes narrowly triangular to linear. Corolla yellow; standard shortly clawed, the limb c. 8 mm in diameter, ovate to subcircular, retuse at the apex, sericeous on the back; wings slightly shorter than the standard, narrowly clawed, the limb oblong, asymmetrical at the base, with series of folds between the upper veins; keel ± as long as the wings, much incurved, obtuse at the apex. Stamens c. 9 mm long; anthers glabrous, the long ones 0.6 mm long and the short ones 0.2 mm long. Ovary densely silky hairy and glandular; style c. 4 mm long, glabrous; ovules 3–8. Pod up to 24 × 7 mm, narrowly oblong, straight with a short downward pointing beak, compressed, sparsely pilose or sometimes glabrescent, rough with glandular tubercles. Seeds 2–8, up to 2 mm in diameter.

Zambia. E: Isoka Distr., Nyika Plateau, upper slopes of Kangampande Mt., 2130 m, fl. 3.v.1952, *White* 2573 (K). **Malawi**. N: Rumphi Distr., Nyika Plateau, Chowo Rock, 2210 m, fl. & fr. 17.v.1970, *Brummitt* 10861 (K; LISC; MAL; PRE; SRGH). S: Mulanje Distr., Mulanje Mt., just below Madzeka (Little Ruo) Hut, 1800 m, fl. 27.vii.1970, *Brummitt* 12261 (K; LISC; MAL; PRE; SRGH).

Also from the highlands of northern Nigeria, Cameroon, Bioko, Dem. Rep. Congo, Rwanda, Sudan, Ethiopia, Uganda, Kenya, Tanzania and Angola. In montane grassland, edges of rainforests and on rocky mountain summits; 1500–2250 m (up to 4000 m in East Africa).

Var. *laevicarpa* Verdc., with eglandular pods, has been described from northern Cameroon.

110. LUPINUS L.

By R.M. Polhill

Lupinus L., Sp. Pl. **2**: 721 (1753); Gen. Pl., ed. 5: 322 (1754).

Herbs or rarely shrubs. Leaves usually digitately 5–11(17)-foliolate, elsewhere sometimes 1–3-foliolate; stipules adnate to the base of the petiole. Flowers usually numerous in terminal and leaf-opposed racemes, alternate to verticillate; bracteoles often attached to the calyx. Calyx deeply 2-lipped; upper lobes separate to largely joined; lower lip entire to shortly 3-toothed. Corolla variously coloured, often variegated; standard with a short claw and broad blade, the sides of which are often partially reflexed at anthesis; wings broad, generally enveloping the keel and often marginally adherent; keel beaked. Stamens all joined in a closed tube; anthers

Tab. 3,7: **57**. ADENOCARPUS MANNII. 1, flowering and fruiting branch (× 1); 2, flower (× 4); 3, calyx, opened out (× 4); 4, standard (× 4); 5, back of standard, upper part (× 4); 6, wing (× 4); 7, keel (× 4); 8, androecium and gynoecium (× 4); 9, androecium, opened out (× 4); 10, gynoecium, with ovary opened to show ovules (× 4); 11, stigma (× 60), 1–11 from *Purseglove* 1604; 12, fruits (× $1^1/_2$); 13, surface of pod showing hairs and papillae (× 8); 14, part of valve of pod showing attachment of seeds (× 4); 15, seed (× 8), 12–15 from *Bally* 4962. Drawn by Margaret Stones. From F.T.E.A.

alternately long and short. Ovary usually sessile, 2–many-ovulate; style incurved, glabrous except often for a ring of hairs below the terminal stigma. Pod laterally flattened to varying degrees, often constricted between the seeds, dehiscent. Seeds prolate, with the small hilum on a short side, the radicular side nearly straight to concave, the opposing side rounded and often bulging below the hilum, the radicular lobe small and receding.

A genus of about 200 species, most numerous in the western parts of North and South America, with a second centre of distribution around the Mediterranean. Two species occur in the highlands of eastern Africa.

Apart from *L. mexicanus*, described below, various other species are commonly grown in the tropics for ornament, forage and green manure and are likely to escape from cultivation in due course. Garden lupins are derived as hybrids from the *Lupinus perennis* L. and *L. polyphyllus* Lindl. group of species. One of the native species, *L. princei* Harms, extends from Ethiopia to the Southern Highlands of Tanzania, south of Iringa, but has not been recorded from the Flora Zambesiaca area.

Lupinus mexicanus Cerv. in Lagasca y Segura, Gen. Pl. Nov.: 22 (1816). —Edwards, Bot. Reg.: t. 457 (1820). Type from Mexico.

Lupinus ehrenbergii Schltr. in Linnaea **12**: 33 (1838). —C.P. Smith, Sp. Lupin.: 43 (1938). —Polhill in F.T.E.A., Leguminosae, Pap.: 1000 (1971). —Lock, Leg. Afr. Check-list: 283 (1989). Type from Mexico.

Lupinus pubescens sensu Drummond in Kirkia **8**: 223 (1972) non Benth.

Erect annual or short-lived perennial, up to 1 m tall. Stem villous with very long spreading hairs over a denser pubescence. Leaflets 7–9, 2.5–7 × 0.5–1.5 cm, oblanceolate to narrowly oblong-elliptic, pointed or rounded at the apex, usually pilose on both surfaces; petiole 7–16 cm long; stipules 1–3 cm long, the free part linear-subulate. Racemes up to 15–50 cm long, with many ± verticillate flowers; bracts up to 20 mm long, linear-caudate to lanceolate-caudate; pedicels 3–6 mm long. Calyx (8)10–12 mm long, tomentose; upper lip bifid; lower lobe entire to minutely 3-fid. Standard subcircular, mostly blue or pink, glabrous; wings ± as long as the keel; keel 1.2–1.4 cm long, narrowly beaked. Pod 2.5–4 × 0.5–0.8 cm, narrowly oblong, with a straight beak, villous, with rather thin valves impressed between the 3–6 rather small seeds.

Zimbabwe. C: Harare, Salisbury Drive, Mabelreign, fl. 18.vii.1971, *Biegel* 3599 (K; SRGH). **Malawi**. S: Zomba Distr., Magomero, fl. 30.xi.1955, *Banda* 144 (K; MAL).

Native of Central America, locally naturalized in Zimbabwe, Malawi and southern Tanzania. Damp places by rivers and seasonally flooded grassland ("vleis") and along roadsides; 900 –1450 m.

INDEX TO BOTANICAL NAMES